Third Edition

Mathematical Proofs
A Transition to
Advanced Mathematics

Gary Chartrand
Western Michigan University

Albert D. Polimeni
State University of New York at Fredonia

Ping Zhang
Western Michigan University

Boston Columbus Indianapolis New York San Francisco Upper Saddle River
Amsterdam Cape Town Dubai London Madrid Milan Munich Paris Montreal Toronto
Delhi Mexico City Sao Paulo Sydney Hong Kong Seoul Singapore Taipei Tokyo

Editor in Chief: Deirdre Lynch
Senior Acquisitions Editor: William Hoffman
Assistant Editor: Brandon Rawnsley
Executive Marketing Manager: Jeff Weidenaar
Marketing Assistant: Caitlin Crain
Senior Production Project Manager: Beth Houston
Manager, Cover Visual Research and Permissions: Jayne Conte
Cover Designer: Suzanne Behnke
Cover Art: Shutterstock.com
Full-Service Project Management: Kailash Jadli, Aptara®, Inc.
Composition: Aptara®, Inc.
Printer/Binder: Courier Westford
Cover Printer: Lehigh/Phoenix

Credits and acknowledgments borrowed from other sources and reproduced, with permission, in this textbook appear on the appropriate page within text.

Library of Congress Cataloging-in-Publication Data

Chartrand, Gary.
  Mathematical proofs : a transition to advanced mathematics / Gary Chartrand,
Albert D. Polimeni, Ping Zhang. – 3rd ed.
    p.   cm.
Includes bibliographical references and index.
ISBN-13: 978-0-321-79709-4
ISBN-10: 0-321-79709-4
  1. Proof theory—Textbooks.   I. Polimeni, Albert D., 1938–   II. Zhang, Ping,
    1957–   III. Title.

QA9.54.C48 2013
511.3′6—dc23   2012012552

10 9 8 7 6 5 4 3 2 —CW—16 15 14 13 12

ISBN-13: 978-0-321-79709-4
ISBN-10:   0-321-79709-4

To

the memory of my mother and father   G.C.

the memory of my uncle Joe and my brothers John and Rocky   A.D.P.

my mother and the memory of my father   P.Z.

Contents

Direct Proof and Proof by Contrapositive   77

More on Direct Proof and Proof by Contrapositive   99

Existence and Proof by Contradiction   120

Mathematical Induction   142

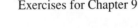

Proofs in Calculus   288

Proofs in Group Theory   322

Proofs in Ring Theory (Online)

Proofs in Linear Algebra (Online)

PREFACE TO THE THIRD EDITION

As we mentioned in the prefaces of the first two editions, because the teaching of calculus in many colleges and universities has become more problem-oriented with added emphasis on the use of calculators and computers, the theoretical gap between the material presented in calculus and the mathematical background expected (or at least hoped for) in more advanced courses such as abstract algebra and advanced calculus has widened. In an attempt to narrow this gap and to better prepare students for the more abstract mathematics courses to follow, many colleges and universities have introduced courses that are now commonly called *transition courses*. In these courses, students are introduced to problems whose solution involves mathematical reasoning and a knowledge of proof techniques and writing clear proofs. Topics such as relations, functions and cardinalities of sets are encountered throughout theoretical mathematics courses. In addition, transition courses often include theoretical aspects of number theory, abstract algebra, and calculus. This textbook has been written for such a course.

The idea for this textbook originated in the early 1980s, long before transition courses became fashionable, during the supervising of undergraduate mathematics research projects. We came to realize that even advanced undergraduates lack a sound understanding of proof techniques and have difficulty writing correct and clear proofs. At that time, a set of notes was developed for these students. This was followed by the introduction of a transition course, for which a more detailed set of notes was written. The first edition of this book emanated from these notes, which in turn has led to a second edition and now this third edition.

While understanding proofs and proof techniques and writing good proofs are major goals here, these are not things that can be accomplished to any great degree in a single course during a single semester. These must continue to be emphasized and practiced in succeeding mathematics courses.

Our Approach

Since this textbook originated from notes that were written exclusively for undergraduates to help them understand proof techniques and to write good proofs, this is the tone in which all editions of this book have been written: to be student-friendly. Numerous examples of proofs are presented in the text. Following common practice, we indicate the end of a proof with the square symbol ∎. Often we precede a proof by a discussion, referred to as a *proof strategy*, where we think through what is needed to present a proof of the result in question. Other times, we find it useful to reflect on a proof we have just presented to point out certain key details. We refer to a discussion of this type as a *proof analysis*. Periodically, problems are presented and solved, and we may find it convenient to discuss some features of the solution, which we refer to simply as an *analysis*. For clarity, we indicate the end of a discussion of a proof strategy, proof analysis, analysis or solution of an example with the diamond symbol ♦.

A major goal of this textbook is to help students learn to construct proofs of their own that are not only mathematically correct but clearly written. More advanced mathematics students should strive to present proofs that are convincing, readable, notationally consistent, and grammatically correct. A secondary goal is to have students gain sufficient knowledge of and confidence with proofs so that they will recognize, understand, and appreciate a proof that is properly written.

As with the first two editions, the third edition of this book is intended to assist the student in making the transition to courses that rely more on mathematical proof and reasoning. We envision students would take a course based on this book after they have had a year of calculus (and possibly another course, such as elementary linear algebra). It is likely that, prior to taking this course, a student's training in mathematics consisted primarily of doing patterned problems; that is, students have been taught methods for solving problems, likely including some explanation as to why these methods worked. Students may very well have had exposure to some proofs in earlier courses but, more than likely, were unaware of the logic involved and of the method of proof being used. There may have even been times when the students were not certain what was being proved.

Outline of the Contents

Since writing good proofs requires a certain degree of competence in writing, we have devoted Chapter 0 to writing mathematics. The emphasis of this chapter is on effective and clear exposition, correct usage of symbols, writing and displaying mathematical expressions, and using key words and phrases. Although every instructor will emphasize writing in his or her own way, we feel that it is useful to read Chapter 0 periodically throughout the course. It will mean more as the student progresses through the course.

Among the additions to and changes in the second edition that resulted in this third edition are the following.

- More than 250 exercises have been added, many of which require more thought to solve.
- New exercises have been added dealing with conjectures to give students practice with this important aspect of more advanced mathematics.
- Additional examples have been provided to assist in understanding and solving new exercises.
- In a number of instances, expanded discussions of a topic have been given to provide added clarity. In particular, the important topic of quantified statements is introduced in Section 2.10 and then reviewed in Section 7.2 to enhance one's understanding of this.

- A discussion of cosets and Lagrange's theorem has been added to Chapter 13 (Proofs in Group Theory).

Each chapter is divided into sections and the exercises for each chapter occur at the end of the chapter, divided into sections in the same way. There is also a final section of exercises for the entire chapter.

Chapter 1 contains a gentle introduction to sets, so that everyone has the same background and is using the same notation as we prepare for what lies ahead. No proofs involving sets occur until Chapter 4. Much of Chapter 1 may very well be a review for many.

Chapter 2 deals exclusively with logic. The goal here is to present what is needed to get into proofs as quickly as possible. Much of the emphasis in Chapter 2 is on statements, implications, and quantified statements, including a discussion of mixed quantifiers. Sets are introduced before logic so that the student's first encounter with mathematics here is a familiar one and because sets are needed to discuss quantified statements properly in Chapter 2.

The two proof techniques of direct proof and proof by contrapositive are introduced in Chapter 3 in the familiar setting of even and odd integers. Proof by cases is discussed in this chapter as well as proofs of "if and only if" statements. Chapter 4 continues this discussion in other settings, namely divisibility of integers, congruence, real numbers, and sets.

The technique of proof by contradiction is introduced in Chapter 5. Since existence proofs and counterexamples have a connection with proof by contradiction, these also occur in Chapter 5. The topic of uniqueness (of an element with specified properties) is also addressed in Chapter 5.

Proof by mathematical induction occurs in Chapter 6. In addition to the Principle of Mathematical Induction and the Strong Principle of Mathematical Induction, this chapter includes proof by minimum counterexample. The main goal of Chapter 7 (Prove or Disprove) concerns the testing of statements of unknown truth value, where it is to be determined, with justification, whether a given statement is true or false. In addition to the challenge of determining whether a statement is true or false, such problems provide added practice with counterexamples and the various proof techniques. Testing statements is preceded in this chapter with an historical discussion of conjectures in mathematics and a review of quantifiers.

Chapter 8 deals with relations, especially equivalence relations. Many examples involving congruence are presented and the set of integers modulo n is described. Chapter 9 involves functions, with emphasis on the properties of one-to-one and onto. This gives rise to a discussion of bijective functions and inverses of functions. The well-defined property of functions is discussed in more detail in this edition. In addition, there is a discussion of images and inverse images of sets with regard to functions and a number of added exercises involving these concepts.

Chapter 10 deals with infinite sets and a discussion of cardinalities of sets. This chapter includes an historical discussion of infinite sets, beginning with Cantor and his fascination and difficulties with the Schröder–Bernstein Theorem, then to Zermelo and the Axiom of Choice, and ending with a proof of the Schröder–Bernstein Theorem.

All of the proof techniques are used in Chapter 11 where numerous results in the area of number theory are introduced and proved. Chapter 12 deals with proofs that occur in calculus. Because these proofs are quite different than those previously encountered but are often more predictable in nature, many illustrations are given that involve limits of sequences and limits of functions and their connections with infinite series, continuity, and differentiability. The final Chapter 13 deals with modern algebra, beginning with binary operations and moving into proofs that are encountered in the area of group theory.

Web Site for Mathematical Proofs

Three additional chapters, Chapters 14–16 (dealing with proofs in ring theory, linear algebra, and topology), can be found on the Web site: http://www.aw.com/info/chartrand.

Teaching a Course from This Text

Although a course using this textbook could be designed in many ways, here are our views on such a course. As we noted earlier, we think it is useful for students to reread (at least portions of) Chapter 0 throughout the course as we feel that with each reading, the chapter becomes more meaningful. The first part of Chapter 1 (Sets) will likely be familiar to most students, although the last part may not. Chapters 2–6 will probably be part of any such course, although certain topics could receive varying degrees of emphasis (with perhaps proof by minimum counterexample in Chapter 6 possibly even omitted). One could spend little or much time on Chapter 7, depending on how much time is used to discuss the large number of "prove or disprove" exercises. We think that most of Chapters 8 and 9 would be covered in such a course. It would be useful to cover some of the fundamental ideas addressed in Chapter 10 (Cardinalities of Sets). As time permits, portions of the later chapters could be covered, especially those of interest to the instructor, including the possibility of going to the Web site for even more variety in the three online chapters.

Exercises

There are numerous exercises for Chapters 1–13 (as well as for Chapters 14–16 on the Web site). The degree of difficulty of the exercises ranges from routine to medium difficulty to moderately challenging. As mentioned earlier, the third edition contains more exercises in the moderately challenging category. There are exercises that present students with statements, asking them to decide whether they are true or false (with justification). There are proposed proofs of statements, asking if the argument is valid. There are proofs without a statement given, asking students to supply a statement of what has been proved. Also, there are exercises that call upon students to make conjectures of their own and possibly to provide proofs of these conjectures.

Chapter 3 is the first chapter in which students will be called upon to write proofs. At such an early stage, we feel that students need to (1) concentrate on constructing a valid proof and not be distracted by unfamiliarity with the mathematics, (2) develop some self-confidence with this process, and (3) learn how to write a proof properly. With this in mind, many of the exercises in Chapter 3 have been intentionally structured so as to be similar to the examples in that chapter.

In general, there are exercises for each section at the end of a chapter (section exercises) and additional exercises for the entire chapter (chapter exercises). Answers or hints to the odd-numbered section exercises appear at the end of text. One should also keep in mind, however, that proofs of results are not unique in general.

Acknowledgments

It is a pleasure to thank the reviewers of the third edition:

Daniel Acosta, Southeastern Louisiana University
Scott Annin, California State University, Fullerton
J. Marshall Ash, DePaul University
Ara Basmajian, Hunter College of CUNY
Matthias Beck, San Francisco State University
Richard Belshoff, Missouri State University
James Brawner, Armstrong Atlantic State University
Manav Das, University of Louisville
David Dempsey, Jacksonville State University
Cristina Domokos, California State University, Sacramento
José D. Flores, University of South Dakota
Eric Gottlieb, Rhodes College
Richard Hammack, Virginia Commonwealth University
Alan Koch, Agnes Scott College
M. Harper Langston, Courant Institute of Mathematical Sciences, New York University
Maria Nogin, California State University, Fresno
Daniel Nucinkis, University of Southampton
Thomas Polaski, Winthrop University
John Randall, Rutgers University
Eileen T. Shugart, Virginia Tech
Brian A. Snyder, Lake Superior State University
Melissa Sutherland, SUNY Geneseo
M.B. Ulmer, University of South Carolina Upstate
Mike Winders, Worcester State University

We also thank Renato Mirollo, Boston College and Tom Weglaitner for giving a final reading of portions of the third edition.

We have been most fortunate to receive the enthusiastic support from many at Pearson. First, we wish to thank the editorial team, as well as others at Pearson who have been so helpful and supportive: Greg Tobin, Publisher, Mathematics and Statistics; William Hoffman, Senior Acquisitions Editor; Jeff Weidenaar, Executive Marketing Manager, Mathematics; and Brandon Rawnsley, Associate Editor, Arts & Sciences, Higher Education. Our thanks to all of you. Finally, thank you as well to Beth Houston, Senior Production Project Manager; Kailash Jadli, Project Manager, Aptara, Inc.; and Mercedes Heston, Copy Editor, for guiding us through the final stages of the third edition.

Gary Chartrand
Albert D. Polimeni
Ping Zhang

0

Communicating Mathematics

Quite likely, the mathematics you have already encountered consists of doing problems using a specific approach or procedure. These may include solving equations in algebra, simplifying algebraic expressions, verifying trigonometric identities, using certain rules to find and simplify the derivatives of functions and setting up and evaluating a definite integral that give the area of a region or the volume of a solid. Accomplishing all of these is often a matter of practice.

Many of the methods that one uses to solve problems in mathematics are based on results in mathematics that were discovered by people and shown to be true. This kind of mathematics may very well be new to you and, as with anything that's new, there are things to be learned. But learning something new can be (in fact should be) fun. There are several steps involved here. The first step is discovering something in mathematics that we believe to be true. How does one discover new mathematics? This usually comes about by considering examples and observing that a pattern seems to be occurring with the examples. This may lead to a guess on our part as to what appears to be happening. We then have to convince ourselves that our guess is correct. In mathematics this involves constructing a proof of what we believe to be true is, in fact, true. But this is not enough. We need to convince others that we are right. So we need to write a proof that is written so clearly and so logically that people who know the methods of mathematics will be convinced. Where mathematics differs from all other scholarly fields is that once a proof has been given of a certain mathematical statement, there is no longer any doubt. This statement is true. Period. There is no other alternative.

Our main emphasis here will be in learning how to construct mathematical proofs and learning to write the proof in such a manner that it will be clear to and understood by others. Even though learning to guess new mathematics is important and can be fun, we will spend only a little time on this as it often requires an understanding of more mathematics than can be discussed at this time. But why would we want to discover new mathematics? While one possible answer is that it comes from the curiosity that most mathematicians possess, a more common explanation is that we have a problem to solve that requires knowing that some mathematical statement is true.

Learning Mathematics

One of the major goals of this book is to assist you as you progress from an individual who uses mathematics to an individual who understands mathematics. Perhaps this will mark the beginning of you becoming someone who actually develops mathematics of your own. This is an attainable goal if you have the desire.

The fact that you've gone this far in your study of mathematics suggests that you have ability in mathematics. This is a real opportunity for you. Much of the mathematics that you will encounter in the future is based on what you are about to learn here. The better you learn the material and the mathematical thought process now, the more you will understand later. To be sure, any area of study is considerably more enjoyable when you understand it. But getting to that point will require effort on your part.

There are probably as many excuses for doing poorly in mathematics as there are strategies for doing well in mathematics. We have all heard students say (sometimes, remarkably, even with pride) that they are not good at mathematics. That's only an alibi. Mathematics can be learned like any other subject. Even some students who have done well in mathematics say that they are not good with proofs. This, too, is unacceptable. What is required is determination and effort. To have done well on an exam with little or no studying is nothing to be proud of. Confidence based on being well-prepared is good, however.

Here is some advice that has worked for several students. First, it is important to understand what goes on in class each day. This means being present and being prepared for every class. After each class, recopy any lecture notes. When recopying the notes, express sentences in your own words and add details so that everything is as clear as possible. If you run into snags (and you will), talk them over with a classmate or your instructor. In fact, it's a good idea (at least in our opinion) to have someone with whom to discuss the material on a regular basis. Not only does it often clarify ideas, it gets you into the habit of using correct terminology and notation.

In addition to learning mathematics from your instructor, solidifying your understanding by careful note-taking and by talking with classmates, your text is (or at least should be) an excellent source as well. Read your text carefully with pen (or pencil) and paper in hand. Make a serious effort to do every homework problem assigned and, eventually, be certain that you know how to solve them. If there are exercises in the text that have not been assigned, you might even try to solve these as well. Another good idea is to try to create your own problems. In fact, when studying for an exam, try creating your own exam. If you start doing this for all of your classes, you might be surprised at how good you become. Creativity is a major part of mathematics. Discovering mathematics not only contributes to your understanding of the subject but has the potential to contribute to mathematics itself. Creativity can come in all forms. The following quote is due to the well-known writer J. K. Rowling (author of the *Harry Potter* novels).

> *Sometimes ideas just come to me. Other times I have to sweat and almost bleed to make ideas come. It's a mysterious process, but I hope I never find out exactly how it works.*

In the book *Defying Gravity: The Creative Career of Stephen Schwartz from Godspell to Wicked*, the author Carol de Giere writes a biography of Stephen Schwartz, one of the most successful composer-lyricists, in which she discusses not only creativity in music but how an idea can lead to something special and interesting and how creative people may have to deal with disappointment. Indeed, de Giere dedicates her book to the *creative spirit within each of us*. While he wrote the music for such famous shows as *Godspell* and *Wicked*, Schwartz discusses creativity head-on in the song "The Spark of Creation" he wrote for the musical *Children of Eden*. In her book, de Giere writes:

> *In many ways, this song expresses the theme of Stephen Schwartz's life—the naturalness and importance of the creative urge within us. At the same time he created an anthem for artists.*

In mathematics our goal is to seek the truth. Finding answers to mathematical questions is important, but we cannot be satisfied with this alone. We must be certain that we are right and that our explanation for why we believe we are correct is convincing to others. The reasoning we use as we proceed from what we know to what we wish to show must be logical. It must make sense to others, not just to ourselves.

There is joint responsibility here. As writers, it is our responsibility to give an accurate clear argument with enough details provided to allow the reader to understand what we have written and to be convinced. It is the reader's responsibility to know the basics of logic and to study the concepts involved so that a well-presented argument will be understood. Consequently, in mathematics writing is important, *very* important. Is it *really* important to write mathematics well? After all, isn't mathematics mainly equations and symbols? Not at all. It is not only important to write mathematics well, it is important to write well. You will be writing the rest of your life, at least reports, letters and e-mail. Many people who never meet you will know you only by what you write and how you write.

Mathematics is a sufficiently complicated subject that we don't need vague, hazy and boring writing to add to it. A teacher has a very positive impression of a student who hands in well-written and well-organized assignments and examinations. You want people to enjoy reading what you've written. It is important to have a good reputation as a writer. It's part of being an educated person. Especially with the large number of e-mail letters that so many of us write, it has become commonplace for writing to be more casual. Although all people would probably subscribe to this (since it is more efficient), we should know how to write well formally and professionally when the situation requires it.

You might think that considering how long you've been writing and that you're set in your ways, it will be very difficult to improve your writing. Not really. If you want to improve, you can and will. Even if you are a good writer, your writing can always be improved. Ordinarily, people don't think much about their writing. Often just thinking about your writing is the first step to writing better.

What Others Have Said about Writing

Many people who are well known in their areas of expertise have expressed their thoughts about writing. Here are quotes by some of these individuals.

Anything that helps communication is good. Anything that hurts it is bad.

I like words more than numbers, and I always did—conceptual more than computational.

Paul Halmos, mathematician

Writing is easy. All you have to do is cross out all the wrong words.

Mark Twain, author (The Adventures of Huckleberry Finn)

You don't write because you want to say something; you write because you've got something to say.

F. Scott Fitzgerald, author (The Great Gatsby)

Writing comes more easily if you have something to say.

Scholem Asch, author

Either write something worth reading or do something worth writing.

Benjamin Franklin, statesman, writer, inventor

What is written without effort is in general read without pleasure.

Samuel Johnson, writer

Easy reading is damned hard writing.

Nathaniel Hawthorne, novelist (The Scarlet Letter)

Everything that is written merely to please the author is worthless.

The last thing one knows when writing a book is what to put first.

I have made this letter longer because I lack the time to make it short.

Blaise Pascal, mathematician and physicist

The best way to become acquainted with a subject is to write a book about it.

Benjamin Disraeli, prime minister of England

In a very real sense, the writer writes in order to teach himself, to understand himself, to satisfy himself; the publishing of his ideas, though it brings gratification, is a curious anticlimax.

Alfred Kazin, literary critic

The skill of writing is to create a context in which other people can think.

Edwin Schlossberg, exhibit designer

A writer needs three things, experience, observation, and imagination, any two of which, at times any one of which, can supply the lack of the other.

William Faulkner, writer (The Sound and the Fury)

If confusion runs rampant in the passage just read,
It may very well be that too much has been said.

So that's what he meant! Then why didn't he say so?

Frank Harary, mathematician

A mathematical theory is not to be considered complete until you have made it so clear that you can explain it to the first man whom you meet on the street.

David Hilbert, mathematician

Everything should be made as simple as possible, but not simpler.

Albert Einstein, physicist

Never let anything you write be published without having had others critique it.

Donald E. Knuth, computer scientist and writer

Some books are to be tasted, others to be swallowed, and some few to be chewed and digested.

Reading maketh a full man, conference a ready man, and writing an exact man.

Francis Bacon, writer and philosopher

Judge an article not by the quality of what is framed and hanging on the wall, but by the quality of what's in the wastebasket.

Anonymous (Quote by Leslie Lamport)

We are all apprentices in a craft where no-one ever becomes a master.

Ernest Hemingway, author (*For Whom the Bell Tolls*)

There are three rules for writing a novel. Unfortunately, no one knows what they are.

W. Somerset Maugham, author (*Of Human Bondage*)

Mathematical Writing

Most of the quotes given above pertain to writing in general, not to mathematical writing in particular. However these suggestions for writing apply as well to writing mathematics. For us, mathematical writing means writing assignments for a mathematics course (particularly a course with proofs). Such an assignment might consist of writing a single proof, writing solutions to a number of problems or perhaps writing a term paper which, more than likely, includes definitions, examples, background *and* proofs. We'll refer to any of these as an *assignment*. Your goal should be to write correctly, clearly and in an interesting manner.

Before you even begin to write, you should have already thought about a number of things. First, you should know what examples and proofs you plan to include if this is appropriate for your assignment. You should not be overly concerned about writing good proofs on your first attempt—but be certain that you do have *proofs*.

As you're writing your assignment, you must be aware of your audience. What is the target group for your assignment? Of course, it should be written for your instructor. But it should be written so that a classmate would understand it. As you grow mathematically, your audience will grow with you and you will adapt your writing to this new audience.

Give yourself enough time to write your assignment. Don't try to put it together just a few minutes before it's due. The disappointing result will be obvious to your instructor. And to you! Find a place to write that is comfortable for you: your room, an office, a study room, the library and sitting at a desk, at a table, in a chair. Do what works best for you. Perhaps you write best when it's quiet or when there is background music.

Now that you're comfortably settled and have allowed enough time to do a good job, let's put a plan together. If the assignment is fairly lengthy, you may need an outline, which, most likely, will include one or more of the following:

1. Background and motivation
2. The definitions to be presented and possibly the notation to be used
3. The examples to include
4. The results to be presented (whose proofs have already been written, probably in rough form)
5. References to other results you intend to use
6. The order of everything mentioned above.

If the assignment is a term paper, it may include extensive background material and may need to be carefully motivated. The subject of the paper should be placed in perspective. Where does it fit in with what we already know?

Many writers write in *spirals*. Even though you have a plan for your assignment which includes an ordered list of things you want to say, it is likely that you will reach some point (perhaps sooner than you think) when you realize that you should have included something earlier—perhaps a definition, a theorem, an example, some notation. (This happened to us many times while writing this textbook.) Insert the missing material, start over again and write until once again you realize that something is missing. It is important, as you reread, that you start at the beginning each time. Then repeat the steps listed above.

We are about to give you some advice, some *pointers*, about writing mathematics. Such advice is necessarily subjective. Not everyone subscribes to these suggestions on writing. Indeed, writing *experts* don't agree on all issues. For the present, your instructor will be your best guide. But writing does not follow a list of rules. As you mature mathematically, perhaps the best advice about your writing is the same advice given by Jiminy Cricket to Disney's Pinocchio: *Always let your conscience be your guide.* You must be yourself. And one additional piece of advice: Be careful about accepting advice on writing. Originality and creativity don't follow rules. Until you reach the stage of being comfortable and confident with your own writing, however, we believe that it is useful to consider a few writing tips.

Since a number of these writing tips may not make sense (since, after all, we don't even have anything to write as yet), it will probably be most useful to return to this chapter periodically as you proceed through the chapters that follow.

Using Symbols

Since mathematics is a symbol-oriented subject, mathematical writing involves a mixture of words and symbols. Here are several guidelines to which a number of mathematicians subscribe.

1. *Never start a sentence with a symbol.*
   Writing mathematics follows the same practice as writing all sentences, namely that the first word should be capitalized. This is confusing if the sentence were to begin with a symbol since the sentence appears to be incomplete. Also, in general, a sentence sounds better if it starts with a word. Instead of writing:

$$x^2 - 6x + 8 = 0 \text{ has two distinct roots.}$$

   write:

$$\text{The equation } x^2 - 6x + 8 = 0 \text{ has two distinct roots.}$$

2. *Separate symbols not in a list by words if possible.*
   Separating symbols by words makes the sentence easier to read and therefore easier to understand. The sentence:

$$\text{Except for } a, b \text{ is the only root of } (x - a)(x - b) = 0.$$

   would be clearer if it were written as:

$$\text{Except for } a, \text{ the number } b \text{ is the only root of } (x - a)(x - b) = 0.$$

3. *Except when discussing logic, avoid writing the following symbols in your assignment:*

$$\Rightarrow, \ \forall, \ \exists, \ \ni, \ etc.$$

   The first four symbols stand for "implies", "for every", "there exists" and "such that", respectively. You may have already seen these symbols and know what they mean. If so, this is good. It is useful when taking notes or writing early drafts of an assignment to use shorthand symbols but many mathematicians avoid such symbols in their professional writing. (We will visit these symbols later.)

4. *Be careful about using i.e. and e.g.*
   These stand for *that is* and *for example*, respectively. There are situations when writing the words is preferable to writing the abbreviations as there may be confusion with nearby symbols. For example, $\sqrt{-1}$ and $\lim\limits_{n \to \infty} \left(1 + \dfrac{1}{n}\right)^n$ are not rational numbers, that is, i and e are not rational numbers.

5. *Write out integers as words when they are used as adjectives and when the numbers are relatively small or are easy to describe in words. Write out numbers numerically when they specify the value of something.*

$$\text{There are exactly two groups of order 4.}$$
$$\text{Fifty million Frenchmen can't be wrong.}$$
$$\text{There are one million positive integers less than 1,000,001.}$$

6. *Don't mix words and symbols improperly.*
   Instead of writing:

$$\text{Every integer} \geq 2 \text{ is a prime or is composite.}$$

it is preferable to write:

> Every integer exceeding 1 is a prime or is composite.

or

> If $n \geq 2$ is an integer, then n is prime or composite.

Although

> Since $(x - 2)(x - 3) = 0$, it follows that $x = 2$ or 3.

sounds correct, it is not written correctly. It should be:

> Since $(x - 2)(x - 3) = 0$, it follows that $x = 2$ or $x = 3$.

7. *Avoid using a symbol in the statement of a theorem when it's not needed.*
   Don't write:

> Theorem *Every bijective function f has an inverse.*

Delete "f". It serves no useful purpose. The theorem does not depend on what the function is called. A symbol should not be used in the statement of a theorem (or in its proof) exactly once. If it is useful to have a name for an arbitrary bijective function in the proof (as it probably will be), then "f" can be introduced there.

8. *Explain the meaning of every symbol that you introduce.*
   Although what you intended may seem clear, don't assume this. For example, if you write $n = 2k + 1$ and k has never appeared before, then say that k is an integer (if indeed k *is* an integer).

9. *Use "frozen symbols" properly.*
   If m and n are typically used for integers (as they probably are), then don't use them for real numbers. If A and B are used for sets, then don't use these as typical elements of a set. If f is used for a function, then don't use this as an integer. Write symbols that the reader would expect. To do otherwise could very well confuse the reader.

10. *Use consistent symbols.*
    Unless there is some special reason to the contrary, use symbols that "fit" together. Otherwise, it is distracting to the reader.
    Instead of writing

> If x and y are even integers, then $x = 2a$
> and $y = 2r$ for some integers a and r.

replace $2r$ by $2b$ (where then, of course, we write "for some integers a and b"). On the other hand, you might prefer to write $x = 2r$ and $y = 2s$.

Writing Mathematical Expressions

There will be numerous occasions when you will want to write mathematical expressions in your assignment, such as algebraic equations, inequalities, and formulas. If these expressions are relatively short, then they should probably be written within the text of the proof or discussion. (We'll explain this in a moment.) If the expressions are rather lengthy, then it is probably preferred for these expressions to be written as *displays*.

For example, suppose that we are discussing the Binomial Theorem. (It's not important if you don't recall what this theorem is.) It's possible that what we are writing includes the following passage:

For example, if we expand $(a + b)^4$, then we obtain $(a + b)^4 = a^4 + 4a^3b + 6a^2b^2 + 4ab^3 + b^4$.

It would probably be better to write the expansion of $(a + b)^4$ as a **display**, where the mathematical expression is placed on a line or lines by itself and is centered. This is illustrated below.

For example, if we expand $(a + b)^4$, then we obtain

$$(a + b)^4 = a^4 + 4a^3b + 6a^2b^2 + 4ab^3 + b^4.$$

If there are several mathematical expressions that are linked by equal signs and inequality symbols, then we would almost certainly write this as a display. For example, suppose that we wanted to write $n^3 + 3n^2 - n + 4$ in terms of k, where $n = 2k + 1$. A possible display is given next:

Since $n = 2k + 1$, it follows that

$$\begin{aligned} n^3 + 3n^2 - n + 4 &= (2k + 1)^3 + 3(2k + 1)^2 - (2k + 1) + 4 \\ &= (8k^3 + 12k^2 + 6k + 1) + 3(4k^2 + 4k + 1) - 2k - 1 + 4 \\ &= 8k^3 + 24k^2 + 16k + 7 = 8k^3 + 24k^2 + 16k + 6 + 1 \\ &= 2(4k^3 + 12k^2 + 8k + 3) + 1. \end{aligned}$$

Notice how the equal signs are lined up. (We wrote two equal signs on one line since that line would have contained very little material otherwise, as well as to balance the lengths of the lines better.)

Let's return to the expression $(a + b)^4 = a^4 + 4a^3b + 6a^2b^2 + 4ab^3 + b^4$ for the moment. If we were to write this expression in the text of a paragraph (as we are doing) and if we find it necessary to write portions of this expression on two separate lines, then this expression should be broken so that the first line ends with an operation or comparative symbol such as $+, -, <, \geq$ or $=$. In other words, the second line should *not* begin with one of these symbols. The reason for doing this is that ending the line with one of these symbols alerts the reader that more will follow; otherwise, the reader might conclude (incorrectly) that the portion of the expression appearing on the first line is the entire expression. Consequently, write

For example, if we expand $(a + b)^4$, then we obtain $(a + b)^4 = a^4 + 4a^3b + 6a^2b^2 + 4ab^3 + b^4$.

and not

For example, if we expand $(a + b)^4$, then we obtain $(a + b)^4 = a^4 + 4a^3b + 6a^2b^2 + 4ab^3 + b^4$.

If there is an occasion to refer to an expression that has already appeared, then this expression should have been written as a display and labeled as below:

$$(a + b)^4 = a^4 + 4a^3b + 6a^2b^2 + 6ab^2 + b^4. \tag{1}$$

Then we can simply refer to expression (1) rather than writing it out each time.

Common Words and Phrases in Mathematics

There are some words and phrases that appear so often in mathematical writing that it is useful to discuss them.

1. *I   We   One   Let's*

> I will now show that n is even.
> We will now show that n is even.
> One now shows that n is even.
> Let's now show that n is even.

These are four ways that we might write a sentence in a proof. Which of these sounds the best to you? It is not considered good practice to use "I" unless you are writing a personal account of something. Otherwise, "I" sounds egotistical and can be annoying. Using "one" is often awkward. Using "we" is standard practice in mathematics. This word also brings the reader into the discussion with the author and gives the impression of a team effort. The word "let's" accomplishes this as well but is much less formal. There is a danger of being *too* casual, however. In general, your writing should be balanced, maintaining a professional style. Of course, there is the possibility of avoiding all of these words:

> The integer n is now shown to be even.

2. *Clearly   Obviously   Of course   Certainly*
   These and similar words can turn a reader off if what's written is not clear to the reader. It can give the impression that the author is putting the reader down. These words should be used sparingly and with caution. If they *are* used, then at least be certain that what you say is true. There is also the possibility that the writer (a student?) has a lack of understanding of the mathematics or is not being careful and is using these words as a cover-up. This gives us even more reasons to avoid these words.

3. *Any   Each   Every*

> This statement is true for any integer n.

Does this mean that the statement is true for *some* integer n or *all* integers n? Since the word *any* can be vague, perhaps it is best to avoid it. If by *any*, we mean *each* or *every*, then use one of these two words instead. When the word *any* is encountered, most of the time the author means *each* or *every*.

4. *Since ⋯,   then ⋯*

A number of people connect these two words. You should use either "If ⋯, then ⋯" (should this be the intended meaning) or "Since ⋯, it follows that ⋯" or, possibly, "Since ⋯, we have ⋯". For example, it is correct to write

If n^2 is even, then n is even.

or

Since n^2 is even, it follows that n is even.

or perhaps

Since n^2 is even, n is even.

but avoid

Since n^2 is even, then n is even.

In this context, the word *since* can be replaced by *because*.

5. *Therefore   Thus   Hence   Consequently   So   It follows that   This implies that*

This is tricky. Mathematicians cannot survive without these words. Often within a proof, we proceed from something we've just learned to something else that can be concluded from it. There are many (many!) openings to sentences which attempt to say this. Although each of the words or phrases

Therefore Thus Hence Consequently So It follows that This implies that

is suitable, it is good to introduce some variety into your writing and not use the same words or phrases any more often than necessary.

6. *That   Which*

These words are often confused with each other. Sometimes they are interchangeable; more often they are not.

The solution to the equation is the number less than 5 that is positive.   **(2)**

The solution to the equation is the number less than 5 which is positive.   **(3)**

Which of these two sentences is correct? The simple answer is: Both are correct—or, at least, both might be correct.

For example, sentence (2) could be the response to the question: Which of the numbers 2, 3, and 5 is the solution of the equation? Sentence (3) could be the response to the question: Which of the numbers 4.9 and 5.0 is the solution of the equation?

The word *that* introduces a *restrictive clause* and, as such, the clause is essential to the meaning of the sentence. That is, if sentence (2) were written only as "The solution to the equation is the number less than 5" then the entire meaning is changed. Indeed, we no longer know what the solution of the equation is.

On the other hand, the word *which* does *not* introduce a restrictive clause. It introduces a nonrestrictive (or parenthetical) clause. A *nonrestrictive clause* only provides additional information that is not essential to the meaning of the sentence. In sentence (3) the phrase "which is positive" simply provides more information about the solution. This clause may have been added because the solution to an earlier equation is negative. In fact, it would be more appropriate to add a comma:

> The solution to the equation is the number less than 5, which is positive.

For another illustration, consider the following two statements:

> I always keep the math text that I like with me.   **(4)**

> I always keep the math text which I like with me.   **(5)**

What is the difference between these two sentences? In (4), the writer of the sentence clearly has more than one math text and is referring to the one that he/she likes. In (5), the writer has only one math text and is providing the added information that he/she likes it. The nonrestrictive clause in (5) should be set off by commas:

> I always keep the math text, which I like, with me.

A possible guideline to follow as you seek to determine whether *that* or *which* is the proper word to use is to ask yourself: Does it sound right if it reads "which, by the way"? In general, *that* is normally used considerably more often than *which*. Hence the advice here is: Beware of wicked which's!

While we are discussing the word *that*, we mention that the words *assume* and *suppose* often precede restrictive clauses and, as such, the word *that* should immediately follow one of these words. Omitting *that* leaves us with an implied *that*. Many mathematicians prefer to include it rather than omit it.

In other words, instead of writing:

> Assume N is a normal subgroup.

many would write

> Assume that N is a normal subgroup.

Some Closing Comments about Writing

1. Use good English. Write in complete sentences, ending each sentence with a period (or a question mark when appropriate) and capitalize the first word of each sentence. (Remember: No sentence begins with a symbol!)

2. Capitalize theorem and lemma as in Theorem 1 and Lemma 4.

3. Many mathematicians do not hyphenate words containing the prefix *non*, such as

> nonempty,   nonnegative,   nondecreasing,   nonzero.

4. Many words that occur often in mathematical writing are commonly misspelled. Among these are:

> commutative (independent of order)
> complement (supplement, balance, remainder)
> consistent (conforming, agreeing)

> feasible (suitable, attainable)
> its (possessive, not "it is")
> occurrence (incident)
> parallel (non-intersecting)
> preceding (foregoing, former)
> principle (postulate, regulation, rule)
> proceed (continue, move on)

and, of course,

> corollary, lemma, theorem.

5. There are many pairs of words that fit together in mathematics (while interchanging words among the pairs do not). For example,

> We ask questions.
> We pose problems.
> We present solutions.
> We prove theorems.
> We solve problems.
>   and
> We conclude this chapter.

1

Sets

In this initial chapter, you will be introduced to, or more than likely be reminded of, a fundamental idea that occurs throughout mathematics: sets. Indeed, a *set* is an object from which every mathematical structure is constructed (as we will often see in the succeeding chapters). Although there is a formal subject called *set theory* in which the properties of sets follow from a number of axioms, this is neither our interest nor our need. It is our desire to keep the discussion of sets informal without sacrificing clarity. It is almost a certainty that portions of this chapter will be familiar to you. Nevertheless, it is important that we understand what is meant by a set, how mathematicians describe sets, the notation used with sets, and several concepts that involve sets.

You've been experiencing sets all your life. In fact, all of the following are examples of sets: the students in a particular class who have an iPod, the items on a shopping list, the integers. As a small child, you learned to say the alphabet. When you did this, you were actually listing the letters that make up the set we call the alphabet. A **set** is a collection of objects. The objects that make up a set are called its **elements** (or **members**). The elements of a softball team are the players; while the elements of the alphabet are letters.

It is customary to use capital (upper case) letters (such as A, B, C, S, X, Y) to designate sets and lower case letters (for example, a, b, c, s, x, y) to represent elements of sets. If a is an element of the set A, then we write $a \in A$; if a does not belong to A, then we write $a \notin A$.

1.1 Describing a Set

There will be many occasions when we (or you) will need to describe a set. The most important requirement when describing a set is that the description makes it clear precisely which elements belong to the set.

If a set consists of a small number of elements, then this set can be described by explicitly listing its elements between braces (curly brackets) where the elements are separated by commas. Thus $S = \{1, 2, 3\}$ is a set, consisting of the numbers 1, 2 and 3. The order in which the elements are listed doesn't matter. Thus the set S just mentioned could be written as $S = \{3, 2, 1\}$ or $S = \{2, 1, 3\}$, for example. They describe the same

set. If a set T consists of the first five letters of the alphabet, then it is not essential that
we write $T = \{a, b, c, d, e\}$; that is, the elements of T need not be listed in alphabet-
ical order. On the other hand, listing the elements of T in any other order may create
unnecessary confusion.

The set A of all people who signed the Declaration of Independence and later became
president of the United States is $A = \{$John Adams, Thomas Jefferson$\}$ and the set B
of all positive even integers less than 20 is $B = \{2, 4, 6, 8, 10, 12, 14, 16, 18\}$. Some
sets contain too many elements to be listed this way. Perhaps even the set B just given
contains too many elements to describe in this manner. In such cases, the ellipsis or
"three dot notation" is often helpful. For example, $X = \{1, 3, 5, \ldots, 49\}$ is the set of all
positive odd integers less than 50, while $Y = \{2, 4, 6, \ldots\}$ is the set of all positive even
integers. The three dots mean "and so on" for Y and "and so on up to" for X.

A set need not contain any elements. Although it may seem peculiar to consider sets
without elements, these kinds of sets occur surprisingly often and in a variety of settings.
For example, if S is the set of real number solutions of the equation $x^2 + 1 = 0$, then S
contains no elements. There is only one set that contains no elements, and it is called the
empty set (or sometimes the **null set** or **void set**). The empty set is denoted by $\emptyset$. We
also write $\emptyset = \{\ \}$. In addition to the example given above, the set of all real numbers x
such that $x^2 < 0$ is also empty.

The elements of a set may in fact be sets themselves. The symbol ♦ below indicates
the conclusion of an example.

Example 1.1   *The set $S = \{1, 2, \{1, 2\}, \emptyset\}$ consists of four elements, two of which are sets, namely,*
$\{1, 2\}$ and $\emptyset$. If we write $C = \{1, 2\}$, then we can also write $S = \{1, 2, C, \emptyset\}$.

The set $T = \{0, \{1, 2, 3\}, 4, 5\}$ also has four elements, namely, the three integers
0, 4 and 5 and the set $\{1, 2, 3\}$. Even though $2 \in \{1, 2, 3\}$, the number 2 is not an element
of T; that is, $2 \notin T$.   ♦

Often sets consist of those elements satisfying some condition or possessing some specified property. In this case, we can define such a set as $S = \{x : p(x)\}$, where, by this, we
mean that S consists of all those elements x satisfying some condition $p(x)$ concerning
x. Some mathematicians write $S = \{x \mid p(x)\}$; that is, some prefer to write a vertical
line rather than a colon (which, by itself here, is understood to mean "such that"). For
example, if we are studying real number solutions of equations, then

$$S = \{x : (x - 1)(x + 2)(x + 3) = 0\}$$

is the set of all real numbers x such that $(x - 1)(x + 2)(x + 3) = 0$; that is, S is the
solution set of the equation $(x - 1)(x + 2)(x + 3) = 0$. We could have written $S = \{1, -2, -3\}$; however, even though this way of expressing S is apparently simpler, it
does not tell us that we are interested in the solutions of a particular equation. The
absolute value $|x|$ of a real number x is x if $x \geq 0$; while $|x| = -x$ if $x < 0$. Therefore,

$$T = \{x : |x| = 2\}$$

is the set of all real numbers having absolute value 2, that is, $T = \{2, -2\}$. In the sets S
and T that we have just described, we understand that "x" refers to a real number x. If

there is a possibility that this wouldn't be clear to the reader, then we should specifically say that x is a real number. We'll say more about this soon. The set

$$P = \{x : x \text{ has been a president of the United States}\}$$

describes, rather obviously, all those individuals who have been president of the United States. So Abraham Lincoln belongs to P but Benjamin Franklin does not.

Example 1.2   *Let $A = \{3, 4, 5, \ldots, 20\}$. If B denotes the set consisting of those elements of A that are less than 8, then we can write*

$$B = \{x \in A : x < 8\} = \{3, 4, 5, 6, 7\}. \qquad \blacklozenge$$

Some sets are encountered so often that they are given special notation. We use **N** to denote the set of all **positive integers** (or **natural numbers**); that is, $\mathbf{N} = \{1, 2, 3, \ldots\}$. The set of all **integers** (positive, negative, and zero) is denoted by **Z**. So $\mathbf{Z} = \{\ldots, -2, -1, 0, 1, 2, \ldots\}$. With the aid of the notation we've just introduced, we can now describe the set $E = \{\ldots, -4, -2, 0, 2, 4, \ldots\}$ of even integers by

$$E = \{y : y \text{ is an even integer}\} \text{ or } E = \{2x : x \text{ is an integer}\}, \text{ or as}$$

$$E = \{y : y = 2x \text{ for some } x \in \mathbf{Z}\} \text{ or } E = \{2x : x \in \mathbf{Z}\}.$$

Also,

$$S = \{x^2 : x \text{ is an integer}\} = \{x^2 : x \in \mathbf{Z}\} = \{0, 1, 4, 9, \ldots\}$$

describes the set of squares of integers.

The set of **real numbers** is denoted by **R**, and the set of positive real numbers is denoted by $\mathbf{R}^+$. A real number that can be expressed in the form $\frac{m}{n}$, where $m, n \in \mathbf{Z}$ and $n \neq 0$, is called a **rational number**. For example, $\frac{2}{3}, \frac{-5}{11}, 17 = \frac{17}{1}$ and $\frac{4}{6}$ are rational numbers. Of course, $4/6 = 2/3$. The set of all rational numbers is denoted by **Q**. A real number that is not rational is called **irrational.** The real numbers $\sqrt{2}, \sqrt{3}, \sqrt[3]{2}, \pi$ and e are known to be irrational; that is, none of these numbers can be expressed as the ratio of two integers. It is also known that the real numbers with (infinite) nonrepeating decimal expansions are precisely the irrational numbers. There is no common symbol to denote the set of irrational numbers. We will often use **I** for the set of all irrational numbers, however. Thus, $\sqrt{2} \in \mathbf{R}$ and $\sqrt{2} \notin \mathbf{Q}$; so $\sqrt{2} \in \mathbf{I}$.

For a set S, we write $|S|$ to denote the number of elements in S. The number $|S|$ is also referred to as the **cardinal number** or **cardinality** of S. If $A = \{1, 2\}$ and $B = \{1, 2, \{1, 2\}, \emptyset\}$, then $|A| = 2$ and $|B| = 4$. Also, $|\emptyset| = 0$. Although the notation is identical for the cardinality of a set and the absolute value of a real number, we should have no trouble distinguishing between the two. A set S is **finite** if $|S| = n$ for some nonnegative integer n. A set S is **infinite** if it is not finite. For the present, we will use the notation $|S|$ only for finite sets S. In Chapter 10, we will discuss the cardinality of infinite sets.

Let's now consider a few examples of sets that are defined in terms of the special sets we have just described.

Example 1.3   *Let $D = \{n \in \mathbf{N} : n \le 9\}$, $E = \{x \in \mathbf{Q} : x \le 9\}$, $H = \{x \in \mathbf{R} : x^2 - 2 = 0\}$ and $J = \{x \in \mathbf{Q} : x^2 - 2 = 0\}$.*

(a)   Describe the set D by listing its elements.

(b)   Give an example of three elements that belong to E but do not belong to D.

(c)   Describe the set H by listing its elements.

(d)   Describe the set J in another manner.

(e)   Determine the cardinality of each set D, H and J.

Solution   (a) $D = \{1, 2, 3, 4, 5, 6, 7, 8, 9\}$.

(b) $\frac{7}{5}, 0, -3$.

(c) $H = \{\sqrt{2}, -\sqrt{2}\}$.

(d) $J = \emptyset$.

(e) $|D| = 9$, $|H| = 2$ and $|J| = 0$.   ◆

Example 1.4   *In which of the following sets is the integer -2 an element?*
$S_1 = \{-1, -2, \{-1\}, \{-2\}, \{-1, -2\}\}$, $S_2 = \{x \in \mathbf{N} : -x \in \mathbf{N}\}$,
$S_3 = \{x \in \mathbf{Z} : x^2 = 2^x\}$, $S_4 = \{x \in \mathbf{Z} : |x| = -x\}$,
$S_5 = \{\{-1, -2\}, \{-2, -3\}, \{-1, -3\}\}$.

Solution   The integer -2 is an element of the sets S_1 and S_4. For S_4, $|-2| = 2 = -(-2)$. The set $S_2 = \emptyset$. Since $(-2)^2 = 4$ and $2^{-2} = 1/4$, it follows that $-2 \notin S_3$. Because each element of S_5 is a set, it contains no integers.   ◆

A **complex number** is a number of the form $a + bi$, where $a, b \in \mathbf{R}$ and $i = \sqrt{-1}$. A complex number $a + bi$ where $b = 0$, can be expressed as $a + 0i$ or, more simply, as a. Hence $a + 0i = a$ is a real number. Thus every real number is a complex number. Let $\mathbf{C}$ denote the set of complex numbers. If $K = \{x \in \mathbf{C} : x^2 + 1 = 0\}$, then $K = \{i, -i\}$. Of course, if $L = \{x \in \mathbf{R} : x^2 + 1 = 0\}$, then $L = \emptyset$. You might have seen that the sum of two complex numbers $a + bi$ and $c + di$ is $(a + c) + (b + d)i$, while their product is

$$(a + bi) \cdot (c + di) = ac + adi + bci + bdi^2 = (ac - bd) + (ad + bc)i.$$

The special sets that we've just described are now summarized below:

symbol	for the set of
N	natural numbers (positive integers)
Z	integers
Q	rational numbers
I	irrational numbers
R	real numbers
C	complex numbers

1.2 Subsets

A set A is called a **subset** of a set B if every element of A also belongs to B. If A is a subset of B, then we write $A \subseteq B$. If A, B and C are sets such that $A \subseteq B$ and $B \subseteq C$, then $A \subseteq C$. To see why this is so, suppose that some element x belongs to A. Because $A \subseteq B$, it follows that $x \in B$. But $B \subseteq C$, which implies that $x \in C$. Therefore, every element that belongs to A also belongs to C and so $A \subseteq C$. This property of subsets might remind you of the property of real numbers where if $a, b, c \in \mathbf{R}$ such that if $a \leq b$ and $b \leq c$, then $a \leq c$. For the sets $X = \{1, 3, 6\}$ and $Y = \{1, 2, 3, 5, 6\}$, we have $X \subseteq Y$. Also, $\mathbf{N} \subseteq \mathbf{Z}$ and $\mathbf{Q} \subseteq \mathbf{R}$. In addition, $\mathbf{R} \subseteq \mathbf{C}$. Since $\mathbf{Q} \subseteq \mathbf{R}$ and $\mathbf{R} \subseteq \mathbf{C}$, it therefore follows that $\mathbf{Q} \subseteq \mathbf{C}$. Moreover, every set is a subset of itself.

Example 1.5   *Find two sets A and B such that A is both an element of and a subset of B.*

Solution   Suppose that we seek two sets A and B such that $A \in B$ and $A \subseteq B$. Let's start with a simple example for A, say $A = \{1\}$. Since we want $A \in B$, the set B must contain the set $\{1\}$ as one of its elements. On the other hand, we also require that $A \subseteq B$, so every element of A must belong to B. Since 1 is the only element of A, it follows that B must also contain the number 1. A possible choice for B is then $B = \{1, \{1\}\}$, although $B = \{1, 2, \{1\}\}$ would also satisfy the conditions.   ◆

In the following example, we will see how we arrive at the answer to a question asked there. This is a prelude to logic, which will be discussed in Chapter 2.

Example 1.6   *Two sets A and B have the property that each is a subset of $\{1, 2, 3, 4, 5\}$ and $|A| = |B| = 3$. Furthermore,*

(a) *1 belongs to A but not to B.*
(b) *2 belongs to B but not to A.*
(c) *3 belongs to exactly one of A and B.*
(d) *4 belongs to exactly one of A and B.*
(e) *5 belongs to at least one of A and B.*

What are the possibilities for the set A?

Solution   By (a) and (b), $1 \in A$ and $1 \notin B$, while $2 \in B$ and $2 \notin A$. By (c), 3 belongs to A or B but not both. By (d), 4 belongs to A or B but not both. If 3 and 4 belong to the same set, then either 3 and 4 both belong to A or 3 and 4 both belong to B. Should it occur that $3 \in A$ and $4 \in A$, then $1 \notin B$, $3 \notin B$ and $4 \notin B$. This means that $|B| \neq 3$. On the other hand, if $3 \in B$ and $4 \in B$, then $3 \notin A$ and $4 \notin A$. Therefore, A contains none of 2, 3 and 4 and so $|A| \neq 3$. We can therefore conclude that 3 and 4 belong to different sets. The only way that $|A| = |B| = 3$ is for 5 to belong to both A and B and so either $A = \{1, 3, 5\}$ or $A = \{1, 4, 5\}$.   ◆

If a set C is *not* a subset of a set D, then we write $C \not\subseteq D$. In this case, there must be some element of C that is not an element of D. One consequence of this is that the empty set $\emptyset$ is a subset of every set. If this were not the case, then there must be some set A such that $\emptyset \nsubseteq A$. But this would mean there is some element, say x, in $\emptyset$ that is not in A. However, $\emptyset$ contains no elements. So $\emptyset \subseteq A$ for *every* set A.

Example 1.7 *Let $S = \{1, \{2\}, \{1, 2\}\}$.*

(a) Determine which of the following are elements of S:
$1, \{1\}, 2, \{2\}, \{1, 2\}, \{\{1, 2\}\}$.

(b) Determine which of the following are subsets of S:
$\{1\}, \{2\}, \{1, 2\}, \{\{1\}, 2\}, \{1, \{2\}\}, \{\{1\}, \{2\}\}, \{\{1, 2\}\}$.

Solution

(a) The following are elements of S: $1, \{2\}, \{1, 2\}$.

(b) The following are subsets of S: $\{1\}, \{1, \{2\}\}, \{\{1, 2\}\}$. ♦

In a typical discussion of sets, we are ordinarily concerned with subsets of some specified set U, called the **universal set.** For example, we may be dealing only with integers, in which case the universal set is **Z**, or we may be dealing only with real numbers, in which case the universal set is **R**. On the other hand, the universal set being considered may be neither **Z** nor **R**. Indeed, U may not even be a set of numbers.

Some frequently encountered subsets of **R** are the so-called "intervals," which you have no doubt encountered often. For $a, b \in \mathbf{R}$ and $a < b$, the **open interval** (a, b) is the set

$$(a, b) = \{x \in \mathbf{R} : a < x < b\}.$$

Therefore, all of the real numbers $\frac{5}{2}, \sqrt{5}, e, 3, \pi, 4.99$ belong to $(2, 5)$, but none of the real numbers $\sqrt{2}, 1.99, 2, 5$ belong to $(2, 5)$.

For $a, b \in \mathbf{R}$ and $a \leq b$, the **closed interval** $[a, b]$ is the set

$$[a, b] = \{x \in \mathbf{R} : a \leq x \leq b\}.$$

While $2, 5 \notin (2, 5)$, we do have $2, 5 \in [2, 5]$. The "interval" $[a, a]$ is therefore $\{a\}$. Thus, for $a < b$, we have $(a, b) \subseteq [a, b]$. For $a, b \in \mathbf{R}$ and $a < b$, the **half-open** or **half-closed intervals** $[a, b)$ and $(a, b]$ are defined as expected:

$$[a, b) = \{x \in \mathbf{R} : a \leq x < b\} \text{ and } (a, b] = \{x \in \mathbf{R} : a < x \leq b\}.$$

For $a \in \mathbf{R}$, the infinite intervals $(-\infty, a), (-\infty, a], (a, \infty)$ and $[a, \infty)$ are defined as

$$(-\infty, a) = \{x \in \mathbf{R} : x < a\}, (-\infty, a] = \{x \in \mathbf{R} : x \leq a\},$$
$$(a, \infty) = \{x \in \mathbf{R} : x > a\}, \quad [a, \infty) = \{x \in \mathbf{R} : x \geq a\}.$$

The interval $(-\infty, \infty)$ is the set **R**. Note that the infinity symbols ∞ and $-\infty$ are not real numbers; they are only used to help describe certain intervals. Therefore, $[1, \infty]$, for example, has no meaning.

Two sets A and B are **equal**, indicated by writing $A = B$, if they have exactly the same elements. Another way of saying $A = B$ is that every element of A is in B and every element of B is in A, that is, $A \subseteq B$ and $B \subseteq A$. In particular, whenever some element x belongs to A, then $x \in B$ because $A \subseteq B$. Also, if y is an element of B, then because $B \subseteq A$, it follows that $y \in A$. That is, whenever an element belongs to one of these sets, it must belong to the other and so $A = B$. This fact will be very useful to us

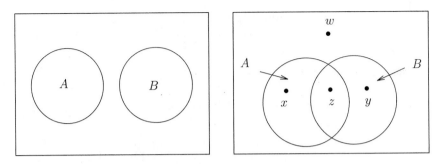

Figure 1.1   Venn diagrams for two sets A and B

in Chapter 4. If $A \neq B$, then there must be some element belonging to one of A and B but not to the other.

It is often convenient to represent sets by diagrams called **Venn diagrams.** For example, Figure 1.1 shows Venn diagrams for two sets A and B. The diagram on the left represents two sets A and B that have no elements in common, while the diagram on the right is more general. The element x belongs to A but not to B, the element y belongs to B but not to A, the element z belongs to both A and B, while w belongs to neither A nor B. In general, the elements of a set are understood to be those displayed within the region that describes the set. A rectangle in a Venn diagram represents the universal set in this case. Since every element under consideration belongs to the universal set, each element in a Venn diagram lies within the rectangle.

A set A is a **proper subset** of a set B if $A \subseteq B$ but $A \neq B$. If A is a proper subset of B, then we write $A \subset B$. For example, if $S = \{4, 5, 7\}$ and $T = \{3, 4, 5, 6, 7\}$, then $S \subset T$. (Although we write $A \subset B$ to indicate that A is a proper subset of B, it should be mentioned that some prefer to write $A \subsetneq B$ to indicate that A is a proper subset of B. Indeed, there are some who write $A \subset B$, rather than $A \subseteq B$, to indicate that A is a subset of B. We will follow the notation introduced above, however.)

The set consisting of all subsets of a given set A is called the **power set** of A and is denoted by $\mathcal{P}(A)$.

Example 1.8   *For each set A below, determine $\mathcal{P}(A)$. In each case, determine $|A|$ and $|\mathcal{P}(A)|$.*

(a) $A = \emptyset$,   (b) $A = \{a, b\}$, · (c) $A = \{1, 2, 3\}$.

Solution   (a) $\mathcal{P}(A) = \{\emptyset\}$. In this case, $|A| = 0$ and $|\mathcal{P}(A)| = 1$.

(b) $\mathcal{P}(A) = \{\emptyset, \{a\}, \{b\}, \{a, b\}\}$. In this case, $|A| = 2$ and $|\mathcal{P}(A)| = 4$.

(c) $\mathcal{P}(A) = \{\emptyset, \{1\}, \{2\}, \{3\}, \{1, 2\}, \{1, 3\}, \{2, 3\}, \{1, 2, 3\}\}$.
      In this case, $|A| = 3$ and $|\mathcal{P}(A)| = 8$.   ♦

Notice that for each set A in Example 1.8, we have $|\mathcal{P}(A)| = 2^{|A|}$. In fact, if A is any finite set, with n elements say, then $\mathcal{P}(A)$ has 2^n elements; that is,

$$|\mathcal{P}(A)| = 2^{|A|}$$

for every finite set A. (Later we will explain why this is true.)

Example 1.9   *If $C = \{\emptyset, \{\emptyset\}\}$, then*

$$\mathcal{P}(C) = \{\emptyset, \{\emptyset\}, \{\{\emptyset\}\}, \{\emptyset, \{\emptyset\}\}\}.$$

It is important to note that no two of the sets $\emptyset$, $\{\emptyset\}$ and $\{\{\emptyset\}\}$ are equal. (An empty box and a box containing an empty box are not the same.) For the set C above, it is therefore correct to write

$$\emptyset \subseteq C, \emptyset \subset C, \emptyset \in C, \{\emptyset\} \subseteq C, \{\emptyset\} \subset C, \{\emptyset\} \in C,$$

as well as

$$\{\{\emptyset\}\} \subseteq C, \{\{\emptyset\}\} \notin C, \{\{\emptyset\}\} \in \mathcal{P}(C). \qquad \blacklozenge$$

1.3   Set Operations

Just as there are several ways of combining two integers to produce another integer (addition, subtraction, multiplication and sometimes division), there are several ways to combine two sets to produce another set. The **union** of two sets A and B, denoted by $A \cup B$, is the set of all elements belonging to A or B, that is,

$$A \cup B = \{x : x \in A \text{ or } x \in B\}.$$

The use of the word "or" here, and in mathematics in general, allows an element of $A \cup B$ to belong to both A and B. That is, x is in $A \cup B$ if x is in A or x is in B or x is in both A and B. A Venn diagram for $A \cup B$ is shown in Figure 1.2.

Example 1.10   *For the sets $A_1 = \{2, 5, 7, 8\}$, $A_2 = \{1, 3, 5\}$ and $A_3 = \{2, 4, 6, 8\}$, we have*

$$A_1 \cup A_2 = \{1, 2, 3, 5, 7, 8\},$$
$$A_1 \cup A_3 = \{2, 4, 5, 6, 7, 8\},$$
$$A_2 \cup A_3 = \{1, 2, 3, 4, 5, 6, 8\}.$$

Also, $\mathbf{N} \cup \mathbf{Z} = \mathbf{Z}$ and $\mathbf{Q} \cup \mathbf{I} = \mathbf{R}$. $\qquad \blacklozenge$

The **intersection** of two sets A and B is the set of all elements belonging to both A and B. The intersection of A and B is denoted by $A \cap B$. In symbols,

$$A \cap B = \{x : x \in A \text{ and } x \in B\}.$$

A Venn diagram for $A \cap B$ is shown in Figure 1.3.

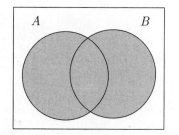

Figure 1.2   A Venn diagram for $A \cup B$

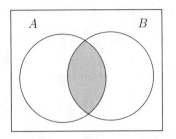

Figure 1.3   A Venn diagram for $A \cap B$

Example 1.11   *For the sets A_1, A_2 and A_3 described in Example 1.10,*

$$A_1 \cap A_2 = \{5\},\ A_1 \cap A_3 = \{2, 8\}\ and\ A_2 \cap A_3 = \emptyset.$$

Also, $\mathbf{N} \cap \mathbf{Z} = \mathbf{N}$ and $\mathbf{Q} \cap \mathbf{R} = \mathbf{Q}$.   ◆

For every two sets A and B, it follows that

$$A \cap B \subseteq A \cup B.$$

To see why this is true, suppose that x is an element belonging to $A \cap B$. Then x belongs to both A and B. Since $x \in A$, for example, $x \in A \cup B$ and so $A \cap B \subseteq A \cup B$.

If two sets A and B have no elements in common, then $A \cap B = \emptyset$ and A and B are said to be **disjoint**. Consequently, the sets A_2 and A_3 described in Example 1.10 are disjoint; however, A_1 and A_3 are not disjoint since 2 and 8 belong to both sets. Also, $\mathbf{Q}$ and $\mathbf{I}$ are disjoint.

The **difference** $A - B$ of two sets A and B (also written as $A \setminus B$ by some mathematicians) is defined as

$$A - B = \{x :\ x \in A \text{ and } x \notin B\}.$$

A Venn diagram for $A - B$ is shown in Figure 1.4.

Example 1.12   *For the sets $A_1 = \{2, 5, 7, 8\}$ and $A_2 = \{1, 3, 5\}$ in Examples 1.10 and 1.11, $A_1 - A_2 = \{2, 7, 8\}$ and $A_2 - A_1 = \{1, 3\}$. Furthermore, $\mathbf{R} - \mathbf{Q} = \mathbf{I}$.*   ◆

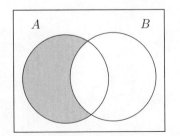

Figure 1.4   A Venn diagram for $A - B$

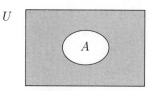

Figure 1.5 A Venn diagram for $\overline{A}$

Example 1.13 *For $A = \{x \in \mathbf{R} : |x| \le 3\}$, $B = \{x \in \mathbf{R} : |x| > 2\}$ and $C = \{x \in \mathbf{R} : |x - 1| \le 4\}$:*

  (a) *Express A, B and C using interval notation.*

  (b) *Determine $A \cap B$, $A - B$, $B \cap C$, $B \cup C$, $B - C$ and $C - B$.*

Solution

  (a) $A = [-3, 3]$, $B = (-\infty, -2) \cup (2, \infty)$ and $C = [-3, 5]$.

  (b) $A \cap B = [-3, -2) \cup (2, 3]$, $A - B = [-2, 2]$, $B \cap C = [-3, -2) \cup (2, 5]$, $B \cup C = (-\infty, \infty)$, $B - C = (-\infty, -3) \cup (5, \infty)$ and $C - B = [-2, 2]$. ◆

Suppose that we are considering a certain universal set U, that is, all sets being discussed are subsets of U. For a set A, its **complement** is

$$\overline{A} = U - A = \{x : x \in U \text{ and } x \notin A\}.$$

If $U = \mathbf{Z}$, then $\overline{\mathbf{N}} = \{0, -1, -2, \ldots\}$; while if $U = \mathbf{R}$, then $\overline{\mathbf{Q}} = \mathbf{I}$. A Venn diagram for $\overline{A}$ is shown in Figure 1.5.

The set difference $A - B$ is sometimes called the **relative complement** of B in A. Indeed, from the definition, $A - B = \{x : x \in A \text{ and } x \notin B\}$. The set $A - B$ can also be expressed in terms of complements, namely, $A - B = A \cap \overline{B}$. This fact will be established later.

Example 1.14 *Let $U = \{1, 2, \ldots, 10\}$ be the universal set, $A = \{2, 3, 5, 7\}$ and $B = \{2, 4, 6, 8, 10\}$. Determine each of the following:*

  (a) $\overline{B}$,   (b) $A - B$,   (c) $A \cap \overline{B}$,   (d) $\overline{\overline{B}}$.

Solution

  (a) $\overline{B} = \{1, 3, 5, 7, 9\}$.

  (b) $A - B = \{3, 5, 7\}$.

  (c) $A \cap \overline{B} = \{3, 5, 7\} = A - B$.

  (d) $\overline{\overline{B}} = B = \{2, 4, 6, 8, 10\}$. ◆

Example 1.15 *Let $A = \{0, \{0\}, \{0, \{0\}\}\}$.*

  (a) Determine which of the following are elements of A: 0, $\{0\}$, $\{\{0\}\}$.

  (b) Determine $|A|$.

(c)   Determine which of the following are subsets of A: 0, $\{0\}$, $\{\{0\}\}$.
       For (d)–(i), determine the indicated sets.

(d) $\{0\} \cap A$

(e) $\{\{0\}\} \cap A$

(f) $\{\{\{0\}\}\} \cap A$

(g) $\{0\} \cup A$

(h) $\{\{0\}\} \cup A$

(i) $\{\{\{0\}\}\} \cup A.$

Solution

(a)   While 0 and $\{0\}$ are elements of A, $\{\{0\}\}$ is not an element of A.

(b)   The set A has three elements: 0, $\{0\}$, $\{0, \{0\}\}$. Therefore, $|A| = 3$.

(c)   The integer 0 is not a set and so cannot be a subset of A (or a subset of any
       other set). Since $0 \in A$ and $\{0\} \in A$, it follows that $\{0\} \subseteq A$ and $\{\{0\}\} \subseteq A$.

(d)   Since 0 is the only element that belongs to both $\{0\}$ and A, it follows that
       $\{0\} \cap A = \{0\}$.

(e)   Since $\{0\}$ is the only element that belongs to both $\{\{0\}\}$ and A, it follows that
       $\{\{0\}\} \cap A = \{\{0\}\}$.

(f)   Since $\{\{0\}\}$ is not an element of A, it follows that $\{\{\{0\}\}\}$ and A are disjoint
       sets and so $\{\{\{0\}\}\} \cap A = \emptyset$.

(g)   Since $0 \in A$, it follows that $\{0\} \cup A = A$.

(h)   Since $\{0\} \in A$, it follows that $\{\{0\}\} \cup A = A$.

(i)   Since $\{\{0\}\} \notin A$, it follows that $\{\{\{0\}\}\} \cup A = \{0, \{0\}, \{\{0\}\}, \{0, \{0\}\}\}$.   ◆

1.4 Indexed Collections of Sets

We will often encounter situations where more than two sets are combined using the set
operations we described earlier. In the case of three sets A, B and C, the standard Venn
diagram is shown in Figure 1.6.

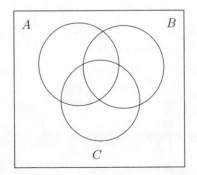

Figure 1.6   A Venn diagram for three sets

The union $A \cup B \cup C$ is defined as

$$A \cup B \cup C = \{x : x \in A \text{ or } x \in B \text{ or } x \in C\}.$$

Thus, in order for an element to belong to $A \cup B \cup C$, the element must belong to at least one of the sets A, B and C. Because it is often useful to consider the union of several sets, additional notation is needed. The union of the $n \geq 2$ sets $A_1, A_2, \ldots, A_n$ is denoted by $A_1 \cup A_2 \cup \cdots \cup A_n$ or $\bigcup_{i=1}^{n} A_i$ and is defined as

$$\bigcup_{i=1}^{n} A_i = \{x : x \in A_i \text{ for some } i, 1 \leq i \leq n\}.$$

Thus, for an element a to belong to $\bigcup_{i=1}^{n} A_i$, it is necessary that a belongs to at least one of the sets $A_1, A_2, \ldots, A_n$.

Example 1.16 *Let* $B_1 = \{1, 2\}$, $B_2 = \{2, 3\}, \ldots,$ $B_{10} = \{10, 11\}$; *that is,* $B_i = \{i, i + 1\}$ *for* $i = 1, 2, \ldots, 10$. *Determine each of the following:*

(a) $\displaystyle\bigcup_{i=1}^{5} B_i$. (b) $\displaystyle\bigcup_{i=1}^{10} B_i$. (c) $\displaystyle\bigcup_{i=3}^{7} B_i$. (d) $\displaystyle\bigcup_{i=j}^{k} B_i$, where $1 \leq j \leq k \leq 10$.

Solution (a) $\displaystyle\bigcup_{i=1}^{5} B_i = \{1, 2, \ldots, 6\}$. (b) $\displaystyle\bigcup_{i=1}^{10} B_i = \{1, 2, \ldots, 11\}$

(c) $\displaystyle\bigcup_{i=3}^{7} B_i = \{3, 4, \ldots, 8\}$. (d) $\displaystyle\bigcup_{i=j}^{k} B_i = \{j, j + 1, \ldots, k + 1\}$. ◆

We are often interested in the intersection of several sets as well. The intersection of the $n \geq 2$ sets $A_1, A_2, \ldots, A_n$ is expressed as $A_1 \cap A_2 \cap \cdots \cap A_n$ or $\bigcap_{i=1}^{n} A_i$ and is defined by

$$\bigcap_{i=1}^{n} A_i = \{x : x \in A_i \text{ for every } i, 1 \leq i \leq n\}.$$

The next example concerns the sets mentioned in Example 1.16.

Example 1.17 *Let* $B_i = \{i, i + 1\}$ *for* $i = 1, 2, \ldots, 10$. *Determine the following:*

(a) $\displaystyle\bigcap_{i=1}^{10} B_i$. (b) $B_i \cap B_{i+1}$. (c) $\displaystyle\bigcap_{i=j}^{j+1} B_i$, where $1 \leq j < 10$.

(d) $\displaystyle\bigcap_{i=j}^{k} B_i$ where $1 \leq j < k \leq 10$.

Solution (a) $\displaystyle\bigcap_{i=1}^{10} B_i = \emptyset$. (b) $B_i \cap B_{i+1} = \{i + 1\}$. (c) $\displaystyle\bigcap_{i=j}^{j+1} B_i = \{j + 1\}$.

(d) $\displaystyle\bigcap_{i=j}^{k} B_i = \{j + 1\}$ if $k = j + 1$; while $\displaystyle\bigcap_{i=j}^{k} B_i = \emptyset$ if $k > j + 1$. ◆

There are instances when the union or intersection of a collection of sets cannot be described conveniently (or perhaps at all) in the manner mentioned above. For this reason, we introduce a (nonempty) set I, called an **index set**, which is used as a mechanism for selecting those sets we want to consider. For example, for an index set I, suppose that there is a set S_α for each $\alpha \in I$. We write $\{S_\alpha\}_{\alpha \in I}$ to describe the collection of all sets S_α, where $\alpha \in I$. Such a collection is called an **indexed collection of sets**. We define the union of the sets in $\{S_\alpha\}_{\alpha \in I}$ by

$$\bigcup_{\alpha \in I} S_\alpha = \{x : x \in S_\alpha \text{ for some } \alpha \in I\},$$

and the intersection of these sets by

$$\bigcap_{\alpha \in I} S_\alpha = \{x : x \in S_\alpha \text{ for all } \alpha \in I\}.$$

Hence an element a belongs to $\bigcup_{\alpha \in I} S_\alpha$ if a belongs to at least one of the sets in the collection $\{S_\alpha\}_{\alpha \in I}$, while a belongs to $\bigcap_{\alpha \in I} S_\alpha$ if a belongs to every set in the collection $\{S_\alpha\}_{\alpha \in I}$. We refer to $\bigcup_{\alpha \in I} S_\alpha$ as the union of the collection $\{S_\alpha\}_{\alpha \in I}$ and $\bigcap_{\alpha \in I} S_\alpha$ as the intersection of the collection $\{S_\alpha\}_{\alpha \in I}$. Just as there is nothing special about our choice of i in $\bigcup_{i=1}^{n} A_i$ (that is, we could just as well describe this set by $\bigcup_{j=1}^{n} A_j$, say), there is nothing special about α in $\bigcup_{\alpha \in I} S_\alpha$. We could also describe this set by $\bigcup_{x \in I} S_x$. The variables i and α above are *dummy variables* and any appropriate symbol could be used. Indeed, we could write J or some other symbol for an index set.

Example 1.18   *For $n \in \mathbf{N}$, define $S_n = \{n, 2n\}$. For example, $S_1 = \{1, 2\}$, $S_2 = \{2, 4\}$ and $S_4 = \{4, 8\}$. Then $S_1 \cup S_2 \cup S_4 = \{1, 2, 4, 8\}$. We can also describe this set by means of an index set. If we let $I = \{1, 2, 4\}$, then*

$$\bigcup_{\alpha \in I} S_\alpha = S_1 \cup S_2 \cup S_4. \qquad \blacklozenge$$

Example 1.19   *For each $n \in \mathbf{N}$, define A_n to be the closed interval $[-\frac{1}{n}, \frac{1}{n}]$ of real numbers; that is,*

$$A_n = \left\{ x \in \mathbf{R} : -\frac{1}{n} \leq x \leq \frac{1}{n} \right\}.$$

So $A_1 = [-1, 1]$, $A_2 = [-\frac{1}{2}, \frac{1}{2}]$, $A_3 = [-\frac{1}{3}, \frac{1}{3}]$ and so on. We have now defined the sets $A_1, A_2, A_3, \ldots$. The union of these sets can be written as $A_1 \cup A_2 \cup A_3 \cup \cdots$ or $\bigcup_{i=1}^{\infty} A_i$. Using $\mathbf{N}$ as an index set, we can also write this union as $\bigcup_{n \in \mathbf{N}} A_n$. Since $A_n \subseteq A_1 = [-1, 1]$ for every $n \in \mathbf{N}$, it follows that $\bigcup_{n \in \mathbf{N}} A_n = [-1, 1]$. Certainly, $0 \in A_n$ for every $n \in \mathbf{N}$; in fact, $\bigcap_{n \in \mathbf{N}} A_n = \{0\}$. $\qquad \blacklozenge$

Example 1.20   *Let A denote the set of the letters of the alphabet, that is, $A = \{a, b, \ldots, z\}$. For $\alpha \in A$, let A_α consist of α and the two letters that follow α. So $A_a = \{a, b, c\}$ and $A_b = \{b, c, d\}$. By A_y, we will mean the set $\{y, z, a\}$ and $A_z = \{z, a, b\}$. Hence $|A_\alpha| = 3$ for every $\alpha \in A$. Therefore $\bigcup_{\alpha \in A} A_\alpha = A$. Indeed, if*

$$B = \{a, d, g, j, m, p, s, v, y\},$$

then $\bigcup_{\alpha \in B} A_\alpha = A$ as well. On the other hand, if $I = \{p, q, r\}$, then $\bigcup_{\alpha \in I} A_\alpha = \{p, q, r, s, t\}$ while $\bigcap_{\alpha \in I} A_\alpha = \{r\}$.   ♦

Example 1.21   *Let $S = \{1, 2, \ldots, 10\}$. Each of the sets*

$$S_1 = \{1, 2, 3, 4\}, \; S_2 = \{4, 5, 6, 7, 8\} \; and \; S_3 = \{7, 8, 9, 10\}$$

is a subset of S. Also, $S_1 \cup S_2 \cup S_3 = S$. This union can be described in a number of ways. Define $I = \{1, 2, 3\}$ and $J = \{S_1, S_2, S_3\}$. Then the union of the three sets belonging to J is precisely $S_1 \cup S_2 \cup S_3$, which can also be written as

$$S = S_1 \cup S_2 \cup S_3 = \bigcup_{i=1}^{3} S_i = \bigcup_{\alpha \in I} S_\alpha = \bigcup_{X \in J} X.$$ ♦

1.5 Partitions of Sets

Recall that two sets are disjoint if their intersection is the empty set. A collection S of subsets of a set A is called **pairwise disjoint** if every two distinct subsets that belong to S are disjoint. For example, let $A = \{1, 2, \ldots, 7\}$, $B = \{1, 6\}$, $C = \{2, 5\}$, $D = \{4, 7\}$ and $S = \{B, C, D\}$. Then S is a pairwise disjoint collection of subsets of A since $B \cap C = B \cap D = C \cap D = \emptyset$. On the other hand, let $A' = \{1, 2, 3\}$, $B' = \{1, 2\}$, $C' = \{1, 3\}$, $D' = \{2, 3\}$ and $S' = \{B', C', D'\}$. Although S' is a collection of subsets of A' and $B' \cap C' \cap D' = \emptyset$, the set S' is *not* a pairwise disjoint collection of sets since $B' \cap C' \neq \emptyset$, for example. Indeed, $B' \cap D'$ and $C' \cap D'$ are also nonempty.

We will often have the occasion (especially in Chapter 8) to encounter, for a nonempty set A, a collection S of pairwise disjoint nonempty subsets of A with the added property that every element of A belongs to some subset in S. Such a collection is called a **partition** of A. A **partition** of A can also be defined as a collection S of nonempty subsets of A such that every element of A belongs to exactly one subset in S. Furthermore, a partition of A can be defined as a collection S of subsets of A satisfying the three properties:

(1) $X \neq \emptyset$ for every set $X \in S$;

(2)   for every two sets $X, Y \in S$, either $X = Y$ or $X \cap Y = \emptyset$;

(3) $\bigcup_{X \in S} X = A$.

Example 1.22   *Consider the following collections of subsets of the set $A = \{1, 2, 3, 4, 5, 6\}$:*

$$S_1 = \{\{1, 3, 6\}, \{2, 4\}, \{5\}\};$$
$$S_2 = \{\{1, 2, 3\}, \{4\}, \emptyset, \{5, 6\}\};$$
$$S_3 = \{\{1, 2\}, \{3, 4, 5\}, \{5, 6\}\};$$
$$S_4 = \{\{1, 4\}, \{3, 5\}, \{2\}\}.$$

Determine which of these sets are partitions of A.

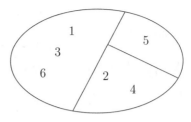

Figure 1.7   A partition of a set

Solution   The set S_1 is a partition of A. The set S_2 is not a partition of A since $\emptyset$ is one of the elements of S_2. The set S_3 is not a partition of A either since the element 5 belongs to two distinct subsets in S_3, namely, $\{3, 4, 5\}$ and $\{5, 6\}$. Finally, S_4 is also not a partition of A because the element 6 belongs to no subset in S_4.   ♦

As the word *partition* probably suggests, a partition of a nonempty set A is a division of A into nonempty subsets. The partition S_1 of the set A in Example 1.22 is illustrated in the diagram shown in Figure 1.7.

For example, the set **Z** of integers can be partitioned into the set of even integers and the set of odd integers. The set **R** of real numbers can be partitioned into the set $\mathbf{R}^+$ of positive real numbers, the set of negative real numbers and the set $\{0\}$ consisting of the number 0. In addition, **R** can be partitioned into the set **Q** of rational numbers and the set **I** of irrational numbers.

Example 1.23   *Let $A = \{1, 2, \ldots, 12\}$.*

(a)   Give an example of a partition S of A such that $|S| = 5$.

(b)   Give an example of a subset T of the partition S in (a) such that $|T| = 3$.

(c)   List all those elements B in the partition S in (a) such that $|B| = 2$.

Solution   (a)   We are seeking a partition S of A consisting of five subsets. One such example is

$$S = \{\{1, 2\}, \{3, 4\}, \{5, 6\}, \{7, 8, 9\}, \{10, 11, 12\}\}.$$

(b)   We are seeking a subset T of S (given in (a)) consisting of three elements. One such example is

$$T = \{\{1, 2\}, \{3, 4\}, \{7, 8, 9\}\}.$$

(c)   We have been asked to list all those elements of S (given in (a)) consisting of two elements of A. These elements are: $\{1, 2\}, \{3, 4\}, \{5, 6\}$.   ♦

1.6 Cartesian Products of Sets

We've already mentioned that when a set A is described by listing its elements, the order in which the elements of A are listed doesn't matter. That is, if the set A consists of two elements x and y, then $A = \{x, y\} = \{y, x\}$. When we speak of the **ordered pair** (x, y), however, this is another story. The ordered pair (x, y) is a single element consisting of a pair of elements in which x is the first element (or first coordinate) of the ordered pair (x, y) and y is the second element (or second coordinate). Moreover, for two ordered pairs (x, y) and (w, z) to be equal, that is, $(x, y) = (w, z)$, we must have $x = w$ and $y = z$. So, if $x \neq y$, then $(x, y) \neq (y, x)$.

The **Cartesian product** (or simply the product) $A \times B$ of two sets A and B is the set consisting of all ordered pairs whose first coordinate belongs to A and whose second coordinate belongs to B. In other words,

$$A \times B = \{(a, b) : a \in A \text{ and } b \in B\}.$$

Example 1.24   *If $A = \{x, y\}$ and $B = \{1, 2, 3\}$, then*

$$A \times B = \{(x, 1), (x, 2), (x, 3), (y, 1), (y, 2), (y, 3)\},$$

while

$$B \times A = \{(1, x), (1, y), (2, x), (2, y), (3, x), (3, y)\}.$$

Since, for example, $(x, 1) \in A \times B$ and $(x, 1) \notin B \times A$, these two sets do not contain the same elements; so $A \times B \neq B \times A$. Also,

$$A \times A = \{(x, x), (x, y), (y, x), (y, y)\}$$

and

$$B \times B = \{(1, 1), (1, 2), (1, 3), (2, 1), (2, 2), (2, 3), (3, 1), (3, 2), (3, 3)\}. \qquad \blacklozenge$$

We also note that if $A = \emptyset$ or $B = \emptyset$, then $A \times B = \emptyset$.

The Cartesian product $\mathbf{R} \times \mathbf{R}$ is the set of all points in the Euclidean plane. For example, the graph of the straight line $y = 2x + 3$ is the set

$$\{(x, y) \in \mathbf{R} \times \mathbf{R} : y = 2x + 3\}.$$

For the sets $A = \{x, y\}$ and $B = \{1, 2, 3\}$ given in Example 1.24, $|A| = 2$ and $|B| = 3$, while $|A \times B| = 6$. Indeed, for all finite sets A and B,

$$|A \times B| = |A| \cdot |B|.$$

Cartesian products will be explored in more detail in Chapter 7.

EXERCISES FOR CHAPTER 1

Section 1.1: Describing a Set

1.1. Which of the following are sets?

  (a) 1, 2, 3
  (b) $\{1, 2\}, 3$
  (c) $\{\{1\}, 2\}, 3$
  (d) $\{1, \{2\}, 3\}$
  (e) $\{1, 2, a, b\}$.

1.2. Let $S = \{-2, -1, 0, 1, 2, 3\}$. Describe each of the following sets as $\{x \in S : p(x)\}$, where $p(x)$ is some condition on x.

   (a) $A = \{1, 2, 3\}$
   (b) $B = \{0, 1, 2, 3\}$
   (c) $C = \{-2, -1\}$
   (d) $D = \{-2, 2, 3\}$

1.3. Determine the cardinality of each of the following sets:

   (a) $A = \{1, 2, 3, 4, 5\}$
   (b) $B = \{0, 2, 4, \ldots, 20\}$
   (c) $C = \{25, 26, 27, \ldots, 75\}$
   (d) $D = \{\{1, 2\}, \{1, 2, 3, 4\}\}$
   (e) $E = \{\emptyset\}$
   (f) $F = \{2, \{2, 3, 4\}\}$

1.4. Write each of the following sets by listing its elements within braces.

   (a) $A = \{n \in \mathbf{Z} : -4 < n \leq 4\}$
   (b) $B = \{n \in \mathbf{Z} : n^2 < 5\}$
   (c) $C = \{n \in \mathbf{N} : n^3 < 100\}$
   (d) $D = \{x \in \mathbf{R} : x^2 - x = 0\}$
   (e) $E = \{x \in \mathbf{R} : x^2 + 1 = 0\}$

1.5. Write each of the following sets in the form $\{x \in \mathbf{Z} : p(x)\}$, where $p(x)$ is a property concerning x.

   (a) $A = \{-1, -2, -3, \ldots\}$
   (b) $B = \{-3, -2, \ldots, 3\}$
   (c) $C = \{-2, -1, 1, 2\}$

1.6. The set $E = \{2x : x \in \mathbf{Z}\}$ can be described by listing its elements, namely $E = \{\ldots, -4, -2, 0, 2, 4, \ldots\}$. List the elements of the following sets in a similar manner.

   (a) $A = \{2x + 1 : x \in \mathbf{Z}\}$
   (b) $B = \{4n : n \in \mathbf{Z}\}$
   (c) $C = \{3q + 1 : q \in \mathbf{Z}\}$

1.7. The set $E = \{\ldots, -4, -2, 0, 2, 4, \ldots\}$ of even integers can be described by means of a defining condition by $E = \{y = 2x : x \in \mathbf{Z}\} = \{2x : x \in \mathbf{Z}\}$. Describe the following sets in a similar manner.

   (a) $A = \{\ldots, -4, -1, 2, 5, 8, \ldots\}$
   (b) $B = \{\ldots, -10, -5, 0, 5, 10, \ldots\}$
   (c) $C = \{1, 8, 27, 64, 125, \ldots\}$

1.8. Let $A = \{n \in \mathbf{Z} : 2 \leq |n| < 4\}$, $B = \{x \in \mathbf{Q} : 2 < x \leq 4\}$, $C = \{x \in \mathbf{R} : x^2 - (2 + \sqrt{2})x + 2\sqrt{2} = 0\}$ and $D = \{x \in \mathbf{Q} : x^2 - (2 + \sqrt{2})x + 2\sqrt{2} = 0\}$.

   (a) Describe the set A by listing its elements.
   (b) Give an example of three elements that belong to B but do not belong to A.
   (c) Describe the set C by listing its elements.
   (d) Describe the set D in another manner.
   (e) Determine the cardinality of each of the sets A, C and D.

1.9. For $A = \{2, 3, 5, 7, 8, 10, 13\}$, let

$$B = \{x \in A : x = y + z, \text{ where } y, z \in A\} \text{ and } C = \{r \in B : r + s \in B \text{ for some } s \in B\}.$$

Determine C.

Section 1.2: Subsets

1.10. Give examples of three sets A, B and C such that

(a) $A \subseteq B \subset C$
(b) $A \in B$, $B \in C$ and $A \notin C$
(c) $A \in B$ and $A \subset C$.

1.11. Let (a, b) be an open interval of real numbers and let $c \in (a, b)$. Describe an open interval I centered at c such that $I \subseteq (a, b)$.

1.12. Which of the following sets are equal?

$A = \{n \in \mathbf{Z} : |n| < 2\}$ $D = \{n \in \mathbf{Z} : n^2 \le 1\}$
$B = \{n \in \mathbf{Z} : n^3 = n\}$ $E = \{-1, 0, 1\}$.
$C = \{n \in \mathbf{Z} : n^2 \le n\}$

1.13. For a universal set $U = \{1, 2, \ldots, 8\}$ and two sets $A = \{1, 3, 4, 7\}$ and $B = \{4, 5, 8\}$, draw a Venn diagram that represents these sets.

1.14. Find $\mathcal{P}(A)$ and $|\mathcal{P}(A)|$ for

(a) $A = \{1, 2\}$.
(b) $A = \{\emptyset, 1, \{a\}\}$.

1.15. Find $\mathcal{P}(A)$ for $A = \{0, \{0\}\}$.

1.16. Find $\mathcal{P}(\mathcal{P}(\{1\}))$ and its cardinality.

1.17. Find $\mathcal{P}(A)$ and $|\mathcal{P}(A)|$ for $A = \{0, \emptyset, \{\emptyset\}\}$.

1.18. For $A = \{x : x = 0 \text{ or } x \in \mathcal{P}(\{0\})\}$, determine $\mathcal{P}(A)$.

1.19. Give an example of a set S such that

(a) $S \subseteq \mathcal{P}(\mathbf{N})$
(b) $S \in \mathcal{P}(\mathbf{N})$
(c) $S \subseteq \mathcal{P}(\mathbf{N})$ and $|S| = 5$
(d) $S \in \mathcal{P}(\mathbf{N})$ and $|S| = 5$

1.20. Determine whether the following statements are true or false.

(a) If $\{1\} \in \mathcal{P}(A)$, then $1 \in A$ but $\{1\} \notin A$.
(b) If A, B and C are sets such that $A \subset \mathcal{P}(B) \subset C$ and $|A| = 2$, then $|C|$ can be 5 but $|C|$ cannot be 4.
(c) If a set B has one more element than a set A, then $\mathcal{P}(B)$ has at least two more elements than $\mathcal{P}(A)$.
(d) If four sets A, B, C and D are subsets of $\{1, 2, 3\}$ such that $|A| = |B| = |C| = |D| = 2$, then at least two of these sets are equal.

1.21. Three subsets A, B and C of $\{1, 2, 3, 4, 5\}$ have the same cardinality. Furthermore,

(a) 1 belongs to A and B but not to C.
(b) 2 belongs to A and C but not to B.
(c) 3 belongs to A and exactly one of B and C.
(d) 4 belongs to an even number of A, B and C.

(e) 5 belongs to an odd number of A, B and C.

(f) The sums of the elements in two of the sets A, B and C differ by 1.

What is B?

Section 1.3: Set Operations

1.22. Let $U = \{1, 3, \ldots, 15\}$ be the universal set, $A = \{1, 5, 9, 13\}$, and $B = \{3, 9, 15\}$. Determine the following:

(a) $A \cup B$ (b) $A \cap B$ (c) $A - B$ (d) $B - A$ (e) $\overline{A}$ (f) $A \cap \overline{B}$.

1.23. Give examples of two sets A and B such that $|A - B| = |A \cap B| = |B - A| = 3$. Draw the accompanying Venn diagram.

1.24. Give examples of three sets A, B and C such that $B \neq C$ but $B - A = C - A$.

1.25. Give examples of three sets A, B and C such that

(a) $A \in B$, $A \subseteq C$ and $B \nsubseteq C$

(b) $B \in A$, $B \subset C$ and $A \cap C \neq \emptyset$

(c) $A \in B$, $B \subseteq C$ and $A \nsubseteq C$.

1.26. Let U be a universal set and let A and B be two subsets of U. Draw a Venn diagram for each of the following sets.

(a) $\overline{A \cup B}$ (b) $\overline{A} \cap \overline{B}$ (c) $\overline{A \cap B}$ (d) $\overline{A} \cup \overline{B}$.

What can you say about parts (a) and (b)? parts (c) and (d)?

1.27. Give an example of a universal set U, two sets A and B and accompanying Venn diagram such that $|A \cap B| = |A - B| = |B - A| = |\overline{A \cup B}| = 2$.

1.28. Let A, B and C be nonempty subsets of a universal set U. Draw a Venn diagram for each of the following set operations.

(a) $(C - B) \cup A$

(b) $C \cap (A - B)$.

1.29. Let $A = \{\emptyset, \{\emptyset\}, \{\{\emptyset\}\}\}$.

(a) Determine which of the following are elements of A: $\emptyset$, $\{\emptyset\}$, $\{\emptyset, \{\emptyset\}\}$.

(b) Determine $|A|$.

(c) Determine which of the following are subsets of A: $\emptyset$, $\{\emptyset\}$, $\{\emptyset, \{\emptyset\}\}$.

For (d)–(i), determine the indicated sets.

(d) $\emptyset \cap A$

(e) $\{\emptyset\} \cap A$

(f) $\{\emptyset, \{\emptyset\}\} \cap A$

(g) $\emptyset \cup A$

(h) $\{\emptyset\} \cup A$

(i) $\{\emptyset, \{\emptyset\}\} \cup A$.

1.30. Let $A = \{x \in \mathbf{R} : |x - 1| \leq 2\}$, $B = \{x \in \mathbf{R} : |x| \geq 1\}$ and $C = \{x \in \mathbf{R} : |x + 2| \leq 3\}$.

(a) Express A, B and C using interval notation.

(b) Determine each of the following sets using interval notation:
$A \cup B$, $A \cap B$, $B \cap C$, $B - C$.

1.31. Give an example of four different sets A, B, C and D such that (1) $A \cup B = \{1, 2\}$ and $C \cap D = \{2, 3\}$ and (2) if B and C are interchanged and $\cup$ and $\cap$ are interchanged, then we get the same result.

1.32. Give an example of four different subsets A, B, C and D of $\{1, 2, 3, 4\}$ such that all intersections of two subsets are different.

1.33. Give an example of two nonempty sets A and B such that $\{A \cup B, A \cap B, A - B, B - A\}$ is the power set of some set.

1.34. Give an example of two subsets A and B of $\{1, 2, 3\}$ such that all of the following sets are different: $A \cup B$, $A \cup \overline{B}$, $\overline{A} \cup B$, $\overline{A} \cup \overline{B}$, $A \cap B$, $A \cap \overline{B}$, $\overline{A} \cap B$, $\overline{A} \cap \overline{B}$.

1.35. Give examples of a universal set U and sets A, B and C such that each of the following sets contains exactly one element: $A \cap B \cap C$, $(A \cap B) - C$, $(A \cap C) - B$, $(B \cap C) - A$, $A - (B \cup C)$, $B - (A \cup C)$, $C - (A \cup B)$, $\overline{A \cup B \cup C}$. Draw the accompanying Venn diagram.

Section 1.4: Indexed Collections of Sets

1.36. For a real number r, define S_r to be the interval $[r - 1, r + 2]$. Let $A = \{1, 3, 4\}$. Determine $\bigcup_{\alpha \in A} S_\alpha$ and $\bigcap_{\alpha \in A} S_\alpha$.

1.37. Let $A = \{1, 2, 5\}$, $B = \{0, 2, 4\}$, $C = \{2, 3, 4\}$ and $S = \{A, B, C\}$. Determine $\bigcup_{X \in S} X$ and $\bigcap_{X \in S} X$.

1.38. For a real number r, define $A_r = \{r^2\}$, B_r as the closed interval $[r - 1, r + 1]$ and C_r as the interval (r, ∞). For $S = \{1, 2, 4\}$, determine

   (a) $\bigcup_{\alpha \in S} A_\alpha$ and $\bigcap_{\alpha \in S} A_\alpha$

   (b) $\bigcup_{\alpha \in S} B_\alpha$ and $\bigcap_{\alpha \in S} B_\alpha$

   (c) $\bigcup_{\alpha \in S} C_\alpha$ and $\bigcap_{\alpha \in S} C_\alpha$.

1.39. Let $A = \{a, b, \ldots, z\}$ be the set consisting of the letters of the alphabet. For $\alpha \in A$, let A_α consist of α and the two letters that follow it, where $A_y = \{y, z, a\}$ and $A_z = \{z, a, b\}$. Find a set $S \subseteq A$ of smallest cardinality such that $\bigcup_{\alpha \in S} A_\alpha = A$. Explain why your set S has the required properties.

1.40. For $i \in \mathbf{Z}$, let $A_i = \{i - 1, i + 1\}$. Determine the following:

   (a) $\displaystyle\bigcup_{i=1}^{5} A_{2i}$ (b) $\displaystyle\bigcup_{i=1}^{5} (A_i \cap A_{i+1})$ (c) $\displaystyle\bigcup_{i=1}^{5} (A_{2i-1} \cap A_{2i+1})$.

1.41. For each of the following, find an indexed collection $\{A_n\}_{n \in \mathbf{N}}$ of distinct sets (that is, no two sets are equal) satisfying the given conditions.

   (a) $\bigcap_{n=1}^{\infty} A_n = \{0\}$ and $\bigcup_{n=1}^{\infty} A_n = [0, 1]$

   (b) $\bigcap_{n=1}^{\infty} A_n = \{-1, 0, 1\}$ and $\bigcup_{n=1}^{\infty} A_n = \mathbf{Z}$.

1.42. For each of the following collections of sets, define a set A_n for each $n \in \mathbf{N}$ such that the indexed collection $\{A_n\}_{n \in \mathbf{N}}$ is precisely the given collection of sets. Then find both the union and intersection of the indexed collection of sets.

   (a) $\{[1, 2 + 1), [1, 2 + 1/2), [1, 2 + 1/3), \ldots\}$

   (b) $\{(-1, 2), (-3/2, 4), (-5/3, 6), (-7/4, 8), \ldots\}$.

1.43. For $r \in \mathbf{R}^+$, let $A_r = \{x \in \mathbf{R} : |x| < r\}$. Determine $\bigcup_{r \in \mathbf{R}^+} A_r$ and $\bigcap_{r \in \mathbf{R}^+} A_r$.

1.44. Each of the following sets is a subset of $A = \{1, 2, \ldots, 10\}$:
$A_1 = \{1, 5, 7, 9, 10\}$, $A_2 = \{1, 2, 3, 8, 9\}$, $A_3 = \{2, 4, 6, 8, 9\}$,
$A_4 = \{2, 4, 8\}$, $A_5 = \{3, 6, 7\}$, $A_6 = \{3, 8, 10\}$, $A_7 = \{4, 5, 7, 9\}$,
$A_8 = \{4, 5, 10\}$, $A_9 = \{4, 6, 8\}$, $A_{10} = \{5, 6, 10\}$,
$A_{11} = \{5, 8, 9\}$, $A_{12} = \{6, 7, 10\}$, $A_{13} = \{6, 8, 9\}$.
Find a set $I \subseteq \{1, 2, \ldots, 13\}$ such that for every two distinct elements $j, k \in I$, $A_j \cap A_k = \emptyset$ and $\left|\bigcup_{i \in I} A_i\right|$ is maximum.

1.45. For $n \in \mathbf{N}$, let $A_n = \left(-\frac{1}{n}, 2 - \frac{1}{n},\right)$. Determine $\bigcup_{n \in \mathbf{N}} A_n$ and $\bigcap_{n \in \mathbf{N}} A_n$.

Section 1.5: Partitions of Sets

1.46. Which of the following are partitions of $A = \{a, b, c, d, e, f, g\}$? For each collection of subsets that is not a partition of A, explain your answer.
   (a) $S_1 = \{\{a, c, e, g\}, \{b, f\}, \{d\}\}$ (b) $S_2 = \{\{a, b, c, d\}, \{e, f\}\}$
   (c) $S_3 = \{A\}$ (d) $S_4 = \{\{a\}, \emptyset, \{b, c, d\}, \{e, f, g\}\}$
   (e) $S_5 = \{\{a, c, d\}, \{b, g\}, \{e\}, \{b, f\}\}$.

1.47. Which of the following sets are partitions of $A = \{1, 2, 3, 4, 5\}$?
   (a) $S_1 = \{\{1, 3\}, \{2, 5\}\}$ (b) $S_2 = \{\{1, 2\}, \{3, 4, 5\}\}$
   (c) $S_3 = \{\{1, 2\}, \{2, 3\}, \{3, 4\}, \{4, 5\}\}$ (d) $S_4 = A$.

1.48. Let $A = \{1, 2, 3, 4, 5, 6\}$. Give an example of a partition S of A such that $|S| = 3$.

1.49. Give an example of a set A with $|A| = 4$ and two disjoint partitions S_1 and S_2 of A with $|S_1| = |S_2| = 3$.

1.50. Give an example of a partition of $\mathbf{N}$ into three subsets.

1.51. Give an example of a partition of $\mathbf{Q}$ into three subsets.

1.52. Give an example of three sets A, S_1 and S_2 such that S_1 is a partition of A, S_2 is a partition of S_1 and $|S_2| < |S_1| < |A|$.

1.53. Give an example of a partition of $\mathbf{Z}$ into four subsets.

1.54. Let $A = \{1, 2, \ldots, 12\}$. Give an example of a partition S of A satisfying the following requirements: (i) $|S| = 5$, (ii) there is a subset T of S such that $|T| = 4$ and $|\cup_{X \in T} X| = 10$ and (iii) there is no element $B \in S$ such that $|B| = 3$.

1.55. A set S is partitioned into two subsets S_1 and S_2. This produces a partition $\mathcal{P}_1$ of S where $\mathcal{P}_1 = \{S_1, S_2\}$ and so $|\mathcal{P}_1| = 2$. One of the sets in $\mathcal{P}_1$ is then partitioned into two subsets, producing a partition $\mathcal{P}_2$ of S with $|\mathcal{P}_2| = 3$. A total of $|\mathcal{P}_1|$ sets in $\mathcal{P}_2$ are partitioned into two subsets each, producing a partition $\mathcal{P}_3$ of S. Next, a total of $|\mathcal{P}_2|$ sets in $\mathcal{P}_3$ are partitioned into two subsets each, producing a partition $\mathcal{P}_4$ of S. This is continued until a partition $\mathcal{P}_6$ of S is produced. What is $|\mathcal{P}_6|$?

1.56. We mentioned that there are three ways that a collection S of subsets of a nonempty set A is defined to be a partition of A.
   Definition 1 The collection S consists of pairwise disjoint nonempty subsets of A and every element of A belongs to a subset in S.
   Definition 2 The collection S consists of nonempty subsets of A and every element of A belongs to exactly one subset in S.
   Definition 3 The collection S consists of subsets of A satisfying the three properties (1) every subset in S is nonempty, (2) every two subsets of A are equal or disjoint and (3) the union of all subsets in S is A.

   (a) Show that any collection S of subsets of A satisfying Definition 1 satisfies Definition 2.
   (b) Show that any collection S of subsets of A satisfying Definition 2 satisfies Definition 3.
   (c) Show that any collection S of subsets of A satisfying Definition 3 satisfies Definition 1.

Section 1.6: Cartesian Products of Sets

1.57. Let $A = \{x, y, z\}$ and $B = \{x, y\}$. Determine $A \times B$.

1.58. Let $A = \{1, \{1\}, \{\{1\}\}\}$. Determine $A \times A$.

1.59. For $A = \{a, b\}$, determine $A \times \mathcal{P}(A)$.

1.60. For $A = \{\emptyset, \{\emptyset\}\}$, determine $A \times \mathcal{P}(A)$.

1.61. For $A = \{1, 2\}$ and $B = \{\emptyset\}$, determine $A \times B$ and $\mathcal{P}(A) \times \mathcal{P}(B)$.

1.62. Describe the graph of the circle whose equation is $x^2 + y^2 = 4$ as a subset of $\mathbf{R} \times \mathbf{R}$.

1.63. List the elements of the set $S = \{(x, y) \in \mathbf{Z} \times \mathbf{Z} : |x| + |y| = 3\}$. Plot the corresponding points in the Euclidean xy- plane.

1.64. For $A = \{1, 2\}$ and $B = \{1\}$, determine $\mathcal{P}(A \times B)$.

1.65. For $A = \{x \in \mathbf{R} : |x - 1| \leq 2\}$ and $B = \{y \in \mathbf{R} : |y - 4| \leq 2\}$, give a geometric description of the points in the xy-plane belonging to $A \times B$.

1.66. For $A = \{a \in \mathbf{R} : |a| \leq 1\}$ and $B = \{b \in \mathbf{R} : |b| = 1\}$, give a geometric description of the points in the xy-plane belonging to $(A \times B) \cup (B \times A)$.

ADDITIONAL EXERCISES FOR CHAPTER 1

1.67. The set $T = \{2k + 1 : k \in \mathbf{Z}\}$ can be described as $T = \{\ldots, -3, -1, 1, 3, \ldots\}$. Describe the following sets in a similar manner.
(a) $A = \{4k + 3 : k \in \mathbf{Z}\}$
(b) $B = \{5k - 1 : k \in \mathbf{Z}\}$.

1.68. Let $S = \{-10, -9, \ldots, 9, 10\}$. Describe each of the following sets as $\{x \in S : p(x)\}$, where $p(x)$ is some condition on x.
(a) $A = \{-10, -9, \ldots, -1, 1, \ldots, 9, 10\}$
(b) $B = \{-10, -9, \ldots, -1, 0\}$
(c) $C = \{-5, -4, \ldots, 0, 1, \ldots, 7\}$
(d) $D = \{-10, -9, \ldots, 4, 6, 7, \ldots, 10\}$.

1.69. Describe each of the following sets by listing its elements within braces.
(a) $\{x \in \mathbf{Z} : x^3 - 4x = 0\}$
(b) $\{x \in \mathbf{R} : |x| = -1\}$
(c) $\{m \in \mathbf{N} : 2 < m \leq 5\}$
(d) $\{n \in \mathbf{N} : 0 \leq n \leq 3\}$
(e) $\{k \in \mathbf{Q} : k^2 - 4 = 0\}$
(f) $\{k \in \mathbf{Z} : 9k^2 - 3 = 0\}$
(g) $\{k \in \mathbf{Z} : 1 \leq k^2 \leq 10\}$.

1.70. Determine the cardinality of each of the following sets.
(a) $A = \{1, 2, 3, \{1, 2, 3\}, 4, \{4\}\}$
(b) $B = \{x \in \mathbf{R} : |x| = -1\}$
(c) $C = \{m \in \mathbf{N} : 2 < m \leq 5\}$
(d) $D = \{n \in \mathbf{N} : n < 0\}$
(e) $E = \{k \in \mathbf{N} : 1 \leq k^2 \leq 100\}$
(f) $F = \{k \in \mathbf{Z} : 1 \leq k^2 \leq 100\}$.

1.71. For $A = \{-1, 0, 1\}$ and $B = \{x, y\}$, determine $A \times B$.

1.72. Let $U = \{1, 2, 3\}$ be the universal set and let $A = \{1, 2\}$, $B = \{2, 3\}$ and $C = \{1, 3\}$. Determine the following.
(a) $(A \cup B) - (B \cap C)$
(b) $\overline{A}$
(c) $\overline{B \cup C}$
(d) $A \times B$.

1.73. Let $A = \{1, 2, \ldots, 10\}$. Give an example of two sets S and B such that $S \subseteq \mathcal{P}(A)$, $|S| = 4$, $B \in S$ and $|B| = 2$.

1.74. For $A = \{1\}$ and $C = \{1, 2\}$, give an example of a set B such that $\mathcal{P}(A) \subset B \subset \mathcal{P}(C)$.

1.75. Give examples of two sets A and B such that $A \cap \mathcal{P}(A) \in B$ and $\mathcal{P}(A) \subseteq A \cup B$.

1.76. Which of the following sets are equal?

$A = \{n \in \mathbf{Z} : -4 \le n \le 4\}$ $\qquad$ $D = \{x \in \mathbf{Z} : x^3 = 4x\}$

$B = \{x \in \mathbf{N} : 2x + 2 = 0\}$ $\qquad$ $E = \{-2, 0, 2\}.$

$C = \{x \in \mathbf{Z} : 3x - 2 = 0\}$

1.77. Let A and B be subsets of some unknown universal set U. Suppose that $\overline{A} = \{3, 8, 9\}$, $A - B = \{1, 2\}$, $B - A = \{8\}$ and $A \cap B = \{5, 7\}$. Determine U, A and B.

1.78. Let I denote the interval $[0, \infty)$. For each $r \in I$, define

$$A_r = \{(x, y) \in \mathbf{R} \times \mathbf{R} : x^2 + y^2 = r^2\}$$
$$B_r = \{(x, y) \in \mathbf{R} \times \mathbf{R} : x^2 + y^2 \le r^2\}$$
$$C_r = \{(x, y) \in \mathbf{R} \times \mathbf{R} : x^2 + y^2 > r^2\}.$$

(a) Determine $\bigcup_{r \in I} A_r$ and $\bigcap_{r \in I} A_r$.

(b) Determine $\bigcup_{r \in I} B_r$ and $\bigcap_{r \in I} B_r$.

(c) Determine $\bigcup_{r \in I} C_r$ and $\bigcap_{r \in I} C_r$.

1.79. Give an example of four sets A_1, A_2, A_3, A_4 such that $|A_i \cap A_j| = |i - j|$ for every two integers i and j with $1 \le i < j \le 4$.

1.80. (a) Give an example of two problems suggested by Exercise 1.79 (above).

(b) Solve one of the problems in (a).

1.81. Let $A = \{1, 2, 3\}$, $B = \{1, 2, 3, 4\}$ and $C = \{1, 2, 3, 4, 5\}$. For the sets S and T described below, explain whether $|S| < |T|$, $|S| > |T|$ or $|S| = |T|$.

(a) Let B be the universal set and let S be the set of all subsets X of B for which $|X| \ne |\overline{X}|$. Let T be the set of 2-element subsets of C.

(b) Let S be the set of all partitions of the set A and let T be the set of 4-element subsets of C.

(c) Let $S = \{(b, a) : b \in B, a \in A, a + b \text{ is odd}\}$ and let T be the set of all nonempty proper subsets of A.

1.82. Give an example of a set $A = \{1, 2, \ldots, k\}$ for a smallest $k \in \mathbf{N}$ containing subsets A_1, A_2, A_3 such that $|A_i - A_j| = |A_j - A_i| = |i - j|$ for every two integers i and j with $1 \le i < j \le 3$.

1.83. (a) For $A = \{-3, -2, \ldots, 4\}$ and $B = \{1, 2, \ldots, 6\}$, determine
$S = \{(a, b) \in A \times B : a^2 + b^2 = 25\}.$

(b) For $C = \{a \in B : (a, b) \in S\}$ and $D = \{b \in A : (a, b) \in S\}$, where A, B, S are the sets in (a), determine $C \times D$.

1.84. For $A = \{1, 2, 3\}$, let B be the set of 2-element sets belonging to $\mathcal{P}(A)$ and let C be the set consisting of the sets that are the intersections of two distinct elements of B. Determine $D = \mathcal{P}(C)$.

1.85. For a real number r, let $A_r = \{r, r + 1\}$. Let $S = \{x \in \mathbf{R} : x^2 + 2x - 1 = 0\}$.

(a) Determine $B = A_s \times A_t$ for the distinct elements $s, t \in S$, where $s < t$.

(b) Let $C = \{ab : (a, b) \in B\}$. Determine the sum of the elements of C.

2

Logic

In mathematics our goal is to seek the truth. Are there connections between two given mathematical concepts? If so, what are they? Under what conditions does an object possess a particular property? Finding answers to questions such as these is important, but we cannot be satisfied only with this. We must be certain that we are right and that our explanation for why we believe we are correct is convincing to others. The reasoning we use as we proceed from what we know to what we wish to show must be logical. It must make sense to others, not just to ourselves.

There is joint responsibility here, however. It is the writer's responsibility to use the rules of logic to give a valid and clear argument with enough details provided to allow the reader to understand what we have written and to be convinced. It is the reader's responsibility to know the basics of logic and to study the concepts involved sufficiently well so that he or she will not only be able to understand a well-presented argument but can decide as well whether it is valid. Consequently, both writer and reader must be familiar with logic.

Although it is possible to spend a great deal of time studying logic, we will present only what we actually need and will instead use the majority of our time putting what we learn into practice.

2.1 Statements

In mathematics we are constantly dealing with statements. By a **statement** we mean a declarative sentence or assertion that is true or false (but not both). Statements therefore declare or assert the truth of something. Of course, the statements in which we will be primarily interested deal with mathematics. For example, the sentences

> The integer 3 is odd.
> The integer 57 is prime.

are statements (only the first of which is true).

Every statement has a **truth value**, namely **true** (denoted by T) or **false** (denoted by F). We often use P, Q and R to denote statements, or perhaps $P_1, P_2, \ldots, P_n$ if there are several statements involved. We have seen that

$$P_1 : \text{The integer 3 is odd.}$$

and

$$P_2 : \text{The integer 57 is prime.}$$

are statements, where P_1 has truth value T and P_2 has truth value F.

Sentences that are imperative (commands) such as

Substitute the number 2 for x.
Find the derivative of $f(x) = e^{-x} \cos 2x$.

or are interrogative (questions) such as

Are these sets disjoint?
What is the derivative of $f(x) = e^{-x} \cos 2x$?

or are exclamatory such as

What an interesting question!
How difficult this problem is!

are not statements since these sentences are not declarative.

It may not be immediately clear whether a statement is true or false. For example, the sentence "The 100th digit in the decimal expansion of π is 7." is a statement, but it may be necessary to find this information in a Web site on the Internet to determine whether this statement is true. Indeed, for a sentence to be a statement, it is not a requirement that we be able to determine its truth value.

The sentence "The real number r is rational." is a statement *provided* we know what real number r is being referred to. Without this additional information, however, it is impossible to assign a truth value to it. This is an example of what is often referred to as an open sentence. In general, an **open sentence** is a declarative sentence that contains one or more variables, each variable representing a value in some prescribed set, called the **domain** of the variable, and which becomes a statement when values from their respective domains are substituted for these variables. For example, the open sentence "$3x = 12$" where the domain of x is the set of integers is a true statement only when $x = 4$.

An open sentence that contains a variable x is typically represented by $P(x)$, $Q(x)$ or $R(x)$. If $P(x)$ is an open sentence, where the domain of x is S, then we say $P(x)$ is an **open sentence over the domain** S. Also, $P(x)$ is a statement for each $x \in S$. For example, the open sentence

$$P(x) : (x - 3)^2 \le 1$$

over the domain **Z** is a true statement when $x \in \{2, 3, 4\}$ and is a false statement otherwise.

Example 2.1   *For the open sentence*

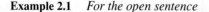

$$P(x, y) : |x + 1| + |y| = 1$$

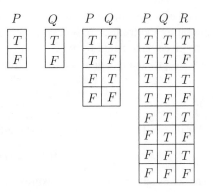

Figure 2.1   Truth tables for one, two and three statements

in two variables, suppose that the domain of the variable x is $S = \{-2, -1, 0, 1\}$ and the domain of the variable y is $T = \{-1, 0, 1\}$. Then

$$P(-1, 1): \ |(-1) + 1| + |1| = 1$$

is a true statement, while

$$P(1, -1): \ |1 + 1| + |-1| = 1$$

is a false statement. In fact, $P(x, y)$ is a true statement when

$$(x, y) \in \{(-2, 0), (-1, -1), (-1, 1), (0, 0)\},$$

while $P(x, y)$ is a false statement for all other elements $(x, y) \in S \times T$.   ◆

The possible truth values of a statement are often listed in a table, called a **truth table.** The truth tables for two statements P and Q are given in Figure 2.1. Since there are two possible truth values for each of P and Q, there are four possible combinations of truth values for P and Q. The truth table showing all these combinations is also given in Figure 2.1. If a third statement R is involved, then there are eight possible combinations of truth values for P, Q and R. This is displayed in Figure 2.1 as well. In general, a truth table involving n statements $P_1, P_2, \cdots, P_n$ contains 2^n possible combinations of truth values for these statements and a truth table showing these combinations would have n columns and 2^n rows. Much of the time, we will be dealing with two statements, usually denoted by P and Q; so the associated truth table will have four rows with the first two columns headed by P and Q. In this case, it is customary to consider the four combinations of the truth values in the order TT, TF, FT, FF, from top to bottom.

2.2 The Negation of a Statement

Much of the interest in integers and other familiar sets of numbers comes not only from the numbers themselves but from properties of the numbers that result by performing operations on them (such as taking their negatives, adding or multiplying them or combinations of these). Similarly, much of our interest in statements comes from investigating the truth or falseness of new statements that can be produced from one or more given statements by performing certain operations on them. Our first example concerns producing a new statement from a single given statement.

The **negation** of a statement P is the statement:

$$\text{not } P.$$

and is denoted by $\sim P$. Although $\sim P$ could always be expressed as

It is not the case that P.

there are usually better ways to express the statement $\sim P$.

Example 2.2   *For the statement*

$$P_1 : \textit{The integer 3 is odd.}$$

described above, we have

$$\sim P_1 : \textit{It is not the case that the integer 3 is odd.}$$

but it would be much preferred to write

$$\sim P_1 : \textit{The integer 3 is not odd.}$$

or better yet to write

$$\sim P_1 : \textit{The integer 3 is even.}$$

Similarly, the negation of the statement

$$P_2 : \textit{The integer 57 is prime.}$$

considered above is

$$\sim P_2 : \textit{The integer 57 is not prime.}$$

Note that $\sim P_1$ is false, while $\sim P_2$ is true.   ◆

Indeed, the negation of a true statement is always false and the negation of a false statement is always true; that is, the truth value of $\sim P$ is opposite to that of P. This is summarized in Figure 2.2, which gives the truth table for $\sim P$ (in terms of the possible truth values of P).

P	$\sim P$
T	F
F	T

Figure 2.2   The truth table for negation

2.3 The Disjunction and Conjunction of Statements

For two given statements P and Q, a common way to produce a new statement from them is by inserting the word "or" or "and" between P and Q. The **disjunction** of the statements P and Q is the statement

$$P \textbf{ or } Q$$

and is denoted by $P \vee Q$. The disjunction $P \vee Q$ is true if at least one of P and Q is true; otherwise, $P \vee Q$ is false. Therefore, $P \vee Q$ is true if exactly one of P and Q is true or if both P and Q are true.

Example 2.3   *For the statements*

$$P_1 : \textit{The integer 3 is odd. and } P_2 : \textit{The integer 57 is prime.}$$

described earlier, the disjunction is the new statement

$$P_1 \vee P_2 \textit{: Either 3 is odd or 57 is prime.}$$

which is true since at least one of P_1 and P_2 is true (namely, P_1 is true). Of course, in this case exactly one of P_1 and P_2 is true.   ◆

For two statements P and Q, the truth table for $P \vee Q$ is shown in Figure 2.3. This truth table then describes precisely when $P \vee Q$ is true (or false).

Although the truth of "P or Q" allows for both P and Q to be true, there are instances when the use of "or" does not allow that possibility. For example, for an integer n, if we were to say "n is even or n is odd," then surely it is not possible for both "n is even" and "n is odd" to be true. When "or" is used in this manner, it is called the **exclusive or**. Suppose, for example, that $\mathcal{P} = \{S_1, S_2, \ldots, S_k\}$, where $k \geq 2$, is a partition of a set S and x is some element of S. If

$$x \in S_1 \text{ or } x \in S_2$$

is true, then it is impossible for both $x \in S_1$ and $x \in S_2$ to be true.

P	Q	$P \vee Q$
T	T	T
T	F	T
F	T	T
F	F	F

Figure 2.3   The truth table for disjunction

P	Q	$P \wedge Q$
T	T	T
T	F	F
F	T	F
F	F	F

Figure 2.4   The truth table for conjunction

The **conjunction** of the statements P and Q is the statement:

$$P \text{ and } Q$$

and is denoted by $P \wedge Q$. The conjunction $P \wedge Q$ is true only when both P and Q are true; otherwise, $P \wedge Q$ is false.

Example 2.4   *For P_1 : The integer 3 is odd. and P_2 : The integer 57 is prime., the statement*

$$P_1 \wedge P_2 : 3 \text{ is odd and 57 is prime.}$$

is false since P_2 is false and so not both P_1 and P_2 are true.   ♦

The truth table for the conjunction of two statements is shown in Figure 2.4.

2.4 The Implication

A statement formed from two given statements in which we will be most interested is the implication (also called the conditional). For statements P and Q, the **implication** (or **conditional**) is the statement

$$\textbf{If } P\textbf{, then } Q\textbf{.}$$

and is denoted by $P \Rightarrow Q$. In addition to the wording "If P, then Q," we also express $P \Rightarrow Q$ in words as

$$P \textbf{ implies } Q\textbf{.}$$

The truth table for $P \Rightarrow Q$ is given in Figure 2.5.

Notice that $P \Rightarrow Q$ is false only when P is true and Q is false ($P \Rightarrow Q$ is true otherwise).

Example 2.5   *For P_1 : The integer 3 is odd. and P_2 : The integer 57 is prime., the implication*

$$P_1 \Rightarrow P_2 : \text{If 3 is an odd integer, then 57 is prime.}$$

P	Q	$P \Rightarrow Q$
T	T	T
T	F	F
F	T	T
F	F	T

Figure 2.5   The truth table for implication

is a false statement. The implication

$$P_2 \Rightarrow P_1 : \text{If 57 is prime, then 3 is odd.}$$

is true, however.   ◆

While the truth tables for the negation $\sim P$, the disjunction $P \vee Q$ and the conjunction $P \wedge Q$ are probably what one would expect, this may not be so for the implication $P \Rightarrow Q$. There is ample justification, however, for the truth values in the truth table of $P \Rightarrow Q$. We illustrate this with an example.

Example 2.6   *A student is taking a math class (let's say this one) and is currently receiving a B+. He visits his instructor a few days before the final examination and asks her, "Is there any chance that I can get an A in this course?" His instructor looks through her grade book and says, "If you earn an A on the final exam, then you will receive an A for your final grade." We now check the truth or falseness of this implication based on the various combinations of truth values of the statements*

$$P : \text{You earn an A on the final exam.}$$

and

$$Q : \text{You receive an A for your final grade.}$$

which make up the implication.

Analysis   Suppose first that P and Q are both true. That is, the student receives an A on his final exam and later learns that he got an A for his final grade in the course. Did his instructor tell the truth? I think we would all agree that she did. So if P and Q are both true, then so too is $P \Rightarrow Q$, which agrees with the first row of the truth table of Figure 2.5.

Second, suppose that P is true and Q is false. So the student got an A on his final exam but did not receive an A as a final grade, say he received a B. Certainly, his instructor did not do as she promised (as she will soon be reminded by her student). What she said was false, which agrees with the second row of the table in Figure 2.5.

Third, suppose that P is false and Q is true. In this case, the student did not get an A on his final exam (say he earned a B) *but* when he received his final grades, he learned (and was pleasantly surprised) that his final grade was an A. How could this happen? Perhaps his instructor was lenient. Perhaps the final exam was unusually difficult and a grade of B on it indicated an exceptionally good performance. Perhaps the instructor made a mistake. In any case, the instructor did not lie; so she told the truth. Indeed, she never promised anything if the student did not get an A on his final exam. This agrees with the third row of the table in Figure2.5.

Finally, suppose that P and Q are both false. That is, suppose the student did not get an A on his final exam and he also did not get an A for a final grade. The instructor did not lie here either. She only promised the student an A *if* he got an A on the final exam. Once again, she did not promise anything if the student did not get an A on the final exam. So the instructor told the truth and this agrees with the fourth and final row of the table.   ◆

In summary then, the only situation for which $P \Rightarrow Q$ is false is when P is true and Q is false (so $\sim Q$ is true). That is, the truth tables for

$$\sim(P \Rightarrow Q) \ \text{ and } \ P \wedge (\sim Q)$$

are the same. We'll revisit this observation again soon.

We have already mentioned that the implication $P \Rightarrow Q$ can be expressed as both "If P, then Q" and "P implies Q." In fact, there are several ways of expressing $P \Rightarrow Q$ in words, namely:

If P, then Q.
Q if P.
P implies Q.
P only if Q.
P is sufficient for Q.
Q is necessary for P.

It is probably not surprising that the first three of these say the same thing, but perhaps not at all obvious that the last three say the same thing as the first three. Consider the statement "P only if Q." This says that P is true only under the condition that Q is true; in other words, it cannot be the case that P is true and Q is false. Thus it says that if P is true, then necessarily Q must be true. We can also see from this that the statement "Q is necessary for P" has the same meaning as "P only if Q." The statement "P is sufficient for Q" states that the truth of P is sufficient for the truth of Q. In other words, the truth of P implies the truth of Q; that is, "P implies Q."

2.5 More on Implications

We have just discussed four ways to create new statements from one or two given statements. In mathematics, however, we are often interested in declarative sentences containing variables and whose truth or falseness is only known once we have assigned values to the variables. The values assigned to the variables come from their respective domains. These sentences are, of course, precisely the sentences we have referred to as open sentences. Just as new statements can be formed from statements P and Q by negation, disjunction, conjunction or implication, new open sentences can be constructed from open sentences in the same manner.

Example 2.7   *Consider the open sentences*

$$P_1(x) : x = -3. \text{ and } P_2(x) : |x| = 3,$$

where $x \in \mathbf{R}$, that is, where the domain of x is $\mathbf{R}$ in each case. We can then form the following open sentences:

$$\sim P_1(x) : \ x \neq -3.$$
$$P_1(x) \vee P_2(x) : \ x = -3 \ or \ |x| = 3.$$
$$P_1(x) \wedge P_2(x) : \ x = -3 \ and \ |x| = 3.$$
$$P_1(x) \Rightarrow P_2(x) : \ If \ x = -3, \ then \ |x| = 3.$$

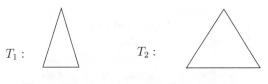

Figure 2.6   Isosceles and equilateral triangles

For a specific real number x, the truth value of each resulting statement can be deter-mined. For example, $\sim P_1(-3)$ is a false statement, while each of the remaining sentences above results in a true statement when $x = -3$. Both $P_1(2) \vee P_2(2)$ and $P_1(2) \wedge P_2(2)$ are false statements. On the other hand, both $\sim P_1(2)$ and $P_1(2) \Rightarrow P_2(2)$ are true state-ments. In fact, for each real number $x \neq -3$, the implication $P_1(x) \Rightarrow P_2(x)$ is a true statement since $P_1(x) : x = -3$ is a false statement. Thus $P_1(x) \Rightarrow P_2(x)$ is true for all $x \in \mathbf{R}$. We will see that open sentences which result in true statements for all values of the domain will be especially interesting to us.

Listed below are various ways of wording the implication $P_1(x) \Rightarrow P_2(x)$:

> *If $x = -3$, then $|x| = 3$.*
> *$|x| = 3$ if $x = -3$.*
> *$x = -3$ implies that $|x| = 3$.*
> *$x = -3$ only if $|x| = 3$.*
> *$x = -3$ is sufficient for $|x| = 3$.*
> *$|x| = 3$ is necessary for $x = -3$.*   ◆

We now consider another example, this time from geometry. You may recall that a triangle is called **equilateral** if the lengths of its three sides are the same, while a triangle is **isosceles** if the lengths of any two of its three sides are the same. Figure 2.6 shows an isosceles triangle T_1 and an equilateral triangle T_2. Actually, since the lengths of any two of the three sides of T_2 are the same, T_2 is isosceles as well. Indeed, this is precisely the fact we wish to discuss.

Example 2.8   *For a triangle T, let*

> $P(T) : T$ *is equilateral. and $Q(T) : T$ is isosceles.*

Thus, $P(T)$ and $Q(T)$ are open sentences over the domain S of all triangles. Consider the implication $P(T) \Rightarrow Q(T)$, where the domain then of the variable T is the set S. For an equilateral triangle T_1, both $P(T_1)$ and $Q(T_1)$ are true statements and so $P(T_1) \Rightarrow Q(T_1)$ is a true statement as well. If T_2 is not an equilateral triangle, then $P(T_2)$ is a false statement and so $P(T_2) \Rightarrow Q(T_2)$ is true. Therefore, $P(T) \Rightarrow Q(T)$ is a true statement for all $T \in S$. We now state $P(T) \Rightarrow Q(T)$ in a variety of ways:

> *If T is an equilateral triangle, then T is isosceles.*
> *A triangle T is isosceles if T is equilateral.*
> *A triangle T being equilateral implies that T is isosceles.*
> *A triangle T is equilateral only if T is isosceles.*
> *For a triangle T to be isosceles, it is sufficient that T be equilateral.*
> *For a triangle T to be equilateral, it is necessary that T be isosceles.*   ◆

Notice that at times we change the wording to make the sentence sound better. In general, the sentence P in the implication $P \Rightarrow Q$ is commonly referred to as the **hypothesis** or **premise** of $P \Rightarrow Q$, while Q is called the **conclusion** of $P \Rightarrow Q$.

It is often easier to deal with an implication when expressed in an "if, then" form. This allows us to identify the hypothesis and conclusion more easily. Indeed, since implications can be stated in a wide variety of ways (even in addition to those mentioned above), being able to reword an implication as "if, then" is especially useful. For example, the implication $P(T) \Rightarrow Q(T)$ described in Example 2.8 can be encountered in many ways, including the following:

- Let T be an equilateral triangle. Then T is isosceles.
- Suppose that T is an equilateral triangle. Then T is isosceles.
- Every equilateral triangle is isosceles.
- Whenever a triangle is equilateral, it is isosceles.

We now investigate the truth or falseness of implications involving open sentences for values of their variables.

Example 2.9   *Let $S = \{2, 3, 5\}$ and let*

$$P(n): \ n^2 - n + 1 \text{ is prime. and } Q(n): \ n^3 - n + 1 \text{ is prime.}$$

be open sentences over the domain S. Determine the truth or falseness of the implication $P(n) \Rightarrow Q(n)$ for each $n \in S$.

Solution   In this case, we have the following:

$$P(2): \ 3 \text{ is prime.} \quad P(3): \ 7 \text{ is prime.} \quad P(5): \ 21 \text{ is prime.}$$
$$Q(2): \ 7 \text{ is prime.} \quad Q(3): \ 25 \text{ is prime.} \quad Q(5): \ 121 \text{ is prime.}$$

Consequently, $P(2) \Rightarrow Q(2)$ and $P(5) \Rightarrow Q(5)$ are true, while $P(3) \Rightarrow Q(3)$ is false. ◆

Example 2.10   *Let $S = \{1, 2\}$ and let $T = \{-1, 4\}$. Also, let*

$$P(x, y): \ ||x + y| - |x - y|| = 2. \text{ and } Q(x, y): \ x^{y+1} = y^x.$$

be open sentences, where the domain of the variable x is S and the domain of y is T. Determine the truth or falseness of the implication $P(x, y) \Rightarrow Q(x, y)$ for all $(x, y) \in S \times T$.

Solution   For $(x, y) = (1, -1)$, we have

$$P(1, -1) \Rightarrow Q(1, -1): \ \text{If } 2 = 2, \text{ then } 1 = -1.$$

which is false. For $(x, y) = (1, 4)$, we have

$$P(1, 4) \Rightarrow Q(1, 4): \ \text{If } 2 = 2, \text{ then } 1 = 4.$$

which is also false. For $(x, y) = (2, -1)$, we have

$$P(2, -1) \Rightarrow Q(2, -1): \ \text{If } 2 = 2, \text{ then } 1 = 1.$$

which is true; while for $(x, y) = (2, 4)$, we have

$$P(2, 4) \Rightarrow Q(2, 4): \ \text{If } 2 = 4, \text{ then } 32 = 16.$$

which is true.   ◆

For statements (or open sentences) P and Q, the implication $Q \Rightarrow P$ is called the **converse** of $P \Rightarrow Q$. The converse of an implication will often be of interest to us, either by itself or in conjunction with the original implication.

Example 2.11   *For the statements*

$$P_1 : 3 \text{ is an odd integer.} \quad P_2 : 57 \text{ is prime.}$$

the converse of the implication

$$P_1 \Rightarrow P_2 : \text{ If } 3 \text{ is an odd integer, then } 57 \text{ is prime.}$$

is the implication

$$P_2 \Rightarrow P_1 : \text{ If } 57 \text{ is prime, then } 3 \text{ is an odd integer.} \qquad \blacklozenge$$

For statements (or open sentences) P and Q, the conjunction

$$(P \Rightarrow Q) \wedge (Q \Rightarrow P)$$

of the implication $P \Rightarrow Q$ and its converse is called the **biconditional** of P and Q and is denoted by $P \Leftrightarrow Q$. For statements P and Q, the truth table for $P \Leftrightarrow Q$ can therefore be determined. This is given in Figure 2.7. From this table, we see that $P \Leftrightarrow Q$ is true whenever the statements P and Q are both true or are both false, while $P \Leftrightarrow Q$ is false otherwise. That is, $P \Leftrightarrow Q$ is true precisely when P and Q have the same truth values.

The biconditional $P \Leftrightarrow Q$ is often stated as

$$P \text{ \textbf{is equivalent to} } Q.$$

P	Q	$P \Rightarrow Q$	$Q \Rightarrow P$	$(P \Rightarrow Q) \wedge (Q \Rightarrow P)$
T	T	T	T	T
T	F	F	T	F
F	T	T	F	F
F	F	T	T	T

P	Q	$P \Leftrightarrow Q$
T	T	T
T	F	F
F	T	F
F	F	T

Figure 2.7   The truth table for a biconditional

or

$$P \text{ if and only if } Q.$$

or as

$$P \text{ is a necessary and sufficient condition for } Q.$$

For statements P and Q, it then follows that the biconditional "P if and only if Q" is true only when P and Q have the same truth values.

Example 2.12   *The biconditional*

$$3 \text{ is an odd integer if and only if } 57 \text{ is prime.}$$

is false; while the biconditional

$$100 \text{ is even if and only if } 101 \text{ is prime.}$$

is true. Furthermore, the biconditional

$$5 \text{ is even if and only if } 4 \text{ is odd.}$$

is also true.   ◆

The phrase "if and only if" occurs often in mathematics and we shall discuss this at greater length later. For the present, we consider two examples involving statements containing the phrase "if and only if."

Example 2.13   *We noted in Example 2.7 that for the open sentences*

$$P_1(x) : x = -3. \text{ and } P_2(x) : |x| = 3.$$

over the domain **R***, the implication*

$$P_1(x) \Rightarrow P_2(x) : \textit{If } x = -3, \textit{ then } |x| = 3.$$

is a true statement for each $x \in$ **R***. However, the converse*

$$P_2(x) \Rightarrow P_1(x) : \textit{If } |x| = 3, \textit{ then } x = -3.$$

is a false statement when $x = 3$ *since* $P_2(3)$ *is true and* $P_1(3)$ *is false. For all other real numbers* x*, the implication* $P_2(x) \Rightarrow P_1(x)$ *is true. Therefore, the biconditional*

$$P_1(x) \Leftrightarrow P_2(x) : \; x = -3 \textit{ if and only if } |x| = 3.$$

is false when $x = 3$ *and is true for all other real numbers* x*.*   ◆

Example 2.14   *For the open sentences*

$$P(T) : T \textit{ is equilateral. and } Q(T) : T \textit{ is isosceles.}$$

over the domain S of all triangles, the converse of the implication

$$P(T) \Rightarrow Q(T) : \text{If } T \text{ is equilateral, then } T \text{ is isosceles.}$$

is the implication

$$Q(T) \Rightarrow P(T) : \text{If } T \text{ is isosceles, then } T \text{ is equilateral.}$$

We noted that $P(T) \Rightarrow Q(T)$ is a true statement for all triangles T, while $Q(T) \Rightarrow P(T)$ is a false statement when T is an isosceles triangle that is not equilateral. On the other hand, the second implication becomes a true statement for all other triangles T. Therefore, the biconditional

$$P(T) \Leftrightarrow Q(T) : T \text{ is equilateral if and only if } T \text{ is isosceles.}$$

is false for all triangles that are isosceles and not equilateral, while it is true for all other triangles T.   ◆

We now investigate the truth or falseness of biconditionals obtained by assigning to a variable each value in its domain.

Example 2.15   *Let $S = \{0, 1, 4\}$. Consider the following open sentences over the domain S:*

$$P(n) : \ \frac{n(n+1)(2n+1)}{6} \ \text{is odd.}$$
$$Q(n) : \ (n+1)^3 = n^3 + 1.$$

Determine three distinct elements a, b, c in S such that $P(a) \Rightarrow Q(a)$ is false, $Q(b) \Rightarrow P(b)$ is false, and $P(c) \Leftrightarrow Q(c)$ is true.

Solution   Observe that

$$P(0) : \ 0 \text{ is odd.} \quad P(1) : \ 1 \text{ is odd.} \quad P(4) : \ 30 \text{ is odd.}$$

$$Q(0) : \ 1 = 1. \quad Q(1) : \ 8 = 2. \quad Q(4) : \ 125 = 65.$$

Thus $P(0)$ and $P(4)$ are false, while $P(1)$ is true. Also, $Q(1)$ and $Q(4)$ are false, while $Q(0)$ is true. Thus $P(1) \Rightarrow Q(1)$ and $Q(0) \Rightarrow P(0)$ are false, while $P(4) \Leftrightarrow Q(4)$ is true. Hence we may take $a = 1$, $b = 0$ and $c = 4$.   ◆

Analysis   Notice in Example 2.15 that both $P(0) \Leftrightarrow Q(0)$ and $P(1) \Leftrightarrow Q(1)$ are false biconditionals. Hence the value 4 in S is the only choice for c.   ◆

2.7 Tautologies and Contradictions

The symbols $\sim$, $\vee$, $\wedge$, $\Rightarrow$ and $\Leftrightarrow$ are sometimes referred to as **logical connectives**. From given statements, we can use these logical connectives to form more intricate statements. For example, the statement $(P \vee Q) \wedge (P \vee R)$ is a statement formed from the given statements P, Q and R and the logical connectives $\vee$ and $\wedge$. We call $(P \vee Q) \wedge (P \vee R)$ a compound statement. More generally, a **compound statement** is a statement composed of one or more given statements (called **component statements** in this context) and at least one logical connective. For example, for a given component statement P, its negation $\sim P$ is a compound statement.

The compound statement $P \vee (\sim P)$, whose truth table is given in Figure 2.8, has the feature that it is true regardless of the truth value of P.

A compound statement S is called a **tautology** if it is true for all possible combinations of truth values of the component statements that comprise S. Hence $P \vee (\sim P)$ is a tautology, as is $(\sim Q) \vee (P \Rightarrow Q)$. This latter fact is verified in the truth table shown in Figure 2.9.

Letting

$$P_1 : 3 \text{ is odd. and } P_2 : 57 \text{ is prime.}$$

we see that not only is

$$57 \text{ is not prime, or } 57 \text{ is prime if } 3 \text{ is odd.}$$

a true statement, but $(\sim P_2) \vee (P_1 \Rightarrow P_2)$ is true regardless of which statements P_1 and P_2 are being considered.

On the other hand, a compound statement S is called a **contradiction** if it is false for all possible combinations of truth values of the component statements that are used to form S. The statement $P \wedge (\sim P)$ is a contradiction, as is shown in Figure 2.10. Hence the statement

$$3 \text{ is odd and } 3 \text{ is not odd.}$$

is false.

Another example of a contradiction is $(P \wedge Q) \wedge (Q \Rightarrow (\sim P))$, which is verified in the truth table shown in Figure 2.11.

Indeed, if a compound statement S is a tautology, then its negation $\sim S$ is a contradiction.

P	$\sim P$	$P \vee (\sim P)$
T	F	T
F	T	T

Figure 2.8   An example of a tautology

P	Q	$\sim Q$	$P \Rightarrow Q$	$(\sim Q) \vee (P \Rightarrow Q)$
T	T	F	T	T
T	F	T	F	T
F	T	F	T	T
F	F	T	T	T

Figure 2.9   Another tautology

P	$\sim P$	$P \wedge (\sim P)$
T	F	*F*
F	T	*F*

Figure 2.10   An example of a contradiction

P	Q	$\sim P$	$P \wedge Q$	$Q \Rightarrow (\sim P)$	$(P \wedge Q) \wedge (Q \Rightarrow (\sim P))$
T	T	F	T	F	*F*
T	F	F	F	T	*F*
F	T	T	F	T	*F*
F	F	T	F	T	*F*

Figure 2.11   Another contradiction

2.8 Logical Equivalence

Figure 2.12 shows a truth table for the two statements $P \Rightarrow Q$ and $(\sim P) \vee Q$. The corresponding columns of these compound statements are identical; in other words, these two compound statements have exactly the same truth value for every combination of truth values of the statements P and Q. Let R and S be two compound statements involving the same component statements. Then R and S are called **logically equivalent** if R and S have the same truth values for all combinations of truth values of their component statements. If R and S are logically equivalent, then this is denoted by $R \equiv S$. Hence $P \Rightarrow Q$ and $(\sim P) \vee Q$ are logically equivalent and so $P \Rightarrow Q \equiv (\sim P) \vee Q$.

Another, even simpler, example of logical equivalence concerns $P \wedge Q$ and $Q \wedge P$. That $P \wedge Q \equiv Q \wedge P$ is verified in the truth table shown in Figure 2.13.

What is the practical significance of logical equivalence? Suppose that R and S are logically equivalent compound statements. Then we know that R and S have the same truth values for all possible combinations of truth values of their component statements. But this means that the biconditional $R \Leftrightarrow S$ is true for all possible combinations of truth values of their component statements and hence $R \Leftrightarrow S$ is a tautology..Conversely, if $R \Leftrightarrow S$ is a tautology, then R and S are logically equivalent.

Let R be a mathematical statement that we would like to show is true and suppose that R and some statement S are logically equivalent. If we can show that S is true, then R is true as well. For example, suppose that we want to verify the truth of an

P	Q	$\sim P$	$P \Rightarrow Q$	$(\sim P) \vee Q$
T	T	F	*T*	*T*
T	F	F	*F*	*F*
F	T	T	*T*	*T*
F	F	T	*T*	*T*

Figure 2.12   Verification of $P \Rightarrow Q \equiv (\sim P) \vee Q$

P	Q	$P \wedge Q$	$Q \wedge P$
T	T	T	T
T	F	F	F
F	T	F	F
F	F	F	F

Figure 2.13   Verification of $P \wedge Q \equiv Q \wedge P$

implication $P \Rightarrow Q$. If we can establish the truth of the statement $(\sim P) \vee Q$, then the logical equivalence of $P \Rightarrow Q$ and $(\sim P) \vee Q$ guarantees that $P \Rightarrow Q$ is true as well.

Example 2.16   *Returning to the mathematics instructor in Example 2.6 and whether she kept her promise that*

> *If you earn an A on the final exam, then you will receive an A for the final grade.*

we need only know that the student did not receive an A on the final exam or the student received an A as a final grade to see that she kept her promise.   ♦

Since the logical equivalence of $P \Rightarrow Q$ and $(\sim P) \vee Q$, verified in Figure 2.12, is especially important and we will have occasion to use this fact often, we state it as a theorem.

Theorem 2.17   *Let P and Q be two statements. Then*

$$P \Rightarrow Q \text{ and } (\sim P) \vee Q$$

are logically equivalent.

Let's return to the truth table in Figure 2.13, where we showed that $P \wedge Q$ and $Q \wedge P$ are logically equivalent for any two statements P and Q. In particular, this says that

$$(P \Rightarrow Q) \wedge (Q \Rightarrow P) \text{ and } (Q \Rightarrow P) \wedge (P \Rightarrow Q)$$

are logically equivalent. Of course, $(P \Rightarrow Q) \wedge (Q \Rightarrow P)$ is precisely what is called the biconditional of P and Q. Since $(P \Rightarrow Q) \wedge (Q \Rightarrow P)$ and $(Q \Rightarrow P) \wedge (P \Rightarrow Q)$ are logically equivalent, $(Q \Rightarrow P) \wedge (P \Rightarrow Q)$ represents the biconditional of P and Q as well. Since $Q \Rightarrow P$ can be written as "P if Q" and $P \Rightarrow Q$ can be expressed as "P only if Q," their conjunction can be written as "P if Q and P only if Q" or, more simply, as

$$P \text{ if and only if } Q.$$

Consequently, expressing $P \Leftrightarrow Q$ as "P if and only if Q" is justified. Furthermore, since $Q \Rightarrow P$ can be phrased as "P is necessary for Q" and $P \Rightarrow Q$ can be expressed as "P is sufficient for Q," writing $P \Leftrightarrow Q$ as "P is necessary and sufficient for Q" is likewise justified.

2.9 Some Fundamental Properties of Logical Equivalence

It probably comes as no surprise that the statements P and $\sim(\sim P)$ are logically equivalent. This fact is verified in Figure 2.14.

We mentioned in Figure 2.13 that, for two statements P and Q, the statements $P \wedge Q$ and $Q \wedge P$ are logically equivalent. There are other fundamental logical equivalences that we often encounter as well.

Theorem 2.18   *For statements P, Q and R,*

(1) *Commutative Laws*
    (a) $P \vee Q \equiv Q \vee P$
    (b) $P \wedge Q \equiv Q \wedge P$
(2) *Associative Laws*
    (a) $P \vee (Q \vee R) \equiv (P \vee Q) \vee R$
    (b) $P \wedge (Q \wedge R) \equiv (P \wedge Q) \wedge R$
(3) *Distributive Laws*
    (a) $P \vee (Q \wedge R) \equiv (P \vee Q) \wedge (P \vee R)$
    (b) $P \wedge (Q \vee R) \equiv (P \wedge Q) \vee (P \wedge R)$
(4) *De Morgan's Laws*
    (a) $\sim(P \vee Q) \equiv (\sim P) \wedge (\sim Q)$
    (b) $\sim(P \wedge Q) \equiv (\sim P) \vee (\sim Q).$

Each part of Theorem 2.18 is verified by means of a truth table. We have already established the commutative law for conjunction (namely $P \wedge Q \equiv Q \wedge P$) in Figure 2.13. In Figure 2.15 $P \vee (Q \wedge R) \equiv (P \vee Q) \wedge (P \vee R)$ is verified by observing that the columns corresponding to the statements $P \vee (Q \wedge R)$ and $(P \vee Q) \wedge (P \vee R)$ are identical.

The laws given in Theorem 2.18, together with other known logical equivalences, can be used to good advantage at times to prove other logical equivalences (without introducing a truth table).

Example 2.19   *Suppose we are asked to verify that*

$$\sim(P \Rightarrow Q) \equiv P \wedge (\sim Q)$$

for every two statements P and Q. Using the logical equivalence of $P \Rightarrow Q$ and $(\sim P) \vee Q$ from Theorem 2.17 and Theorem 2.18(4a), we see that

$$\sim(P \Rightarrow Q) \equiv \sim((\sim P) \vee Q) \equiv (\sim(\sim P)) \wedge (\sim Q) \equiv P \wedge (\sim Q), \tag{2.1}$$

P	$\sim P$	$\sim(\sim P)$
T	F	T
F	T	F

Figure 2.14   Verification of $P \equiv \sim(\sim P)$

P	Q	R	$Q \wedge R$	$P \vee Q$	$P \vee R$	$P \vee (Q \wedge R)$	$(P \vee Q) \wedge (P \vee R)$
T	T	T	T	T	T	T	T
T	T	F	F	T	T	T	T
T	F	T	F	T	T	T	T
T	F	F	F	T	T	T	T
F	T	T	T	T	T	T	T
F	T	F	F	T	F	F	F
F	F	T	F	F	T	F	F
F	F	F	F	F	F	F	F

Figure 2.15   Verification of the distributive law $P \vee (Q \wedge R) \equiv (P \vee Q) \wedge (P \vee R)$

implying that the statements $\sim (P \Rightarrow Q)$ and $P \wedge (\sim Q)$ are logically equivalent, which we alluded to earlier.   ◆

It is important to keep in mind what we have said about logical equivalence. For example, the logical equivalence of $P \wedge Q$ and $Q \wedge P$ allows us to replace a statement of the type $P \wedge Q$ by $Q \wedge P$ without changing its truth value. As an additional example, according to De Morgan's Laws in Theorem 2.18, if it is not the case that an integer a is even or an integer b is even, then it follows that a and b are both odd.

Example 2.20   *Using the second of De Morgan's Laws and (2.1), we can establish a useful logically equivalent form of the negation of $P \Leftrightarrow Q$ by the following string of logical equivalences:*

$$\sim (P \Leftrightarrow Q) \equiv \; \sim ((P \Rightarrow Q) \wedge (Q \Rightarrow P))$$
$$\equiv (\sim (P \Rightarrow Q)) \vee (\sim (Q \Rightarrow P))$$
$$\equiv (P \wedge (\sim Q)) \vee (Q \wedge (\sim P)).$$ ◆

What we have observed about the negation of an implication and a biconditional is repeated in the following theorem.

Theorem 2.21   *For statements P and Q,*

(a) $\sim (P \Rightarrow Q) \equiv P \wedge (\sim Q)$
(b) $\sim (P \Leftrightarrow Q) \equiv (P \wedge (\sim Q)) \vee (Q \wedge (\sim P))$.

Example 2.22   *Once again, let's return to what the mathematics instructor in Example 2.6 said:*

If you earn an A on the final exam, then you will receive an A for your final grade.

If this instructor was not truthful, then it follows by Theorem 2.21(a) that

You earned an A on the final exam and did not receive A as your final grade.

Suppose, on the other hand, that the mathematics instructor had said:

> *If you earn an A on the final exam, then you will receive an A for the final grade—and that's the only way that you will get an A for a final grade.*

If this instructor was not truthful, then it follows by Theorem 2.21(b) that

> *Either you earned an A on the final exam and didn't receive A as your final grade or you received an A for your final grade and you didn't get an A on the final exam.*   ♦

2.10 Quantified Statements

We have mentioned that if $P(x)$ is an open sentence over a domain S, then $P(x)$ is a statement for each $x \in S$. We illustrate this again.

Example 2.23   *If $S = \{1, 2, \cdots, 7\}$, then*

$$P(n): \quad \frac{2n^2 + 5 + (-1)^n}{2} \text{ is prime.}$$

is a statement for each $n \in S$. Therefore,

$$P(1): \ 3 \text{ is prime.}$$
$$P(2): \ 7 \text{ is prime.}$$
$$P(3): \ 11 \text{ is prime.}$$
$$P(4): \ 19 \text{ is prime.}$$

are true statements; while

$$P(5): \ 27 \text{ is prime.}$$
$$P(6): \ 39 \text{ is prime.}$$
$$P(7): \ 51 \text{ is prime.}$$

are false statements.   ♦

There are other ways that an open sentence can be converted into a statement, namely by a method called **quantification**. Let $P(x)$ be an open sentence over a domain S. Adding the phrase "For every $x \in S$" to $P(x)$ produces a statement called a **quantified statement**. The phrase "for every" is referred to as the **universal quantifier** and is denoted by the symbol $\forall$. Other ways to express the universal quantifier are "for each" and "for all." This quantified statement is expressed in symbols by

$$\forall x \in S, P(x) \tag{2.2}$$

and is expressed in words by

$$\text{For every } x \in S, P(x). \tag{2.3}$$

The quantified statement (2.2) (or (2.3)) is true if $P(x)$ is true for every $x \in S$, while the quantified statement (2.2) is false if $P(x)$ is false for at least one element $x \in S$.

Another way to convert an open sentence $P(x)$ over a domain S into a statement through quantification is by the introduction of a quantifier called an existential quantifier. Each of the phrases *there exists, there is, for some* and *for at least one* is referred to as an **existential quantifier** and is denoted by the symbol ∃. The quantified statement

$$\exists x \in S, P(x) \tag{2.4}$$

can be expressed in words by

$$\text{There exists } x \in S \text{ such that } P(x). \tag{2.5}$$

The quantified statement (2.4) (or (2.5)) is true if $P(x)$ is true for at least one element $x \in S$, while the quantified statement (2.4) is false if $P(x)$ is false for all $x \in S$.

We now consider two quantified statements constructed from the open sentence we saw in Example 2.23.

Example 2.24   *For the open sentence*

$$P(n) : \frac{2n^2 + 5 + (-1)^n}{2} \text{ is prime.}$$

over the domain $S = \{1, 2, \cdots, 7\}$, the quantified statement

$$\forall n \in S, P(n) : \text{ For every } n \in S, \frac{2n^2 + 5 + (-1)^n}{2} \text{ is prime.}$$

is false since $P(5)$ is false, for example; while the quantified statement

$$\exists n \in S, P(n) : \text{ There exists } n \in S \text{ such that } \frac{2n^2 + 5 + (-1)^n}{2} \text{ is prime.}$$

is true since $P(1)$ is true, for example.   ◆

The quantified statement $\forall x \in S, P(x)$ can also be expressed as

$$\text{If } x \in S, \text{ then } P(x).$$

Consider the open sentence $P(x) : x^2 \geq 0.$ over the set **R** of real numbers. Then

$$\forall x \in \mathbf{R}, P(x)$$

or, equivalently,

$$\forall x \in \mathbf{R}, x^2 \geq 0$$

can be expressed as

$$\text{For every real number } x, x^2 \geq 0.$$

or

$$\text{If } x \text{ is a real number, then } x^2 \geq 0.$$

as well as

$$\text{The square of every real number is nonnegative.}$$

In general, the universal quantifier is used to claim that the statement resulting from a given open sentence is true when each value of the domain of the variable is assigned to the variable. Consequently, the statement $\forall x \in \mathbf{R}, x^2 \geq 0$ is true since $x^2 \geq 0$ is true for every real number x.

Suppose now that we were to consider the open sentence $Q(x) : x^2 \leq 0$. The statement $\forall x \in \mathbf{R}, Q(x)$ (that is, for every real number x, we have $x^2 \leq 0$) is false since, for example, $Q(1)$ is false. Of course, this means that its negation is true. If it were not the case that for every real number x, we have $x^2 \leq 0$, then there must exist some real number x such that $x^2 > 0$. This negation

> There exists a real number x such that $x^2 > 0$.

can be written in symbols as

$$\exists x \in \mathbf{R}, x^2 > 0 \text{ or } \exists x \in \mathbf{R}, \sim Q(x).$$

More generally, if we are considering an open sentence $P(x)$ over a domain S, then

$$\sim (\forall x \in S, P(x)) \equiv \exists x \in S, \sim P(x).$$

Example 2.25 *Suppose that we are considering the set $A = \{1, 2, 3\}$ and its power set $\mathcal{P}(A)$, the set of all subsets of A. Then the quantified statement*

$$\text{For every set } B \in \mathcal{P}(A), A - B \neq \emptyset. \tag{2.6}$$

is false since for the subset $B = A = \{1, 2, 3\}$, we have $A - B = \emptyset$. The negation of the statement (2.6) is

$$\text{There exists } B \in \mathcal{P}(A) \text{ such that } A - B = \emptyset. \tag{2.7}$$

The statement (2.7) is therefore true since for $B = A \in \mathcal{P}(A)$, we have $A - B = \emptyset$. The statement (2.6) can also be written as

$$\text{If } B \subseteq A, \text{ then } A - B \neq \emptyset. \tag{2.8}$$

Consequently, the negation of (2.8) can be expressed as

$$\text{There exists some subset } B \text{ of } A \text{ such that } A - B = \emptyset. \qquad \blacklozenge$$

The existential quantifier is used to claim that at least one statement resulting from a given open sentence is true when the values of a variable are assigned from its domain. We know that for an open sentence $P(x)$ over a domain S, the quantified statement $\exists x \in S, P(x)$ is true provided $P(x)$ is a true statement for at least one element $x \in S$. Thus the statement $\exists x \in \mathbf{R}, x^2 > 0$ is true since, for example, $x^2 > 0$ is true for $x = 1$.

The quantified statement

$$\exists x \in \mathbf{R}, \ 3x = 12$$

is therefore true since there is some real number x for which $3x = 12$, namely $x = 4$ has this property. (Indeed, $x = 4$ is the *only* real number for which $3x = 12$.) On the other hand, the quantified statement

$$\exists n \in \mathbf{Z}, \ 4n - 1 = 0$$

is false as there is no integer n for which $4n - 1 = 0$. (Of course, $4n - 1 = 0$ when $n = 1/4$, but $1/4$ is not an integer.)

Suppose that $Q(x)$ is an open sentence over a domain S. If the statement $\exists x \in S$, $Q(x)$ is *not* true, then it must be the case that for every $x \in S$, $Q(x)$ is false. That is,

$$\sim (\exists x \in S, Q(x)) \equiv \forall x \in S, \sim Q(x)$$

is true. We illustrate this with a specific example.

Example 2.26   *The following statement contains the existential quantifier:*

$$\text{There exists a real number } x \text{ such that } x^2 = 3. \tag{2.9}$$

If we let $P(x) : x^2 = 3$, then (2.9) can be rewritten as $\exists x \in \mathbf{R}, P(x)$. The statement (2.9) is true since $P(x)$ is true when $x = \sqrt{3}$ (or when $x = -\sqrt{3}$). Hence the negation of (2.9) is:

$$\text{For every real number } x, x^2 \neq 3. \tag{2.10}$$

The statement (2.10) is therefore false.   ♦

Let $P(x, y)$ be an open sentence, where the domain of the variable x is S and the domain of the variable y is T. Then the quantified statement

$$\text{For all } x \in S \text{ and } y \in T, P(x, y).$$

can be expressed symbolically as

$$\forall x \in S, \forall y \in T, \ P(x, y). \tag{2.11}$$

The negation of the statement (2.11) is

$$\sim (\forall x \in S, \forall y \in T, \ P(x, y)) \equiv \exists x \in S, \ \sim (\forall y \in T, \ P(x, y))$$
$$\equiv \exists x \in S, \exists y \in T, \ \sim P(x, y). \tag{2.12}$$

We now consider examples of quantified statements involving two variables.

Example 2.27   *Consider the statement*

$$\text{For every two real numbers } x \text{ and } y, x^2 + y^2 \geq 0. \tag{2.13}$$

If we let

$$P(x, y) : \ x^2 + y^2 \geq 0$$

where the domain of both x and y is $\mathbf{R}$, then statement (2.13) can be expressed as

$$\forall x \in \mathbf{R}, \forall y \in \mathbf{R}, \ P(x, y) \tag{2.14}$$

or as

$$\forall x, y \in \mathbf{R}, \ P(x, y).$$

Since $x^2 \geq 0$ and $y^2 \geq 0$ for all real numbers x and y, it follows that $x^2 + y^2 \geq 0$ and so $P(x, y)$ is true for all real numbers x and y. Thus the quantified statement (2.14) is true.

The negation of statement (2.14) is therefore

$$\sim (\forall x \in \mathbf{R}, \forall y \in \mathbf{R}, \ P(x, y)) \equiv \exists x \in \mathbf{R}, \exists y \in \mathbf{R}, \ \sim P(x, y) \equiv \exists x, y \in \mathbf{R}, \sim P(x, y),$$
$$\text{(2.15)}$$

which, in words, is

$$\text{There exist real numbers } x \text{ and } y \text{ such that } x^2 + y^2 < 0. \qquad \text{(2.16)}$$

The statement (2.16) is therefore false. ◆

For an open sentence containing two variables, the domains of the variables need not be the same.

Example 2.28 *Consider the statement*

$$\text{For every } s \in S \text{ and } t \in T, st + 2 \text{ is a prime.} \qquad \text{(2.17)}$$

where the domain of the variable s is $S = \{1, 3, 5\}$ and the domain of the variable t is $T = \{3, 9\}$. If we let

$$Q(s, t): \ st + 2 \text{ is a prime.}$$

then the statement (2.17) can be expressed as

$$\forall s \in S, \forall t \in T, \ Q(s, t). \qquad \text{(2.18)}$$

Since all of the statements

$Q(1, 3)$: $1 \cdot 3 + 2$ *is a prime.* $Q(3, 3)$: $3 \cdot 3 + 2$ *is a prime.*
$Q(5, 3)$: $5 \cdot 3 + 2$ *is a prime.*

$Q(1, 9)$: $1 \cdot 9 + 2$ *is a prime.* $Q(3, 9)$: $3 \cdot 9 + 2$ *is a prime.*
$Q(5, 9)$: $5 \cdot 9 + 2$ *is a prime.*

are true, the quantified statement (2.18) is true.
As we saw in (2.12), the negation of the quantified statement (2.18) is

$$\sim (\forall s \in S, \forall t \in T, \ Q(s, t)) \equiv \exists s \in S, \exists t \in T, \ \sim Q(s, t)$$

and so the negation of (2.17) is

$$\text{There exist } s \in S \text{ and } t \in T \text{ such that } st + 2 \text{ is not a prime.} \qquad \text{(2.19)}$$

The statement (2.19) is therefore false. ◆

Again, let $P(x, y)$ be an open sentence, where the domain of the variable x is S and the domain of the variable y is T. The quantified statement

$$\text{There exist } x \in S \text{ and } y \in T \text{ such that } P(x, y)$$

can be expressed in symbols as

$$\exists x \in S, \exists y \in T, \ P(x, y). \qquad \text{(2.20)}$$

The negation of the statement (2.20) is then

$$\sim(\exists x \in S, \exists y \in T, \ P(x, y)) \equiv \forall x \in S, \ \sim(\exists y \in T, \ P(x, y))$$
$$\equiv \forall x \in S, \forall y \in T, \ \sim P(x, y). \tag{2.21}$$

We now illustrate this situation.

Example 2.29   *Consider the open sentence*

$$R(s, t): \ |s - 1| + |t - 2| \le 2.$$

where the domain of the variable s is the set S of even integers and the domain of the variable t is the set T of odd integers. Then the quantified statement

$$\exists s \in S, \exists t \in T, R(s, t). \tag{2.22}$$

can be expressed in words as

> *There exist an even integer s and an odd integer t such that $|s - 1| + |t - 2| \le 2$.*
> $\qquad\qquad$ **(2.23)**

Since $R(2, 3): \ 1 + 1 \le 2$ is true, the quantified statement (2.23) is true.
$\qquad$ *The negation of (2.22) is therefore*

$$\sim(\exists s \in S, \exists t \in T, \ R(s, t)) \equiv \forall s \in S, \forall t \in T, \ \sim R(s, t). \tag{2.24}$$

and so the negation of (2.22), in words, is

> *For every even integer s and every odd integer t, $|s - 1| + |t - 2| > 2$.* $\qquad$ **(2.25)**

The quantified statement (2.25) is therefore false. $\qquad\qquad\qquad\qquad\qquad$ ♦

In the next two examples of negations of quantified statements, De Morgan's laws are also used.

Example 2.30   *The negation of*

> *For all integers a and b, if ab is even, then a is even and b is even.*

is

> *There exist integers a and b such that ab is even and a or b is odd.* $\qquad$ ♦

Example 2.31   *The negation of*

> *There exists a rational number r such that $r \in A = \{\sqrt{2}, \pi\}$ or*
> $\qquad\qquad\qquad r \in B = \{-\sqrt{2}, \sqrt{3}, e\}.$

is

> *For every rational number r, both $r \notin A$ and $r \notin B$.* $\qquad\qquad\qquad$ ♦

Quantified statements may contain both universal and existential quantifiers. While we present examples of these now, we will discuss these in more detail in Section 7.2.

Example 2.32   *Consider the open sentence*

$$P(a, b): \ ab = 1.$$

where the domain of both a and b is the set $\mathbf{Q}^+$ of positive rational numbers. Then the quantified statement

$$\forall a \in \mathbf{Q}^+, \exists b \in \mathbf{Q}^+, P(a, b) \tag{2.26}$$

can be expressed in words as

> *For every positive rational number a, there exists*
> *a positive rational number b such that ab = 1.*

It turns out that the quantified statement (2.26) is true. If we replace $\mathbf{Q}^+$ by $\mathbf{R}$, then we have

$$\forall a \in \mathbf{R}, \exists b \in \mathbf{R}, P(a, b). \tag{2.27}$$

The negation of this statement is

$$\sim (\forall a \in \mathbf{R}, \exists b \in \mathbf{R}, P(a, b)) \equiv \exists a \in \mathbf{R}, \sim (\exists b \in \mathbf{R}, P(a, b))$$
$$\equiv \exists a \in \mathbf{R}, \forall b \in \mathbf{R}, \sim P(a, b),$$

which, in words, says that

> *There exists a real number a such that for every real number b, $ab \neq 1$.*

This negation is true since for $a = 0$ and every real number b, $ab = 0 \neq 1$. Thus the quantified statement (2.27) is false. ◆

Example 2.33   *Consider the open sentence*

$$Q(a, b): \ ab \text{ is odd.}$$

where the domain of both a and b is the set $\mathbf{N}$ of positive integers. Then the quantified statement

$$\exists a \in \mathbf{N}, \forall b \in \mathbf{N}, Q(a, b), \tag{2.28}$$

expressed in words, is

> *There exists a positive integer a such that for every positive integer b, ab is odd.*

The statement (2.28) turns out to be false. The negation of (2.28), in symbols, is

$$\sim (\exists a \in \mathbf{N}, \forall b \in \mathbf{N}, Q(a, b)) \equiv \forall a \in \mathbf{N}, \sim (\forall b \in \mathbf{N}, Q(a, b))$$
$$\equiv \forall a \in \mathbf{N}, \exists b \in \mathbf{N}, \sim Q(a, b).$$

In words, this says

> *For every positive integer a, there exists a positive integer b such that ab is even.*

This statement, therefore, is true. ◆

Suppose that $P(x, y)$ is an open sentence, where the domain of x is S and the domain of y is T. Then the quantified statement

$$\forall x \in S, \exists y \in T, P(x, y)$$

is true if $\exists y \in T, P(x, y)$ is true for each $x \in S$. This means that for every $x \in S$, there is some $y \in T$ for which $P(x, y)$ is true.

Example 2.34   *Consider the open sentence*

$$P(x, y):\ x + y \text{ is prime.}$$

where the domain of x is $S = \{2, 3\}$ and the domain of y is $T = \{3, 4\}$. The quantified statement

$$\forall x \in S, \exists y \in T, P(x, y),$$

expressed in words, is

For every $x \in S$, there exists $y \in T$ such that $x + y$ is prime.

This statement is true. For $x = 2$, $P(2, 3)$ is true and for $x = 3$, $P(3, 4)$ is true.   ◆

Suppose that $Q(x, y)$ is an open sentence, where S is the domain of x and T is the domain of y. The quantified statement

$$\exists x \in S, \forall y \in T, Q(x, y)$$

is true if $\forall y \in T, Q(x, y)$ is true for some $x \in S$. This means that for some element x in S, the open sentence $Q(x, y)$ is true for all $y \in T$.

Example 2.35   *Consider the open sentence*

$$Q(x, y):\ x + y \text{ is prime.}$$

where the domain of x is $S = \{3, 5, 7\}$ and the domain of y is $T = \{2, 6, 8, 12\}$. The quantified statement

$$\exists x \in S, \forall y \in T, Q(x, y), \tag{2.29}$$

expressed in words, is

There exists some $x \in S$ such that for every $y \in T$, $x + y$ is prime.

For $x = 5$, all of the numbers $5 + 2, 5 + 6, 5 + 8$, and $5 + 12$ are prime. Consequently, the quantified statement (2.29) is true.   ◆

Let's review symbols that we have introduced in this chapter:

$\sim$	negation (not)
$\vee$	disjunction (or)
$\wedge$	conjunction (and)
$\Rightarrow$	implication
$\Leftrightarrow$	biconditional
$\forall$	universal quantifier (for every)
$\exists$	existential quantifier (there exists)

2.11 Characterizations of Statements

Let's return to the biconditional $P \Leftrightarrow Q$. Recall that $P \Leftrightarrow Q$ represents the compound statement $(P \Rightarrow Q) \wedge (Q \Rightarrow P)$. Earlier, we described how this compound statement can be expressed as

$$P \text{ if and only if } Q.$$

Many mathematicians abbreviate the phrase "if and only if" by writing "iff." Although "iff" is informal and, of course, is not a word, its use is common and you should be familiar with it.

Recall that whenever you see

$$P \text{ if and only if } Q.$$

or

$$P \text{ is necessary and sufficient for } Q.$$

this means

$$\text{If } P \text{ then } Q \text{ and if } Q \text{ then } P.$$

Example 2.36   *Suppose that*

$$P(x) : x = -3. \text{ and } Q(x) : |x| = 3.$$

where $x \in \mathbf{R}$. Then the biconditional $P(x) \Leftrightarrow Q(x)$ can be expressed as

$$x = -3 \text{ if and only if } |x| = 3.$$

or

$$x = -3 \text{ is necessary and sufficient for } |x| = 3.$$

or, perhaps better, as

$$x = -3 \text{ is a necessary and sufficient condition for } |x| = 3.$$

Let's now consider the quantified statement $\forall x \in \mathbf{R}, P(x) \Leftrightarrow Q(x)$. This statement is false because $P(3) \Leftrightarrow Q(3)$ is false. ♦

Suppose that some concept (or object) is expressed in an open sentence $P(x)$ over a domain S and $Q(x)$ is another open sentence over the domain S concerning this concept. We say that this concept is **characterized** by $Q(x)$ if $\forall x \in S$, $P(x) \Leftrightarrow Q(x)$ is a true statement. The statement $\forall x \in S$, $P(x) \Leftrightarrow Q(x)$ is then called a **characterization** of this concept. For example, *irrational numbers* are defined as real numbers that are not rational and are characterized as real numbers whose decimal expansions are nonrepeating. This provides a characterization of irrational numbers:

A real number r is irrational if and only if r has a nonrepeating decimal expansion.

We saw that equilateral triangles are defined as triangles whose sides are equal. They are characterized however as triangles whose angles are equal. Therefore, we have the characterization:

A triangle T is equilateral if and only if T has three equal angles.

You might think that equilateral triangles are also characterized as those triangles having three equal sides but the associated biconditional:

A triangle T is equilateral if and only if T has three equal sides.

is not a characterization of equilateral triangles. Indeed, this is the definition we gave of equilateral triangles. A characterization of a concept then gives an alternative, but equivalent, way of looking at this concept. Characterizations are often valuable in studying concepts or in proving other results. We will see examples of this in future chapters.

We mentioned that the following biconditional, though true, is not a characterization: A triangle T is equilateral if and only if T has three equal sides. Although this is the definition of equilateral triangles, mathematicians rarely use the phrase "if and only if" in a definition since this is what is meant in a definition. That is, a triangle is defined to be equilateral if it has three equal sides. Consequently, a triangle with three equal sides is equilateral but a triangle that does not have three equal sides is not equilateral.

EXERCISES FOR CHAPTER 2

Section 2.1: Statements

2.1. Which of the following sentences are statements? For those that are, indicate the truth value.

(a) The integer 123 is prime.

(b) The integer 0 is even.

(c) Is $5 \times 2 = 10$?

(d) $x^2 - 4 = 0$.

(e) Multiply $5x + 2$ by 3.

(f) $5x + 3$ is an odd integer.

(g) What an impossible question!

2.2. Consider the sets A, B, C and D below. Which of the following statements are true? Give an explanation for each false statement.

$$A = \{1, 4, 7, 10, 13, 16, \ldots\} \quad C = \{x \in \mathbf{Z} : x \text{ is prime and } x \neq 2\}$$
$$B = \{x \in \mathbf{Z} : x \text{ is odd}\} \quad D = \{1, 2, 3, 5, 8, 13, 21, 34, 55, \ldots\}$$

(a) $25 \in A$ (b) $33 \in D$ (c) $22 \notin A \cup D$ (d) $C \subseteq B$ (e) $\emptyset \in B \cap D$ (f) $53 \notin C$.

2.3. Which of the following statements are true? Give an explanation for each false statement.
(a) $\emptyset \in \emptyset$ (b) $\emptyset \in \{\emptyset\}$ (c) $\{1, 3\} = \{3, 1\}$
(d) $\emptyset = \{\emptyset\}$ (e) $\emptyset \subset \{\emptyset\}$ (f) $1 \subseteq \{1\}$.

2.4. Consider the open sentence $P(x) : x(x - 1) = 6$ over the domain $\mathbf{R}$.

(a) For what values of x is $P(x)$ a true statement?
(b) For what values of x is $P(x)$ a false statement?

2.5. For the open sentence $P(x) : 3x - 2 > 4$ over the domain $\mathbf{Z}$, determine:

(a) the values of x for which $P(x)$ is true.
(b) the values of x for which $P(x)$ is false.

2.6. For the open sentence $P(A) : A \subseteq \{1, 2, 3\}$ over the domain $S = \mathcal{P}(\{1, 2, 4\})$, determine:

(a) all $A \in S$ for which $P(A)$ is true.
(b) all $A \in S$ for which $P(A)$ is false.
(c) all $A \in S$ for which $A \cap \{1, 2, 3\} = \emptyset$.

2.7. Let $P(n)$: n and $n + 2$ are primes. be an open sentence over the domain $\mathbf{N}$. Find six positive integers n for which $P(n)$ is true. If $n \in \mathbf{N}$ such that $P(n)$ is true, then the two integers n, $n + 2$ are called **twin primes**. It has been conjectured that there are infinitely many twin primes.

2.8. Let $P(n) :$ $\frac{n^2+5n+6}{2}$ is even.

(a) Find a set S_1 of three integers such that $P(n)$ is an open sentence over the domain S_1 and $P(n)$ is true for each $n \in S_1$.
(b) Find a set S_2 of three integers such that $P(n)$ is an open sentence over the domain S_2 and $P(n)$ is false for each $n \in S_2$.

2.9. Find an open sentence $P(n)$ over the domain $S = \{3, 5, 7, 9\}$ such that $P(n)$ is true for half of the integers in S and false for the other half.

2.10. Find two open sentences $P(n)$ and $Q(n)$, both over the domain $S = \{2, 4, 6, 8\}$, such that $P(2)$ and $Q(2)$ are both true, $P(4)$ and $Q(4)$ are both false, $P(6)$ is true and $Q(6)$ is false, while $P(8)$ is false and $Q(8)$ is true.

Section 2.2: The Negation of a Statement

2.11. State the negation of each of the following statements.

(a) $\sqrt{2}$ is a rational number.
(b) 0 is not a negative integer.
(c) 111 is a prime number.

2.12. Complete the truth table in Figure 2.16.

2.13. State the negation of each of the following statements.

(a) The real number r is at most $\sqrt{2}$.
(b) The absolute value of the real number a is less than 3.
(c) Two angles of the triangle are $45°$.

P	Q	~ P	~ Q
T	T		
T	F		
F	T		
F	F		

Figure 2.16   The truth table for Exercise 2.12.

(d) The area of the circle is at least 9π.
(e) Two sides of the triangle have the same length.
(f) The point P in the plane lies outside of the circle C.

2.14. State the negation of each of the following statements.

(a) At least two of my library books are overdue.
(b) One of my two friends misplaced his homework assignment.
(c) No one expected that to happen.
(d) It's not often that my instructor teaches that course.
(e) It's surprising that two students received the same exam score.

Section 2.3: The Disjunction and Conjunction of Statements

2.15. Complete the truth table in Figure 2.17.

P	Q	~ Q	$P \wedge (\sim Q)$
T	T		
T	F		
F	T		
F	F		

Figure 2.17   The truth table for Exercise 2.15

2.16. For the sets $A = \{1, 2, \cdots, 10\}$ and $B = \{2, 4, 6, 9, 12, 25\}$, consider the statements

$$P: A \subseteq B. \quad Q: |A - B| = 6.$$

Determine which of the following statements are true.
(a) $P \vee Q$ (b) $P \vee (\sim Q)$ (c) $P \wedge Q$
(d) $(\sim P) \wedge Q$ (e) $(\sim P) \vee (\sim Q)$.

2.17. Let P: 15 is odd. and Q : 21 is prime. State each of the following in words, and determine whether they are true or false.
(a) $P \vee Q$ (b) $P \wedge Q$ (c) $(\sim P) \vee Q$ (d) $P \wedge (\sim Q)$.

2.18. Let $S = \{1, 2, \ldots, 6\}$ and let

$$P(A) : A \cap \{2, 4, 6\} = \emptyset. \text{ and } Q(A) : A \neq \emptyset.$$

be open sentences over the domain $\mathcal{P}(S)$.

(a) Determine all $A \in \mathcal{P}(S)$ for which $P(A) \wedge Q(A)$ is true.

(b) Determine all $A \in \mathcal{P}(S)$ for which $P(A) \vee (\sim Q(A))$ is true.

(c) Determine all $A \in \mathcal{P}(S)$ for which $(\sim P(A)) \wedge (\sim Q(A))$ is true.

Section 2.4: The Implication

2.19. Consider the statements P: 17 is even. and Q: 19 is prime. Write each of the following statements in words and indicate whether it is true or false.
   (a) $\sim P$ (b) $P \vee Q$ (c) $P \wedge Q$ (d) $P \Rightarrow Q$.

2.20. For statements P and Q, construct a truth table for $(P \Rightarrow Q) \Rightarrow (\sim P)$.

2.21. Consider the statements $P : \sqrt{2}$ is rational. and $Q : 22/7$ is rational. Write each of the following statements in words and indicate whether it is true or false.
   (a) $P \Rightarrow Q$ (b) $Q \Rightarrow P$ (c) $(\sim P) \Rightarrow (\sim Q)$ (d) $(\sim Q) \Rightarrow (\sim P)$.

2.22. Consider the statements:

$$P: \sqrt{2} \text{ is rational.} \quad Q: \tfrac{2}{3} \text{ is rational.} \quad R: \sqrt{3} \text{ is rational.}$$

Write each of the following statements in words and indicate whether the statement is true or false.
   (a) $(P \wedge Q) \Rightarrow R$ (b) $(P \wedge Q) \Rightarrow (\sim R)$
   (c) $((\sim P) \wedge Q) \Rightarrow R$ (d) $(P \vee Q) \Rightarrow (\sim R)$.

2.23. Suppose that $\{S_1, S_2\}$ is a partition of a set S and $x \in S$. Which of the following are true?

   (a) If we know that $x \notin S_1$, then x must belong to S_2.
   (b) It's possible that $x \notin S_1$ and $x \notin S_2$.
   (c) Either $x \notin S_1$ or $x \notin S_2$.
   (d) Either $x \in S_1$ or $x \in S_2$.
   (e) It's possible that $x \in S_1$ and $x \in S_2$.

2.24. Two sets A and B are nonempty disjoint subsets of a set S. If $x \in S$, then which of the following are true?

   (a) It's possible that $x \in A \cap B$.
   (b) If x is an element of A, then x can't be an element of B.
   (c) If x is not an element of A, then x must be an element of B.
   (d) It's possible that $x \notin A$ and $x \notin B$.
   (e) For each nonempty set C, either $x \in A \cap C$ or $x \in B \cap C$.
   (f) For some nonempty set C, both $x \in A \cup C$ and $x \in B \cup C$.

2.25. A college student makes the following statement:

> If I receive an A in both Calculus I and Discrete Mathematics this semester, then I'll take either Calculus II or Computer Programming this summer.

For each of the following, determine whether this statement is true or false.

   (a) The student doesn't get an A in Calculus I but decides to take Calculus II this summer anyway.
   (b) The student gets an A in both Calculus I and Discrete Mathematics but decides not to take any class this summer.
   (c) The student does not get an A in Calculus I and decides not to take Calculus II but takes Computer Programming this summer.
   (d) The student gets an A in both Calculus I and Discrete Mathematics and decides to take both Calculus II and Computer Programming this summer.
   (e) The student gets an A in neither Calculus I nor Discrete Mathematics and takes neither Calculus II nor Computer Programming this summer.

2.26. A college student makes the following statement:

   If I don't see my advisor today, then I'll see her tomorrow.

For each of the following, determine whether this statement is true or false.

   (a) The student doesn't see his advisor either day.
   (b) The student sees his advisor both days.
   (c) The student sees his advisor on one of the two days.
   (d) The student doesn't see his advisor today and waits until next week to see her.

2.27. The instructor of a computer science class announces to her class that there will be a well-known speaker on campus later that day. Four students in the class are Alice, Ben, Cindy and Don. Ben says that he'll attend the lecture if Alice does. Cindy says that she'll attend the talk if Ben does. Don says that he will go to the lecture if Cindy does. That afternoon exactly two of the four students attend the talk. Which two students went to the lecture?

2.28. Consider the statement (implication):
   If Bill takes Sam to the concert, then Sam will take Bill to dinner.
Which of the following implies that this statement is true?

   (a) Sam takes Bill to dinner only if Bill takes Sam to the concert.
   (b) Either Bill doesn't take Sam to the concert or Sam takes Bill to dinner.
   (c) Bill takes Sam to the concert.
   (d) Bill takes Sam to the concert and Sam takes Bill to dinner.
   (e) Bill takes Sam to the concert and Sam doesn't take Bill to dinner.
   (f) The concert is canceled.
   (g) Sam doesn't attend the concert.

2.29. Let P and Q be statements. Which of the following implies that $P \vee Q$ is false?
   (a) $(\sim P) \vee (\sim Q)$ is false.   (b) $(\sim P) \vee Q$ is true.
   (c) $(\sim P) \wedge (\sim Q)$ is true.   (d) $Q \Rightarrow P$ is true.   (e) $P \wedge Q$ is false.

Section 2.5: More on Implications

2.30. Consider the open sentences $P(n) : 5n + 3$ is prime. and $Q(n) : 7n + 1$ is prime., both over the domain **N**.

   (a) State $P(n) \Rightarrow Q(n)$ in words.
   (b) State $P(2) \Rightarrow Q(2)$ in words. Is this statement true or false?
   (c) State $P(6) \Rightarrow Q(6)$ in words. Is this statement true or false?

2.31. In each of the following, two open sentences $P(x)$ and $Q(x)$ over a domain S are given. Determine the truth value of $P(x) \Rightarrow Q(x)$ for each $x \in S$.

   (a) $P(x) : |x| = 4$; $Q(x) : x = 4$; $S = \{-4, -3, 1, 4, 5\}$.
   (b) $P(x) : x^2 = 16$; $Q(x) : |x| = 4$; $S = \{-6, -4, 0, 3, 4, 8\}$.
   (c) $P(x) : x > 3$; $Q(x) : 4x - 1 > 12$; $S = \{0, 2, 3, 4, 6\}$.

2.32. In each of the following, two open sentences $P(x)$ and $Q(x)$ over a domain S are given. Determine all $x \in S$ for which $P(x) \Rightarrow Q(x)$ is a true statement.

   (a) $P(x) : x - 3 = 4$; $Q(x) : x \geq 8$; $S = \mathbf{R}$.
   (b) $P(x) : x^2 \geq 1$; $Q(x) : x \geq 1$; $S = \mathbf{R}$.
   (c) $P(x) : x^2 \geq 1$; $Q(x) : x \geq 1$; $S = \mathbf{N}$.
   (d) $P(x) : x \in [-1, 2]$; $Q(x) : x^2 \leq 2$; $S = [-1, 1]$.

2.33. In each of the following, two open sentences $P(x, y)$ and $Q(x, y)$ are given, where the domain of both x and y is $\mathbf{Z}$. Determine the truth value of $P(x, y) \Rightarrow Q(x, y)$ for the given values of x and y.

(a) $P(x, y)$: $x^2 - y^2 = 0$. and $Q(x, y)$: $x = y$.
$(x, y) \in \{(1, -1), (3, 4), (5, 5)\}$.

(b) $P(x, y)$: $|x| = |y|$. and $Q(x, y)$: $x = y$.
$(x, y) \in \{(1, 2), (2, -2), (6, 6)\}$.

(c) $P(x, y)$: $x^2 + y^2 = 1$. and $Q(x, y)$: $x + y = 1$.
$(x, y) \in \{(1, -1), (-3, 4), (0, -1), (1, 0)\}$.

2.34. Each of the following describes an implication. Write the implication in the form "if, then."

(a) Any point on the straight line with equation $2y + x - 3 = 0$ whose x-coordinate is an integer also has an integer for its y-coordinate.

(b) The square of every odd integer is odd.

(c) Let $n \in \mathbf{Z}$. Whenever $3n + 7$ is even, n is odd.

(d) The derivative of the function $f(x) = \cos x$ is $f'(x) = -\sin x$.

(e) Let C be a circle of circumference 4π. Then the area of C is also 4π.

(f) The integer n^3 is even only if n is even.

Section 2.6: The Biconditional

2.35. Let P : 18 is odd. and Q : 25 is even. State $P \Leftrightarrow Q$ in words. Is $P \Leftrightarrow Q$ true or false?

2.36. Let $P(x)$: x is odd. and $Q(x)$: x^2 is odd. be open sentences over the domain $\mathbf{Z}$. State $P(x) \Leftrightarrow Q(x)$ in two ways: (1) using "if and only if" and (2) using "necessary and sufficient."

2.37. For the open sentences $P(x)$: $|x - 3| < 1$. and $Q(x)$: $x \in (2, 4)$. over the domain $\mathbf{R}$, state the biconditional $P(x) \Leftrightarrow Q(x)$ in two different ways.

2.38. Consider the open sentences:

$$P(x) : x = -2. \text{ and } Q(x) : x^2 = 4.$$

over the domain $S = \{-2, 0, 2\}$. State each of the following in words and determine all values of $x \in S$ for which the resulting statements are true.

(a) $\sim P(x)$ (b) $P(x) \vee Q(x)$ (c) $P(x) \wedge Q(x)$ (d) $P(x) \Rightarrow Q(x)$
(e) $Q(x) \Rightarrow P(x)$ (f) $P(x) \Leftrightarrow Q(x)$.

2.39. For the following open sentences $P(x)$ and $Q(x)$ over a domain S, determine all values of $x \in S$ for which the biconditional $P(x) \Leftrightarrow Q(x)$ is true.

(a) $P(x)$: $|x| = 4$; $Q(x)$: $x = 4$; $S = \{-4, -3, 1, 4, 5\}$.

(b) $P(x)$: $x \geq 3$; $Q(x)$: $4x - 1 > 12$; $S = \{0, 2, 3, 4, 6\}$.

(c) $P(x)$: $x^2 = 16$; $Q(x)$: $x^2 - 4x = 0$; $S = \{-6, -4, 0, 3, 4, 8\}$.

2.40. In each of the following, two open sentences $P(x, y)$ and $Q(x, y)$ are given, where the domain of both x and y is $\mathbf{Z}$. Determine the truth value of $P(x, y) \Leftrightarrow Q(x, y)$ for the given values of x and y.

(a) $P(x, y)$: $x^2 - y^2 = 0$ and; $Q(x, y)$: $x = y$.
$(x, y) \in \{(1, -1), (3, 4), (5, 5)\}$.

(b) $P(x, y)$: $|x| = |y|$ and; $Q(x, y)$: $x = y$.
$(x, y) \in \{(1, 2), (2, -2), (6, 6)\}$.

(c) $P(x, y)$: $x^2 + y^2 = 1$ and; $Q(x, y)$: $x + y = 1$.
$(x, y) \in \{(1, -1), (-3, 4), (0, -1), (1, 0)\}$.

2.41. Determine all values of n in the domain $S = \{1, 2, 3\}$ for which the following is a true statement:
A necessary and sufficient condition for $\frac{n^3 + n}{2}$ to be even is that $\frac{n^2 + n}{2}$ is odd.

2.42. Determine all values of n in the domain $S = \{2, 3, 4\}$ for which the following is a true statement: The integer $\frac{n(n-1)}{2}$ is odd if and only if $\frac{n(n+1)}{2}$ is even.

2.43. Let $S = \{1, 2, 3\}$. Consider the following open sentences over the domain S:

$$P(n): \frac{(n+4)(n+5)}{2} \text{ is odd.}$$
$$Q(n): 2^{n-2} + 3^{n-2} + 6^{n-2} > (2.5)^{n-1}.$$

Determine three distinct elements a, b, c in S such that $P(a) \Rightarrow Q(a)$ is false, $Q(b) \Rightarrow P(b)$ is false, and $P(c) \Leftrightarrow Q(c)$ is true.

2.44. Let $S = \{1, 2, 3, 4\}$. Consider the following open sentences over the domain S:

$$P(n): \frac{n(n-1)}{2} \text{ is even.}$$
$$Q(n): 2^{n-2} - (-2)^{n-2} \text{ is even.}$$
$$R(n): 5^{n-1} + 2^n \text{ is prime.}$$

Determine four distinct elements a, b, c, d in S such that
(i) $P(a) \Rightarrow Q(a)$ is false;   (ii) $Q(b) \Rightarrow P(b)$ is true;
(iii) $P(c) \Leftrightarrow R(c)$ is true;   (iv) $Q(d) \Leftrightarrow R(d)$ is false.

2.45. Let $P(n): 2^n - 1$ is a prime. and $Q(n): n$ is a prime. be open sentences over the domain $S = \{2, 3, 4, 5, 6, 11\}$. Determine all values of $n \in S$ for which $P(n) \Leftrightarrow Q(n)$ is a true statement.

Section 2.7: Tautologies and Contradictions

2.46. For statements P and Q, show that $P \Rightarrow (P \vee Q)$ is a tautology.

2.47. For statements P and Q, show that $(P \wedge (\sim Q)) \wedge (P \wedge Q)$ is a contradiction.

2.48. For statements P and Q, show that $(P \wedge (P \Rightarrow Q)) \Rightarrow Q$ is a tautology. Then state $(P \wedge (P \Rightarrow Q)) \Rightarrow Q$ in words. (This is an important logical argument form, called **modus ponens**.)

2.49. For statements P, Q and R, show that $((P \Rightarrow Q) \wedge (Q \Rightarrow R)) \Rightarrow (P \Rightarrow R)$ is a tautology. Then state this compound statement in words. (This is another important logical argument form, called **syllogism**.)

2.50. Let R and S be compound statements involving the same component statements. If R is a tautology and S is a contradiction, then what can be said of the following?
(a) $R \vee S$ (b) $R \wedge S$ (c) $R \Rightarrow S$ (d) $S \Rightarrow R$.

Section 2.8: Logical Equivalence

2.51. For statements P and Q, the implication $(\sim P) \Rightarrow (\sim Q)$ is called the **inverse** of the implication $P \Rightarrow Q$.

(a) Use a truth table to show that these statements are not logically equivalent.
(b) Find another implication that is logically equivalent to $(\sim P) \Rightarrow (\sim Q)$ and verify your answer.

2.52. Let P and Q be statements.

(a) Is $\sim (P \vee Q)$ logically equivalent to $(\sim P) \vee (\sim Q)$? Explain.
(b) What can you say about the biconditional $\sim (P \vee Q) \Leftrightarrow ((\sim P) \vee (\sim Q))$?

2.53. For statements P, Q and R, use a truth table to show that each of the following pairs of statements is logically equivalent.

(a) $(P \wedge Q) \Leftrightarrow P$ and $P \Rightarrow Q$.
(b) $P \Rightarrow (Q \vee R)$ and $(\sim Q) \Rightarrow ((\sim P) \vee R)$.

2.54. For statements P and Q, show that $(\sim Q) \Rightarrow (P \wedge (\sim P))$ and Q are logically equivalent.

2.55. For statements P, Q and R, show that $(P \vee Q) \Rightarrow R$ and $(P \Rightarrow R) \wedge (Q \Rightarrow R)$ are logically equivalent.

2.56. Two compound statements S and T are composed of the same component statements P, Q and R. If S and T are not logically equivalent, then what can we conclude from this?

2.57. Five compound statements S_1, S_2, S_3, S_4 and S_5 are all composed of the same component statements P and Q and whose truth tables have identical first and fourth rows. Show that at least two of these five statements are logically equivalent.

Section 2.9: Some Fundamental Properties of Logical Equivalence

2.58. Verify the following laws stated in Theorem 2.18:

(a) Let P, Q and R be statements. Then
$$P \vee (Q \wedge R) \text{ and } (P \vee Q) \wedge (P \vee R) \text{ are logically equivalent.}$$

(b) Let P and Q be statements. Then
$$\sim(P \vee Q) \text{ and } (\sim P) \wedge (\sim Q) \text{ are logically equivalent.}$$

2.59. Write negations of the following open sentences:

(a) Either $x = 0$ or $y = 0$.
(b) The integers a and b are both even.

2.60. Consider the implication: If x and y are even, then xy is even.

(a) State the implication using "only if."
(b) State the converse of the implication.
(c) State the implication as a disjunction (see Theorem 2.17).
(d) State the negation of the implication as a conjunction (see Theorem 2.21(a)).

2.61. For a real number x, let $P(x) : x^2 = 2$. and $Q(x) : x = \sqrt{2}$. State the negation of the biconditional $P \Leftrightarrow Q$ in words (see Theorem 2.21(b)).

2.62. Let P and Q be statements. Show that $[(P \vee Q) \wedge \sim (P \wedge Q)] \equiv \sim (P \Leftrightarrow Q)$.

2.63. Let $n \in \mathbf{Z}$. For which implication is its negation the following?
The integer $3n + 4$ is odd and $5n - 6$ is even.

2.64. For which biconditional is its negation the following?
n^3 and $7n + 2$ are odd or n^3 and $7n + 2$ are even.

Section 2.10: Quantified Statements

2.65. Let S denote the set of odd integers and let
$$P(x) : x^2 + 1 \text{ is even.} \quad \text{and} \quad Q(x) : x^2 \text{ is even.}$$
be open sentences over the domain S. State $\forall x \in S$, $P(x)$ and $\exists x \in S$, $Q(x)$ in words.

2.66. Define an open sentence $R(x)$ over some domain S and then state $\forall x \in S$, $R(x)$ and $\exists x \in S$, $R(x)$ in words.

2.67. State the negations of the following quantified statements, where all sets are subsets of some universal set U:

(a) For every set A, $A \cap \overline{A} = \emptyset$.
(b) There exists a set A such that $\overline{A} \subseteq A$.

2.68. State the negations of the following quantified statements:

(a) For every rational number r, the number $1/r$ is rational.
(b) There exists a rational number r such that $r^2 = 2$.

2.69. Let $P(n)$: $(5n - 6)/3$ is an integer. be an open sentence over the domain $\mathbf{Z}$. Determine, with explanations, whether the following statements are true:
(a) $\forall n \in \mathbf{Z}, P(n)$.   (b) $\exists n \in \mathbf{Z}, P(n)$.

2.70. Determine the truth value of each of the following statements.
(a) $\exists x \in \mathbf{R}, x^2 - x = 0$.
(b) $\forall n \in \mathbf{N}, n + 1 \geq 2$.
(c) $\forall x \in \mathbf{R}, \sqrt{x^2} = x$.
(d) $\exists x \in \mathbf{Q}, 3x^2 - 27 = 0$.
(e) $\exists x \in \mathbf{R}, \exists y \in \mathbf{R}, x + y + 3 = 8$.
(f) $\forall x, y \in \mathbf{R}, x + y + 3 = 8$.
(g) $\exists x, y \in \mathbf{R}, x^2 + y^2 = 9$.
(h) $\forall x \in \mathbf{R}, \forall y \in \mathbf{R}, x^2 + y^2 = 9$.

2.71. The statement
$$\text{For every integer } m, \text{ either } m \leq 1 \text{ or } m^2 \geq 4.$$
can be expressed using a quantifier as:
$$\forall m \in \mathbf{Z}, m \leq 1 \text{ or } m^2 \geq 4.$$
Do this for the following two statements.
(a) There exist integers a and b such that both $ab < 0$ and $a + b > 0$.
(b) For all real numbers x and y, $x \neq y$ implies that $x^2 + y^2 > 0$.
(c) Express in words the negations of the statements in (a) and (b).
(d) Using quantifiers, express in symbols the negations of the statements in both (a) and (b).

2.72. Let $P(x)$ and $Q(x)$ be open sentences where the domain of the variable x is S. Which of the following implies that $(\sim P(x)) \Rightarrow Q(x)$ is false for some $x \in S$?

(a) $P(x) \wedge Q(x)$ is false for all $x \in S$.
(b) $P(x)$ is true for all $x \in S$.
(c) $Q(x)$ is true for all $x \in S$.
(d) $P(x) \vee Q(x)$ is false for some $x \in S$.
(e) $P(x) \wedge (\sim Q(x))$ is false for all $x \in S$.

2.73. Let $P(x)$ and $Q(x)$ be open sentences where the domain of the variable x is T. Which of the following implies that $P(x) \Rightarrow Q(x)$ is true for all $x \in T$?

(a) $P(x) \wedge Q(x)$ is false for all $x \in T$.
(b) $Q(x)$ is true for all $x \in T$.
(c) $P(x)$ is false for all $x \in T$.
(d) $P(x) \wedge (\sim (Q(x))$ is true for some $x \in T$.
(e) $P(x)$ is true for all $x \in T$.
(f) $(\sim P(x)) \wedge (\sim Q(x))$ is false for all $x \in T$.

2.74. Consider the open sentence
$$P(x, y, z): (x - 1)^2 + (y - 2)^2 + (z - 2)^2 > 0.$$
where the domain of each of the variables x, y and z is $\mathbf{R}$.

(a) Express the quantified statement $\forall x \in \mathbf{R}, \forall y \in \mathbf{R}, \forall z \in \mathbf{R}, P(x, y, z)$ in words.
(b) Is the quantified statement in (a) true or false? Explain.
(c) Express the negation of the quantified statement in (a) in symbols.
(d) Express the negation of the quantified statement in (a) in words.
(e) Is the negation of the quantified statement in (a) true or false? Explain.

2.75. Consider quantified statement
$$\text{For every } s \in S \text{ and } t \in S, st - 2 \text{ is prime.}$$

where the domain of the variables s and t is $S = \{3, 5, 11\}$.

(a) Express this quantified statement in symbols.

(b) Is the quantified statement in (a) true or false? Explain.

(c) Express the negation of the quantified statement in (a) in symbols.

(d) Express the negation of the quantified statement in (a) in words.

(e) Is the negation of the quantified statement in (a) true or false? Explain.

2.76. Let A be the set of circles in the plane with center $(0, 0)$ and let B be the set of circles in the plane with center $(1, 1)$. Furthermore, let

$$P(C_1, C_2): \ C_1 \text{ and } C_2 \text{ have exactly two points in common.}$$

be an open sentence where the domain of C_1 is A and the domain of C_2 is B.

(a) Express the following quantified statement in words:

$$\forall C_1 \in A, \exists C_2 \in B, P(C_1, C_2). \tag{2.30}$$

(b) Express the negation of the quantified statement in (2.30) in symbols.

(c) Express the negation of the quantified statement in (2.30) in words.

2.77. For a triangle T, let $r(T)$ denote the ratio of the length of the longest side of T to the length of the smallest side of T. Let A denote the set of all triangles and let

$$P(T_1, T_2): \ r(T_2) \ge r(T_1).$$

be an open sentence where the domain of both T_1 and T_2 is A.

(a) Express the following quantified statement in words:

$$\exists T_1 \in A, \forall T_2 \in A, P(T_1, T_2). \tag{2.31}$$

(b) Express the negation of the quantified statement in (2.31) in symbols.

(c) Express the negation of the quantified statement in (2.31) in words.

2.78. Consider the open sentence $P(a, b): \ a/b < 1$. where the domain of a is $A = \{2, 3, 5\}$ and the domain of b is $B = \{2, 4, 6\}$.

(a) State the quantified statement $\forall a \in A, \exists b \in B, P(a, b)$ in words.

(b) Show the quantified statement in (a) is true.

2.79. Consider the open sentence $Q(a, b): \ a - b < 0$. where the domain of a is $A = \{3, 5, 8\}$ and the domain of b is $B = \{3, 6, 10\}$.

(a) State the quantified statement $\exists b \in B, \forall a \in A, Q(a, b)$ in words.

(b) Show the quantified statement in (a) is true.

Section 2.11: Characterizations of Statements

2.80. Give a definition of each of the following and then state a characterization of each.

(a) Two lines in the plane are perpendicular.

(b) A rational number.

2.81. Define an integer n to be odd if n is not even. State a characterization of odd integers.

2.82. Define a triangle to be isosceles if it has two equal sides. Which of the following statements are characterizations of isosceles triangles? If a statement is not a characterization of isosceles triangles, then explain why.

(a) If a triangle is equilateral, then it is isosceles.

(b) A triangle T is isosceles if and only if T has two equal sides.

(c) If a triangle has two equal sides, then it is isosceles.

(d) A triangle T is isosceles if and only if T is equilateral.

(e) If a triangle has two equal angles, then it is isosceles.

(f) A triangle T is isosceles if and only if T has two equal angles.

2.83. By definition, a right triangle is a triangle one of whose angles is a right angle. Also, two angles in a triangle are complementary if the sum of their degrees is $90°$. Which of the following statements are characterizations of a right triangle? If a statement is not a characterization of a right triangle, then explain why.

(a) A triangle is a right triangle if and only if two of its sides are perpendicular.

(b) A triangle is a right triangle if and only if it has two complementary angles.

(c) A triangle is a right triangle if and only if its area is half of the product of the lengths of some pair of its sides.

(d) A triangle is a right triangle if and only if the square of the length of its longest side equals to the sum of the squares of the lengths of the two smallest sides.

(e) A triangle is a right triangle if and only if twice of the area of the triangle equals the area of some rectangle.

2.84. Two distinct lines in the plane are defined to be parallel if they don't intersect. Which of the following is a characterization of parallel lines?

(a) Two distinct lines ℓ_1 and ℓ_2 are parallel if and only if any line ℓ_3 that is perpendicular to ℓ_1 is also perpendicular to ℓ_2.

(b) Two distinct lines ℓ_1 and ℓ_2 are parallel if and only if any line distinct from ℓ_1 and ℓ_2 that doesn't intersect ℓ_1 also doesn't intersect ℓ_2.

(c) Two distinct lines ℓ_1 and ℓ_2 are parallel if and only if whenever a line ℓ intersects ℓ_1 in an acute angle α, then ℓ also intersects ℓ_2 in an acute angle α.

(d) Two distinct lines ℓ_1 and ℓ_2 are parallel if and only if whenever a point P is not on ℓ_1, the point P is not on ℓ_2.

ADDITIONAL EXERCISES FOR CHAPTER 2

2.85. Construct a truth table for $P \wedge (Q \Rightarrow (\sim P))$.

2.86. Given that the implication $(Q \vee R) \Rightarrow (\sim P)$ is false and Q is false, determine the truth values of R and P.

2.87. Find a compound statement involving the component statements P and Q that has the truth table given in Figure 2.18.

$(Q \Rightarrow \sim Q) \vee P$

P	Q	$\sim Q$	
T	T	F	T
T	F	T	T
F	T	F	F
F	F	T	T

Figure 2.18   Truth table for Exercise 2.87.

2.88. Determine the truth value of each of the following quantified statements:
  (a) $\exists x \in \mathbf{R}, x^3 + 2 = 0$.   (b) $\forall n \in \mathbf{N}, 2 \geq 3 - n$.
  (c) $\forall x \in \mathbf{R}, |x| = x$.   (d) $\exists x \in \mathbf{Q}, x^4 - 4 = 0$.
  (e) $\exists x, y \in \mathbf{R}, x + y = \pi$.   (f) $\forall x, y \in \mathbf{R}, x + y = \sqrt{x^2 + y^2}$.

2.89. Rewrite each of the implications below using (1) only if and (2) sufficient.

  (a) If a function f is differentiable, then f is continuous.
  (b) If $x = -5$, then $x^2 = 25$.

2.90. Let $P(n)$: $n^2 - n + 5$ is a prime. be an open sentence over a domain S.

  (a) Determine the truth values of the quantified statements $\forall n \in S, P(n)$ and $\exists n \in S, \sim P(n)$ for
      $S = \{1, 2, 3, 4\}$.
  (b) Determine the truth values of the quantified statements $\forall n \in S, P(n)$ and $\exists n \in S, \sim P(n)$ for
      $S = \{1, 2, 3, 4, 5\}$.
  (c) How are the statements in (a) and (b) related?

2.91. (a) For statements P, Q and R, show that

$$((P \wedge Q) \Rightarrow R) \equiv ((P \wedge (\sim R)) \Rightarrow (\sim Q)).$$

  (b) For statements P, Q and R, show that

$$((P \wedge Q) \Rightarrow R) \equiv ((Q \wedge (\sim R)) \Rightarrow (\sim P)).$$

2.92. For a fixed integer n, use Exercise 2.91 to restate the following implication in two different ways:

$$\text{If } n \text{ is a prime and } n > 2, \text{ then } n \text{ is odd.}$$

2.93. For fixed integers m and n, use Exercise 2.91 to restate the following implication in two different ways:

$$\text{If } m \text{ is even and } n \text{ is odd, then } m + n \text{ is odd.}$$

2.94. For a real-valued function f and a real number x, use Exercise 2.91 to restate the following implication in two different ways:

$$\text{If } f'(x) = 3x^2 - 2x \text{ and } f(0) = 4, \text{ then } f(x) = x^3 - x^2 + 4.$$

2.95. For the set $S = \{1, 2, 3\}$, give an example of three open sentences $P(n), Q(n)$ and $R(n)$, each over the domain S, such that (1) each of $P(n), Q(n)$ and $R(n)$ is a true statement for exactly two elements of S, (2) all of the implications $P(1) \Rightarrow Q(1), Q(2) \Rightarrow R(2)$ and $R(3) \Rightarrow P(3)$ are true, and (3) the converse of each implication in (2) is false.

2.96. Do there exist a set S of cardinality 2 and a set $\{P(n), Q(n), R(n)\}$ of three open sentences over the domain S such that (1) the implications $P(a) \Rightarrow Q(a), Q(b) \Rightarrow R(b)$ and $R(c) \Rightarrow P(c)$ are true, where $a, b, c \in S$ and (2) the converses of the implications in (1) are false? Necessarily, at least two of these elements a, b and c of S are equal.

2.97. Let $A = \{1, 2, \ldots, 6\}$ and $B = \{1, 2, \ldots, 7\}$. For $x \in A$, let $P(x)$: $7x + 4$ is odd. For $y \in B$, let $Q(y)$: $5y + 9$ is odd. Let

$$S = \{(P(x), Q(y)) : x \in A, y \in B, P(x) \Rightarrow Q(y) \text{ is false}\}.$$

What is $|S|$?

2.98. Let $P(x, y, z)$ be an open sentence, where the domains of x, y and z are A, B and C, respectively.

  (a) State the quantified statement $\forall x \in A, \forall y \in B, \exists z \in C, P(x, y, z)$ in words.
  (b) State the quantified statement $\forall x \in A, \forall y \in B, \exists z \in C, P(x, y, z)$ in words for $P(x, y, z)$: $x = yz$.

(c) Determine whether the quantified statement in (b) is true when $A = \{4, 8\}$, $B = \{2, 4\}$ and $C = \{1, 2, 4\}$.

2.99. Let $P(x, y, z)$ be an open sentence, where the domains of x, y and z are A, B and C, respectively.

(a) Express the negation of $\forall x \in A, \forall y \in B, \exists z \in C, P(x, y, z)$ in symbols.
(b) Express $\sim (\forall x \in A, \forall y \in B, \exists z \in C, P(x, y, z))$ in words.
(c) Determine whether $\sim (\forall x \in A, \forall y \in B, \exists z \in C, P(x, y, z))$ is true when $P(x, y, z): x + z = y$. for $A = \{1, 3\}$, $B = \{3, 5, 7\}$ and $C = \{0, 2, 4, 6\}$.

2.100. Write each of the following using "if, then."

(a) A sufficient condition for a triangle to be isosceles is that it has two equal angles.
(b) Let C be a circle of diameter $\sqrt{2/\pi}$. Then the area of C is $1/2$.
(c) The 4th power of every odd integer is odd.
(d) Suppose that the slope of a line ℓ is 2. Then the equation of ℓ is $y = 2x + b$ for some real number b.
(e) Whenever a and b are nonzero rational numbers, a/b is a nonzero rational number.
(f) For every three integers, there exist two of them whose sum is even.
(g) A triangle is a right triangle if the sum of two of its angles is $90°$.
(h) The number $\sqrt{3}$ is irrational.

3

Direct Proof and Proof by Contrapositive

We are now prepared to begin discussing our main topic: mathematical proofs. Initially, we will be primarily concerned with one question: For a given true mathematical statement, how can we show that it is true? In this chapter, you will be introduced to two important proof techniques.

A true mathematical statement whose truth is accepted without proof is referred to as an **axiom**. For example, an axiom of Euclid in geometry states that for every line ℓ and point P not on ℓ, there is a unique line containing P that is parallel to ℓ. A true mathematical statement whose truth can be verified is often referred to as a **theorem**, although many mathematicians reserve the word *theorem* for such statements that are especially significant or interesting. For example, the mathematical statement "$2 + 3 = 5$" is true but few, if any, would consider this to be a theorem under this latter interpretation. In addition to the word *theorem,* other common terms for such statements include proposition, result, observation and fact, the choice often depending on the significance of the statement or the degree of difficulty of its proof. We will use the word *theorem* sparingly, however, primarily reserving it for true mathematical statements that will be used to verify other mathematical statements that we will encounter later. Otherwise, we will simply use the word *result*. For the most part then, our results are examples used to illustrate proof techniques and our goal is to prove these results.

A **corollary** is a mathematical result that can be deduced from, and is thereby a consequence of, some earlier result. A **lemma** is a mathematical result that is useful in establishing the truth of some other result. Some people like to think of a lemma as a "helping result." Indeed, the German word for lemma is *hilfsatz*, whose English translation is "helping theorem." Ordinarily then, a lemma is not of primary importance itself. Indeed, its very existence is due only to its usefulness in proving another (more interesting) result.

Most theorems (or results) are stated as implications. We now begin our study of proofs of such mathematical statements.

3.1 Trivial and Vacuous Proofs

In nearly all of the implications $P \Rightarrow Q$ that we will encounter, P and Q are open sentences; that is, we will actually be considering $P(x) \Rightarrow Q(x)$ or $P(n) \Rightarrow Q(n)$ or some related implication, depending on which variable is being used. The variables x or n (or some other symbols) are used to represent elements of some set S being discussed; that is, S is the domain of the variable. As we have seen, for each value of a variable from its domain, a statement results. (It is possible, of course, that P and Q are expressed in terms of two or more variables.) Whether $P(x)$ (or $Q(x)$) is true ordinarily depends on which element $x \in S$ we are considering; that is, it is rarely the case that $P(x)$ is true for all $x \in S$ (or that $P(x)$ is false for all $x \in S$). For example, for

$$P(n) : 3n^2 - 4n + 1 \text{ is even}$$

where $n \in \mathbf{Z}$, $P(1)$ is a true statement while $P(2)$ is a false statement. Likewise, it is seldom the case that $Q(x)$ is true for all $x \in S$ or that $Q(x)$ is false for all $x \in S$.

When the quantified statement $\forall x \in S$, $P(x) \Rightarrow Q(x)$ is expressed as a result or theorem, we often write such a statement as

For $x \in S$, if $P(x)$ then $Q(x)$.

or as

Let $x \in S$. If $P(x)$, then $Q(x)$. **(3.1)**

Thus (3.1) is true if $P(x) \Rightarrow Q(x)$ is a true statement for each $x \in S$, while (3.1) is false if $P(x) \Rightarrow Q(x)$ is false for at least one element $x \in S$.

For a given element $x \in S$, let's recall (see the truth table in Figure 3.1) the conditions under which $P(x) \Rightarrow Q(x)$ has a particular truth value.

Accordingly, if $Q(x)$ is true for all $x \in S$ or $P(x)$ is false for all $x \in S$, then determining the truth or falseness of (3.1) becomes considerably easier. Indeed, if it can be shown that $Q(x)$ is true for all $x \in S$ (regardless of the truth value of $P(x)$), then, according to the truth table for the implication (shown in Figure 3.1), (3.1) is true. This constitutes a proof of (3.1) and is called a **trivial proof**. Accordingly, the statement

Let $n \in \mathbf{Z}$. If $n^3 > 0$, then 3 is odd.

is true and a (trivial) proof consists only of observing that 3 is an odd integer. The following provides a more interesting example of a trivial proof.

$P(x)$	$Q(x)$	$P(x) \Rightarrow Q(x)$
T	T	T
T	F	F
F	T	T
F	F	T

Figure 3.1   The truth table for the implication
$P(x) \Rightarrow Q(x)$ for an element x in its domain

Result 3.1   *Let $x \in \mathbf{R}$. If $x < 0$, then $x^2 + 1 > 0$.*

Proof   Since $x^2 \geq 0$ for each real number x, it follows that

$$x^2 + 1 > x^2 \geq 0.$$

Hence $x^2 + 1 > 0$.   ■
  Consider

$$P(x) : x < 0 \text{ and } Q(x) : x^2 + 1 > 0$$

where $x \in \mathbf{R}$. Then Result 3.1 asserts the truth of: For all $x \in \mathbf{R}$, $P(x) \Rightarrow Q(x)$. Since we verified that $Q(x)$ is true for every $x \in \mathbf{R}$, it follows that $P(x) \Rightarrow Q(x)$ is true for all $x \in \mathbf{R}$ and so Result 3.1 is true. In this case, when considered over the domain $\mathbf{R}$, $Q(x)$ is actually a true statement. It is this fact that allowed us to give a trivial proof of Result 3.1.

The proof of Result 3.1 does not depend on $x < 0$. Indeed, provided that $x \in \mathbf{R}$, we could have replaced "$x < 0$" by any hypothesis (including the more satisfying "$x \in \mathbf{R}$") and the result would still be true. In fact, this new result has the same proof. To be sure, it is rare indeed when a trivial proof is used to verify an implication; nevertheless, this is an important reminder of the truth table in Figure 3.1.

The symbol ■ that occurs at the end of the proof of Result 3.1 indicates that the proof is complete. There are definite advantages to using ■ (or some other symbol) to indicate the conclusion of a proof. First, as you start reading a proof, you can look ahead for this symbol to determine the length of the proof. Also, without this symbol, you may continue to read past the end of the proof, still thinking that you're reading a proof of the result. When you reach this symbol, you are *supposed* to be convinced that the result is true. If you are, this is good! Everything happened as planned. On the other hand, if you're not convinced, then, to you, the writer hasn't presented a proof. This may not be the writer's fault, however.

In the past, the most common way to indicate that a proof has concluded was to write Q.E.D., which stands for the Latin phrase "quod erat demonstrandum," whose English translation is "which was to be demonstrated." Some still use it.

Let $P(x)$ and $Q(x)$ be open sentences over a domain S. Then $\forall x \in S$, $P(x) \Rightarrow Q(x)$ is a true statement if it can be shown that $P(x)$ is false for all $x \in S$ (regardless of the truth value of $Q(x)$), according to the truth table for implication. Such a proof is called a **vacuous proof** of $\forall x \in S$, $P(x) \Rightarrow Q(x)$. Therefore,

Let $n \in \mathbf{Z}$. If 3 is even, then $n^3 > 0$.

is a true statement. Let's take a look, however, at a more interesting example of a vacuous proof.

Result 3.2   *Let $x \in \mathbf{R}$. If $x^2 - 2x + 2 \leq 0$, then $x^3 \geq 8$.*

Proof   First observe that

$$x^2 - 2x + 1 = (x - 1)^2 \geq 0.$$

Therefore, $x^2 - 2x + 2 = (x - 1)^2 + 1 \geq 1 > 0$. Thus $x^2 - 2x + 2 \leq 0$ is false for all $x \in \mathbf{R}$ and the implication is true.   ■

For

$$P(x) : x^2 - 2x + 2 \le 0 \quad \text{and} \quad Q(x) : x^3 \ge 8$$

over the domain $\mathbf{R}$, Result 3.2 asserts the truth of $\forall x \in \mathbf{R}, P(x) \Rightarrow Q(x)$. Since we verified that $P(x)$ is false for every $x \in \mathbf{R}$, it follows that $P(x) \Rightarrow Q(x)$ is true for each $x \in \mathbf{R}$. Hence Result 3.2 is true. In this case, $P(x)$ is a false statement for each $x \in \mathbf{R}$. This is what permitted us to give a vacuous proof of Result 3.2.

In the proof of Result 3.2, the truth or falseness of $x^3 \ge 8$ played no role whatsoever. Indeed, had we replaced $x^3 \ge 8$ by $x^3 \le 8$, for example, then neither the truth nor the proof of Result 3.2 would be affected. Whenever there is a vacuous proof of a result, we often say that the result follows **vacuously**. As we mentioned, a trivial proof is almost never encountered in mathematics; however, the same cannot be said of vacuous proofs, as we will see later.

We consider one additional example.

Result 3.3   *Let $S = \{n \in \mathbf{Z} : n \ge 2\}$ and let $n \in S$. If $2n + \frac{2}{n} < 5$, then $4n^2 + \frac{4}{n^2} < 25$.*

Proof   First, we observe that if $n = 2$, then $2n + \frac{2}{n} = 5$. Of course, $5 < 5$ is false. If $n \ge 3$, then $2n + \frac{2}{n} > 2n \ge 6$. So, when $n \ge 3$, $2n + \frac{2}{n} < 5$ is false as well. Thus $2n + \frac{2}{n} < 5$ is false for all $n \in S$. Hence the implication is true.   ∎

In two of the examples that we presented to illustrate trivial and vacuous proofs, we used the fact (and assumed it was known) that 3 is odd. Also, in the proofs of Results 3.1 and 3.2, we used the fact that if r is any real number, then $r^2 \ge 0$. Although you are certainly familiar with this property of real numbers, it is essential that any facts used within a proof are known to and likely to be recalled by the reader. Facts used within a proof should not come as a surprise to the reader. This subject will be discussed in more detail shortly.

Even though the trivial and vacuous proofs are rarely encountered in mathematics, they are important reminders of the truth table for implication. We are now prepared to be introduced to the first major proof technique in mathematics.

3.2 Direct Proofs

Typically, when we are discussing an implication $P(x) \Rightarrow Q(x)$ over a domain S, there is some connection between $P(x)$ and $Q(x)$. That is, the truth value of $Q(x)$ for a particular $x \in S$ often depends on the truth value of $P(x)$ for that same element x, or the truth value of $P(x)$ depends on the truth value of $Q(x)$. These are the kinds of implications in which we are primarily interested and it is the proofs of these types of results that will occupy much of our attention. We begin with the first major proof technique, which occurs more often in mathematics than any other technique.

Let $P(x)$ and $Q(x)$ be open sentences over a domain S. Suppose that our goal is to show that $P(x) \Rightarrow Q(x)$ is true for every $x \in S$, that is, our goal is to show that the quantified statement $\forall x \in S, P(x) \Rightarrow Q(x)$ is true. If $P(x)$ is false for some $x \in S$, then $P(x) \Rightarrow Q(x)$ is true for this element x. Hence we need only be concerned with showing that $P(x) \Rightarrow Q(x)$ is true for all $x \in S$ for which $P(x)$ is true. In a **direct proof** of $P(x) \Rightarrow Q(x)$ for all $x \in S$, we consider an arbitrary element $x \in S$ for which $P(x)$ is true and show that $Q(x)$ is true for this element x. To summarize then, to give a direct proof of $P(x) \Rightarrow Q(x)$ for all $x \in S$, we assume that $P(x)$ is true for an arbitrary element $x \in S$ and show that $Q(x)$ must be true as well for this element x.

In order to illustrate this type of proof (and others as well), we need to deal with mathematical topics with which we're all familiar. Let's first consider the integers and some of their elementary properties. We assume that you are familiar with the integers and the following properties of integers:

1.   The negative of every integer is an integer.

2.   The sum (and difference) of every two integers is an integer.

3.   The product of every two integers is an integer.

We will agree that we can use any of these properties. No justification is required or expected. Initially, we will use even and odd integers to illustrate our proof techniques. In this case, however, any properties of even and odd integers must be verified before they can be used. For example, you probably know that the sum of every two even integers is even but this must first be proved to be used. We need to lay some groundwork before any examples of direct proofs are given.

Since we will be working with even and odd integers, it is essential that we have precise definitions of these kinds of numbers. An integer n is defined to be **even** if $n = 2k$ for some integer k. For example, 10 is even since $10 = 2 \cdot 5$ (where, of course, 5 is an integer). Also, $-14 = 2(-7)$ is even, as is $0 = 2 \cdot 0$. The integer 17 is not even since there is no *integer* k for which $17 = 2k$. Thus we see that the set of all even integers is the set

$$S = \{2k : k \in \mathbf{Z}\} = \{\cdots, -4, -2, 0, 2, 4, \cdots\}.$$

We could define an integer n to be odd if it's not even but it would be difficult to work with this definition. Instead, we define an integer n to be **odd** if $n = 2k + 1$ for some integer k. Now 17 is odd since $17 = 2 \cdot 8 + 1$. Also, -5 is odd because $-5 = 2(-3) + 1$. On the other hand, 26 is not odd since there is no *integer* k such that $26 = 2k + 1$. In fact, 26 is even. Hence, according to the definition of odd integers that we have just given, we see that the set of all odd integers is precisely the set

$$T = \{2k + 1 : k \in \mathbf{Z}\} = \{\cdots, -5, -3, -1, 1, 3, 5, \cdots\}.$$

Observe that S and T are disjoint sets and $S \cup T = \mathbf{Z}$; that is, $\mathbf{Z}$ is partitioned into S and T. Therefore, every integer is either even or odd.

From time to time, we will find ourselves in a position where we have a result to prove, and it may not be entirely clear how to proceed. In such a case, we need to consider our options and develop a plan, which we refer to as a **proof strategy**. The idea is to discuss a proof strategy for the result and, from it, construct a proof. At other times, we may wish to reflect on a proof that we have just given in order to understand it better. Such a discussion will be referred to as a **proof analysis**. As with examples, we conclude both a proof strategy and a proof analysis with the symbol ♦.

We are now prepared to illustrate the direct proof technique. We follow the proof by a proof analysis.

Result 3.4   *If n is an odd integer, then $3n + 7$ is an even integer.*

Proof   Assume that n is an odd integer. Since n is odd, we can write $n = 2k + 1$ for some integer k. Now

$$3n + 7 = 3(2k + 1) + 7 = 6k + 3 + 7 = 6k + 10 = 2(3k + 5).$$

Since $3k + 5$ is an integer, $3n + 7$ is even.   ∎

<u>PROOF ANALYSIS</u>   First, notice that Result 3.4 could have been stated as either

For every odd integer n, the integer $3n + 7$ is even.

or

Let n be an odd integer. Then $3n + 7$ is even.

Thus the domain of the variable n in Result 3.4 is the set of odd integers. In the proof of Result 3.4, the expression $2k + 1$ was substituted for n in $3n + 7$ and simplified as $6k + 10$. Since our goal was to show that $3n + 7$ is even, we needed to show that $3n + 7$ can be expressed as twice an integer. Consequently, we factored 2 from $6k + 10$ and wrote it as $2(3k + 5)$. Since 3 and k are integers, so is $3k$ (the product of two integers is an integer). Since $3k$ and 5 are integers, so is $3k + 5$ (the sum of two integers is an integer). Therefore, $3n + 7$ satisfies the definition of an even integer.

One other remark deserves mention here. In the second sentence, we wrote:

Since n is odd, we can write $n = 2k + 1$ for some integer k.

It would be incorrect to write: "*If n is odd*" rather than "Since n is odd" because we have already assumed that n is odd and therefore n is now known to be odd.   ◆

We defined an integer n to be odd if we can write n as $2k + 1$ for some integer k. This means that whenever we want to show that an integer, say m, is odd, we must follow this definition; that is, we must show that $m = 2k + 1$ for some integer k. (Of course, the use of the symbol k is not important. For example, an odd integer n can be written as $n = 2\ell + 1$ for some integer ℓ.) We could have defined an integer n to be odd if it is possible to write $n = 2k - 1$ for some integer k, but we *didn't*. However, if we could prove that an integer n is odd if and only if n can be expressed as $2k - 1$ for some integer k, then we could use this characterization of odd integers to show that an integer is odd. This, however, would require additional work on our part with no obvious benefit. Similarly, we could have defined an integer n to be even if we can write $n = 2k + 2$, or $n = 2k - 2$ or perhaps $n = 2k + 100$ for some integer k. The definitions of even and odd integers that we chose are probably the most commonly used. Any other definitions that could have been given provide no special advantage to us.

The proof given of Result 3.4 is an example of a direct proof. Let

$Q(n)$: $3n + 7$ is an even integer.

over the domain of odd integers. We verified Result 3.4 by assuming that n is an arbitrary odd integer and then showing that $Q(n)$ is true for this element n. Showing that $Q(n)$ is true essentially required one step on our part. As we venture further into proofs, we will see that we can't always establish the truth of the desired conclusion so quickly. It may be necessary to establish the truth of some other mathematical statements along the way that can then be used to establish the truth of $Q(n)$. We will see examples of this later.

Let's consider another example. For variety, we use an alternative opening sentence and different symbols in the proof of the following result.

Result 3.5 *If n is an even integer, then $-5n - 3$ is an odd integer.*

Proof Let n be an even integer. Then $n = 2x$, where x is an integer. Therefore,

$$-5n - 3 = -5(2x) - 3 = -10x - 3 = -10x - 4 + 1 = 2(-5x - 2) + 1.$$

Since $-5x - 2$ is an integer, $-5n - 3$ is an odd integer. ∎

We now consider another example, which may have a surprise ending.

Result 3.6 *If n is an odd integer, then $4n^3 + 2n - 1$ is odd.*

Proof Assume that n is odd. Then $n = 2y + 1$ for some integer y. Therefore,

$$\begin{aligned} 4n^3 + 2n - 1 &= 4(2y + 1)^3 + 2(2y + 1) - 1 \\ &= 4(8y^3 + 12y^2 + 6y + 1) + 4y + 2 - 1 \\ &= 32y^3 + 48y^2 + 28y + 5 \\ &= 2(16y^3 + 24y^2 + 14y + 2) + 1. \end{aligned}$$

Since $16y^3 + 24y^2 + 14y + 2$ is an integer, $4n^3 + 2n - 1$ is odd. ∎

PROOF ANALYSIS   Although the direct proof of Result 3.6 that we gave is correct, this is *not* the desired proof. Indeed, had we observed that

$$4n^3 + 2n - 1 = 4n^3 + 2n - 2 + 1 = 2(2n^3 + n - 1) + 1$$

and that $2n^3 + n - 1 \in \mathbf{Z}$, we could have concluded immediately that $4n^3 + 2n - 1$ is odd for *every* integer n. Hence a trivial proof of Result 3.6 could be given and, in fact, is preferred. The fact that $4n^3 + 2n - 1$ is odd does not depend on n being odd. Indeed, it would be far better to replace the statement of Result 3.6 by

If n is an integer, then $4n^3 + 2n - 1$ is odd. ♦

We give an additional example of a somewhat different type.

Result 3.7 *Let $S = \{1, 2, 3\}$ and let $n \in S$. If $\dfrac{n(n + 3)}{2}$ is even, then $\dfrac{(n + 2)(n - 5)}{2}$ is even.*

Proof Let $n \in S$ such that $n(n + 3)/2$ is even. Since $n(n+3)/2 = 2$ when $n = 1$, $n(n + 3)/2 = 5$ when $n = 2$ and $n(n + 3)/2 = 9$ when $n = 3$, it follows that $n = 1$. When $n = 1$, $(n + 2)(n - 5)/2 = -6$, which is even. Therefore, the implication is true. ∎

PROOF ANALYSIS   In the proof of Result 3.7, we were only concerned with those elements $n \in S$ for which $n(n + 3)/2$ is even. Furthermore, it is not initially clear for which elements n of S the integer $n(n + 3)/2$ is even. Since S consists only of three elements, this can be determined rather quickly, which is what we did. We saw that only $n = 1$ has the desired property and this is the only element we needed to consider.   ◆

If our goal is to establish the truth of $P(x) \Rightarrow Q(x)$ for all x in a domain S by means of a direct proof, then the proof begins by assuming that $P(x)$ is true for an arbitrary element $x \in S$. It is often common in this situation, however, to omit the initial assumption that $P(x)$ is true for an arbitrary element $x \in S$. It is then understood that we are giving a direct proof. We illustrate this with a short example.

Result 3.8   *If n is an even integer, then $3n^5$ is an even integer.*

Proof   Since n is an even integer, $n = 2x$ for some integer x. Therefore,

$$3n^5 = 3(2x)^5 = 3(32x^5) = 96x^5 = 2(48x^5).$$

Since $48x^5 \in \mathbf{Z}$, the integer $3n^5$ is even.   ■

For the present, when giving a direct proof of $P(x) \Rightarrow Q(x)$ for all x in a domain S, we will often include the initial assumption that $P(x)$ is true for an arbitrary element $x \in S$ in order to solidify this technique in your mind.

3.3 Proof by Contrapositive

For statements P and Q, the **contrapositive** of the implication $P \Rightarrow Q$ is the implication $(\sim Q) \Rightarrow (\sim P)$. For example, for P_1 : 3 is odd and P_2 : 57 is prime, the contrapositive of the implication

$$P_1 \Rightarrow P_2 : \text{If 3 is odd, then 57 is prime.}$$

is the implication

$$(\sim P_2) \Rightarrow (\sim P_1): \text{If 57 is not prime, then 3 is even.}$$

The most important feature of the contrapositive $(\sim Q) \Rightarrow (\sim P)$ is that it is logically equivalent to $P \Rightarrow Q$. This fact is stated formally as a theorem and is verified in the truth table shown in Figure 3.2.

Theorem 3.9   *For every two statements P and Q, the implication $P \Rightarrow Q$ and its contrapositive are logically equivalent; that is,*

$$P \Rightarrow Q \equiv (\sim Q) \Rightarrow (\sim P).$$

Let

$$P(x) : x = 2. \text{ and } Q(x) : x^2 = 4.$$

where $x \in \mathbf{R}$. The contrapositive of the implication

$$P(x) \Rightarrow Q(x) : \text{If } x = 2, \text{ then } x^2 = 4.$$

is the implication

P	Q	$P \Rightarrow Q$	$\sim P$	$\sim Q$	$(\sim Q) \Rightarrow (\sim P)$
T	T	T	F	F	T
T	F	F	F	T	F
F	T	T	T	F	T
F	F	T	T	T	T

Figure 3.2   The logical equivalence of an implication and its contrapositive

$$(\sim Q(x)) \Rightarrow (\sim P(x)) : \text{If } x^2 \neq 4, \text{ then } x \neq 2.$$

Suppose that we wish to prove a result (or theorem) which is expressed as

$$\text{Let } x \in S. \text{ If } P(x), \text{ then } Q(x). \tag{3.2}$$

or as

$$\text{For all } x \in S, \text{ if } P(x), \text{ then } Q(x). \tag{3.3}$$

We have seen that a proof of such a result consists of establishing the truth of the implication $P(x) \Rightarrow Q(x)$ for all $x \in S$. If it can be shown that $(\sim Q(x)) \Rightarrow (\sim P(x))$ is true for all $x \in S$, then $P(x) \Rightarrow Q(x)$ is true for all $x \in S$. A **proof by contrapositive** of the result (3.2) (or of (3.3)) is a direct proof of its contrapositive:

$$\text{Let } x \in S. \text{ If } \sim Q(x), \text{ then } \sim P(x).$$

or

$$\text{For all } x \in S, \text{ if } \sim Q(x), \text{ then } \sim P(x).$$

Thus to give a proof by contrapositive of (3.2) (or of (3.3)), we assume that $\sim Q(x)$ is true for an arbitrary element $x \in S$ and show that $\sim P(x)$ is true for this element x.

There are certain types of results where a proof by contrapositive is preferable or perhaps even essential. We now give some examples to illustrate this method of proof.

Result 3.10   *Let $x \in \mathbf{Z}$. If $5x - 7$ is even, then x is odd.*

Proof   Assume that x is even. Then $x = 2a$ for some integer a. So

$$5x - 7 = 5(2a) - 7 = 10a - 7 = 10a - 8 + 1 = 2(5a - 4) + 1.$$

Since $5a - 4 \in \mathbf{Z}$, the integer $5x - 7$ is odd.   ∎

PROOF ANALYSIS   Some comments are now in order. The goal of Result 3.10 was to prove $P(x) \Rightarrow Q(x)$ for all $x \in \mathbf{Z}$, where $P(x) : 5x - 7$ is even. and $Q(x) : x$ is odd. Since we chose to give a proof by contrapositive, we gave a direct proof of $(\sim Q(x)) \Rightarrow (\sim P(x))$ for all $x \in \mathbf{Z}$. Hence the proof began by assuming that x is not odd; that is, x is even. The object then was to show that $5x - 7$ is odd.

If we had attempted to prove Result 3.10 with a direct proof, then we would have begun by assuming that $5x - 7$ is even for an arbitrary integer x. Then $5x - 7 = 2a$ for some integer a. So $x = (2a + 7)/5$. We then would want to show that x is odd. With the expression we have for x, it is not even clear that x is an integer, much less that x is an *odd* integer, although, of course, we were told in the statement of Result 3.10 that the domain of x is the set of integers. Therefore, it is not only that a proof by contrapositive provides us with a rather simple method of proving Result 3.10, it may not be immediately clear how or whether a direct proof can be used.

How did we know beforehand that it is a proof by contrapositive that we should use here? This is not as difficult as it may appear. If we use a direct proof, then we begin by assuming that $5x - 7$ is even for an arbitrary integer x; while if we use a proof by contrapositive, then we begin by assuming that x is even. Therefore, using a proof by contrapositive allows us to work with x initially rather than the more complicated expression $5x - 7$.   ◆

In all of the examples that we have seen so far, we have considered only implications. Now we look at a biconditional.

Result 3.11   *Let $x \in \mathbf{Z}$. Then $11x - 7$ is even if and only if x is odd.*

Proof   There are two implications to prove here, namely,

    (1) if x is odd, then $11x - 7$ is even and
    (2) if $11x - 7$ is even, then x is odd.

We begin with (1). In this case, a direct proof is appropriate. Assume that x is odd. Then $x = 2r + 1$, where $r \in \mathbf{Z}$. So

$$11x - 7 = 11(2r + 1) - 7 = 22r + 11 - 7 = 22r + 4 = 2(11r + 2).$$

Since $11r + 2$ is an integer, $11x - 7$ is even.

We now prove (2), which is the converse of (1). We use a proof by contrapositive here. Assume that x is even. Then $x = 2s$, where $s \in \mathbf{Z}$. Therefore,

$$11x - 7 = 11(2s) - 7 = 22s - 7 = 22s - 8 + 1 = 2(11s - 4) + 1.$$

Since $11s - 4$ is an integer, $11x - 7$ is odd.   ∎

A comment concerning the statements of Results 3.10 and 3.11 bears repeating here. These results begin with the sentence: Let $x \in \mathbf{Z}$. This, of course, is informing us that the domain in this case is $\mathbf{Z}$. That is, we are being told that x represents an integer. We need not state this assumption in the proof. The sentence "Let $x \in \mathbf{Z}$." is commonly called an "overriding" assumption or hypothesis and so x is assumed to be an integer throughout the proofs of Results 3.10 and 3.11.

In the proof of Result 3.11, we discussed our plan of attack. Namely, we stated that there were two implications to prove, and we specifically stated each. Ordinarily we don't include such information within the proof—unless the proof is quite long, in which case a roadmap indicating the steps we plan to take may be helpful. We give an additional example of this type, where this time a more conventional condensed proof is presented. The following example will be useful to us in the future, thus we refer to it as a theorem.

Theorem 3.12   *Let $x \in \mathbf{Z}$. Then x^2 is even if and only if x is even.*

Proof   Assume that x is even. Then $x = 2a$ for some integer a. Therefore,

$$x^2 = (2a)^2 = 4a^2 = 2(2a^2).$$

Because $2a^2 \in \mathbf{Z}$, the integer x^2 is even.

For the converse, assume that x is odd. So $x = 2b + 1$, where $b \in \mathbf{Z}$. Then

$$x^2 = (2b + 1)^2 = 4b^2 + 4b + 1 = 2(2b^2 + 2b) + 1.$$

Since $2b^2 + 2b$ is an integer, x^2 is odd.   ∎

Suppose now that you were asked to prove the following result:

Let $x \in \mathbf{Z}$. Then x^2 is odd if and only if x is odd.   **(3.4)**

How would you do this? You might think of proving the implication "If x is odd, then x^2 is odd." by a direct proof and its converse "If x^2 is odd, then x is odd." by a proof by contrapositive, where, of course, the domain of x is $\mathbf{Z}$. If we look at what is happening here, we see that we are duplicating the proof of Theorem 3.12. This is no surprise whatsoever. Theorem 3.12 states that if x is even, then x^2 is even; and if x^2 is even, then x is even. The contrapositive of the first implication is "If x^2 is odd, then x is odd," while the contrapositive of the second implication is "If x is odd, then x^2 is odd." In other words, (3.4) simply restates Theorem 3.12 in terms of contrapositives. Thus (3.4) requires no proof at all. It is essentially a restatement of Theorem 3.12. And speaking of restatements of Theorem 3.12, we need to recognize that this theorem can be restated in other ways. For example, we could restate

If x is an even integer, then x^2 is even.

as

The square of every even integer is even.

Hence Theorem 3.12 could be stated as:

An integer is even if and only if its square is even.

It is not only useful to sometimes restate results in different manners for variety, it is important to recognize what a result is saying regardless of the manner in which it may be stated.

At this point, it is convenient to pause and discuss how theorems (or results) can be used and why it is that we may be interested in proving a particular theorem. First, it is only by providing a proof of a theorem that we know for certain that the theorem is true and therefore have the right to call it a theorem. A fundamental reason why mathematicians may want to give a proof of some mathematical statement is that they consider this a challenge—this is what mathematicians do.

This, in fact, brings up a question that many mathematicians consider of greater importance. Where do such statements come from? Of course, the answer is that they come from mathematicians or students. How these people arrive at such questions does not follow any set rule. But this deals with the creative aspect of mathematics. Some people are curious and imaginative. Perhaps while proving some theorem, it is realized that the method of proof used could be applied to prove something even more interesting. (What is interesting, of course, is quite subjective.) More than likely however, a person has observed some relationship that exists in an example being considered that appears to occur in a more general setting. This individual then attempts to show that this is the case by giving a proof. This entire process involves the idea of conjectures (guesses) and trying to show the accuracy of a conjecture. We'll discuss this at greater length later.

Suppose that we have been successful in proving $P(x) \Rightarrow Q(x)$ for all x in some domain S (by whatever method). We therefore know that for every $x \in S$ for which the statement $P(x)$ is true, the statement $Q(x)$ is true. Also, for any $x \in S$ for which the statement $Q(x)$ is false, the statement $P(x)$ is false. For example, since we know that Result 3.10 is true, if we should ever encounter an integer n for which $5n - 7$ is even, then we know that n is odd. Furthermore, if we should encounter an integer n for which n^2 is odd, then we can conclude by statement (3.4) or, better yet, by Theorem 3.12, that n itself must be odd.

It is not only knowing that a particular theorem might be useful to us in the future, it is perhaps that a theorem seems surprising, interesting or even beautiful. (Yes—to mathematicians, and hopefully to you as well, a theorem can be beautiful.)

We next describe a type of result that we have not yet encountered. Consider the following result, which we would like to prove.

Result to Prove   Let $x \in \mathbf{Z}$. If $5x - 7$ is odd, then $9x + 2$ is even.

PROOF STRATEGY   This result doesn't seem to fit into the kinds of results we've been proving. (This is not unusual. After learning how to prove certain statements, we encounter new statements that require us to ... think.) If we attempt to give either a direct proof or a proof by contrapositive of this result, we may be headed for difficulties. There is, however, another approach. Even though we must be very careful about what we are assuming, from what we know about even and odd integers, it appears that if $5x - 7$ is odd, then x must be even. In fact, if we *knew* that whenever $5x - 7$ is odd then x is even, this fact would be extremely helpful. We illustrate this next. Don't forget that our goal is to prove the following result, which we will refer to as Result 3.14: Let $x \in \mathbf{Z}$. If $5x - 7$ is odd, then $9x + 2$ is even. The (unusual) numbering of this result is because we will first state and prove a lemma (Lemma 3.13) that will aid us in the proof of Result 3.14.   ◆

In order to verify the truth of Result 3.14, we first prove the following lemma.

Lemma 3.13   *Let $x \in \mathbf{Z}$. If $5x - 7$ is odd, then x is even.*

Proof   Assume that x is odd. Then $x = 2y + 1$, where $y \in \mathbf{Z}$. Therefore,

$$5x - 7 = 5(2y + 1) - 7 = 10y - 2 = 2(5y - 1).$$

Since $5y - 1$ is an integer, $5x - 7$ is even.   ∎

We are now prepared to give a proof of Result 3.14.

Result 3.14   *Let $x \in \mathbf{Z}$. If $5x - 7$ is odd, then $9x + 2$ is even.*

Proof   Let $5x - 7$ be an odd integer. By Lemma 3.13, the integer x is even. Since x is even, $x = 2z$ for some integer z. Thus

$$9x + 2 = 9(2z) + 2 = 18z + 2 = 2(9z + 1).$$

Because $9z + 1$ is an integer, $9x + 2$ is even.   ∎

So, with the aid of Lemma 3.13, we have produced a very uncomplicated (and, hopefully, easy-to-follow) proof of Result 3.14.

The main reason for presenting Result 3.14 was to show how helpful a lemma can be in producing a proof of another result. However, having just said this, we now show how we can prove Result 3.14 without the aid of a lemma, by performing a bit of algebraic manipulation.

Alternative Proof   Assume that $5x - 7$ is odd. Then $5x - 7 = 2n + 1$ for some integer n. Observe that
of Result 3.14

$$9x + 2 = (5x - 7) + (4x + 9) = 2n + 1 + 4x + 9$$
$$= 2n + 4x + 10 = 2(n + 2x + 5).$$

Because $n + 2x + 5$ is an integer, $9x + 2$ is even.   ∎

You may prefer one proof of Result 3.14 over the other. Whether you do or not, it is important to know that two different methods can be used. These methods might prove to be useful for future results you encounter. Also, you might think we used a trick to give the second proof of Result 3.14; but, as we will see, if the same "trick" can be used often, then it becomes a technique.

3.4 Proof by Cases

While attempting to give a proof of a mathematical statement concerning an element x in some set S, it is sometimes useful to observe that x possesses one of two or more properties. A common property which x may possess is that of belonging to a particular subset of S. If we can verify the truth of the statement for each property that x may have, then we have a proof of the statement. Such a proof is then divided into parts called **cases**, one case for each property that x may possess or for each subset to which x may belong. This method is called **proof by cases**. Indeed, it may be useful in a proof by cases to further divide a case into other cases, called **subcases**.

For example, in a proof of $\forall n \in \mathbf{Z}$, $R(n)$, it might be convenient to use a proof by cases whose proof is divided into the two cases

   Case 1. n is even. and *Case 2. n is odd.*

Other possible proofs by cases might involve proving $\forall x \in \mathbf{R}$, $P(x)$ using the cases

   Case 1. $x = 0$, Case 2. $x < 0$ and *Case 3. $x > 0$.*

Also, we might attempt to prove $\forall n \in \mathbf{N}$, $P(n)$ using the cases

$Case$ 1. $n = 1$ and $Case$ 2. $n \geq 2$.

Furthermore, for $S = \mathbf{Z} - \{0\}$, we might try to prove $\forall x, y \in S$, $P(x, y)$ by using the cases:

$Case$ 1. $xy > 0$ and $Case$ 2. $xy < 0$.

Case 1 could, in fact, be divided into two subcases:

$Subcase$ 1.1. $x > 0$ and $y > 0$. and $Subcase$ 1.2. $x < 0$ and $y < 0$,

while Case 2 could be divided into the two subcases:

$Subcase$ 2.1. $x > 0$ and $y < 0$. and $Subcase$ 2.2. $x < 0$ and $y > 0$.

Let's look at an example of a proof by cases.

Result 3.15 *If $n \in \mathbf{Z}$, then $n^2 + 3n + 5$ is an odd integer.*

Proof We proceed by cases, according to whether n is even or odd.

Case 1. n is even. Then $n = 2x$ for some $x \in \mathbf{Z}$. So

$$n^2 + 3n + 5 = (2x)^2 + 3(2x) + 5 = 4x^2 + 6x + 5 = 2(2x^2 + 3x + 2) + 1.$$

Since $2x^2 + 3x + 2 \in \mathbf{Z}$, the integer $n^2 + 3n + 5$ is odd.

Case 2. n is odd. Then $n = 2y + 1$, where $y \in \mathbf{Z}$. Thus

$$n^2 + 3n + 5 = (2y + 1)^2 + 3(2y + 1) + 5 = 4y^2 + 10y + 9$$
$$= 2(2y^2 + 5y + 4) + 1.$$

Because $2y^2 + 5y + 4 \in \mathbf{Z}$, the integer $n^2 + 3n + 5$ is odd. ∎

Two integers x and y are said to be **of the same parity** if x and y are both even or are both odd. The integers x and y are **of opposite parity** if one of x and y is even and the other is odd. For example, 5 and 13 are of the same parity, while 8 and 11 are of opposite parity. Because the definition of two integers having the same (or opposite) parity requires the two integers to satisfy one of two properties, any result containing these terms is likely to be proved by cases. The following theorem presents a characterization of two integers that are of the same parity.

Theorem 3.16 *Let $x, y \in \mathbf{Z}$. Then x and y are of the same parity if and only if $x + y$ is even.*

Proof First, assume that x and y are of the same parity. We consider two cases.

Case 1. x and y are even. Then $x = 2a$ and $y = 2b$ for some integers a and b. So $x + y = 2a + 2b = 2(a + b)$. Since $a + b \in \mathbf{Z}$, the integer $x + y$ is even.

Case 2. x and y are odd. Then $x = 2a + 1$ and $y = 2b + 1$, where $a, b \in \mathbf{Z}$. Therefore,

$$x + y = (2a + 1) + (2b + 1) = 2a + 2b + 2 = 2(a + b + 1).$$

Since $a + b + 1$ is an integer, $x + y$ is even.

For the converse, assume that x and y are of opposite parity. Again, we consider two cases.

Case 1. x is even and y is odd. Then $x = 2a$ and $y = 2b + 1$, where $a, b \in \mathbf{Z}$. Then

$$x + y = 2a + (2b + 1) = 2(a + b) + 1.$$

Since $a + b \in \mathbf{Z}$, the integer $x + y$ is odd.

Case 2. x is odd and y is even. The proof is similar to the proof of the preceding case and is therefore omitted.   ■

PROOF ANALYSIS   A comment concerning the proof of Theorem 3.16 is useful here. Although there is always some concern when omitting steps or proofs, it should be clear that it is truly a waste of effort by writer and reader alike to give a proof of the case when x is odd and y is even in Theorem 3.16. Indeed, there is an alternative when the converse is considered.

For the converse, assume that x and y are of opposite parity. Without loss of generality, assume that x is even and y is odd. Then $x = 2a$ and $y = 2b + 1$, where $a, b \in \mathbf{Z}$. Then

$$x + y = 2a + (2b + 1) = 2(a + b) + 1.$$

Since $a + b \in \mathbf{Z}$, the integer $x + y$ is odd.   ◆

We used the phrase **without loss of generality** (some abbreviate this as *WOLOG* or *WLOG*) to indicate that the proofs of the two situations are similar, so the proof of only one of these is needed. Sometimes it is rather subjective to say that two situations are similar. We present one additional example to illustrate this.

Theorem to Prove   Let a and b be integers. Then ab is even if and only if a is even or b is even.

PROOF STRATEGY   Before we begin a proof of this result (Theorem 3.17 below), let's see what we will be required to show. We need to prove two implications, namely, (1) If a is even or b is even, then ab is even and (2) if ab is even, then a is even or b is even. We consider (1) first. A direct proof seems appropriate. Here, we will assume that a is even or b is even. We could give a proof by cases: (i) a is even, (ii) b is even. On the other hand, since the proofs of these cases will certainly be similar, we could say, without loss of generality, that a is even. We will see that it is unnecessary to make any assumption about b.

If we were to give a direct proof of (2), then we would begin by assuming that ab is even, say $ab = 2k$ for some integer k. But how could we deduce any information about a and b individually? Let's try another approach. If we use a proof by contrapositive, then we would begin by assuming that it is not the case that a is even or b is even. This is exactly the situation covered by one of De Morgan's laws:

$$\sim (P \vee Q) \text{ is logically equivalent to } (\sim P) \wedge (\sim Q).$$

It is important not to forget this. In this case, we have $P : a$ is even. and $Q : b$ is even. So the negation of "a is even or b is even" is "a is odd *and* b is odd."   ◆

Let's now prove this result.

Theorem 3.17   *Let a and b be integers. Then ab is even if and only if a is even or b is even.*

Proof   First, assume that a is even or b is even. Without loss of generality, let a be even. Then $a = 2x$ for some integer x. Thus $ab = (2x)b = 2(xb)$. Since xb is an integer, ab is even.

For the converse, assume that a is odd and b is odd. Then $a = 2x + 1$ and $b = 2y + 1$, where $x, y \in \mathbf{Z}$. Hence

$$ab = (2x + 1)(2y + 1) = 4xy + 2x + 2y + 1 = 2(2xy + x + y) + 1.$$

Since $2xy + x + y$ is an integer, ab is odd.   ∎

3.5 Proof Evaluations

We have now stated several results and have given a proof of each result (sometimes preceding a proof by a proof strategy or following the proof with a proof analysis). Let's reverse this process by giving an example of a proof of a result but not stating the result being proved. We will follow the proof with several options for the statements of the result being proved.

Example 3.18   Given below is a proof of a result.

Proof   Assume that n is an odd integer. Then $n = 2k + 1$ for some integer k. Then

$$3n - 5 = 3(2k + 1) - 5 = 6k + 3 - 5 = 6k - 2 = 2(3k - 1).$$

Since $3k - 1$ is an integer, $3n - 5$ is even.   ∎

Which of the following is proved above?

    (1) $3n - 5$ is an even integer.
    (2) If n is an odd integer, then $3n - 5$ is an even integer.
    (3) Let n be an integer. If $3n - 5$ is an even integer, then n is an odd integer.
    (4) Let n be an integer. If $3n - 5$ is an odd integer, then n is an even integer.

The correct answers are (2) and (4). The proof given is a direct proof of (2) and a proof by contrapositive of (4). The sentence (1) is an open sentence, not a statement, and is only the conclusion of (2). Statement (3) is the converse of (2).   ◆

When learning any mathematical subject, it is not the least bit unusual to make mistakes along the way. In fact, part of learning mathematics is to learn from your mistakes and those of others. For this reason, you will see a few exercises at the end of most chapters (beginning with this chapter) where you are asked to evaluate the proof of a result. That is, a result and a proposed proof of this result will be given. You are then asked to read this proposed proof and determine whether, in your opinion, it is, in fact, a proof. If you don't believe that the given argument provides a proof of the result, then you should point out the (or a) mistake. We give two examples of this.

Problem 3.19   *Evaluate the proposed proof of the following result.*

Result   *If x and y are integers of the same parity, then $x - y$ is even.*

Proof   Let x and y be two integers of the same parity. We consider two cases, according to whether x and y are both even or are both odd.

Case 1. x and y are both even. Let $x = 6$ and $y = 2$, which are both even. Then $x - y = 4$, which is even.

Case 2. x and y are both odd. Let $x = 7$ and $y = 1$, which are both odd. Then $x - y = 6$, which is even.   ∎

Proof Evaluation   Although the proof started correctly, assuming that x and y are two integers of the same parity and dividing the proof into these two cases, the proof of each case is incorrect. When we assume that x and y are both even, for example, x and y must represent arbitrary even integers, not specific even integers.   ◆

Problem 3.20   *Evaluate the proposed proof of the following result.*

Result   *If m is an even integer and n is an odd integer, then $3m + 5n$ is odd.*

Proof   Let m be an even integer and n an odd integer. Then $m = 2k$ and $n = 2k + 1$, where $k \in \mathbf{Z}$. Therefore,

$$3m + 5n = 3(2k) + 5(2k + 1) = 6k + 10k + 5$$
$$= 16k + 5 = 2(8k + 2) + 1.$$

Since $8k + 2$ is an integer, $3m + 5n$ is odd.   ∎

Proof Evaluation   There is a mistake in the second sentence of the proposed proof, where it is written that $m = 2k$ and $n = 2k + 1$, where $k \in \mathbf{Z}$. Since the same symbol k is used for both m and n, we have inadvertently added the assumption that $n = m + 1$. This is incorrect, however, as it was never stated that m and n must be consecutive integers. In other words, we should write $m = 2k$ and $n = 2\ell + 1$, say, where $k, \ell \in \mathbf{Z}$.   ◆

EXERCISES FOR CHAPTER 3

Section 3.1: Trivial and Vacuous Proofs

3.1. Let $x \in \mathbf{R}$. Prove that if $0 < x < 1$, then $x^2 - 2x + 2 \neq 0$.

3.2. Let $n \in \mathbf{N}$. Prove that if $|n - 1| + |n + 1| \leq 1$, then $|n^2 - 1| \leq 4$.

3.3. Let $r \in \mathbf{Q}^+$. Prove that if $\frac{r^2+1}{r} \leq 1$, then $\frac{r^2+2}{r} \leq 2$.

3.4. Let $x \in \mathbf{R}$. Prove that if $x^3 - 5x - 1 \geq 0$, then $(x - 1)(x - 3) \geq -2$.

3.5. Let $n \in \mathbf{N}$. Prove that if $n + \frac{1}{n} < 2$, then $n^2 + \frac{1}{n^2} < 4$.

3.6. Prove that if a, b and c are odd integers such that $a + b + c = 0$, then $abc < 0$. (You are permitted to use well-known properties of integers here.)

3.7. Prove that if x, y and z are three real numbers such that $x^2 + y^2 + z^2 < xy + xz + yz$, then $x + y + z > 0$.

Section 3.2: Direct Proofs

3.8. Prove that if x is an odd integer, then $9x + 5$ is even.

3.9. Prove that if x is an even integer, then $5x - 3$ is an odd integer.

3.10. Prove that if a and c are odd integers, then $ab + bc$ is even for every integer b.

3.11. Let $n \in \mathbf{Z}$. Prove that if $1 - n^2 > 0$, then $3n - 2$ is an even integer.

3.12. Let $x \in \mathbf{Z}$. Prove that if 2^{2x} is an odd integer, then 2^{-2x} is an odd integer.

3.13. Let $S = \{0, 1, 2\}$ and let $n \in S$. Prove that if $(n + 1)^2(n + 2)^2/4$ is even, then $(n + 2)^2(n + 3)^2/4$ is even.

3.14. Let $S = \{1, 5, 9\}$. Prove that if $n \in S$ and $\frac{n^2+n-6}{2}$ is odd, then $\frac{2n^3+3n^2+n}{6}$ is even.

3.15. Let $A = \{n \in \mathbf{Z} : n > 2 \text{ and } n \text{ is odd}\}$ and $B = \{n \in \mathbf{Z} : n < 11\}$. Prove that if $n \in A \cap B$, then $n^2 - 2$ is prime.

Section 3.3: Proof by Contrapositive

3.16. Let $x \in \mathbf{Z}$. Prove that if $7x + 5$ is odd, then x is even.

3.17. Let $n \in \mathbf{Z}$. Prove that if $15n$ is even, then $9n$ is even.

3.18. Let $x \in \mathbf{Z}$. Prove that $5x - 11$ is even if and only if x is odd.

3.19. Let $x \in \mathbf{Z}$. Use a lemma to prove that if $7x + 4$ is even, then $3x - 11$ is odd.

3.20. Let $x \in \mathbf{Z}$. Prove that $3x + 1$ is even if and only if $5x - 2$ is odd.

3.21. Let $n \in \mathbf{Z}$. Prove that $(n + 1)^2 - 1$ is even if and only if n is even.

3.22. Let $S = \{2, 3, 4\}$ and let $n \in S$. Use a proof by contrapositive to prove that if $n^2(n - 1)^2/4$ is even, then $n^2(n + 1)^2/4$ is even.

3.23. Let $A = \{0, 1, 2\}$ and $B = \{4, 5, 6\}$ be subsets of $S = \{0, 1, \ldots, 6\}$. Let $n \in S$. Prove that if $\frac{n(n-1)(n-2)}{6}$ is even, then $n \in A \cup B$.

3.24. Let $n \in \mathbf{Z}$. Prove that $2n^2 + n$ is odd if and only if $\cos \frac{n\pi}{2}$ is even.

3.25. Let $\{A, B\}$ be a partition of the set of $S = \{1, 2, \ldots, 7\}$, where $A = \{1, 4, 5\}$ and $B = \{2, 3, 6, 7\}$. Let $n \in S$. Prove that if $\frac{n^2+3n-4}{2}$ is even, then $n \in A$.

Section 3.4: Proof by Cases

3.26. Prove that if $n \in \mathbf{Z}$, then $n^2 - 3n + 9$ is odd.

3.27. Prove that if $n \in \mathbf{Z}$, then $n^3 - n$ is even.

3.28. Let $x, y \in \mathbf{Z}$. Prove that if xy is odd, then x and y are odd.

3.29. Let $a, b \in \mathbf{Z}$. Prove that if ab is odd, then $a^2 + b^2$ is even.

3.30. Let $x, y \in \mathbf{Z}$. Prove that $x - y$ is even if and only if x and y are of the same parity.

3.31. Let $a, b \in \mathbf{Z}$. Prove that if $a + b$ and ab are of the same parity, then a and b are even.

3.32. (a) Let x and y be integers. Prove that $(x + y)^2$ is even if and only if x and y are of the same parity.
(b) Restate the result in (a) in terms of odd integers.

3.33. Let $A = \{1, 2, 3\}$ and $B = \{2, 3, 4\}$ be subsets of $S = \{1, 2, 3, 4\}$. Let $n \in S$. Prove that $2n^2 - 5n$ is either (a) positive and even or (b) negative and odd if and only if $n \notin A \cap B$.

3.34. Let $A = \{3, 4\}$ be a subset of $S = \{1, 2, \ldots, 6\}$. Let $n \in S$. Prove that if $\frac{n^2(n+1)^2}{4}$ is even, then $n \in A$.

3.35. Prove for every nonnegative integer n that $2^n + 6^n$ is an even integer.

3.36. A collection of nonempty subsets of a nonempty set S is called a **cover** of S if every element of S belongs to at least one of the subsets. (A cover is a partition of S if every element of S belongs to exactly one of the subsets.) Consider the following.

Result   Let $a, b \in \mathbf{Z}$. If a is even or b is even, then ab is even.

Proof   Assume that a is even or b is even. We consider the following cases.

Case 1. a is even. Then $a = 2k$, where $k \in \mathbf{Z}$. Thus $ab = (2k)b = 2(kb)$. Since $kb \in \mathbf{Z}$, it follows that ab is even.

Case 2. b is even. Then $b = 2\ell$, where $\ell \in \mathbf{Z}$. Thus $ab = a(2\ell) = 2(a\ell)$. Since $a\ell \in \mathbf{Z}$, it follows that ab is even.   ∎

Since the domain is $\mathbf{Z}$ for both a and b, we might think of $\mathbf{Z} \times \mathbf{Z}$ being the domain of (a, b). Consider the following subsets of $\mathbf{Z} \times \mathbf{Z}$:

$S_1 = \{(a, b) \in \mathbf{Z} \times \mathbf{Z} : a \text{ and } b \text{ are odd}\}$

$S_2 = \{(a, b) \in \mathbf{Z} \times \mathbf{Z} : a \text{ is even}\}$

$S_3 = \{(a, b) \in \mathbf{Z} \times \mathbf{Z} : b \text{ is even}\}.$

(a) Why is $\{S_1, S_2, S_3\}$ a cover of $\mathbf{Z} \times \mathbf{Z}$ and not a partition of $\mathbf{Z} \times \mathbf{Z}$?

(b) Why does the set S_1 not appear in the proof above?

(c) Give a proof by cases of the result above where the cases are determined by a partition and not a cover.

Section 3.5: Proof Evaluations

3.37. Below is a proof of a result.

Proof   We consider two cases.

Case 1. a and b are even. Then $a = 2r$ and $b = 2s$ for integers r and s. Thus

$$a^2 - b^2 = (2r)^2 - (2s)^2 = 4r^2 - 4s^2 = 2(2r^2 - 2s^2).$$

Since $2r^2 - 2s^2$ is an integer, $a^2 - b^2$ is even.

Case 2. a and b are odd. Then $a = 2r + 1$ and $b = 2s + 1$ for integers r and s. Thus

$$a^2 - b^2 = (2r + 1)^2 - (2s + 1)^2 = (4r^2 + 4r + 1) - (4s^2 + 4s + 1)$$
$$= 4r^2 + 4r - 4s^2 - 4s = 2(2r^2 + 2r - 2s^2 - 2s).$$

Since $2r^2 + 2r - 2s^2 - 2s$ is an integer, $a^2 - b^2$ is even.   ∎

Which of the following is proved?

(1) Let $a, b \in \mathbf{Z}$. Then a and b are of the same parity if and only if $a^2 - b^2$ is even.

(2) Let $a, b \in \mathbf{Z}$. Then $a^2 - b^2$ is even.

(3) Let $a, b \in \mathbf{Z}$. If a and b are of the same parity, then $a^2 - b^2$ is even.

(4) Let $a, b \in \mathbf{Z}$. If $a^2 - b^2$ is even, then a and b are of the same parity.

3.38. Below is given a proof of a result. What result is being proved?

Proof   Assume that x is even. Then $x = 2a$ for some integer a. So

$$3x^2 - 4x - 5 = 3(2a)^2 - 4(2a) - 5 = 12a^2 - 8a - 5 = 2(6a^2 - 4a - 3) + 1.$$

Since $6a^2 - 4a - 3$ is an integer, $3x^2 - 4x - 5$ is odd.

For the converse, assume that x is odd. So $x = 2b + 1$, where $b \in \mathbf{Z}$. Therefore,

$$3x^2 - 4x - 5 = 3(2b + 1)^2 - 4(2b + 1) - 5 = 3(4b^2 + 4b + 1) - 8b - 4 - 5$$
$$= 12b^2 + 4b - 6 = 2(6b^2 + 2b - 3).$$

Since $6b^2 + 2b - 3$ is an integer, $3x^2 - 4x - 5$ is even.   ∎

3.39. Evaluate the proof of the following result.

Result   Let $n \in \mathbf{Z}$. If $3n - 8$ is odd, then n is odd.

Proof   Assume that n is odd. Then $n = 2k + 1$ for some integer k. Then $3n - 8 = 3(2k + 1) - 8 = 6k + 3 - 8 = 6k - 5 = 2(3k - 3) + 1$. Since $3k - 3$ is an integer, $3n - 8$ is odd.   ∎

3.40. Evaluate the proof of the following result.

Result   Let $a, b \in \mathbf{Z}$. Then $a - b$ is even if and only if a and b are of the same parity.

Proof   We consider two cases.

Case 1. a and b are of the same parity. We now consider two subcases.

Subcase 1.1. a and b are both even. Then $a = 2x$ and $b = 2y$, where $x, y \in \mathbf{Z}$. Then $a - b = 2x - 2y = 2(x - y)$. Since $x - y$ is an integer, $a - b$ is even.

Subcase 1.2. a and b are both odd. Then $a = 2x + 1$ and $b = 2y + 1$, where $x, y \in \mathbf{Z}$. Then $a - b = (2x + 1) - (2y + 1) = 2(x - y)$. Since $x - y$ is an integer, $a - b$ is even.

Case 2. a and b are of opposite parity. We again have two subcases.

Subcase 2.1. a is odd and b is even. Then $a = 2x + 1$ and $b = 2y$, where $x, y \in \mathbf{Z}$. Then $a - b = (2x + 1) - 2y = 2(x - y) + 1$. Since $x - y$ is an integer, $a - b$ is odd.

Subcase 2.2. a is even and b is odd. Then $a = 2x$ and $b = 2y + 1$, where $x, y \in \mathbf{Z}$. Then $a - b = 2x - (2y + 1) = 2x - 2y - 1 = 2(x - y - 1) + 1$. Since $x - y - 1$ is an integer, $a - b$ is odd.   ∎

3.41. The following is an attempted proof of a result. What is the result and is the attempted proof correct?

Proof   Assume, without loss of generality, that x is even. Then $x = 2a$ for some integer a. Thus

$$xy^2 = (2a)y^2 = 2(ay^2).$$

Since ay^2 is an integer, xy^2 is even.   ∎

3.42. Given below is a proof of a result. What is the result?

Proof   Assume, without loss of generality, that x and y are even. Then $x = 2a$ and $y = 2b$ for integers a and b. Therefore,

$$xy + xz + yz = (2a)(2b) + (2a)z + (2b)z = 2(2ab + az + bz).$$

Since $2ab + az + bz$ is an integer, $xy + xz + yz$ is even.   ∎

3.43. What result is being proved below, and what procedure is being used to verify the result? First, we present the following proof.

Proof   Assume that x is even. Then $x = 2a$ for some integer a. Thus
$$7x - 3 = 7(2a) - 3 = 14a - 3 = 14a - 4 + 1 = 2(7a - 2) + 1.$$
Since $7a - 2$ is an integer, $7x - 3$ is odd.   ∎

We are now prepared to prove our main result.

Proof   Assume that $7x - 3$ is even. From the result above, x is odd. So $x = 2b + 1$ for some integer b. Thus
$$3x + 8 = 3(2b + 1) + 8 = 6b + 11 = 2(3b + 5) + 1.$$
Since $3b + 5$ is an integer, $3x + 8$ is odd.   ∎

3.44. Consider the following statement.
Let $n \in \mathbf{Z}$. Then $(n - 5)(n + 7)(n + 13)$ is odd if and only if n is even.
Which of the following would be an appropriate way to begin a proof of this statement?
(a) Assume that $(n - 5)(n + 7)(n + 13)$ is odd.
(b) Assume that $(n - 5)(n + 7)(n + 13)$ is even.
(c) Assume that n is even.
(d) Assume that n is odd.
(e) We consider two cases, according to whether n is even or n is odd.

ADDITIONAL EXERCISES FOR CHAPTER 3

3.45. Let $x \in \mathbf{Z}$. Prove that if $7x - 8$ is even, then x is even.

3.46. Let $x \in \mathbf{Z}$. Prove that x^3 is even if and only if x is even.

3.47. Let $x \in \mathbf{Z}$. Use one or two lemmas to prove that $3x^3$ is even if and only if $5x^2$ is even.

3.48. Give a direct proof of the following: Let $x \in \mathbf{Z}$. If $11x - 5$ is odd, then x is even.

3.49. Let $x, y \in \mathbf{Z}$. Prove that if $x + y$ is odd, then x and y are of opposite parity.

3.50. Let $x, y \in \mathbf{Z}$. Prove that if $3x + 5y$ is even, then x and y are of the same parity.

3.51. Let $x, y \in \mathbf{Z}$. Prove that $(x + 1)y^2$ is even if and only if x is odd or y is even.

3.52. Let $x, y \in \mathbf{Z}$. Prove that if xy and $x + y$ are even, then both x and y are even.

3.53. Prove, for every integer x, that the integers $3x + 1$ and $5x + 2$ are of opposite parity.

3.54. Prove the following two results:
(a) Result A: Let $n \in \mathbf{Z}$. If n^3 is even, then n is even.
(b) Result B: If n is an odd integer, then $5n^9 + 13$ is even.

3.55. Prove for every two distinct real numbers a and b, either $\frac{a+b}{2} > a$ or $\frac{a+b}{2} > b$.

3.56. Let $x, y \in \mathbf{Z}$. Prove that if a and b are even integers, then $ax + by$ is even.

3.57. Evaluate the proof of the following result.

Result   Let $x, y \in \mathbf{Z}$ and let a and b be odd integers. If $ax + by$ is even, then x and y are of the same parity.

Proof   Assume that x and y are of opposite parity. Then $x = 2p$ and $y = 2q + 1$ for some integers p and q. Since a and b are odd integers, $a = 2r + 1$ and $b = 2s + 1$ for integers r and s. Hence
$$ax + by = (2r + 1)(2p) + (2s + 1)(2q + 1)$$
$$= 4pr + 2p + 4qs + 2s + 2q + 1$$
$$= 2(2pr + p + 2qs + s + q) + 1.$$
Since $2pr + p + 2qs + s + q$ is an integer, $ax + by$ is odd.   ∎

3.58. Let $S = \{a, b, c, d\}$ be a set of four distinct integers. Prove that if either (1) for each $x \in S$, the integer x and the sum of any two of the remaining three integers of S are of the same parity or (2) for each $x \in S$, the integer x and the sum of any two of the remaining three integers of S are of opposite parity, then every pair of integers of S is of the same parity.

3.59. Prove that if a and b are two positive integers, then $a^2(b + 1) + b^2(a + 1) \geq 4ab$.

3.60. Let $a, b \in \mathbf{Z}$. Prove that if $ab = 4$, then $(a - b)^3 - 9(a - b) = 0$.

3.61. Let a, b and c be the lengths of the sides of a triangle T where $a \leq b \leq c$. Prove that if T is a right triangle, then

$$(abc)^2 = \frac{c^6 - a^6 - b^6}{3}.$$

3.62. Consider the following statement.
Let $n \in \mathbf{Z}$. Then $3n^3 + 4n^2 + 5$ is even if and only if n is even.
Which of the following would be an appropriate way to begin a proof of this statement?

    (a) Assume that $3n^3 + 4n^2 + 5$ is odd.
    (b) Assume that $3n^3 + 4n^2 + 5$ is even.
    (c) Assume that n is even.
    (d) Assume that n is odd.
    (e) We consider two cases, according to whether n is even or n is odd.

3.63. Let $\mathcal{P} = \{A, B, C\}$ be a partition of a set S of integers, where $A = \{n \in S : n \text{ is odd and } n > 0\}$, $B = \{n \in S : n \text{ is odd and } n < 0\}$ and $C = \{n \in S : n \text{ is even and } n > 0\}$. Prove that if x and y are elements of S belonging to distinct subsets in $\mathcal{P}$, then xy is either odd, even and greater than 1, or even and less than -1.

3.64. Let $n \in \mathbf{N}$. Prove that if $n^3 - 5n - 10 > 0$, then $n \geq 3$.

3.65. Prove for every odd integer a that $(a^2 + 3)(a^2 + 7) = 32b$ for some integer b.

3.66. Prove for every two positive integers a and b that

$$(a + b)\left(\frac{1}{a} + \frac{1}{b}\right) \geq 4.$$

3.67. Which result is being proved below, and what procedure is being used to verify the result?
We begin with the following proof.

Proof   First, assume that x is even. Then $x = 2a$, where $a \in \mathbf{Z}$. Thus

$$3x - 2 = 3(2a) - 2 = 6a - 2 = 2(3a - 1).$$

Since $3a - 1$ is an integer, $3x - 2$ is even.
    Next, suppose that x is odd. Then $x = 2b + 1$ for some integer b. So
$$3x - 2 = 3(2b + 1) - 2 = 6b + 1 = 2(3a) + 1.$$

Since $3a$ is an integer, $3x - 2$ is odd.   ■
    We can now give the following proof.

Proof   First, assume that $3x - 2$ is even. From the preceding result, x is even and so $x = 2a$, where $a \in \mathbf{Z}$. Thus

$$5x + 1 = 5(2a) + 1 = 2(5a) + 1.$$

Since $5a$ is an integer, $5x + 1$ is odd.
    Next, assume that $3x - 2$ is odd. Again, by the preceding result, x is odd. Hence $x = 2b + 1$ for some integer b. Therefore,

$$5x + 1 = 5(2b + 1) + 1 = 10b + 6 = 2(5b + 3).$$

Since $5b + 3$ is an integer, $5x + 1$ is even.   ■

4

More on Direct Proof and Proof by Contrapositive

The vast majority of the examples illustrating direct proof and proof by contrapositive that we have seen involve properties of even and odd integers. In this chapter, we will give additional examples of direct proofs and proofs by contrapositive concerning integers but in new surroundings. First, we will see how even and odd integers can be studied in a more general setting, through divisibility of integers. We will then explore some properties of real numbers and, finally, look at some properties of set operations.

4.1 Proofs Involving Divisibility of Integers

We have now seen many examples of integers that can be written as $2x$ for some integer x. These are precisely the even integers, of course. However, some integers can also be expressed as $3x$ or $4x$, or as $-5x$ for some integer x. In general, for integers a and b with $a \neq 0$, we say that a **divides** b if there is an integer c such that $b = ac$. In this case, we write $a \mid b$. Hence if n is an even integer, then $2 \mid n$; moreover, if 2 divides some integer n, then n is even. That is, an integer n is even if and only if $2 \mid n$. Theorem 3.17 (which states for integers a and b, that ab is even if and only if a or b is even) can therefore be restated for integers a and b as: $2 \mid ab$ if and only if $2 \mid a$ or $2 \mid b$.

If $a \mid b$, then we also say that b is a **multiple** of a and that a is a **divisor** of b. Thus every even integer is a multiple of 2. If a does not divide b, then we write $a \nmid b$. For example, $4 \mid 48$ since $48 = 4 \cdot 12$ and $-3 \mid 57$ since $57 = (-3) \cdot (-19)$. On the other hand, $4 \nmid 66$ as there is no integer c such that $66 = 4c$.

We now apply the techniques we've learned to prove some results concerning divisibility properties of integers.

Result to Prove Let a, b and c be integers with $a \neq 0$ and $b \neq 0$. If $a \mid b$ and $b \mid c$, then $a \mid c$.

PROOF STRATEGY It seems reasonable here to use a direct proof and to begin by assuming that $a \mid b$ and $b \mid c$. This means that $b = ax$ and $c = by$ for some integers x and y. Since our goal is to show that $a \mid c$, we need to show that c can be written as the product of a and some other integer. Hence it is logical to consider c and determine how we can express it. ◆

Result 4.1   *Let a, b and c be integers with $a \neq 0$ and $b \neq 0$. If $a \mid b$ and $b \mid c$, then $a \mid c$.*

Proof   Assume that $a \mid b$ and $b \mid c$. Then $b = ax$ and $c = by$, where $x, y \in \mathbf{Z}$. Therefore, $c = by = (ax)y = a(xy)$. Since xy is an integer, $a \mid c$.   ∎

We now verify two other divisibility properties of integers.

Result 4.2   *Let a, b, c and d be integers with $a \neq 0$ and $b \neq 0$. If $a \mid c$ and $b \mid d$, then $ab \mid cd$.*

Proof   Let $a \mid c$ and $b \mid d$. Then $c = ax$ and $d = by$, where $x, y \in \mathbf{Z}$. Then

$$cd = (ax)(by) = (ab)(xy).$$

Since xy is an integer, $ab \mid cd$.   ∎

Result 4.3   *Let a, b, c, x, $y \in \mathbf{Z}$, where $a \neq 0$. If $a \mid b$ and $a \mid c$, then $a \mid (bx + cy)$.*

Proof   Assume that $a \mid b$ and $a \mid c$. Then $b = ar$ and $c = as$, where $r, s \in \mathbf{Z}$. Then

$$bx + cy = (ar)x + (as)y = a(rx + sy).$$

Since $rx + sy$ is an integer, $a \mid (bx + cy)$.   ∎

The examples that we have presented thus far concern general properties of divisibility of integers. We now look at some specialized properties of divisibility.

Result 4.4   *Let $x \in \mathbf{Z}$. If $2 \mid (x^2 - 1)$, then $4 \mid (x^2 - 1)$.*

Proof   Assume that $2 \mid (x^2 - 1)$. So $x^2 - 1 = 2y$ for some integer y. Thus $x^2 = 2y + 1$ is an odd integer. It then follows by Theorem 3.12 that x too is odd. Hence $x = 2z + 1$ for some integer z. Then

$$x^2 - 1 = (2z + 1)^2 - 1 = (4z^2 + 4z + 1) - 1 = 4z^2 + 4z = 4(z^2 + z).$$

Since $z^2 + z$ is an integer, $4 \mid (x^2 - 1)$.   ∎

For each of the Results 4.1–4.4, a direct proof worked very well. For the following result, however, the situation is quite different.

Result to Prove   Let $x, y \in \mathbf{Z}$. If $3 \nmid xy$, then $3 \nmid x$ and $3 \nmid y$.

PROOF STRATEGY   If we let

$$P: 3 \nmid xy, \quad Q: 3 \nmid x \quad \text{and} \quad R: 3 \nmid y,$$

then we wish to prove that $P \Rightarrow Q \wedge R$ for all integers x and y. (It should be clear that P, Q and R are open sentences in this case, but we omit the variables here for simplicity.) If we use a direct proof, then we would assume that $3 \nmid xy$ and attempt to show that $3 \nmid x$ and $3 \nmid y$. Thus we would know that xy *cannot* be expressed as 3 times an integer. On the other hand, if we use a proof by contrapositive, then we are considering the implication $(\sim(Q \wedge R)) \Rightarrow (\sim P)$, which, by De Morgan's Law, is logically equivalent to

$((\sim Q) \vee (\sim R)) \Rightarrow (\sim P)$ and which, in words, is: If $3 \mid x$ or $3 \mid y$, then $3 \mid xy$. This method looks more promising. ◆

Result 4.5   *Let $x, y \in \mathbf{Z}$. If $3 \nmid xy$, then $3 \nmid x$ and $3 \nmid y$.*

Proof   Assume that $3 \mid x$ or $3 \mid y$. Without loss of generality, assume that 3 divides x. Then $x = 3z$ for some integer z. Hence $xy = (3z)y = 3(zy)$. Since zy is an integer, $3 \mid xy$. ∎

We have already mentioned that if an integer n is not a multiple of 2, then we can write $n = 2q + 1$ for some integer q (that is, if an integer n is not even, then it is odd). This is a consequence of knowing that 0 and 1 are the only possible remainders when an integer is divided by 2. Along the same lines, if an integer n is not a multiple of 3, then we can write $n = 3q + 1$ or $n = 3q + 2$ for some integer q; that is, every integer can be expressed as $3q, 3q + 1$ or $3q + 2$ for some integer q since 0, 1 and 2 are the only remainders that can result when an integer is divided by 3. Similarly, if an integer n is not a multiple of 4, then n can be expressed as $4q + 1, 4q + 2$ or $4q + 3$ for some integer q. This topic concerns a well-known theorem called the Division Algorithm, which will be explored in more detail in Chapter 11.

Result to Prove   Let $x \in \mathbf{Z}$. If $3 \nmid (x^2 - 1)$, then $3 \mid x$.

PROOF STRATEGY   We have two options here, namely (1) use a direct proof and begin a proof by assuming that $3 \nmid (x^2 - 1)$ or (2) use a proof by contrapositive and begin a proof by assuming that $3 \nmid x$. Certainly, we cannot avoid assuming that 3 does *not* divide *some* integer. However, it appears far easier to know that $3 \nmid x$ and attempt to show that $3 \mid (x^2 - 1)$ than to know that $3 \nmid (x^2 - 1)$ and show that $3 \mid x$. Also, if $3 \nmid x$, then we now know that $x = 3q + 1$ or $x = 3q + 2$ for some integer q, which suggests a proof by cases. ◆

Result 4.6   *Let $x \in \mathbf{Z}$. If $3 \nmid (x^2 - 1)$, then $3 \mid x$.*

Proof   Assume that $3 \nmid x$. Then either $x = 3q + 1$ for some integer q or $x = 3q + 2$ for some integer q. We consider these two cases.

Case 1. $x = 3q + 1$ for some integer q. Then

$$x^2 - 1 = (3q + 1)^2 - 1 = (9q^2 + 6q + 1) - 1$$
$$= 9q^2 + 6q = 3(3q^2 + 2q).$$

Since $3q^2 + 2q$ is an integer, $3 \mid (x^2 - 1)$.

Case 2. $x = 3q + 2$ for some integer q. Then

$$x^2 - 1 = (3q + 2)^2 - 1 = (9q^2 + 12q + 4) - 1$$
$$= 9q^2 + 12q + 3 = 3(3q^2 + 4q + 1).$$

Since $3q^2 + 4q + 1$ is an integer, $3 \mid (x^2 - 1)$. ∎

We now consider a biconditional involving divisibility.

Result 4.7   *Let $x, y \in \mathbf{Z}$. Then $4 \mid (x^2 - y^2)$ if and only if x and y are of the same parity.*

Proof   Assume first that x and y are of the same parity. We show that $4 \mid (x^2 - y^2)$. There are two cases.

Case 1. x and y are both even. Thus $x = 2a$ and $y = 2b$ for some integers a and b. Then

$$x^2 - y^2 = (2a)^2 - (2b)^2 = 4a^2 - 4b^2 = 4(a^2 - b^2).$$

Since $a^2 - b^2$ is an integer, $4 \mid (x^2 - y^2)$.

Case 2. x and y are both odd. So $x = 2c + 1$ and $y = 2d + 1$ for some integers c and d. Then

$$x^2 - y^2 = (2c + 1)^2 - (2d + 1)^2 = (4c^2 + 4c + 1) - (4d^2 + 4d + 1)$$
$$= 4c^2 + 4c - 4d^2 - 4d = 4(c^2 + c - d^2 - d).$$

Since $c^2 + c - d^2 - d$ is an integer, $4 \mid (x^2 - y^2)$.

For the converse, assume that x and y are of opposite parity. We show that $4 \nmid (x^2 - y^2)$. We consider two cases.

Case 1. x is even and y is odd. Thus $x = 2a$ and $y = 2b + 1$ for some integers a and b. Then

$$x^2 - y^2 = (2a)^2 - (2b + 1)^2 = 4a^2 - [4b^2 + 4b + 1]$$
$$= 4a^2 - 4b^2 - 4b - 1 = 4a^2 - 4b^2 - 4b - 4 + 3$$
$$= 4(a^2 - b^2 - b - 1) + 3.$$

Since $a^2 - b^2 - b - 1$ is an integer, it follows that there is a remainder of 3 when $x^2 - y^2$ is divided by 4 and so $4 \nmid (x^2 - y^2)$.

Case 2. x is odd and y is even. The proof of this case is similar to that of Case 1 and is therefore omitted.   ∎

We consider a result of a somewhat different nature.

Result to Prove   For every integer $n \geq 7$, there exist positive integers a and b such that $n = 2a + 3b$.

PROOF STRATEGY   First, notice that we can write $7 = 2 \cdot 2 + 3 \cdot 1$, $8 = 2 \cdot 1 + 3 \cdot 2$ and $9 = 2 \cdot 3 + 3 \cdot 1$. So the result is certainly true for $n = 7, 8, 9$. On the other hand, there is no pair a, b of positive integers such that $6 = 2a + 3b$. Of course, this observation only shows that we cannot replace $n \geq 7$ by $n \geq 6$.

Suppose that n is an integer such that $n \geq 7$. We could bring the integer 2 into the discussion by observing that we can write $n = 2q$ or $n = 2q + 1$, where $q \in \mathbf{Z}$. Actually, if $n = 2q$, then $q \geq 4$ since $n \geq 7$; while if $n = 2q + 1$, then $q \geq 3$ since $n \geq 7$. This is a useful observation.   ◆

Result 4.8   *For every integer $n \geq 7$, there exist positive integers a and b such that $n = 2a + 3b$.*

Proof   Let n be an integer such that $n \geq 7$. Then $n = 2q$ or $n = 2q + 1$ for some integer q. We consider these two cases.

Case 1. $n = 2q$. Since $n \geq 7$, it follows that $q \geq 4$. Thus

$$n = 2q = 2(q - 3) + 6 = 2(q - 3) + 3 \cdot 2.$$

Since $q \geq 4$, it follows that $q - 3 \in \mathbf{N}$.

Case 2. $n = 2q + 1$. Since $n \geq 7$, it follows that $q \geq 3$. Thus

$$n = 2q + 1 = 2(q - 1) + 2 + 1 = 2(q - 1) + 3 \cdot 1.$$

Since $q \geq 3$, it follows that $q - 1 \in \mathbf{N}$.   ■

4.2 Proofs Involving Congruence of Integers

We know that an integer x is even if $x = 2q$ for some integer q, while x is odd if $x = 2q + 1$ for some integer q. Furthermore, two integers x and y are of the same parity if they are both even or are both odd. From this, it follows that x and y are of the same parity if and only if $2 \mid (x - y)$. Consequently, $2 \mid (x - y)$ if and only if x and y have the same remainder when divided by 2. We also know that an integer x can be expressed as $3q, 3q + 1$ or $3q + 2$ for some integer q, according whether the remainder is 0, 1 or 2 when x is divided by 3. If integers x and y are both of the form $3q + 1$, then $x = 3s + 1$ and $y = 3t + 1$, where $s, t \in \mathbf{Z}$, and so $x - y = 3(s - t)$. Since $s - t$ is an integer, $3 \mid (x - y)$. Similarly, if x and y are both of the form $3q$ or are both of the form $3q + 2$, then $3 \mid (x - y)$ as well. Hence if x and y have the same remainder when divided by 3, then $3 \mid (x - y)$. The converse of this implication is true as well. This suggests a special interest in pairs x, y of integers such that $2 \mid (x - y)$ or $3 \mid (x - y)$ or, in fact, in pairs x, y of integers such that $n \mid (x - y)$ for some integer $n \geq 2$.

For integers a, b and $n \geq 2$, we say that a is **congruent to** b **modulo** n, written $a \equiv b \pmod{n}$, if $n \mid (a - b)$. For example, $15 \equiv 7 \pmod{4}$ since $4 \mid (15 - 7)$ and $3 \equiv -15 \pmod{9}$ since $9 \mid (3 - (-15))$. On the other hand, 14 is not congruent to 4 modulo 6, written $14 \not\equiv 4 \pmod{6}$, since $6 \nmid (14 - 4)$.

Since we know that every integer x can be expressed as $x = 2q$ or as $x = 2q + 1$ for some integer q, it follows that either $2 \mid (x - 0)$ or $2 \mid (x - 1)$, that is, $x \equiv 0 \pmod{2}$ or $x \equiv 1 \pmod{2}$. Also, since each integer x can be expressed as $x = 3q, x = 3q + 1$ or $x = 3q + 2$ for some integer q, it follows that $3 \mid (x - 0), 3 \mid (x - 1)$ or $3 \mid (x - 2)$. Hence

$$x \equiv 0 \pmod{3}, \quad x \equiv 1 \pmod{3} \quad \text{or} \quad x \equiv 2 \pmod{3}.$$

Moreover, for each integer x, exactly one of

$$x \equiv 0 \pmod{4}, \quad x \equiv 1 \pmod{4}, \quad x \equiv 2 \pmod{4}, \quad x \equiv 3 \pmod{4}$$

holds, according to whether the remainder is 0, 1, 2 or 3, respectively, when x is divided by 4. Similar statements can also be made when x is divided by n for each integer $n \geq 5$. We now consider some properties of congruence of integers.

Result to Prove   Let a, b, k and n be integers where $n \geq 2$. If $a \equiv b \pmod{n}$, then $ka \equiv kb \pmod{n}$.

PROOF STRATEGY   A direct proof seems reasonable here. So, we begin by assuming that $a \equiv b \pmod{n}$. Our goal is to show that $ka \equiv kb \pmod{n}$. Because we know that $a \equiv b \pmod{n}$, it follows from the definition that $n \mid (a - b)$, which implies that $a - b = nx$ for some integer x. We need to show that $ka \equiv kb \pmod{n}$, which means that we need to show that $n \mid (ka - kb)$. Thus, we must show that $ka - kb = nt$ for some integer t. This suggests considering the expression $ka - kb$.   ◆

Result 4.9   *Let a, b, k and n be integers where $n \geq 2$. If $a \equiv b \pmod{n}$, then $ka \equiv kb \pmod{n}$.*

Proof   Assume that $a \equiv b \pmod{n}$. Then $n \mid (a - b)$. Hence $a - b = nx$ for some integer x. Therefore,

$$ka - kb = k(a - b) = k(nx) = n(kx).$$

Since kx is an integer, $n \mid (ka - kb)$ and so $ka \equiv kb \pmod{n}$.   ∎

Result 4.10   *Let $a, b, c, d, n \in \mathbf{Z}$ where $n \geq 2$. If $a \equiv b \pmod{n}$ and $c \equiv d \pmod{n}$, then $a + c \equiv b + d \pmod{n}$.*

Proof   Assume that $a \equiv b \pmod{n}$ and $c \equiv d \pmod{n}$. Then $a - b = nx$ and $c - d = ny$ for some integers x and y. Adding these two equations, we obtain

$$(a - b) + (c - d) = nx + ny$$

and so

$$(a + c) - (b + d) = n(x + y).$$

Since $x + y$ is an integer, $n \mid [(a + c) - (b + d)]$. Hence $a + c \equiv b + d \pmod{n}$.   ∎

The next result parallels that of Result 4.10 in terms of multiplication.

Result to Prove   Let $a, b, c, d, n \in \mathbf{Z}$ where $n \geq 2$. If $a \equiv b \pmod{n}$ and $c \equiv d \pmod{n}$, then $ac \equiv bd \pmod{n}$.

<u>**PROOF STRATEGY**</u>   This result and Result 4.10 have the same hypothesis. In the proof of Result 4.10, we arrived at the equations $a - b = nx$ and $c - d = ny$ and needed only to add them to complete the proof. This suggests that in the current result, it would be reasonable to multiply these two equations. However, if we multiply them, we obtain $(a - b)(c - d) = (nx)(ny)$, which does not give us the desired conclusion that $ac - bd$ is a multiple of n. It is essential though that we work $ac - bd$ into the proof. By rewriting $a - b = nx$ and $c - d = ny$ as $a = b + nx$ and $c = d + ny$, respectively, and *then* multiplying, we can accomplish this, however.   ◆

Result 4.11   *Let $a, b, c, d, n \in \mathbf{Z}$ where $n \geq 2$. If $a \equiv b \pmod{n}$ and $c \equiv d \pmod{n}$, then $ac \equiv bd \pmod{n}$.*

Proof   Assume that $a \equiv b \pmod{n}$ and $c \equiv d \pmod{n}$. Then $a - b = nx$ and $c - d = ny$, where $x, y \in \mathbf{Z}$. Thus $a = b + nx$ and $c = d + ny$. Multiplying these two equations, we obtain

$$ac = (b + nx)(d + ny) = bd + dnx + bny + n^2xy$$
$$= bd + n(dx + by + nxy)$$

and so $ac - bd = n(dx + by + nxy)$. Since $dx + by + nxy$ is an integer, $ac \equiv bd \pmod{n}$. ∎

The proofs of the preceding three results use a direct proof. This is not a convenient proof technique for the next result, however.

Result to Prove   Let $n \in \mathbf{Z}$. If $n^2 \not\equiv n \pmod 3$, then $n \not\equiv 0 \pmod 3$ and $n \not\equiv 1 \pmod 3$.

PROOF STRATEGY   Let

$$P(n) : n^2 \not\equiv n \pmod 3, \ Q(n) : n \not\equiv 0 \pmod 3 \text{ and } R(n) : n \not\equiv 1 \pmod 3.$$

Our goal is then to show that $P(n) \Rightarrow (Q(n) \wedge R(n))$ is true for every integer n. A direct proof does not appear to be a good choice. However, a proof by contrapositive would lead us to the implication $\sim(Q(n) \wedge R(n)) \Rightarrow (\sim P(n))$, which, by De Morgan's Law, is logically equivalent to

$$((\sim Q(n)) \vee (\sim R(n))) \Rightarrow (\sim P(n)).$$

In words, we then have: If $n \equiv 0 \pmod 3$ or $n \equiv 1 \pmod 3$, then $n^2 \equiv n \pmod 3$. ♦

Result 4.12   *Let $n \in \mathbf{Z}$. If $n^2 \not\equiv n \pmod 3$, then $n \not\equiv 0 \pmod 3$ and $n \not\equiv 1 \pmod 3$.*

Proof   Let n be an integer such that $n \equiv 0 \pmod 3$ or $n \equiv 1 \pmod 3$. We consider these two cases.

Case 1. $n \equiv 0 \pmod 3$. Then $n = 3k$ for some integer k. Hence

$$n^2 - n = (3k)^2 - (3k) = 9k^2 - 3k = 3(3k^2 - k).$$

Since $3k^2 - k$ is an integer, $3 \mid (n^2 - n)$. Thus $n^2 \equiv n \pmod 3$.

Case 2. $n \equiv 1 \pmod 3$. So $n = 3\ell + 1$ for some integer ℓ and

$$n^2 - n = (3\ell + 1)^2 - (3\ell + 1) = (9\ell^2 + 6\ell + 1) - (3\ell + 1)$$
$$= 9\ell^2 + 3\ell = 3(3\ell^2 + \ell).$$

Since $3\ell^2 + \ell$ is an integer, $3 \mid (n^2 - n)$ and so $n^2 \equiv n \pmod 3$. ∎

As a consequence of Result 4.12, if an integer n and its square n^2 have different remainders when divided by 3, then the remainder for n (when divided by 3) is 2.

4.3 Proofs Involving Real Numbers

We now apply the proof techniques we have introduced to verify some mathematical statements involving real numbers. To be certain that we are working under the same set of rules, let us recall some facts about real numbers whose truth we accept without justification. We have already mentioned that $a^2 \geq 0$ for every real number a. Indeed, $a^n \geq 0$ for every real number a if n is a positive even integer. If $a < 0$ and n is a positive odd integer, then $a^n < 0$. Of course, the product of two real numbers is positive if and only if both numbers are positive or both are negative.

Now let $a, b, c \in \mathbf{R}$. If $a \geq b$ and $c \geq 0$, then the inequality $ac \geq bc$ holds. Indeed, if $c > 0$, then $a/c \geq b/c$.

$$\text{If } a > b \text{ and } c > 0, \text{ then } ac > bc \text{ and } a/c > b/c. \tag{4.1}$$

If $c < 0$, then the inequalities in (4.1) are reversed; namely:

$$\text{If } a > b \text{ and } c < 0, \text{ then } ac < bc \text{ and } a/c < b/c. \tag{4.2}$$

Another important and well-known property of real numbers is that if the product of two real numbers is 0, then at least one of these numbers is 0.

Theorem to Prove   If x and y are real numbers such that $xy = 0$, then $x = 0$ or $y = 0$.

<u>**PROOF STRATEGY**</u>   If we use a direct proof, then we begin by assuming that $xy = 0$. If $x = 0$, then we already have the desired result. On the other hand, if $x \neq 0$, then we are required to show that $y = 0$. However, if $x \neq 0$, then $1/x$ is a real number. This suggests multiplying $xy = 0$ by $1/x$.   ◆

Theorem 4.13   *Let $x, y \in \mathbf{R}$. If $xy = 0$, then $x = 0$ or $y = 0$.*

Proof   Assume that $xy = 0$. We consider two cases, according to whether $x = 0$ or $x \neq 0$.

Case 1. $x = 0$. Then we have the desired result.

Case 2. $x \neq 0$. Multiplying $xy = 0$ by the number $1/x$, we obtain $\dfrac{1}{x}(xy) = \dfrac{1}{x} \cdot 0$.

Since

$$\frac{1}{x}(xy) = \left(\frac{1}{x}x\right) y = 1 \cdot y = y,$$

it follows that $y = 0$.   ∎

We now use Theorem 4.13 to prove the next result.

Result 4.14   *Let $x \in \mathbf{R}$. If $x^3 - 5x^2 + 3x = 15$, then $x = 5$.*

Proof   Assume that $x^3 - 5x^2 + 3x = 15$. Thus $x^3 - 5x^2 + 3x - 15 = 0$. Observe that

$$x^3 - 5x^2 + 3x - 15 = x^2(x - 5) + 3(x - 5) = (x^2 + 3)(x - 5).$$

Since $x^3 - 5x^2 + 3x - 15 = 0$, it follows that $(x^2 + 3)(x - 5) = 0$. By Theorem 4.13, $x^2 + 3 = 0$ or $x - 5 = 0$. Since $x^2 + 3 > 0$, it follows that $x - 5 = 0$ and so $x = 5$.   ∎

Next we consider an example of a proof by contrapositive involving an inequality.

Result 4.15   *Let $x \in \mathbf{R}$. If $x^5 - 3x^4 + 2x^3 - x^2 + 4x - 1 \geq 0$, then $x \geq 0$.*

Proof   Assume that $x < 0$. Then $x^5 < 0$, $2x^3 < 0$ and $4x < 0$. In addition, $-3x^4 < 0$ and $-x^2 < 0$. Thus

$$x^5 - 3x^4 + 2x^3 - x^2 + 4x - 1 < 0 - 1 < 0,$$

as desired.   ∎

On occasion we may encounter problems that involve the verification of a certain equality or inequality and where it is convenient to find an equivalent formulation of the equality or inequality whose truth is clear. This then becomes the starting point of a proof. We now verify an inequality whose proof uses this common approach.

Result to Prove If $x, y \in \mathbf{R}$, then

$$\frac{1}{3}x^2 + \frac{3}{4}y^2 \geq xy.$$

<u>**PROOF STRATEGY**</u> Let's first eliminate fractions from the expression. Showing that $\frac{1}{3}x^2 + \frac{3}{4}y^2 \geq xy$ is equivalent to showing that

$$12\left(\frac{1}{3}x^2 + \frac{3}{4}y^2\right) \geq 12xy,$$

that is,

$$4x^2 + 9y^2 \geq 12xy,$$

which, in turn, is equivalent to

$$4x^2 - 12xy + 9y^2 \geq 0.$$

That is, if we could show that $4x^2 - 12xy + 9y^2 \geq 0$, then we would be able to show that

$$\frac{1}{3}x^2 + \frac{3}{4}y^2 \geq xy.$$

Making a simple observation about $4x^2 - 12xy + 9y^2$ leads to a proof. ◆

Result 4.16 *If $x, y \in \mathbf{R}$, then*

$$\frac{1}{3}x^2 + \frac{3}{4}y^2 \geq xy.$$

Proof Since $(2x - 3y)^2 \geq 0$, it follows that $4x^2 - 12xy + 9y^2 \geq 0$ and so $4x^2 + 9y^2 \geq 12xy$. Dividing this inequality by 12, we obtain

$$\frac{1}{3}x^2 + \frac{3}{4}y^2 \geq xy$$

producing the desired inequality. ∎

Recall that for a real number x, its **absolute value** $|x|$ is defined as

$$|x| = \begin{cases} x & \text{if } x \geq 0 \\ -x & \text{if } x < 0. \end{cases}$$

A well-known property of absolute values is that $|xy| = |x||y|$ for every two real numbers x and y (see Exercise 30). The following theorem gives another familiar property of absolute values of real numbers (called the triangle inequality) that has numerous applications. Since the definition of $|x|$ is essentially a definition by cases, proofs involving $|x|$ are often by cases.

Theorem 4.17 (**The Triangle Inequality**) *For every two real numbers x and y,*

$$|x + y| \leq |x| + |y|.$$

Proof   Since $|x + y| = |x| + |y|$ if either x or y is 0, we can assume that x and y are nonzero. We proceed by cases.

Case 1. $x > 0$ and $y > 0$. Then $x + y > 0$ and

$$|x + y| = x + y = |x| + |y|.$$

Case 2. $x < 0$ and $y < 0$. Since $x + y < 0$,

$$|x + y| = -(x + y) = (-x) + (-y) = |x| + |y|.$$

Case 3. One of x and y is positive and the other is negative. Assume, without loss of generality, that $x > 0$ and $y < 0$. We consider two subcases.

Subcase 3.1. $x + y \geq 0$. Then

$$|x| + |y| = x + (-y) = x - y > x + y = |x + y|.$$

Subcase 3.2. $x + y < 0$. Here

$$|x| + |y| = x + (-y) = x - y > -x - y = -(x + y) = |x + y|.$$

Therefore, $|x + y| \leq |x| + |y|$ for every two real numbers.   ∎

Example 4.18   *Show that if $|x - 1| < 1$ and $|x - 1| < r/4$, where $r \in \mathbf{R}^+$, then $|x^2 + x - 2| < r$.*

Solution   First, observe that

$$|x^2 + x - 2| = |(x + 2)(x - 1)| = |x + 2||x - 1|.$$

By Theorem 4.17,

$$|x + 2| = |(x - 1) + 3| \leq |x - 1| + |3| < 1 + 3 = 4.$$

Therefore,

$$|x^2 + x - 2| = |x + 2||x - 1| < 4 \left(\frac{r}{4} \right) = r.$$ ◆

4.4 Proofs Involving Sets

We now turn our attention to proofs concerning properties of sets. Recall, for sets A and B contained in some universal set U, that the **intersection** of A and B is

$$A \cap B = \{x \; : \; x \in A \text{ and } x \in B\},$$

the **union** of A and B is

$$A \cup B = \{x \; : \; x \in A \text{ or } x \in B\}$$

and the **difference** of A and B is

$$A - B = \{x \; : \; x \in A \text{ and } x \notin B\}.$$

The set $A - B$ is also called the **relative complement** of B in A and the relative complement of A in U is called simply the **complement** of A and is denoted by $\overline{A}$. Thus, $\overline{A} = U - A$. In what follows, we will always assume that the sets under discussion are subsets of some universal set U.

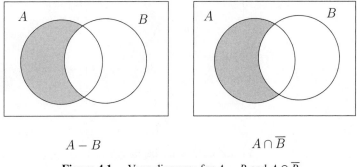

$A - B$ $A \cap \overline{B}$

Figure 4.1   Venn diagrams for $A - B$ and $A \cap \overline{B}$

Figure 4.1 shows Venn diagrams of $A - B$ and $A \cap \overline{B}$ for arbitrary sets A and B. The diagrams suggest that these two sets are equal. This is, in fact, the case. Recall that to show the equality of two sets C and D, we can verify the two set inclusions $C \subseteq D$ and $D \subseteq C$. To establish the inclusion $C \subseteq D$, we show that every element of C is also an element of D; that is, if $x \in C$ then $x \in D$. This is accomplished with a direct proof by letting x be an (arbitrary) element of C and showing that x must belong to D as well. Recall that we need not be concerned if C contains no elements; for in this case $x \in C$ is false for every element x and so the implication "If $x \in C$, then $x \in D$." is true for all $x \in U$ and in this case the statement follows vacuously. As a consequence of this observation, if $C = \emptyset$, then C contains no elements and it follows that $C \subseteq D$.

Result 4.19   *For every two sets A and B,*
$$A - B = A \cap \overline{B}.$$

Proof   First we show that $A - B \subseteq A \cap \overline{B}$. Let $x \in A - B$. Then $x \in A$ and $x \notin B$. Since $x \notin B$, it follows that $x \in \overline{B}$. Therefore, $x \in A$ and $x \in \overline{B}$; so $x \in A \cap \overline{B}$. Hence $A - B \subseteq A \cap \overline{B}$.

   Next we show that $A \cap \overline{B} \subseteq A - B$. Let $y \in A \cap \overline{B}$. Then $y \in A$ and $y \in \overline{B}$. Since $y \in \overline{B}$, we see that $y \notin B$. Now because $y \in A$ and $y \notin B$, we conclude that $y \in A - B$. Thus, $A \cap \overline{B} \subseteq A - B$.   ∎

<u>PROOF ANALYSIS</u>   In the second paragraph of the proof of Result 4.19, we used y (rather than x) to denote an arbitrary element of $A \cap \overline{B}$. We did this only for variety. We could have used x twice. Once we decided to use distinct symbols, y was the logical choice since x was used in the first paragraph of the proof. This keeps our use of symbols consistent. Another possibility would have been to use a in the first paragraph and b in the second. This has some disadvantages, however. Since the sets are being called A and B, we might have a tendency to think that $a \in A$ and $b \in B$, which may confuse the reader. For this reason, we chose x and y over a and b.

   Before leaving the proof of Result 4.19, we have one other remark. At one point in the second paragraph, we learned that $y \in A$ and $y \notin B$. From this we could have concluded (correctly) that $y \notin A \cap B$, but this is not what we wanted. Instead, we wrote that $y \in A - B$. It is always a good idea to keep our goal in sight. We wanted to show that $y \in A - B$; so it was important to keep in mind that it was the set $A - B$ in which we were interested, not $A \cap B$.   ◆

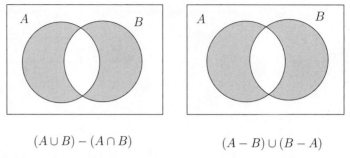

$$(A \cup B) - (A \cap B) \qquad\qquad (A - B) \cup (B - A)$$

Figure 4.2   Venn diagrams for $(A \cup B) - (A \cap B)$ and $(A - B) \cup (B - A)$

Next, let's consider the Venn diagrams for $(A \cup B) - (A \cap B)$ and $(A - B) \cup (B - A)$, which are shown in Figure 4.2. From these two diagrams, we might conclude (correctly) that the two sets $(A \cup B) - (A \cap B)$ and $(A - B) \cup (B - A)$ are equal. Indeed, all that is lacking is a *proof* that these two sets are equal. That is, Venn diagrams can be useful in suggesting certain results concerning sets, but they are only drawings and do not constitute a proof.

Result 4.20   *For every two sets A and B,*

$$(A \cup B) - (A \cap B) = (A - B) \cup (B - A).$$

Proof   First we show that $(A \cup B) - (A \cap B) \subseteq (A - B) \cup (B - A)$. Let $x \in (A \cup B) - (A \cap B)$. Then $x \in A \cup B$ and $x \notin A \cap B$. Since $x \in A \cup B$, it follows that $x \in A$ or $x \in B$. Without loss of generality, we can assume $x \in A$. Since $x \notin A \cap B$, the element $x \notin B$. Therefore, $x \in A - B$ and so $x \in (A - B) \cup (B - A)$. Hence

$$(A \cup B) - (A \cap B) \subseteq (A - B) \cup (B - A).$$

Next we show that $(A - B) \cup (B - A) \subseteq (A \cup B) - (A \cap B)$. Let $x \in (A - B) \cup (B - A)$. Then $x \in A - B$ or $x \in B - A$, say the former. So $x \in A$ and $x \notin B$. Thus $x \in A \cup B$ and $x \notin A \cap B$. Consequently, $x \in (A \cup B) - (A \cap B)$. Therefore,

$$(A - B) \cup (B - A) \subseteq (A \cup B) - (A \cap B),$$

as desired. ∎

PROOF ANALYSIS   In the proof of Result 4.20, when we were verifying the set inclusion

$$(A \cup B) - (A \cap B) \subseteq (A - B) \cup (B - A),$$

we concluded that $x \in A$ or $x \in B$. At that point, we could have divided the proof into two cases (*Case* 1. $x \in A$ and *Case* 2. $x \in B$); however, the proofs of the two cases would be identical, except that A and B would be interchanged. Therefore, we decided to consider only one of these. Since it really didn't matter which case we handled, we simply chose the case where $x \in A$. This was accomplished by writing:

Without loss of generality, assume that $x \in A$.

In the proof of the reverse set containment, we found ourselves in a similar situation, namely, $x \in A - B$ or $x \in B - A$. Again, these two situations were basically identical and we simply chose to work with the first (former) situation. (Had we decided to assume that $x \in B - A$, we would have considered the *latter* case.)   ♦

We now look at an example of a biconditional concerning sets.

Result 4.21   *Let A and B be sets. Then $A \cup B = A$ if and only if $B \subseteq A$.*

Proof   First we prove that if $A \cup B = A$, then $B \subseteq A$. We use a proof by contrapositive. Assume that B is not a subset of A. Then there must be some element $x \in B$ such that $x \notin A$. Since $x \in B$, it follows that $x \in A \cup B$. However, since $x \notin A$, we have $A \cup B \neq A$.

Next we verify the converse, namely, if $B \subseteq A$, then $A \cup B = A$. We use a direct proof here. Assume that $B \subseteq A$. To verify that $A \cup B = A$, we show that $A \subseteq A \cup B$ and $A \cup B \subseteq A$. The set inclusion $A \subseteq A \cup B$ is immediate (if $x \in A$, then $x \in A \cup B$). It remains only to show then that $A \cup B \subseteq A$. Let $y \in A \cup B$. Thus $y \in A$ or $y \in B$. If $y \in A$, then we already have the desired result. If $y \in B$, then since $B \subseteq A$, it follows that $y \in A$. Thus $A \cup B \subseteq A$.   ∎

PROOF ANALYSIS   In the first paragraph of the proof of Result 4.21 we indicated that we were using a proof by contrapositive, while in the second paragraph we mentioned that we were using a direct proof. This really wasn't necessary as the assumptions we made would inform the reader what technique we were applying. Also, in the proof of Result 4.21, we used a proof by contrapositive for one implication and a direct proof for its converse. This wasn't necessary either. Indeed, it is quite possible to interchange the techniques we used (see Exercise 41).   ♦

4.5 Fundamental Properties of Set Operations

Many results concerning sets follow from some very fundamental properties of sets, which, in turn, follow from corresponding results about logical statements that were described in Chapter 2. For example, we know that if P and Q are two statements, then $P \vee Q$ and $Q \vee P$ are logically equivalent. Similarly, if A and B are two sets, then $A \cup B = B \cup A$. We list some of the fundamental properties of set operations in the following theorem.

Theorem 4.22   *For sets A, B and C,*

(1) *Commutative Laws*

    (*a*) $A \cup B = B \cup A$
    (*b*) $A \cap B = B \cap A$

(2) *Associative Laws*

    (*a*) $A \cup (B \cup C) = (A \cup B) \cup C$
    (*b*) $A \cap (B \cap C) = (A \cap B) \cap C$

(3) *Distributive Laws*

    (*a*) $A \cup (B \cap C) = (A \cup B) \cap (A \cup C)$
    (*b*) $A \cap (B \cup C) = (A \cap B) \cup (A \cap C)$

(4) *De Morgan's Laws*

    (a) $\overline{A \cup B} = \overline{A} \cap \overline{B}$
    (b) $\overline{A \cap B} = \overline{A} \cup \overline{B}$.

We present proofs of only three parts of Theorem 4.22, beginning with the commutative law of the union of two sets.

Proof of
Theorem 4.22(1a)
We show that $A \cup B \subseteq B \cup A$. Assume that $x \in A \cup B$. Then $x \in A$ or $x \in B$. Applying the commutative law for disjunction of statements, we conclude that $x \in B$ or $x \in A$; so $x \in B \cup A$. Thus, $A \cup B \subseteq B \cup A$. The proof of the reverse set inclusion $B \cup A \subseteq A \cup B$ is similar and is therefore omitted. ∎

Next we verify one of the distributive laws.

Proof of
Theorem 4.22(3a)
First we show that $A \cup (B \cap C) \subseteq (A \cup B) \cap (A \cup C)$. Let $x \in A \cup (B \cap C)$. Then $x \in A$ or $x \in B \cap C$. If $x \in A$, then $x \in A \cup B$ and $x \in A \cup C$. Thus $x \in (A \cup B) \cap (A \cup C)$, as desired. On the other hand, if $x \in B \cap C$, then $x \in B$ and $x \in C$; and again, $x \in A \cup B$ and $x \in A \cup C$. So $x \in (A \cup B) \cap (A \cup C)$. Therefore, $A \cup (B \cap C) \subseteq (A \cup B) \cap (A \cup C)$.

To verify the reverse set inclusion, let $x \in (A \cup B) \cap (A \cup C)$. Then $x \in A \cup B$ and $x \in A \cup C$. If $x \in A$, then $x \in A \cup (B \cap C)$. So we may assume that $x \notin A$. Then the fact that $x \in A \cup B$ and $x \notin A$ implies that $x \in B$. By the same reasoning, $x \in C$. Thus $x \in B \cap C$ and so $x \in A \cup (B \cap C)$. Therefore, $(A \cup B) \cap (A \cup C) \subseteq A \cup (B \cap C)$. ∎

As a final example, we prove one of De Morgan's laws.

Proof of
Theorem 4.22(4a)
First, we show that $\overline{A \cup B} \subseteq \overline{A} \cap \overline{B}$. Let $x \in \overline{A \cup B}$. Then $x \notin A \cup B$. Hence $x \notin A$ and $x \notin B$. Therefore, $x \in \overline{A}$ and $x \in \overline{B}$, so $x \in \overline{A} \cap \overline{B}$; Consequently, $\overline{A \cup B} \subseteq \overline{A} \cap \overline{B}$.

Next we show that $\overline{A} \cap \overline{B} \subseteq \overline{A \cup B}$. Let $x \in \overline{A} \cap \overline{B}$. Then $x \in \overline{A}$ and $x \in \overline{B}$. Thus, $x \notin A$ and $x \notin B$; so $x \notin A \cup B$. Therefore, $x \in \overline{A \cup B}$. Hence $\overline{A} \cap \overline{B} \subseteq \overline{A \cup B}$. ∎

PROOF ANALYSIS
In the proof of the De Morgan law that we just presented, we arrived at the step $x \notin A \cup B$ at one point and then next wrote $x \notin A$ *and* $x \notin B$. Since $x \in A \cup B$ implies that $x \in A$ *or* $x \in B$, you might have expected us to write that $x \notin A$ *or* $x \notin B$ after writing $x \notin A \cup B$, but this would not be the correct conclusion. When we say that $x \notin A \cup B$, this is equivalent to writing $\sim (x \in A \cup B)$, which is logically equivalent to $\sim ((x \in A) \text{ or } (x \in B))$. By the De Morgan law for the negation of the disjunction of two statements (or two open sentences), we have that $\sim ((x \in A) \text{ or } (x \in B))$ is logically equivalent to $\sim (x \in A)$ and $\sim (x \in B)$; that is, $x \notin A$ *and* $x \notin B$. ◆

Proofs of some other parts of Theorem 4.22 are left as exercises.

4.6 Proofs Involving Cartesian Products of Sets

Recall that the **Cartesian product** (or simply the **product**) $A \times B$ of two sets A and B is defined as

$$A \times B = \{(a, b) \, : \, a \in A \text{ and } b \in B\}.$$

If $A = \emptyset$ or $B = \emptyset$, then $A \times B = \emptyset$.

     Before looking at several examples of proofs concerning Cartesian products of sets, it is important to keep in mind that an arbitrary element of the Cartesian product $A \times B$ of two sets A and B is of the form (a, b), where $a \in A$ and $b \in B$.

Result 4.23   *Let A, B, C and D be sets. If $A \subseteq C$ and $B \subseteq D$, then $A \times B \subseteq C \times D$.*

Proof   Let $(x, y) \in A \times B$. Then $x \in A$ and $y \in B$. Since $A \subseteq C$ and $B \subseteq D$, it follows that $x \in C$ and $y \in D$. Hence $(x, y) \in C \times D$.   ∎

Result 4.24   *For sets A, B and C,*

$$A \times (B \cup C) = (A \times B) \cup (A \times C).$$

Proof   We first show that $A \times (B \cup C) \subseteq (A \times B) \cup (A \times C)$. Let $(x, y) \in A \times (B \cup C)$. Then $x \in A$ and $y \in B \cup C$. Thus $y \in B$ or $y \in C$, say the former. Then $(x, y) \in A \times B$ and so $(x, y) \in (A \times B) \cup (A \times C)$. Consequently, $A \times (B \cup C) \subseteq (A \times B) \cup (A \times C)$.

     Next we show that $(A \times B) \cup (A \times C) \subseteq A \times (B \cup C)$. Let $(x, y) \in (A \times B) \cup (A \times C)$. Then $(x, y) \in A \times B$ or $(x, y) \in A \times C$, say the former. Then $x \in A$ and $y \in B \subseteq B \cup C$. Hence $(x, y) \in A \times (B \cup C)$, implying that $(A \times B) \cup (A \times C) \subseteq A \times (B \cup C)$.   ∎

     We give one additional example of a proof involving the Cartesian products of sets.

Result 4.25   *For sets A, B and C,*

$$A \times (B - C) = (A \times B) - (A \times C).$$

Proof   First we show that $A \times (B - C) \subseteq (A \times B) - (A \times C)$. Let $(x, y) \in A \times (B - C)$. Then $x \in A$ and $y \in B - C$. Since $y \in B - C$, it follows that $y \in B$ and $y \notin C$. Because $x \in A$ and $y \in B$, we have $(x, y) \in A \times B$. Since $y \notin C$, however, $(x, y) \notin A \times C$. Therefore, $(x, y) \in (A \times B) - (A \times C)$. Hence $A \times (B - C) \subseteq (A \times B) - (A \times C)$.

     We now show that $(A \times B) - (A \times C) \subseteq A \times (B - C)$. Let $(x, y) \in (A \times B) - (A \times C)$. Then $(x, y) \in A \times B$ and $(x, y) \notin A \times C$. Since $(x, y) \in A \times B$, it follows that $x \in A$ and $y \in B$. Also, since $x \in A$ and $(x, y) \notin A \times C$, it follows that $y \notin C$. So $y \in B - C$. Thus $(x, y) \in A \times (B - C)$ and $(A \times B) - (A \times C) \subseteq A \times (B - C)$.   ∎

PROOF ANALYSIS   We add one comment concerning the preceding proof. During the proof of $(A \times B) - (A \times C) \subseteq A \times (B - C)$, we needed to show that $y \notin C$. We learned that $(x, y) \notin A \times C$. However, this information alone did not allow us to conclude that $y \notin C$. Indeed, if

$(x, y) \notin A \times C$, then $x \notin A$ or $y \notin C$. Since we knew, however, that $x \in A$ and $(x, y) \notin A \times C$, we were able to conclude that $y \notin C$. ♦

EXERCISES FOR CHAPTER 4

Section 4.1: Proofs Involving Divisibility of Integers

4.1. Let a and b be integers, where $a \neq 0$. Prove that if $a \mid b$, then $a^2 \mid b^2$.

4.2. Let $a, b \in \mathbf{Z}$, where $a \neq 0$ and $b \neq 0$. Prove that if $a \mid b$ and $b \mid a$, then $a = b$ or $a = -b$.

4.3. Let $m \in \mathbf{Z}$.

   (a) Give a direct proof of the following: If $3 \mid m$, then $3 \mid m^2$.
   (b) State the contrapositive of the implication in (a).
   (c) Give a direct proof of the following: If $3 \nmid m$, then $3 \nmid m^2$.
   (d) State the contrapositive of the implication in (c).
   (e) State the conjunction of the implications in (a) and (c) using "if and only if."

4.4. Let $x, y \in \mathbf{Z}$. Prove that if $3 \nmid x$ and $3 \nmid y$, then $3 \mid (x^2 - y^2)$.

4.5. Let $a, b, c \in \mathbf{Z}$, where $a \neq 0$. Prove that if $a \nmid bc$, then $a \nmid b$ and $a \nmid c$.

4.6. Let $a \in \mathbf{Z}$. Prove that if $3 \mid 2a$, then $3 \mid a$.

4.7. Let $n \in \mathbf{Z}$. Prove that $3 \mid (2n^2 + 1)$ if and only if $3 \nmid n$.

4.8. In Result 4.4, it was proved for an integer x that if $2 \mid (x^2 - 1)$, then $4 \mid (x^2 - 1)$. Prove that if $2 \mid (x^2 - 1)$, then $8 \mid (x^2 - 1)$.

4.9. (a) Let $x \in \mathbf{Z}$. Prove that if $2 \mid (x^2 - 5)$, then $4 \mid (x^2 - 5)$.
   (b) Give an example of an integer x such that $2 \mid (x^2 - 5)$ but $8 \nmid (x^2 - 5)$.

4.10. Let $n \in \mathbf{Z}$. Prove that $2 \mid (n^4 - 3)$ if and only if $4 \mid (n^2 + 3)$.

4.11. Prove that for every integer $n \geq 8$, there exist nonnegative integers a and b such that $n = 3a + 5b$.

4.12. In Result 4.7, it was proved for integers x and y that $4 \mid (x^2 - y^2)$ if and only if x and y are of the same parity. In particular, this says that if x and y are both even, then $4 \mid (x^2 - y^2)$; while if x and y are both odd, then $4 \mid (x^2 - y^2)$. Prove that if x and y are both odd, then $8 \mid (x^2 - y^2)$.

4.13. Prove that if $a, b, c \in \mathbf{Z}$ and $a^2 + b^2 = c^2$, then $3 \mid ab$.

Section 4.2: Proofs Involving Congruence of Integers

4.14. Let $a, b, n \in \mathbf{Z}$, where $n \geq 2$. Prove that if $a \equiv b \pmod{n}$, then $a^2 \equiv b^2 \pmod{n}$.

4.15. Let $a, b, c, n \in \mathbf{Z}$, where $n \geq 2$. Prove that if $a \equiv b \pmod{n}$ and $a \equiv c \pmod{n}$, then $b \equiv c \pmod{n}$.

4.16. Let $a, b \in \mathbf{Z}$. Prove that if $a^2 + 2b^2 \equiv 0 \pmod{3}$, then either a and b are both congruent to 0 modulo 3 or neither is congruent to 0 modulo 3.

4.17. (a) Prove that if a is an integer such that $a \equiv 1 \pmod{5}$, then $a^2 \equiv 1 \pmod{5}$.
   (b) Given that b is an integer such that $b \equiv 1 \pmod{5}$, what can we conclude from (a)?

4.18. Let $m, n \in \mathbf{N}$ such that $m \geq 2$ and $m \mid n$. Prove that if a and b are integers such that $a \equiv b \pmod{n}$, then $a \equiv b \pmod{m}$.

4.19. Let $a, b \in \mathbf{Z}$. Show that if $a \equiv 5 \pmod{6}$ and $b \equiv 3 \pmod{4}$, then $4a + 6b \equiv 6 \pmod{8}$.

4.20. (a) Result 4.12 states: Let $n \in \mathbf{Z}$. If $n^2 \not\equiv n \pmod 3$, then $n \not\equiv 0 \pmod 3$ and $n \not\equiv 1 \pmod 3$. State and prove the converse of this result.

(b) State the conjunction of Result 4.12 and its converse using "if and only if."

4.21. Let $a \in \mathbf{Z}$. Prove that $a^3 \equiv a \pmod 3$.

4.22. Let $n \in \mathbf{Z}$. Prove each of the statements (a)–(f).

(a) If $n \equiv 0 \pmod 7$, then $n^2 \equiv 0 \pmod 7$.

(b) If $n \equiv 1 \pmod 7$, then $n^2 \equiv 1 \pmod 7$.

(c) If $n \equiv 2 \pmod 7$, then $n^2 \equiv 4 \pmod 7$.

(d) If $n \equiv 3 \pmod 7$, then $n^2 \equiv 2 \pmod 7$.

(e) For each integer n, $n^2 \equiv (7-n)^2 \pmod 7$.

(f) For every integer n, n^2 is congruent to exactly one of 0, 1, 2 or 4 modulo 7.

4.23. Prove for any set $S = \{a, a+1, \ldots, a+5\}$ of six integers where $6 \mid a$ that $24 \mid (x^2 - y^2)$ for distinct odd integers x and y in S if and only if one of x and y is congruent to 1 modulo 6 while the other is congruent to 5 modulo 6.

4.24. Let x and y be even integers. Prove that $x^2 \equiv y^2 \pmod{16}$ if and only if either (1) $x \equiv 0 \pmod 4$ and $y \equiv 0 \pmod 4$ or (2) $x \equiv 2 \pmod 4$ and $y \equiv 2 \pmod 4$.

Section 4.3: Proofs Involving Real Numbers

4.25. Let $x, y \in \mathbf{R}$. Prove that if $x^2 - 4x = y^2 - 4y$ and $x \neq y$, then $x + y = 4$.

4.26. Let a, b and m be integers. Prove that if $2a + 3b \geq 12m + 1$, then $a \geq 3m + 1$ or $b \geq 2m + 1$.

4.27. Let $x \in \mathbf{R}$. Prove that if $3x^4 + 1 \leq x^7 + x^3$, then $x > 0$.

4.28. Prove that if r is a real number such that $0 < r < 1$, then $\frac{1}{r(1-r)} \geq 4$.

4.29. Prove that if r is a real number such that $|r - 1| < 1$, then $\frac{4}{r(4-r)} \geq 1$.

4.30. Let $x, y \in \mathbf{R}$. Prove that $|xy| = |x| \cdot |y|$.

4.31. Prove for every two real numbers x and y that $|x + y| \geq |x| - |y|$.

4.32. (a) Recall that $\sqrt{r} > 0$ for every positive real number r. Prove that if a and b are positive real numbers, then $0 < \sqrt{ab} \leq \frac{a+b}{2}$. (The number $\sqrt{ab}$ is called the **geometric mean** of a and b, while $(a + b)/2$ is called the **arithmetic mean** or **average** of a and b.)

(b) Under what conditions does $\sqrt{ab} = (a + b)/2$ for positive real numbers a and b? Justify your answer.

4.33. The **geometric mean** of three positive real numbers a, b and c is $\sqrt[3]{abc}$ and the **arithmetic mean** is $(a + b + c)/3$. Prove that $\sqrt[3]{abc} \leq (a + b + c)/3$. [Note: The numbers a, b and c can be expressed as $a = r^3, b = s^3$ and $c = t^3$ for positive numbers r, s and t.]

4.34. Prove for every three real numbers x, y and z that $|x - z| \leq |x - y| + |y - z|$.

4.35. Prove that if x is a real number such that $x(x + 1) > 2$, then $x < -2$ or $x > 1$.

4.36. Prove for every positive real number x that $1 + \frac{1}{x^4} \geq \frac{1}{x} + \frac{1}{x^3}$.

4.37. Prove for $x, y, z \in \mathbf{R}$ that $x^2 + y^2 + z^2 \geq xy + xz + yz$.

4.38. Let $a, b, x, y \in \mathbf{R}$ and $r \in \mathbf{R}^+$. Prove that if $|x - a| < r/2$ and $|y - b| < r/2$, then $|(x + y) - (a + b)| < r$.

4.39. Prove that if $a, b, c, d \in \mathbf{R}$, then $(ab + cd)^2 \leq (a^2 + c^2)(b^2 + d^2)$.

Section 4.4: Proofs Involving Sets

4.40. Let A and B be sets. Prove that $A \cup B = (A - B) \cup (B - A) \cup (A \cap B)$.

4.41. In Result 4.21, it was proved for sets A and B that $A \cup B = A$ if and only if $B \subseteq A$. Provide another proof of this result by giving a direct proof of the implication "If $A \cup B = A$, then $B \subseteq A$" and a proof by contrapositive of its converse.

4.42. Let A and B be sets. Prove that $A \cap B = A$ if and only if $A \subseteq B$.

4.43.
(a) Give an example of three sets A, B and C such that $A \cap B = A \cap C$ but $B \neq C$.
(b) Give an example of three sets A, B and C such that $A \cup B = A \cup C$ but $B \neq C$.
(c) Let A, B and C be sets. Prove that if $A \cap B = A \cap C$ *and* $A \cup B = A \cup C$, then $B = C$.

4.44. Prove that if A and B are sets such that $A \cup B \neq \emptyset$, then $A \neq \emptyset$ or $B \neq \emptyset$.

4.45. Let $A = \{n \in \mathbf{Z} : n \equiv 1 \ (\mathrm{mod}\ 2)\}$ and $B = \{n \in \mathbf{Z} : n \equiv 3 \ (\mathrm{mod}\ 4)\}$. Prove that $B \subseteq A$.

4.46. Let A and B be sets. Prove that $A \cup B = A \cap B$ if and only if $A = B$.

4.47. Let $A = \{n \in \mathbf{Z} : n \equiv 2 \ (\mathrm{mod}\ 3)\}$ and $B = \{n \in \mathbf{Z} : n \equiv 1 \ (\mathrm{mod}\ 2)\}$.

(a) Describe the elements of the set $A - B$.
(b) Prove that if $n \in A \cap B$, then $n^2 \equiv 1 \ (\mathrm{mod}\ 12)$.

4.48. Let $A = \{n \in \mathbf{Z} : 2 \mid n\}$ and $B = \{n \in \mathbf{Z} : 4 \mid n\}$. Let $n \in \mathbf{Z}$. Prove that $n \in A - B$ if and only if $n = 2k$ for some odd integer k.

4.49. Prove for every two sets A and B that $A = (A - B) \cup (A \cap B)$.

4.50. Prove for every two sets A and B that $A - B$, $B - A$ and $A \cap B$ are pairwise disjoint.

4.51. Let A and B be subsets of a universal set. Which of the following is a necessary condition for A and B to be disjoint?

(a) Either $A = \emptyset$ or $B = \emptyset$.
(b) Whenever $x \notin A$, it must occur that $x \in B$.
(c) Whenever $x \notin A$, it must occur that $x \notin B$.
(d) Whenever $x \in A$, it must occur that $x \in B$.
(e) Whenever $x \in A$, it must occur that $x \notin B$.

Section 4.5: Fundamental Properties of Set Operations

4.52. Prove that $A \cap B = B \cap A$ for every two sets A and B (Theorem 4.22(1b)).

4.53. Prove that $A \cap (B \cup C) = (A \cap B) \cup (A \cap C)$ for every three sets A, B and C (Theorem 4.22(3b)).

4.54. Prove that $\overline{A \cap B} = \overline{A} \cup \overline{B}$ for every two sets A and B (Theorem 4.22(4b)).

4.55. Let A, B and C be sets. Prove that $(A - B) \cap (A - C) = A - (B \cup C)$.

4.56. Let A, B and C be sets. Prove that $(A - B) \cup (A - C) = A - (B \cap C)$.

4.57. Let A, B and C be sets. Use Theorem 4.22 to prove that $\overline{A} \cup (\overline{B} \cap C) = (A \cap B) \cup (A - C)$.

4.58. Let A, B and C be sets. Prove that $A \cap (B \cap \overline{C}) = \overline{(\overline{A} \cup B)} \cap (\overline{A} \cup \overline{C})$.

4.59. Show for every three sets A, B and C that $A - (B - C) = (A \cap C) \cup (A - B)$.

Section 4.6: Proofs Involving Cartesian Products of Sets

4.60. For $A = \{x, y\}$, determine $A \times \mathcal{P}(A)$.

4.61. For $A = \{1\}$ and $B = \{2\}$, determine $\mathcal{P}(A \times B)$ and $\mathcal{P}(A) \times \mathcal{P}(B)$.

4.62. Let A and B be sets. Prove that $A \times B = \emptyset$ if and only if $A = \emptyset$ or $B = \emptyset$.

4.63. For sets A and B, find a necessary and sufficient condition for $A \times B = B \times A$.

4.64. For sets A and B, find a necessary and sufficient condition for $(A \times B) \cap (B \times A) = \emptyset$. Verify that this condition *is* necessary and sufficient.

4.65. Let A, B and C be nonempty sets. Prove that $A \times C \subseteq B \times C$ if and only if $A \subseteq B$.

4.66. Result 4.23 states that if A, B, C and D are sets such that $A \subseteq C$ and $B \subseteq D$, then $A \times B \subseteq C \times D$.

    (a) Show that the converse of Result 4.23 is false.
    (b) Under what added hypothesis is the converse true? Prove your assertion.

4.67. Let A, B and C be sets. Prove that $A \times (B \cap C) = (A \times B) \cap (A \times C)$.

4.68. Let A, B, C and D be sets. Prove that $(A \times B) \cap (C \times D) = (A \cap C) \times (B \cap D)$.

4.69. Let A, B, C and D be sets. Prove that $(A \times B) \cup (C \times D) \subseteq (A \cup C) \times (B \cup D)$.

4.70. Let A and B be sets. Show, in general, that $\overline{A \times B} \neq \overline{A} \times \overline{B}$.

ADDITIONAL EXERCISES FOR CHAPTER 4

4.71. Let $n \in \mathbf{Z}$. Prove that $5 \mid n^2$ if and only if $5 \mid n$.

4.72. Prove for integers a and b that $3 \mid ab$ if and only if $3 \mid a$ or $3 \mid b$.

4.73. Prove that if n is an odd integer, then $8 \mid [n^2 + (n + 6)^2 + 6]$.

4.74. Prove that if n is an odd integer, then $8 \mid (n^4 + 4n^2 + 11)$.

4.75. Let $n, m \in \mathbf{Z}$. Prove that if $n \equiv 1 \pmod 2$ and $m \equiv 3 \pmod 4$, then $n^2 + m \equiv 0 \pmod 4$.

4.76. Find two distinct positive integer values of a for which the following is true and give a proof in each case:
    For every integer n, $a \nmid (n^2 + 1)$.

4.77. Prove for every two real numbers a and b that $ab \leq \sqrt{a^2}\sqrt{b^2}$.

4.78. Prove for every two positive real numbers a and b that $\frac{a}{b} + \frac{b}{a} \geq 2$.

4.79. Prove the following: Let $x \in \mathbf{R}$. If $x(x - 5) = -4$, then $\sqrt{5x^2 - 4} = 1$ implies that $x + \frac{1}{x} = 2$.

4.80. Let $x, y \in \mathbf{R}$. Prove that if $x < 0$, then $x^3 - x^2 y \leq x^2 y - xy^2$.

4.81. Prove that $3 \mid (n^3 - 4n)$ for every integer n.

4.82. Evaluate the proposed proof of the following result.

    Result   Let $x, y \in \mathbf{Z}$. If $x \equiv 2 \pmod 3$ and $y \equiv 2 \pmod 3$, then $xy \equiv 1 \pmod 3$.

    Proof   Let $x \equiv 2 \pmod 3$ and $y \equiv 2 \pmod 3$. Then $x = 3k + 2$ and $y = 3k + 2$ for some integer k. Hence

$$xy = (3k + 2)(3k + 2) = 9k^2 + 12k + 4 = 9k^2 + 12k + 3 + 1$$
$$= 3(3k^2 + 4k + 1) + 1.$$

    Since $3k^2 + 4k + 1$ is an integer, $xy \equiv 1 \pmod 3$. ∎

4.83. Below is given a proof of a result. What result is proved?

Proof Assume that $x \equiv 1 \pmod 5$ and $y \equiv 2 \pmod 5$. Then $5 \mid (x - 1)$ and $5 \mid (y - 2)$. Hence $x - 1 = 5a$ and $y - 2 = 5b$ for some integers a and b. So $x = 5a + 1$ and $y = 5b + 2$. Therefore,

$$x^2 + y^2 = (5a + 1)^2 + (5b + 2)^2 = (25a^2 + 10a + 1) + (25b^2 + 20b + 4)$$
$$= 25a^2 + 10a + 25b^2 + 20b + 5 = 5(5a^2 + 2a + 5b^2 + 4b + 1).$$

Since $5a^2 + 2a + 5b^2 + 4b + 1$ is an integer, $5 \mid (x^2 + y^2)$ and so $x^2 + y^2 \equiv 0 \pmod 5$. ∎

4.84. A proof of the following result is given.

Result   Let $n \in \mathbf{Z}$. If n^4 is even, then $3n + 1$ is odd.

Proof Assume that $n^4 = (n^2)^2$ is even. Since n^4 is even, n^2 is even. Furthermore, since n^2 is even, n is even. Because n is even, $n = 2k$ for some integer k. Then
$$3n + 1 = 3(2k) + 1 = 6k + 1 = 2(3k) + 1.$$

Since $3k$ is an integer, $3n + 1$ is odd. ∎

Answer the following questions.
(1) Which proof technique is being used?
(2) What is the starting assumption?
(3) What must be shown to give a complete proof?
(4) Give a reason for each of the following steps in the proof.

    (a) Since n^4 is even, n^2 is even.
    (b) Furthermore, since n^2 is even, n is even.
    (c) Because n is even, $n = 2k$ for some integer k.
    (d) Then $3n + 1 = 3(2k) + 1 = 6k + 1 = 2(3k) + 1$.
    (e) Since $3k$ is an integer, $3n + 1$ is odd.

4.85. Given below is an attempted proof of a result.

Proof First, we show that $A \subseteq (A \cup B) - B$. Let $x \in A$. Since $A \cap B = \emptyset$, it follows that $x \notin B$. Therefore, $x \in A \cup B$ and $x \notin B$; so $x \in (A \cup B) - B$. Thus $A \subseteq (A \cup B) - B$.
    Next, we show that $(A \cup B) - B \subseteq A$. Let $x \in (A \cup B) - B$. Then $x \in A \cup B$ and $x \notin B$. From this, it follows that $x \in A$. Hence $(A \cup B) - B \subseteq A$. ∎

(a) What result is being proved above?
(b) What change (or changes) in this proof would make it better (from your point of view)?

4.86. Evaluate the proposed proof of the following result.

Result   Let $x, y \in \mathbf{Z}$ such that $3 \mid x$. If $3 \mid (x + y)$, then $3 \mid y$.

Proof Since $3 \mid x$, it follows that $x = 3a$, where $a \in \mathbf{Z}$. Assume that $3 \mid (x + y)$. Then $x + y = 3b$ for some integer b. Hence $y = 3b - x = 3b - 3a = 3(b - a)$. Since $b - a$ is an integer, $3 \mid y$.
    For the converse, assume that $3 \mid y$. Therefore, $y = 3c$, where $c \in \mathbf{Z}$. Thus $x + y = 3a + 3c = 3(a + c)$. Since $a + c$ is an integer, $3 \mid (x + y)$. ∎

4.87. Evaluate the proposed proof of the following result.

Result Let $x, y \in \mathbf{Z}$. If $x \equiv 1 \pmod 3$ and $y \equiv 1 \pmod 3$, then $xy \equiv 1 \pmod 3$.
Proof   Assume that $x \equiv 1 \pmod 3$ and $y \equiv 1 \pmod 3$. Then $3 \mid (x - 1)$ and $3 \mid (y - 1)$. Hence $x - 1 = 3q$ and $y - 1 = 3q$ for some integer q and so $x = 3q + 1$ and $y = 3q + 1$. Thus

$$xy = (3q + 1)(3q + 1) = 9q^2 + 6q + 1 = 3(3q^2 + 2q) + 1$$

and so $xy - 1 = 3(3q^2 + 2q)$. Since $3q^2 + 2q$ is an integer, $3 \mid (xy - 1)$. Hence $xy \equiv 1 \pmod 3$. ∎

4.88. Evaluate the proposed proof of the following result.

> **Result**   For every three sets A, B and C, $(A \times C) - (B \times C) \subseteq (A - B) \times C$.
> **Proof**   Let $(x, y) \in (A \times C) - (B \times C)$. Then $(x, y) \in A \times C$ and $(x, y) \notin B \times C$. Since $(x, y) \in A \times C$, it follows that $x \in A$ and $y \in C$. Since $(x, y) \notin B \times C$, we have $x \notin B$. Thus $x \in A - B$. Hence $(x, y) \in (A - B) \times C$.   ∎

4.89. Prove that for every three integers a, b and c, the sum $|a - b| + |a - c| + |b - c|$ is an even integer.

4.90. Prove for every four real numbers a, b, c and d that $ac + bd \leq \sqrt{a^2 + b^2}\sqrt{c^2 + d^2}$.

4.91. Prove that for every real number x, $\sin^6 x + 3\sin^2 x \cos^2 x + \cos^6 x = 1$.

4.92. Let $a \in \mathbf{Z}$. Prove that if $6 \mid a$ and $10 \mid a$, then $15 \mid a$.

4.93. Let $A = \{x\}$. Give an example of a set concerning the set A to which each of the following elements belong.
   (a) $(x, \{x\})$ (b) $(\{x\}, x)$ (c) (x, x) (d) $(\{x\}, \{x\})$
   (e) x (f) $\{x\}$ (g) $\{(x, x)\}$ (h) $(\{x\}, \{\{x\}\})$.

4.94. Let $a, b \in \mathbf{Z}$. Prove that if $a \equiv b \pmod 2$ and $b \equiv a \pmod 3$, then $a \equiv b \pmod 6$.

4.95. Let $a, b, c \in \mathbf{R}$. Prove that $\frac{3}{2}(a^2 + b^2 + c^2 + 1) \geq a(b + 1) + b(c + 1) + c(a + 1)$.

4.96. Prove that if a, b and c are positive real numbers, then $(a + b + c)\left(\frac{1}{a} + \frac{1}{b} + \frac{1}{c}\right) \geq 9$.

4.97. Let $T = \{1, 2, \ldots, 8\}$.

   (a) Determine the elements of the set $A = \{a \in T : 2^m \equiv a \pmod 9 \text{ for some } m \in \mathbf{N}\}$.
   (b) Determine the elements of the set $B = \{a \in T : 5^m \equiv a \pmod 9 \text{ for some } m \in \mathbf{N}\}$.
   (c) What property do the sets A and B have in (a) and (b)?

4.98. Consider the open sentence
$$P(m) : 5m + 1 = a^2 \text{ for some } a \in \mathbf{Z},$$
where $m \in \mathbf{N}$. That is, $P(m)$ is the open sentence: $5m + 1$ is a perfect square.

   (a) Determine four distinct solutions t of $t^2 \equiv 4 \pmod 5$. For each solution t, determine $m = \frac{4(t^2 - 4)}{5} + 3$ and show that $P(m)$ is a true statement.
   (b) Show that the set $S = \{t \in \mathbf{Z} : t^2 \equiv 4 \pmod 5\}$ contains infinitely many elements.
   (c) Let t be an element of the set S in (b). Prove that if $m = \frac{4(t^2 - 4)}{5} + 3$, then $5m + 1$ is a perfect square.
   (d) As a consequence of the results established in (a)–(c), what can be concluded about the set $M = \{m \in \mathbf{N} : 5m + 1 \text{ is a perfect square}\}$?

4.99. Let $a_1, a_2, \ldots, a_n$ $(n \geq 3)$ be n integers such that $|a_{i+1} - a_i| \leq 1$ for $1 \leq i \leq n - 1$. Prove that if k is any integer that lies strictly between a_1 and a_n, then there is an integer j with $1 < j < n$ such that $a_j = k$.

5

Existence and Proof by Contradiction

Thus far, we have been primarily concerned with quantified statements involving universal quantifiers, namely statements of the type $\forall x \in S, R(x)$. We now consider problems that involve, either directly or indirectly, quantified statements involving existential quantifiers, that is, statements of the type $\exists x \in S, R(x)$.

5.1 Counterexamples

It must certainly come as no surprise that some quantified statements of the type $\forall x \in S, R(x)$ are false. We have seen that

$$\sim (\forall x \in S, R(x)) \equiv \exists x \in S, \sim R(x),$$

that is, if the statement $\forall x \in S, R(x)$ is false, then there exists some element $x \in S$ for which $R(x)$ is false. Such an element x is called a **counterexample** of the (false) statement $\forall x \in S, R(x)$. Finding a counterexample verifies that $\forall x \in S, R(x)$ is false.

Example 5.1 *Consider the statement:*

$$\text{If } x \in \mathbf{R}, \text{ then } (x^2 - 1)^2 > 0. \tag{5.1}$$

or, equivalently,

$$\text{For every real number } x, (x^2 - 1)^2 > 0.$$

Show that the statement (5.1) is false by exhibiting a counterexample.

Solution For $x = 1$, $(x^2 - 1)^2 = (1^2 - 1)^2 = 0$. Thus $x = 1$ is a counterexample. ♦

It might be noticed that the number $x = -1$ is also a counterexample. In fact, $x = 1$ and $x = -1$ are the only two counterexamples of the statement (5.1). That is, the statement

$$\text{If } x \in \mathbf{R} - \{1, -1\}, \text{ then } (x^2 - 1)^2 > 0. \tag{5.2}$$

is true.

If a statement P is shown to be false in some manner, then P is said to be **disproved**. The counterexample $x = 1$ therefore disproves the statement (5.1).

Example 5.2   *Disprove the statement:*

$$\text{If } x \text{ is a real number, then } \tan^2 x + 1 = \sec^2 x. \tag{5.3}$$

Solution   Since $\tan x$ and $\sec x$ are not defined when $x = \pi/2$, it follows that $\tan^2 x + 1$ and $\sec^2 x$ have no numerical value when $x = \pi/2$ and, consequently, $\tan^2 x + 1$ and $\sec^2 x$ are not equal when $x = \pi/2$. That is, $x = \pi/2$ is a counterexample to the statement (5.3).   ◆

Although $\tan^2 x + 1 = \sec^2 x$ is a well-known identity from trigonometry, statement (5.3), as presented, is false. The following is true, however:

$$\begin{array}{c} \text{If } x \text{ is a real number for which } \tan x \text{ and } \sec x \text{ are defined,} \\ \text{then } \tan^2 x + 1 = \sec^2 x. \end{array} \tag{5.4}$$

Indeed, it is probably statement (5.4) that was intended in Example 5.2, rather than statement (5.3). Since $\tan x$ and $\sec x$ are defined for precisely the same real numbers x (namely, those numbers x such that $\cos x \neq 0$), we can restate (5.4) as

$$\text{If } x \in \mathbf{R} - \left\{ n\pi + \tfrac{\pi}{2} : n \in \mathbf{Z} \right\}, \text{ then } \tan^2 x + 1 = \sec^2 x.$$

Example 5.3   *Disprove the statement:*

$$\text{If } x \in \mathbf{Z}, \text{ then } \frac{x^2 + x}{x^2 - x} = \frac{x + 1}{x - 1}. \tag{5.5}$$

Solution   If $x = 0$, then $x^2 - x = 0$ and so $\dfrac{x^2 + x}{x^2 - x}$ is not defined. On the other hand, if $x = 0$, then $\dfrac{x + 1}{x - 1} = -1$; so the expressions $\dfrac{x^2 + x}{x^2 - x}$ and $\dfrac{x + 1}{x - 1}$ are certainly not equal when $x = 0$. Thus $x = 0$ is a counterexample to the statement (5.5).   ◆

Since neither $\dfrac{x^2 + x}{x^2 - x}$ nor $\dfrac{x + 1}{x - 1}$ is defined when $x = 1$, it follows that $x = 1$ is also a counterexample of statement (5.5). Indeed, $x = 0$ and $x = 1$ are the only counterexamples of statement (5.5) and so the statement

$$\text{If } x \in \mathbf{Z} - \{0, 1\}, \text{ then } \frac{x^2 + x}{x^2 - x} = \frac{x + 1}{x - 1}.$$

is true.

The three preceding examples illustrate the fact that an open sentence $R(x)$ that is false over some domain S may very well be true over a subset of S. Therefore, the truth (or falseness) of a statement $\forall x \in S, R(x)$ depends not only on the open sentence $R(x)$ but on its domain as well.

Example 5.4   *Disprove the statement:*

$$\text{For every odd positive integer } n, 3 \mid (n^2 - 1). \tag{5.6}$$

Solution   Since $3 \nmid (3^2 - 1)$, it follows that $n = 3$ is a counterexample.   ◆

You might have noticed that even though $3 \nmid (3^2 - 1)$, it is the case that $3 \mid (n^2 - 1)$ for *some* odd positive integers. For example, $3 \mid (n^2 - 1)$ if $n = 1, 5, 7, 11, 13, 17$, while $3 \nmid (n^2 - 1)$ if $n = 3, 9, 15, 21$. This should make you wonder for which odd positive integers n, the open sentence $3 \mid (n^2 - 1)$ is true. (See Result 4.6.)

We have seen that a quantified statement of the type

$$\forall x \in S, \ R(x)$$

is false if

$$\exists x \in S, \ \sim R(x)$$

is true, that is, if there exists some element $x \in S$ for which $R(x)$ is false. There will be many instances when $R(x)$ is an implication $P(x) \Rightarrow Q(x)$. Therefore, the quantified statement

$$\forall x \in S, \ P(x) \Rightarrow Q(x) \tag{5.7}$$

is false if

$$\exists x \in S, \ \sim (P(x) \Rightarrow Q(x)) \tag{5.8}$$

is true. By Theorem 2.4(a), the statement (5.8) can be expressed as

$$\exists x \in S, \ (P(x) \wedge (\sim Q(x))).$$

That is, to show that the statement (5.7) is false, we need to exhibit a counterexample, which is then an element $x \in S$ for which $P(x)$ is true and $Q(x)$ is false.

Example 5.5   *Disprove the statement:*

$$\text{Let } n \in \mathbf{Z}. \text{ If } n^2 + 3n \text{ is even, then } n \text{ is odd.}$$

Solution   If $n = 2$, then $n^2 + 3n = 2^2 + 3 \cdot 2 = 10$ is even and 2 is even. Thus $n = 2$ is a counterexample.   ◆

In the preceding example, not only is 2 a counterexample, *every* even integer is a counterexample.

Example 5.6   *Disprove the statement:*

$$\text{If } n \text{ is an odd integer, then } n^2 - n \text{ is odd.} \tag{5.9}$$

Solution   For the odd integer $n = 1$, the integer $n^2 - n = 1^2 - 1 = 0$ is even. Thus $n = 1$ is a counterexample.   ◆

Actually, it is not difficult to prove that the statement   ·

If n is an odd integer, then $n^2 - n$ is even.

is true. Although it may very well be of interest to know this, to show that statement (5.9) is false requires exhibiting only a single counterexample. It does *not* require proving some other result. One should know the difference between these two.

Example 5.7   *Show that the statement:*

$$\text{Let } n \in \mathbf{Z}. \text{ If } 4 \mid (n^2 - 1), \text{ then } 4 \mid (n - 1).$$

is false.

Solution   Since $4 \mid (3^2 - 1)$ but $4 \nmid (3 - 1)$, it follows that $n = 3$ is a counterexample.   ◆

Example 5.8   *Show that the statement*

$$\text{For positive integers } a, b, c, a^{b^c} = \left(a^b\right)^c.$$

is false.

Solution   Let $a = 2$, $b = 2$ and $c = 3$. Then $a^{b^c} = 2^{2^3} = 2^8 = 256$, while $\left(a^b\right)^c = \left(2^2\right)^3 = 4^3 = 64$. Since $256 \neq 64$, the positive integers $a = 2$, $b = 2$ and $c = 3$ constitute a counterexample.   ◆

Example 5.9   *Show that the statement:*

$$\text{Let } a \text{ and } b \text{ be nonzero real numbers. If } x, y \in \mathbf{R}^+, \text{ then}$$

$$\frac{a^2}{2b^2}x^2 + \frac{b^2}{2a^2}y^2 > xy. \tag{5.10}$$

is false.

Solution   Let $x = b^2$ and $y = a^2$. Then

$$\frac{a^2}{2b^2}x^2 + \frac{b^2}{2a^2}y^2 = \frac{a^2b^2}{2} + \frac{a^2b^2}{2} = a^2b^2 = xy.$$

Thus $x = b^2$ and $y = a^2$ is a counterexample and so the inequality is false.   ◆

Analysis   After reading the solution of Example 5.9, the only question that may occur to you is where the counterexample $x = b^2$ and $y = a^2$ came from. Multiplying the inequality (5.10) by $2a^2b^2$ (which eliminates all fractions) produces the equivalent inequality

$$a^4x^2 + b^4y^2 > 2a^2b^2xy$$

and so
$$a^4x^2 - 2a^2b^2xy + b^4y^2 > 0,$$

which can be expressed as
$$(a^2x - b^2y)^2 > 0.$$

Of course, $(a^2x - b^2y)^2 \geq 0$. Thus any values of x and y for which $a^2x - b^2y = 0$ produce a counterexample. Although there are many choices for x and y, one such choice is $x = b^2$ and $y = a^2$.   ◆

5.2 Proof by Contradiction

Suppose, as usual, that we would like to show that a certain mathematical statement R is true. If R is expressed as the quantified statement $\forall\ x \in S,\ P(x) \Rightarrow Q(x)$, then we have already introduced two proof techniques, namely direct proof and proof by contrapositive, that could be used to establish the truth of R. We now introduce a third method that can be used to establish the truth of R, regardless of whether R is expressed in terms of an implication.

Suppose that we assume R is a false statement and, from this assumption, we are able to arrive at or deduce a statement that contradicts some assumption we made in the proof or some known fact. (The known fact might be a definition, an axiom or a theorem.) If we denote this assumption or known fact by P, then what we have deduced is $\sim P$ and have thus produced the contradiction $C : P \wedge (\sim P)$. We have therefore established the truth of the implication

$$(\sim R) \Rightarrow C.$$

However, because $(\sim R) \Rightarrow C$ is true and C is false, it follows that $\sim R$ is false and so R is true, as desired. This technique is called **proof by contradiction**.

If R is the quantified statement $\forall x \in S,\ P(x) \Rightarrow Q(x)$, then a proof by contradiction of this statement consists of verifying the implication

$$\sim (\forall\ x \in S,\ P(x) \Rightarrow Q(x)) \Rightarrow C$$

for some contradiction C. However, since

$$\sim (\forall\ x \in S,\ P(x) \Rightarrow Q(x)) \equiv \exists x \in S,\ \sim (P(x) \Rightarrow Q(x))$$
$$\equiv \exists x \in S,\ (P(x) \wedge (\sim Q(x))),$$

it follows that a proof by contradiction of $\forall\ x \in S,\ P(x) \Rightarrow Q(x)$ would begin by assuming the existence of some element $x \in S$ such that $P(x)$ is true and $Q(x)$ is false. That is, a proof by contradiction of $\forall\ x \in S,\ P(x) \Rightarrow Q(x)$ begins by assuming the existence of a counterexample of this quantified statement. Often the reader is alerted that a proof by contradiction is being used by saying (or writing)

Suppose that R is false.

or

Assume, to the contrary, that R is false.

Therefore, if R is the quantified statement $\forall\ x \in S,\ P(x) \Rightarrow Q(x)$, then a proof by contradiction might begin with:

Assume, to the contrary, that there exists some element $x \in S$ for which $P(x)$ is true and $Q(x)$ is false.

(or something along these lines). The remainder of the proof then consists of showing that this assumption leads to a contradiction.

Let's now look at some examples of proof by contradiction. We begin by establishing a fact about positive real numbers.

Result to Prove   There is no smallest positive real number.

PROOF STRATEGY   In a proof by contradiction, we begin by assuming that the statement is false and attempt to show that this leads us to a contradiction. Hence we begin by assuming that there *is* a smallest positive real number. It is useful to represent this number by a symbol, say r. Our goal is to produce a contradiction. How do we go about doing this? Of course, if we could think of a positive real number that is less then r, then this would give us a contradiction.   ◆

Result 5.10   *There is no smallest positive real number.*

Proof   Assume, to the contrary, that there is a smallest positive real number, say r. Since $0 < r/2 < r$, it follows that $r/2$ is a positive real number that is smaller than r. This, however, is a contradiction.   ∎

PROOF ANALYSIS   The contradiction referred to in the proof of Result 5.10 is the statement: r is the smallest positive real number *and* $r/2$ is a positive real number that is less than r. This statement is certainly false. We have assumed that the reader understands what contradiction has been obtained. If we think that the reader may not see this, then, of course, we should specifically state (in the proof) what the contradiction is.

There is another point concerning Result 5.10 that should be made. This result states that "there is *no* smallest positive real number." This is a negative-sounding result. In the vast majority of cases, proofs of negative-sounding results are given by contradiction. Thus the proof technique used in Result 5.10 is not unexpected.   ◆

Let's consider two additional examples.

Result 5.11   *No odd integer can be expressed as the sum of three even integers.*

Proof   Assume, to the contrary, that there exists an odd integer n which can be expressed as the sum of three even integers x, y and z. Then $x = 2a$, $y = 2b$ and $z = 2c$ with $a, b, c \in \mathbf{Z}$. Therefore,

$$n = x + y + z = 2a + 2b + 2c = 2(a + b + c).$$

Since $a + b + c$ is an integer, n is even. This is a contradiction.   ∎

PROOF ANALYSIS   Consider the statement:

R: No odd integer can be expressed as the sum of three even integers.

Obviously, Result 5.11 states that R is a true statement. In order to give a proof by contradiction of Result 5.11, we attempted to prove an implication of the type $(\sim R) \Rightarrow C$ for some contradiction C. The negation $\sim R$ is

$\sim R$: There exists an odd integer that can be expressed as the sum of three even integers.

The proof we gave of Result 5.11 began by assuming the truth of $\sim R$. We introduced symbols for the four integers involved to make it easier to explain the proof. Eventually, we were able to show that n is an even integer. On the other hand, we knew that n is odd. Hence n was both even and odd. This was our contradiction C.   ◆

In the two examples of proof by contradiction that we have given, neither statement to be proved is expressed as an implication. For our next example, we consider an implication.

Result 5.12   *If a is an even integer and b is an odd integer, then $4 \nmid (a^2 + 2b^2)$.*

Proof   Assume, to the contrary, that there exist an even integer a and an odd integer b such that $4 \mid (a^2 + 2b^2)$. Thus $a = 2x$, $b = 2y + 1$ and $a^2 + 2b^2 = 4z$ for some integers x, y and z. Hence

$$a^2 + 2b^2 = (2x)^2 + 2(2y + 1)^2 = 4z.$$

Simplifying, we obtain $4x^2 + 8y^2 + 8y + 2 = 4z$ or, equivalently,

$$2 = 4z - 4x^2 - 8y^2 - 8y = 4(z - x^2 - 2y^2 - 2y).$$

Since $z - x^2 - 2y^2 - 2y$ is an integer, $4 \mid 2$, which is impossible.   ∎

PROOF ANALYSIS   Let S be the set of even integers and T the set of odd integers. In Result 5.12, our goal was to prove that

$$\forall a \in S, \forall b \in T, \; P(a, b). \tag{5.11}$$

is true, where

$$P(a, b): \; 4 \nmid (a^2 + 2b^2).$$

Since we were attempting to prove (5.11) by contradiction, we wanted to establish the truth of

$$\sim (\forall a \in S, \forall b \in T, \; P(a, b)) \Rightarrow C.$$

for some contradiction C or, equivalently, the truth of

$$\exists a \in S, \exists b \in T, \; (\sim P(a, b))) \Rightarrow C.$$

Hence we began by assuming that there exist an even integer a and an odd integer b such that $4 \mid (a^2 + 2b^2)$. We eventually deduced that $4 \mid 2$, which is a false statement and thereby produced a desired contradiction.

Using some facts we discussed earlier, we could have given a direct proof of Result 5.12. Once we wrote $a = 2x$ and $b = 2y + 1$, we have

$$a^2 + 2b^2 = (2x)^2 + 2(2y + 1)^2 = 4x^2 + 8y^2 + 8y + 2$$
$$= 4(x^2 + 2y^2 + 2y) + 2.$$

Hence we have expressed $a^2 + 2b^2$ as $4q + 2$, where $q = x^2 + 2y^2 + 2y$. That is, dividing $a^2 + 2b^2$ by 4 results in a remainder of 2 and so $4 \nmid (a^2 + 2b^2)$. At this stage, however, a proof by contradiction of Result 5.12 is probably preferred, in order to both practice and understand this proof technique.   ◆

Let's consider two other negative–sounding results.

Result 5.13   *The integer* 100 *cannot be written as the sum of three integers, an odd number of which are odd.*

Proof   Assume, to the contrary, that 100 *can* be written as the sum of three integers a, b and c, an odd number of which are odd. We consider two cases.

Case 1. Exactly one of a, b and c is odd, say a. Then $a = 2x + 1$, $b = 2y$ and $c = 2z$, where $x, y, z \in \mathbf{Z}$. So

$$100 = a + b + c = (2x + 1) + 2y + 2z = 2(x + y + z) + 1.$$

Since $x + y + z \in \mathbf{Z}$, the integer 100 is odd, producing a contradiction.

Case 2. All of a, b and c are odd. Then $a = 2x + 1$, $b = 2y + 1$ and $c = 2z + 1$, where $x, y, z \in \mathbf{Z}$. So

$$100 = a + b + c = (2x + 1) + (2y + 1) + (2z + 1) = 2(x + y + z + 1) + 1.$$

Since $x + y + z + 1 \in \mathbf{Z}$, the integer 100 is odd, again a contradiction.   ■

PROOF ANALYSIS   Observe that the proof of Result 5.13 begins by assuming that 100 can be written as the sum of three integers, an odd number of which are odd (as expected). However, by introducing symbols for these integers, namely a, b and c, this made for an easier and clearer proof.   ◆

Result 5.14   *For every integer m such that $2 \mid m$ and $4 \nmid m$, there exist no integers x and y for which $x^2 + 3y^2 = m$.*

Proof   Assume, to the contrary, that there exist an integer m such that $2 \mid m$ and $4 \nmid m$ and integers x and y for which $x^2 + 3y^2 = m$. Since $2 \mid m$, it follows that m is even. By Theorem 3.16, x^2 and $3y^2$ are of the same parity. We consider two cases.

Case 1. x^2 *and* $3y^2$ *are even.* Since $3y^2$ is even and 3 is odd, it follows by Theorem 3.17 that y^2 is even. Because x^2 and y^2 are both even, we have by Theorem 3.12 that x and y are even. Thus $x = 2a$ and $y = 2b$, where $a, b \in \mathbf{Z}$. Therefore,

$$x^2 + 3y^2 = (2a)^2 + 3(2b)^2 = 4a^2 + 12b^2$$
$$= 4(a^2 + 3b^2) = m.$$

Since $a^2 + 3b^2 \in \mathbf{Z}$, it follows that $4 \mid m$, producing a contradiction.

Case 2. x^2 *and* $3y^2$ *are odd.* Since $3y^2$ is odd and 3 is odd, it follows by (the contrapositive formulation of) Theorem 3.17 that y^2 is odd. By (the contrapositive formulation of) Theorem 3.12, x and y are both odd. Then $x = 2a + 1$ and $y = 2b + 1$, where $a, b \in \mathbf{Z}$. Thus

$$x^2 + 3y^2 = (2a + 1)^2 + 3(2b + 1)^2 = (4a^2 + 4a + 1) + 3(4b^2 + 4b + 1)$$
$$= 4a^2 + 4a + 12b^2 + 12b + 4 = 4(a^2 + a + 3b^2 + 3b + 1) = m.$$

Since $a^2 + a + 3b^2 + 3b + 1 \in \mathbf{Z}$, it follows that $4 \mid m$, producing a contradiction.   ∎

The next result concerns irrational numbers. Recall that a real number is rational if it can be expressed as m/n for some $m, n \in \mathbf{Z}$, where $n \neq 0$. Since "irrational" means "not rational," it is not surprising that proof by contradiction is the proof technique we will use.

Result 5.15   *The sum of a rational number and an irrational number is irrational.*

Proof   Assume, to the contrary, that there exist a rational number x and an irrational number y whose sum is a rational number z. Thus, $x + y = z$, where $x = a/b$ and $z = c/d$ for some integers $a, b, c, d \in \mathbf{Z}$ and $b, d \neq 0$. This implies that

$$y = \frac{c}{d} - \frac{a}{b} = \frac{bc - ad}{bd}.$$

Since $bc - ad$ and bd are integers and $bd \neq 0$, it follows that y is rational, which is a contradiction.   ∎

Result 5.15 concerns the irrationality of numbers. One of the best known irrational numbers is $\sqrt{2}$. Although we have never verified that this number is irrational, we establish this fact now.

Theorem to Prove   The real number $\sqrt{2}$ is irrational.

PROOF STRATEGY   In the proof of this result, we will use Theorem 3.12 which states that an integer x is even if and only if x^2 is even. Also, in the proof, it will be useful to express a rational number m/n, where $m, n \in \mathbf{Z}$ and $n \neq 0$, in lowest terms, which means that m and n contain no common divisor greater than 1.   ♦

Theorem 5.16   *The real number $\sqrt{2}$ is irrational.*

Proof   Assume, to the contrary, that $\sqrt{2}$ is rational. Then $\sqrt{2} = a/b$, where $a, b \in \mathbf{Z}$ and $b \neq 0$. We may further assume that a/b has been expressed in (or reduced to) lowest terms.

Then $2 = a^2/b^2$; so $a^2 = 2b^2$. Since b^2 is an integer, a^2 is even. By Theorem 3.12, a is even. So $a = 2c$, where $c \in \mathbf{Z}$. Thus, $(2c)^2 = 2b^2$ and so $4c^2 = 2b^2$. Therefore, $b^2 = 2c^2$. Because c^2 is an integer, b^2 is even, which implies by Theorem 3.12 that b is even. Since a and b are even, each has 2 as a divisor, which is a contradiction since a/b has been reduced to lowest terms.   ∎

The Three Prisoners Problem

We now take a brief diversion from our discussion of proof by contradiction to present a "story" problem. Three prisoners (see Figure 5.1) have been sentenced to long terms in prison, but due to overcrowded conditions, one prisoner must be released.

The warden devises a scheme to determine which prisoner is to be released. He tells the prisoners that he will blindfold them and then paint a red dot or a blue dot on each forehead. After he paints the dots, he will remove the blindfolds and a prisoner should raise his hand if he sees a red dot on at least one of the other two prisoners. The first prisoner to identify the color of the dot on his own forehead will be released. Of course, the prisoners agree to this. (What do they have to lose?)

The warden blindfolds the prisoners, as promised, and then paints a red dot on the foreheads of *all three prisoners*. He removes the blindfolds and, since each prisoner sees a red dot (in fact two red dots), each prisoner raises his hand. Some time passes when one of the prisoners exclaims, "I know what color my dot is! It's red!" This prisoner is then released. Although the story of the three prisoners is over, there is a lingering question: How did this prisoner correctly identify the color of the dot painted on his forehead?

The solution is given next but try to determine the answer for yourself before reading on.

Solution of the Three Prisoners Problem

Let's assume (without loss of generality) that it's prisoner #1 (see Figure 5.1) who determined that he had a red dot painted on his forehead. How did he come to this conclusion? Perhaps you think he just guessed since he had nothing to lose anyway. But this is not the answer we were looking for.

Prisoner #1 knows that the color of his dot is either red or blue. He thinks, "Assume, to the contrary, that my dot is blue. Then, of course, #2 knows this and he knows that #3 has a red dot. (That's why #2 raised his hand.) But #2 also knows that #3 raised his

#1   #2   #3

Figure 5.1   The three prisoners

hand. So if my dot is blue, #2 knows his dot is red. Similarly, if my dot is blue, then #3 knows his dot is red. In other words, if my dot is blue, then both #2 and #3 should be able to identify the colors of their dots quite quickly. But time has passed and they haven't determined the colors of their dots. So my dot can't be blue." Therefore, #1 exclaims, "I know what color my dot is! It's red!"

What you probably noticed is that the reasoning #1 used to conclude that his dot is red is proof by contradiction. It seems as if there is more to know about prisoner #1. But that's another story. ♦

5.3 A Review of Three Proof Techniques

We have seen that we're often in the situation where we want to prove the truth of a statement $\forall x \in S,\ P(x) \Rightarrow Q(x)$. You have now been introduced to three proof techniques: direct proof, proof by contrapositive, proof by contradiction. For each of these three techniques, you should be aware of how to start a proof and what your goal should be. You should also know what *not* to do. Figure 5.2 gives several ways that we *might* start a proof. Only some of these can lead to a proof, however.

Let's now compare the three proof techniques with two examples.

	First Step of "Proof"	Remarks/Goal
1.	**Assume that there exists $x \in S$ such that $P(x)$ is true.**	**A direct proof is being used.** **Show that $Q(x)$ is true for the element x.**
2.	Assume that there exists $x \in S$ such that $P(x)$ is false.	A mistake has been made.
3.	Assume that there exists $x \in S$ such that $Q(x)$ is true.	A mistake has been made.
4.	**Assume that there exists $x \in S$ such that $Q(x)$ is false.**	**A proof by contrapositive is being used.** **Show that $P(x)$ is false for the element x.**
5.	Assume that there exists $x \in S$ such that $P(x)$ and $Q(x)$ are true.	A mistake has been made.
6.	**Assume that there exists $x \in S$ such that $P(x)$ is true and $Q(x)$ is false.**	**A proof by contradiction is being used.** **Produce a contradiction.**
7.	Assume that there exists $x \in S$ such that $P(x)$ is false and $Q(x)$ is true.	A mistake has been made.
8.	Assume that there exists $x \in S$ such that $P(x)$ and $Q(x)$ are false.	A mistake has been made.
9.	Assume that there exists $x \in S$ such that $P(x) \Rightarrow Q(x)$ is true.	A mistake has been made.
10.	**Assume that there exists $x \in S$ such that $P(x) \Rightarrow Q(x)$ is false.**	**A proof by contradiction is being used.** **Produce a contradiction.**

Figure 5.2   How to prove (and not to prove) that $\forall x \in S,\ P(x) \Rightarrow Q(x)$ is true

Result 5.17   *If n is an even integer, then 3n + 7 is odd.*

Direct Proof   Let n be an even integer. Then $n = 2x$ for some integer x. Therefore,

$$3n + 7 = 3(2x) + 7 = 6x + 7 = 2(3x + 3) + 1.$$

Since $3x + 3$ is an integer, $3n + 7$ is odd.   ∎

Proof by
Contrapositive   Assume that $3n + 7$ is even. Then $3n + 7 = 2y$ for some integer y. Hence

$$n = (3n + 7) + (-2n - 7) = 2y - 2n - 7 = 2(y - n - 4) + 1.$$

Since $y - n - 4$ is an integer, n is odd.   ∎

Proof by
Contradiction   Assume, to the contrary, that there exists an even integer n such that $3n + 7$ is even. Since n is even, $n = 2x$ for some integer x. Hence

$$3n + 7 = 3(2x) + 7 = 6x + 7 = 2(3x + 3) + 1.$$

Since $3x + 3$ is an integer, $3n + 7$ is odd, which is a contradiction.   ∎

Although a direct proof of Result 5.17 is certainly the preferred proof technique in this case, it is useful to compare all three techniques. The following example is more intricate.

Result 5.18   *Let x be a nonzero real number. If $x + \dfrac{1}{x} < 2$, then $x < 0$.*

Direct Proof   Assume that $x + \dfrac{1}{x} < 2$. Since $x \neq 0$, we know that $x^2 > 0$. Multiplying both sides of the inequality $x + \dfrac{1}{x} < 2$ by x^2, we obtain $x^2 \left(x + \dfrac{1}{x} \right) < 2x^2$. Simplifying this inequality, we have $x^3 + x - 2x^2 < 0$; so

$$x(x^2 - 2x + 1) = x(x - 1)^2 < 0.$$

Since $(x - 1)^2 \geq 0$ and $x(x - 1)^2 \neq 0$, we must have $(x - 1)^2 > 0$. Since $x(x - 1)^2 < 0$ and $(x - 1)^2 > 0$, it follows that $x < 0$, as desired.   ∎

Proof Strategy
for Proof by
Contrapositive   For a proof by contrapositive, we will begin by assuming that $x \geq 0$ and attempt to show that $x + \dfrac{1}{x} \geq 2$. This inequality can be simplified by multiplying through by x, obtaining $x^2 + 1 \geq 2x$. Subtracting $2x$ from both sides, we have $x^2 - 2x + 1 = (x - 1)^2 \geq 0$, which, of course, we know to be true. A proof is suggested then by reversing the order of these steps:

$$x + \frac{1}{x} \geq 2$$
$$x^2 + 1 \geq 2x$$
$$x^2 - 2x + 1 = (x - 1)^2 \geq 0.$$

This method is common when dealing with inequalities.   ◆

Proof by Contrapositive   Assume that $x \geq 0$. Since $x \neq 0$, it follows that $x > 0$. Since $(x - 1)^2 \geq 0$, we have $(x - 1)^2 = x^2 - 2x + 1 \geq 0$. Adding $2x$ to both sides of this inequality, we obtain $x^2 + 1 \geq 2x$. Dividing both sides of the inequality $x^2 + 1 \geq 2x$ by the positive number x, we obtain $x + \dfrac{1}{x} \geq 2$, as desired.   ∎

Proof by Contradiction   Assume, to the contrary, that there exists a nonzero real number x such that $x + \dfrac{1}{x} < 2$ and $x \geq 0$. Since $x \neq 0$, it follows that $x > 0$. Multiplying both sides of the inequality $x + \dfrac{1}{x} < 2$ by x, we obtain $x^2 + 1 < 2x$. Subtracting $2x$ from both sides, we have $x^2 - 2x + 1 < 0$. It then follows that $(x - 1)^2 < 0$, which is a contradiction.   ∎

Many mathematicians feel that if a result can be verified by a direct proof, then this is the proof technique that should be used, as it is normally easier to understand. This is only a general guideline, however; it is *not* a hard and fast rule.

5.4 Existence Proofs

In an **existence theorem** the existence of an object (or objects) possessing some specified property or properties is asserted. Typically then, an existence theorem concerning an open sentence $R(x)$ over a domain S can be expressed as a quantified statement

$$\exists x \in S, \, R(x) : \text{ There exists } x \in S \text{ such that } R(x). \tag{5.12}$$

We have seen that such a statement (5.12) is true provided that $R(x)$ is true for some $x \in S$. A proof of an existence theorem is called an **existence proof**. An existence proof may then consist of displaying or constructing an example of such an object or perhaps, with the aid of known results, verifying that such objects must exist without ever producing a single example of the desired type. For example, there are theorems in mathematics that tell us that every polynomial of odd degree with real coefficients has at least one real number as a solution, but we don't know how to find a real number solution for every such polynomial. Indeed, we quote the great mathematician David Hilbert, who used the following example in his lectures to illustrate the idea of an existence proof:

> There is at least one student in this class . . . let us name him 'X' . . . for whom the following statement is true: No other student in the class has more hairs on his head than X. Which student is it? That we shall never know; but of his existence we can be absolutely certain.

Let's now see some examples of existence proofs.

Result to Prove   There exists an integer whose cube equals its square.

PROOF STRATEGY   Since this result is only asserting the existence of an integer whose cube equals its square, we have a proof once we can think of an example. The integer 1 has this property.   ♦

Result 5.19   *There exists an integer whose cube equals its square.*

Proof   Since $1^3 = 1^2 = 1$, the integer 1 has the desired property.   ∎

Suppose that we didn't notice that the integer 1 satisfied the required condition in the preceding theorem. Then an alternate proof may go something like this: Let $x \in \mathbf{Z}$ such that $x^3 = x^2$. Then $x^3 - x^2 = 0$ or $x^2(x - 1) = 0$. Thus, there are only two possible integers with this property, namely 1 and 0, and in fact both integers have the desired property.

A common error in elementary algebra is to write $(a + b)^2 = a^2 + b^2$. Can this *ever* be true?

Result 5.20   *There exist real numbers a and b such that $(a + b)^2 = a^2 + b^2$.*

Proof   Let $a, b \in \mathbf{R}$ such that $(a + b)^2 = a^2 + b^2$. Then $a^2 + 2ab + b^2 = a^2 + b^2$, so $2ab = 0$. Since $a = 1, b = 0$ is a solution to this equation, we have

$$(a + b)^2 = (1 + 0)^2 = 1^2 = 1^2 + 0^2 = a^2 + b^2.$$ ∎

The proof presented of Result 5.20 is longer than necessary. We could have written the following proof:

Proof   Let $a = 1$ and $b = 0$. Then

$$(a + b)^2 = (1 + 0)^2 = 1^2 = 1^2 + 0^2 = a^2 + b^2.$$ ∎

In the first proof, we actually presented an argument for how we thought of $a = 1$ and $b = 0$. In a proof, we are not required to explain where we got the idea for the proof, although it may very well be interesting to know this. If we feel that such information might be interesting or valuable, it may be worthwhile to include this in a discussion preceding or following the proof. The first proof we gave of Result 5.20 actually informs us of all real numbers a and b for which $(a + b)^2 = a^2 + b^2$, namely $(a + b)^2 = a^2 + b^2$ if and only if at least one of a and b is 0. This is more than what was requested of us but, nevertheless, it seems interesting.

We saw in Section 5.2 that $\sqrt{2}$ is irrational. Since $\sqrt{2} = 2^{1/2}$, it follows that there exist rational numbers a and b such that a^b is irrational, namely $a = 2$ and $b = 1/2$ have this property. Let's reverse this question. That is, do there exist *irrational numbers a* and b such that a^b is rational? Although there are many irrational numbers (in fact, an infinite number), we have only verified that $\sqrt{2}$ is irrational. (On the other hand, we know from the exercises for this section that $r + \sqrt{2}$ is irrational for every rational number r and that both $r\sqrt{2}$ and $r/\sqrt{2}$ are irrational for every nonzero rational number r.)

Result to Prove   There exist irrational numbers a and b such that a^b is rational.

PROOF STRATEGY   As we mentioned, there are only certain numbers that we know to be irrational, the simplest being $\sqrt{2}$. This might suggest considering the (real) number $\sqrt{2}^{\sqrt{2}}$. If this number is rational, then this answers our question. But perhaps $\sqrt{2}^{\sqrt{2}}$ is irrational. Then what do we do? This discussion suggests two cases.   ◆

Result 5.21   *There exist irrational numbers a and b such that a^b is rational.*

Proof   Consider the number $\sqrt{2}^{\sqrt{2}}$. Of course, this number is either rational or irrational. We consider these possibilities separately.

Case 1. $\sqrt{2}^{\sqrt{2}}$ is rational. Then we can take $a = b = \sqrt{2}$ and we have the desired result.
Case 2. $\sqrt{2}^{\sqrt{2}}$ is irrational. In this case, consider the number obtained by raising the (irrational) number $\sqrt{2}^{\sqrt{2}}$ to the (irrational) power $\sqrt{2}$; that is, consider a^b, where $a = \sqrt{2}^{\sqrt{2}}$ and $b = \sqrt{2}$. Observe that

$$a^b = \left(\sqrt{2}^{\sqrt{2}}\right)^{\sqrt{2}} = \sqrt{2}^{\sqrt{2}\cdot\sqrt{2}} = \sqrt{2}^2 = 2,$$

which is rational.   ■

The proof of Theorem 5.21 may seem unsatisfactory to you since we still don't know two specific irrational numbers a and b such that a^b is rational. We only know that two such numbers exist. We actually do know a bit more, namely either (1) $\sqrt{2}^{\sqrt{2}}$ is rational or (2) $\sqrt{2}^{\sqrt{2}}$ is irrational and $\left(\sqrt{2}^{\sqrt{2}}\right)^{\sqrt{2}}$ is rational. (Actually it has been proved that $\sqrt{2}^{\sqrt{2}}$ is an irrational number. Hence there are also irrational numbers of the form of a^b, where a and b are both irrational.)

In the next result, we want to show that the equation $x^5 + 2x - 5 = 0$ has a real number solution between $x = 1$ and $x = 2$. It is not easy to find a number that satisfies this equation. Instead, we use a well-known theorem from calculus to show that such a solution exists. You may not remember all the terms used in the following theorem but this is not crucial.

Intermediate
Value Theorem
of Calculus

If f is a function that is continuous on the closed interval $[a, b]$ and k is a number between $f(a)$ and $f(b)$, then there exists a number $c \in (a, b)$ such that $f(c) = k$.

We now give an example to show how this theorem can be used.

Result 5.22   *The equation $x^5 + 2x - 5 = 0$ has a real number solution between $x = 1$ and $x = 2$.*

Proof   Let $f(x) = x^5 + 2x - 5$. Since f is a polynomial function, it is continuous on the set of all real numbers and so f is continuous on the interval $[1, 2]$. Now $f(1) = -2$ and $f(2) = 31$. Since 0 is between $f(1)$ and $f(2)$, it follows by the Intermediate Value Theorem of Calculus that there is a number c between 1 and 2 such that $f(c) = c^5 + 2c - 5 = 0$. Hence c is a solution.   ■

As we just saw, the equation $x^5 + 2x - 5 = 0$ has a real number solution between $x = 1$ and $x = 2$. Actually, the equation $x^5 + 2x - 5 = 0$ has exactly one real number solution between $x = 1$ and $x = 2$. This brings up the topic of uniqueness. An element belonging to some prescribed set A and possessing a certain property P is **unique** if it is the only element of A having property P. Typically, to prove that only one element of A has property P, we proceed in one of two ways:

(1) We assume that a and b are elements of A possessing property P and show that $a = b$.

(2) We assume that a and b are distinct elements of A possessing property P and show that $a = b$.

Although (1) results in a direct proof and (2) results in a proof by contradiction, it is often the case that either proof technique can be used.

As an illustration, we return to Result 5.22 and show, in fact, that the equation $x^5 + 2x - 5 = 0$ has a unique real number solution between $x = 1$ and $x = 2$.

Result 5.23   *The equation $x^5 + 2x - 5 = 0$ has a unique real number solution between $x = 1$ and $x = 2$.*

Proof   Assume, to the contrary, that the equation $x^5 + 2x - 5 = 0$ has two distinct real number solutions a and b between $x = 1$ and $x = 2$. We may assume that $a < b$. Since $1 < a < b < 2$, it follows that $a^5 + 2a - 5 < b^5 + 2b - 5$. On the other hand, $a^5 + 2a - 5 = 0$ and $b^5 + 2b - 5 = 0$. Thus

$$0 = a^5 + 2a - 5 < b^5 + 2b - 5 = 0,$$

which produces a contradiction. ∎

Actually, we could have omitted Result 5.22 altogether and replaced it by Result 5.23 only (renumbering this result by Result 5.22), including the proofs of both Results 5.22 and 5.23.

We now present another result concerning uniqueness.

Result to Prove   For an irrational number r, let

$$S = \{sr + t : s, t \in \mathbf{Q}\}.$$

For every $x \in S$, there exist unique rational numbers a and b such that $x = ar + b$.

PROOF STRATEGY   To verify that a and b are unique, we assume that x can be expressed in two ways, say as $ar + b$ and $cr + d$, where $a, b, c, d \in \mathbf{Q}$, and then show that $a = c$ and $b = d$. Hence $ar + b = cr + d$. If $a \neq c$, then we can show that r is a rational number, producing a contradiction. Thus $a = c$. Subtracting ar from both sides of $ar + b = cr + d$, we obtain $b = d$ as well. ♦

We now give a complete proof.

Result 5.24   *For an irrational number r, let*

$$S = \{sr + t : s, t \in \mathbf{Q}\}.$$

> *For every $x \in S$, there exist unique rational numbers a and b such that $x = ar + b$.*

Proof Let $x \in S$ and suppose that $x = ar + b$ and $x = cr + d$, where $a, b, c, d \in \mathbf{Q}$. Then $ar + b = cr + d$. If $a \neq c$, then $(a - c)r = d - b$ and so

$$r = \frac{d - b}{a - c}.$$

Since $\frac{d-b}{a-c}$ is a rational number, this is impossible. So $a = c$. Subtracting $ar = cr$ from both sides of $ar + b = cr + d$, we obtain $b = d$.   ∎

Example 5.25 (a) *Show that the equation $6x^3 + x^2 - 2x = 0$ has a root in the interval $[-1, 1]$.*
(b) *Does this equation have a unique root in the interval $[-1, 1]$?*

Solution (a) By inspection, we can see that $x = 0$ is a root of the equation.
(b) Observe that

$$6x^3 + x^2 - 2x = x(6x^2 + x - 2) = x(3x + 2)(2x - 1).$$

Thus $x = -2/3$ and $x = 1/2$ are also roots of the equation $6x^3 + x^2 - 2x = 0$ and so this equation does not have a unique root in the interval $[-1, 1]$.   ◆

5.5 Disproving Existence Statements

Let $R(x)$ be an open sentence where the domain of x is S. We have already seen that to disprove a quantified statement of the type $\forall x \in S, R(x)$, it suffices to produce a counterexample (that is, an element x in S for which $R(x)$ is false). However, disproving a quantified statement of the type $\exists x \in S, R(x)$ requires a totally different approach. Since

$$\sim (\exists x \in S, R(x)) \equiv \forall x \in S, \sim R(x),$$

it follows that the statement $\exists x \in S, R(x)$ is false if $R(x)$ is false for *every* $x \in S$. Let's look at some examples of disproving existence statements.

Example 5.26 *Disprove the statement: There exists an odd integer n such that $n^2 + 2n + 3$ is odd.*

Solution We show that if n is an odd integer, then $n^2 + 2n + 3$ is even. Let n be an odd integer. Then $n = 2k + 1$ for some integer k. Thus

$$n^2 + 2n + 3 = (2k + 1)^2 + 2(2k + 1) + 3 = 4k^2 + 4k + 1 + 4k + 2 + 3$$
$$= 4k^2 + 8k + 6 = 2(2k^2 + 4k + 3).$$

Since $2k^2 + 4k + 3$ is an integer, $n^2 + 2n + 3$ is even.   ◆

Example 5.27 *Disprove the statement: There is a real number x such that $x^6 + 2x^4 + x^2 + 2 = 0$.*

Solution Let $x \in \mathbf{R}$. Since x^6, x^4 and x^2 are all even powers of the real number x, it follows that $x^6 \geq 0$, $x^4 \geq 0$ and $x^2 \geq 0$. Therefore, $x^6 + 2x^4 + x^2 + 2 \geq 0 + 0 + 0 + 2 = 2$ and so

$x^6 + 2x^4 + x^2 + 2 \neq 0$. Hence the equation $x^6 + 2x^4 + x^2 + 2 = 0$ has no real number solution.   ♦

Example 5.28   *Disprove the statement: There exists an integer n such that $n^3 - n + 1$ is even.*

Solution   Let $n \in \mathbf{Z}$. We consider two cases.

Case 1. n is even. Then $n = 2a$, where $a \in \mathbf{Z}$. So

$$n^3 - n + 1 = (2a)^3 - (2a) + 1 = 8a^3 - 2a + 1 = 2(4a^3 - a) + 1.$$

Since $4a^3 - a$ is an integer, $n^3 - n + 1$ is odd and so it is not even.
Case 2. n is odd. Then $n = 2b + 1$, where $b \in \mathbf{Z}$. Hence

$$
\begin{aligned}
n^3 - n + 1 &= (2b + 1)^3 - (2b + 1) + 1 \\
&= 8b^3 + 12b^2 + 6b + 1 - 2b - 1 + 1 \\
&= 8b^3 + 12b^2 + 4b + 1 = 2(4b^3 + 6b^2 + 2b) + 1.
\end{aligned}
$$

Since $4b^3 + 6b^2 + 2b$ is an integer, $n^3 - n + 1$ is odd and so it is not even.   ♦

If we had replaced Example 5.26 by

For every odd integer n, $n^2 + 2n + 3$ is even.

replaced Example 5.27 by

For every real number x, $x^6 + 2x^4 + x^2 + 2 \neq 0$.

and replaced Example 5.28 by

For every integer n, $n^3 - n + 1$ is odd.

then we would have a true statement in each case and the solutions of Examples 5.26–5.28 would become proofs.

EXERCISES FOR CHAPTER 5

Section 5.1: Counterexamples

5.1. Disprove the statement: If a and b are any two real numbers, then $\log(ab) = \log(a) + \log(b)$.

5.2. Disprove the statement: If $n \in \{0, 1, 2, 3, 4\}$, then $2^n + 3^n + n(n - 1)(n - 2)$ is prime.

5.3. Disprove the statement: If $n \in \{1, 2, 3, 4, 5\}$, then $3 \mid (2n^2 + 1)$.

5.4. Disprove the statement: Let $n \in \mathbf{N}$. If $\frac{n(n+1)}{2}$ is odd, then $\frac{(n+1)(n+2)}{2}$ is odd.

5.5. Disprove the statement: For every two positive integers a and b, $(a + b)^3 = a^3 + 2a^2b + 2ab + 2ab^2 + b^3$.

5.6. Let $a, b \in \mathbf{Z}$. Disprove the statement: If ab and $(a + b)^2$ are of opposite parity, then a^2b^2 and $a + ab + b$ are of opposite parity.

5.7. For positive real numbers a and b, it can be shown that $(a + b)\left(\frac{1}{a} + \frac{1}{b}\right) \geq 4$. If $a = b$, then this inequality is an equality. Consider the following statement:
If a and b are positive real numbers such that $(a + b)\left(\frac{1}{a} + \frac{1}{b}\right) = 4$, then $a = b$.
Is there a counterexample to this statement?

5.8. In Exercise 5.7, it is stated that $(a + b)\left(\frac{1}{a} + \frac{1}{b}\right) \geq 4$ for every two positive real numbers a and b. Does it therefore follow that $(c^2 + d^2)\left(\frac{1}{c^2} + \frac{1}{d^2}\right) \geq 4^2$ for every two positive real numbers c and d?

5.9. Disprove the statement: For every positive integer x and every integer $n \geq 2$, the equation $x^n + (x + 1)^n = (x + 2)^n$ has no solution.

Section 5.2: Proof by Contradiction

5.10. Prove that there is no largest negative rational number.

5.11. Prove that there is no smallest positive irrational number.

5.12. Prove that 200 cannot be written as the sum of an odd integer and two even integers.

5.13. Use proof by contradiction to prove that if a and b are odd integers, then $4 \nmid (a^2 + b^2)$.

5.14. Prove that if $a \geq 2$ and b are integers, then $a \nmid b$ or $a \nmid (b + 1)$.

5.15. Prove that 1000 cannot be written as the sum of three integers, an even number of which are even.

5.16. Prove that the product of an irrational number and a nonzero rational number is irrational.

5.17. Prove that when an irrational number is divided by a (nonzero) rational number, the resulting number is irrational.

5.18. Let a be an irrational number and r a nonzero rational number. Prove that if s is a real number, then either $ar + s$ or $ar - s$ is irrational.

5.19. Prove that $\sqrt{3}$ is irrational. [Hint: First prove for an integer a that $3 \mid a^2$ if and only if $3 \mid a$. Recall that every integer can be written as $3q, 3q + 1$ or $3q + 2$ for some integer q. Also, see Exercise 3 of Chapter 4.]

5.20. Prove that $\sqrt{2} + \sqrt{3}$ is an irrational number.

5.21. (a) Prove that $\sqrt{6}$ is an irrational number.
     (b) Prove that there are infinitely many positive integers n such that $\sqrt{n}$ is irrational.

5.22. Let $S = \{p + q\sqrt{2} : p, q \in \mathbf{Q}\}$ and $T = \{r + s\sqrt{3} : r, s \in \mathbf{Q}\}$. Prove that $S \cap T = \mathbf{Q}$.

5.23. Prove that there is no integer a such that $a \equiv 5 \pmod{14}$ and $a \equiv 3 \pmod{21}$.

5.24. Prove that there exists no positive integer x such that $2x < x^2 < 3x$.

5.25. Prove that there do not exist three distinct positive integers a, b and c such that each integer divides the difference of the other two.

5.26. Prove that the sum of the squares of two odd integers cannot be the square of an integer.

5.27. Prove that if x and y are positive real numbers, then $\sqrt{x + y} \neq \sqrt{x} + \sqrt{y}$.

5.28. Prove that there do not exist positive integers m and n such that $m^2 - n^2 = 1$.

5.29. Let m be a positive integer of the form $m = 2s$, where s is an odd integer. Prove that there do not exist positive integers x and y such that $x^2 - y^2 = m$.

5.30. Prove that there do not exist three distinct real numbers a, b and c such that all of the numbers $a + b + c$, ab, ac, bc, abc are equal.

5.31. Use a proof by contradiction to prove the following. Let $m \in \mathbf{Z}$. If $3 \nmid (m^2 - 1)$, then $3 \mid m$. (A proof by contrapositive of this result is given in Result 4.6.)

5.32. Prove that there exist no positive integers m and n for which $m^2 + m + 1 = n^2$.

5.33. (a) Prove that there is no rational number solution of the equation $x^2 - 3x + 1 = 0$.
     (b) The problem in (a) should suggest a more general problem. State and outline a proof of this.

Section 5.3: A Review of Three Proof Techniques

5.34. Prove that if n is an odd integer, then $7n - 5$ is even by
(a) a direct proof, (b) a proof by contrapositive and (c) a proof by contradiction.

5.35. Let x be a positive real number. Prove that if $x - \frac{2}{x} > 1$, then $x > 2$ by
(a) a direct proof, (b) a proof by contrapositive and (c) a proof by contradiction.

5.36. Let $a, b \in \mathbf{R}$. Prove that if $ab \neq 0$, then $a \neq 0$ by using as many of the three proof techniques as possible.

5.37. Let $x, y \in \mathbf{R}^+$. Prove that if $x \leq y$, then $x^2 \leq y^2$ by
(a) a direct proof, (b) a proof by contrapositive and (c) a proof by contradiction.

5.38. Prove the following statement using more than one method of proof.
Let $a, b \in \mathbf{Z}$. If a is odd and $a + b$ is even, then b is odd and ab is odd.

5.39. Prove the following statement using more than one method of proof.
For every three integers a, b and c, exactly two of the integers ab, ac and bc cannot be odd.

Section 5.4: Existence Proofs

5.40. Show that there exist a rational number a and an irrational number b such that a^b is rational.

5.41. Show that there exist a rational number a and an irrational number b such that a^b is irrational.

5.42. Show that there exist two distinct irrational numbers a and b such that a^b is rational.

5.43. Show that there exist no nonzero real numbers a and b such that $\sqrt{a^2 + b^2} = \sqrt[3]{a^3 + b^3}$.

5.44. Prove that there exists a unique real number solution to the equation $x^3 + x^2 - 1 = 0$ between $x = 2/3$ and $x = 1$.

5.45. Let $R(x)$ be an open sentence over a domain S. Suppose that $\forall x \in S$, $R(x)$ is a false statement and that the set T of counterexamples is a proper subset of S. Show that there exists a subset W of S such that $\forall x \in W$, $R(x)$ is true.

5.46. (a) Prove that there exist four distinct positive integers such that each integer divides the sum of the remaining integers.
(b) The problem in (a) should suggest another problem to you. State and solve such a problem.

5.47. Let S be a set of three integers. For a nonempty subset A of S, let σ_A be the sum of the elements in A. Prove that there exist two distinct nonempty subsets B and C of S such that $\sigma_B \equiv \sigma_C \pmod{6}$.

5.48. Prove that the equation $\cos^2(x) - 4x + \pi = 0$ has a real number solution in the interval $[0, 4]$. (You may assume that $\cos^2(x)$ is continuous on $[0, 4]$.)

Section 5.5: Disproving Existence Statements

5.49. Disprove the statement: There exist odd integers a and b such that $4 \mid (3a^2 + 7b^2)$.

5.50. Disprove the statement: There is a real number x such that $x^6 + x^4 + 1 = 2x^2$.

5.51. Disprove the statement: There is an integer n such that $n^4 + n^3 + n^2 + n$ is odd.

5.52. The integers $1, 2, 3$ have the property that each divides the sum of the other two. Indeed, for each positive integer a, the integers $a, 2a, 3a$ have the property that each divides the sum of the other two. Show that the following statement is false.
There exists an example of three distinct positive integers different from $a, 2a, 3a$ for some $a \in \mathbf{N}$ having the property that each divides the sum of the other two.

ADDITIONAL EXERCISES FOR CHAPTER 5

5.53. Show that the following statement is false.
    If A and B are two sets of positive integers with $|A| = |B| = 3$ such that whenever an integer s is the sum of the elements of some subset of A, then s is also the sum of the elements of some subset of B, then $A = B$.

5.54. (a) Prove that if $a \geq 2$ and $n \geq 1$ are integers such that $a^2 + 1 = 2^n$, then a is odd.
    (b) Prove that there are no integers $a \geq 2$ and $n \geq 1$ such that $a^2 + 1 = 2^n$.

5.55. Prove that there do not exist positive integers a and n such that $a^2 + 3 = 3^n$.

5.56. Let $x, y \in \mathbf{R}^+$. Use a proof by contradiction to prove that if $x < y$, then $\sqrt{x} < \sqrt{y}$.

5.57. The king's daughter had three suitors and couldn't decide which one to marry. So the king said, "I have three gold crowns and two silver ones. I will put either a gold or silver crown on each of your heads. The suitor who can tell me which crown he has will marry my daughter." The first suitor looked around and said he could not tell. The second did the same. The third suitor said: "I have a gold crown." He is correct, but the daughter was puzzled: This suitor was blind. How did he know? (Reference: Ask Marilyn, *Parade Magazine*, July 6, 2003).

5.58. Prove that if a, b, c, d are four real numbers, then at most four of the numbers ab, ac, ad, bc, bd, cd are negative.

5.59. Evaluate the proposed proof of the following result.

Result   The number 25 cannot be written as the sum of three integers, an even number of which are odd.

Proof   Assume, to the contrary, that 25 can be written as the sum of three integers, an even number of which are odd. Then $25 = x + y + z$, where $x, y, z \in \mathbf{Z}$. We consider two cases.

Case 1. x and y are odd. Then $x = 2a + 1$, $y = 2b + 1$ and $z = 2c$, where $a, b, c \in \mathbf{Z}$. Therefore,

$$25 = x + y + z = (2a + 1) + (2b + 1) + 2c$$
$$= 2a + 2b + 2c + 2 = 2(a + b + c + 1).$$

Since $a + b + c + 1$ is an integer, 25 is even, a contradiction.

Case 2. x, y and z are even. Then $x = 2a$, $y = 2b$ and $z = 2c$, where $a, b, c \in \mathbf{Z}$. Hence

$$25 = x + y + z = 2a + 2b + 2c = 2(a + b + c).$$

Since $a + b + c$ is an integer, 25 is even, again a contradiction.   ∎

5.60. (a) Let n be a positive integer. Show that every integer m with $1 \leq m \leq 2n$ can be expressed as $2^\ell k$, where ℓ is a nonnegative integer and k is an odd integer with $1 \leq k < 2n$.
    (b) Prove for every positive integer n and every subset S of $\{1, 2, \ldots, 2n\}$ with $|S| = n + 1$ that there exist integers $a, b \in S$ such that $a \mid b$.

5.61. Prove that the sum of the irrational numbers $\sqrt{2}, \sqrt{3}$ and $\sqrt{5}$ is also irrational.

5.62. Let $a_1, a_2, \ldots, a_r$ be odd integers where $a_i > 1$ for $i = 1, 2, \ldots, r$. Prove that if $n = a_1 a_2 \cdots a_r + 2$, then $a_i \nmid n$ for each integer i ($1 \leq i \leq r$).

5.63. Below is given a proof of a result. What result is proved?

Proof   Let $a, b, c \in \mathbf{Z}$ such that $a^2 + b^2 = c^2$. Assume, to the contrary, that a, b and c are all odd. Then $a = 2r + 1$, $b = 2s + 1$ and $c = 2t + 1$, where $r, s, t \in \mathbf{Z}$. Thus,

$$a^2 + b^2 = (4r^2 + 4r + 1) + (4s^2 + 4s + 1)$$
$$= 2(2r^2 + 2r + 2s^2 + 2s + 1).$$

Since $2r^2 + 2r + 2s^2 + 2s + 1$ is an integer, it follows that $a^2 + b^2$ is even. On the other hand,

$$c^2 = (2t + 1)^2 = 4t^2 + 4t + 1 = 2(2t^2 + 2t) + 1.$$

Since $2t^2 + 2t$ is an integer, it follows that c^2 is odd. Therefore, $a^2 + b^2$ is even and c^2 is odd, contradicting the fact that $a^2 + b^2 = c^2$.   ∎

5.64. Evaluate the proposed proof of the following result.

Result   If x is an irrational number and y is a rational number, then $z = x - y$ is irrational.

Proof   Assume, to the contrary, that $z = x - y$ is rational. Then $z = a/b$, where $a, b \in \mathbf{Z}$ and $b \neq 0$. Since $\sqrt{2}$ is irrational, we let $x = \sqrt{2}$. Since y is rational, $y = c/d$, where $c, d \in \mathbf{Z}$ and $d \neq 0$. Therefore,

$$\sqrt{2} = x = y + z = \frac{c}{d} + \frac{a}{b} = \frac{ad + bc}{bd}.$$

Since $ad + bc$ and bd are integers, where $bd \neq 0$, it follows that $\sqrt{2}$ is rational, producing a contradiction.   ∎

5.65. Prove that there exist four distinct real numbers a, b, c, d such that exactly four of the numbers ab, ac, ad, bc, bd, cd are irrational.

5.66. Below is given a proof of a result. What result is proved?

Proof   Let $a \equiv 2 \pmod 4$ and $b \equiv 1 \pmod 4$ and assume, to the contrary, that $4 \mid (a^2 + 2b)$. Since $a \equiv 2 \pmod 4$ and $b \equiv 1 \pmod 4$, it follows that $a = 4r + 2$ and $b = 4s + 1$, where $r, s \in \mathbf{Z}$. Therefore,

$$a^2 + 2b = (4r + 2)^2 + 2(4s + 1) = (16r^2 + 16r + 4) + (8s + 2)$$
$$= 16r^2 + 16r + 8s + 6.$$

Since $4 \mid (a^2 + 2b)$, we have $a^2 + 2b = 4t$, where $t \in \mathbf{Z}$. So $16r^2 + 16r + 8s + 6 = 4t$ and

$$6 = 4t - 16r^2 - 16r - 8s = 4(t - 4r^2 - 4r - 2s).$$

Since $t - 4r^2 - 4r - 2s$ is an integer, $4 \mid 6$, which is a contradiction.   ∎

6

Mathematical Induction

We have seen three proof techniques which could be used to prove that a quantified statement $\forall x \in S, P(x)$ is true: direct proof, proof by contrapositive, proof by contradiction. For certain sets S, however, there is another possible method of proof: mathematical induction.

6.1 The Principle of Mathematical Induction

Let A be a nonempty set of real numbers. A number $m \in A$ is called a **least element** (or a **minimum** or **smallest element**) of A if $x \geq m$ for every $x \in A$. Some nonempty sets of real numbers have a least element; others do not. The set **N** has a smallest element, namely 1, while **Z** has no least element. The closed interval $[2, 5]$ has the minimum element 2, but the open interval $(2, 5)$ has no minimum element. The set

$$A = \left\{ \frac{1}{n} : n \in \mathbf{N} \right\}$$

also has no least element.

If a nonempty set A of real numbers has a least element, then this element is necessarily unique. We will verify this fact. Recall that when attempting to prove that an element possessing a certain property is unique, it is customary to assume that there are two elements with this property. We then show that these elements are equal, implying that there is exactly one such element.

Theorem 6.1 *If a set A of real numbers has a least element, then A has a unique least element.*

Proof Let m_1 and m_2 be least elements of A. Since m_1 is a least element, $m_2 \geq m_1$. Also, since m_2 is a least element, $m_1 \geq m_2$. Therefore, $m_1 = m_2$. ∎

The proof we gave of Theorem 6.1 is a direct proof. Suppose that we had replaced the first sentence of this proof by

Assume, to the contrary, that A contains distinct least elements m_1 and m_2.

If the remainder of the proof of Theorem 6.1 were the same except for adding a concluding sentence that we have a contradiction, then this too would be a proof of Theorem 6.1. That is, with a small change, the proof technique used to verify Theorem 6.1 can be transformed from a direct proof to a proof by contradiction.

There is a property possessed by some sets of real numbers that will be of great interest to us here. A nonempty set S of real numbers is said to be **well-ordered** if every nonempty subset of S has a least element. Let $S = \{-7, -1, 2\}$. The nonempty subsets of S are

$$\{-7, -1, 2\}, \{-7, -1\}, \{-7, 2\}, \{-1, 2\}, \{-7\}, \{-1\} \text{ and } \{2\}.$$

Since each of these subsets has a least element, S is well-ordered. Indeed, it should be clear that *every* nonempty finite set of real numbers is well-ordered. (See Exercise 6.20 in Exercises for Section 6.2.) The open interval $(0, 1)$ is *not* well-ordered, since, for example, $(0, 1)$ itself has no least element. The closed interval $[0, 1]$ has the least element 0; however, $[0, 1]$ is *not* well-ordered since the open interval $(0, 1)$ is a (nonempty) subset of $[0, 1]$ without a least element. Because none of the sets $\mathbf{Z}, \mathbf{Q}$, and $\mathbf{R}$ have a least element, none of these sets are well-ordered. Hence, having a least element is a necessary condition for a nonempty set to be well-ordered, but it is not sufficient.

Although it may appear evident that the set $\mathbf{N}$ of positive integers is well-ordered, this statement cannot be proved from the properties of positive integers that we have used and derived thus far. Consequently, this statement is accepted as an axiom, which we state below.

The Well-Ordering Principle

The set $\mathbf{N}$ of positive integers is well-ordered.

A consequence of the Well-Ordering Principle is another principle, which serves as the foundation of another important proof technique.

Theorem 6.2 (**The Principle of Mathematical Induction**) *For each positive integer n, let $P(n)$ be a statement. If*

(1) *$P(1)$ is true and*
(2) *the implication*

$$\textit{If } P(k), \textit{ then } P(k + 1).$$

is true for every positive integer k,

then $P(n)$ is true for every positive integer n.

Proof Assume, to the contrary, that the theorem is false. Then conditions (1) and (2) are satisfied, but there exist some positive integers n for which $P(n)$ is a false statement. Let

$$S = \{n \in \mathbf{N} : \ P(n) \text{ is false}\}.$$

Since S is a nonempty subset of **N**, it follows by the Well-Ordering Principle that S contains a least element s. Since $P(1)$ is true, $1 \notin S$. Thus $s \geq 2$ and $s - 1 \in \mathbf{N}$. Therefore, $s - 1 \notin S$ and so $P(s - 1)$ is a true statement. By condition (2), $P(s)$ is also true and so $s \notin S$. This, however, contradicts our assumption that $s \in S$. ∎

The Principle of Mathematical Induction is stated more symbolically below.

The Principle of Mathematical Induction   *For each positive integer n, let $P(n)$ be a statement. If*

(1) $P(1)$ *is true and*
(2) $\forall k \in \mathbf{N}, \; P(k) \Rightarrow P(k + 1)$ *is true,*

then $\forall n \in \mathbf{N}, \; P(n)$ is true.

As a consequence of the Principle of Mathematical Induction, the quantified statement $\forall n \in \mathbf{N}, \; P(n)$ can be proved to be true if

(1)   we can show that the statement $P(1)$ is true and
(2)   we can establish the truth of the implication

$$\text{If } P(k), \text{ then } P(k + 1).$$

for every positive integer k.

A proof using the Principle of Mathematical Induction is called an **induction proof** or a **proof by induction**. The verification of the truth of $P(1)$ in an induction proof is called the **base step**, **basis step** or the **anchor** of the induction. In the implication

$$\text{If } P(k), \text{ then } P(k + 1).$$

for an arbitrary positive integer k, the statement $P(k)$ is called the **inductive (or induction) hypothesis**. Often we use a direct proof to verify

$$\forall k \in \mathbf{N}, \; P(k) \Rightarrow P(k + 1), \tag{6.1}$$

although any proof technique is acceptable. That is, we typically assume that the inductive hypothesis $P(k)$ is true for an arbitrary positive integer k and attempt to show that $P(k + 1)$ is true. Establishing the truth of (6.1) is called the **inductive step** in the induction proof.

We illustrate this proof technique by showing that the sum of the first n positive integers is given by $n(n + 1)/2$ for every positive integer n, that is,

$$1 + 2 + 3 + \cdots + n = \frac{n(n + 1)}{2}.$$

Result 6.3   *Let*

$$P(n): \; 1 + 2 + 3 + \cdots + n = \frac{n(n + 1)}{2}$$

where $n \in \mathbf{N}$. Then $P(n)$ is true for every positive integer n.

Proof We employ induction. Since $1 = (1 \cdot 2)/2$, the statement $P(1)$ is true. Assume that $P(k)$ is true for an arbitrary positive integer k, that is, assume that

$$1 + 2 + 3 + \cdots + k = \frac{k(k+1)}{2}.$$

We show that $P(k+1)$ is true, that is, we show that

$$1 + 2 + 3 + \cdots + (k+1) = \frac{(k+1)(k+2)}{2}.$$

Thus

$$1 + 2 + 3 + \cdots + (k+1) = (1 + 2 + 3 + \cdots + k) + (k+1)$$
$$= \frac{k(k+1)}{2} + (k+1) = \frac{k(k+1) + 2(k+1)}{2}$$
$$= \frac{(k+1)(k+2)}{2},$$

as desired.

By the Principle of Mathematical Induction, $P(n)$ is true for every positive integer n. ∎

Typically, a statement to be proved by induction is not presented in terms of $P(n)$ or some other symbols. In order to illustrate this, we give an alternative statement and proof of Result 6.3, as it is to be understood what $P(n)$ would represent.

Result 6.4 *For every positive integer n,*

$$1 + 2 + 3 + \cdots + n = \frac{n(n+1)}{2}.$$

Proof We employ induction. Since $1 = (1 \cdot 2)/2$, the statement is true for $n = 1$. Assume that

$$1 + 2 + 3 + \cdots + k = \frac{k(k+1)}{2},$$

where k is a positive integer. We show that

$$1 + 2 + 3 + \cdots + (k+1) = \frac{(k+1)(k+2)}{2}.$$

Thus

$$1 + 2 + 3 + \cdots + (k+1) = (1 + 2 + 3 + \cdots + k) + (k+1)$$
$$= \frac{k(k+1)}{2} + (k+1) = \frac{k(k+1) + 2(k+1)}{2}$$
$$= \frac{(k+1)(k+2)}{2}.$$

By the Principle of Mathematical Induction,

$$1 + 2 + 3 + \cdots + n = \frac{n(n+1)}{2}$$

for every positive integer n. ∎

The proof of Result 6.3 (or of Result 6.4) began by stating that induction was being used. This alerts the reader of what to expect in the proof. Also, in the proof of the inductive step, it is assumed that

$$1 + 2 + 3 + \cdots + k = \frac{k(k+1)}{2}$$

for a positive integer k, that is, for an *arbitrary* positive integer k. We do *not* assume that

$$1 + 2 + 3 + \cdots + k = \frac{k(k+1)}{2}$$

for *every* positive integer k as this would be assuming what we are attempting to prove in Result 6.3 (and in Result 6.4). ♦

Carl Friedrich Gauss (1777–1855) is considered to be one of the most brilliant mathematicians of all time. The story goes that when he was very young (in grade school in Germany) his teacher gave him and his classmates the supposedly unpleasant task of adding the integers from 1 to 100. He obtained the correct result of 5050 quickly. It is believed that he considered both the sum $1 + 2 + \cdots + 100$ and its reverse sum $100 + 99 + \cdots + 1$ and added these to obtain the sum $101 + 101 + \cdots + 101$, which has 100 terms and so equals 10,100. Since this is twice the required sum, $1 + 2 + \cdots + 100 = 10100/2 = 5050$. This, of course, can be quite easily generalized to find a formula for $1 + 2 + 3 + \cdots + n$, where $n \in \mathbf{N}$. Let

$$S = 1 + 2 + 3 + \cdots + n. \tag{6.2}$$

If we reverse the order of the terms on the right side of (6.2), then we obtain

$$S = n + (n - 1) + (n - 2) + \cdots + 1. \tag{6.3}$$

Adding (6.2) and (6.3), we have

$$2S = (n + 1) + (n + 1) + (n + 1) + \cdots + (n + 1). \tag{6.4}$$

Since there are n terms on the right side of (6.4), we conclude that $2S = n(n + 1)$ or $S = n(n + 1)/2$. Hence

$$1 + 2 + 3 + \cdots + n = \frac{n(n+1)}{2}.$$

You might think that the proof of Result 6.3 (and Result 6.4) that we gave by mathematical induction is longer (and more complicated) than the one we just gave and this may very well be true. But, in general, mathematical induction is a technique that can be used to prove a wide range of statements. In this chapter, we will see a variety of statements where mathematical induction is a natural technique used in verifying their truth. We begin with an example that leads to a problem involving mathematical induction.

Suppose that an $n \times n$ square S is composed of n^2 1×1 squares. For all integers k with $1 \le k \le n$, how many different $k \times k$ squares does S contain? (See Figure 6.1 for the case where $n = 3$.) For $n = 3$, the square S contains the 3×3 square S itself, four 2×2 squares and nine 1×1 squares (see Figure 6.1). Therefore, the number of different squares that S contains is $1 + 4 + 9 = 1^2 + 2^2 + 3^2 = 14$.

$S:$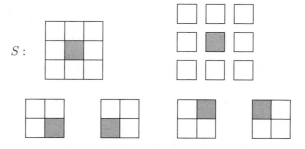

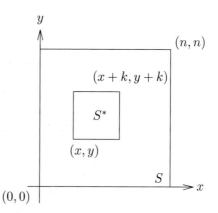

Figure 6.1 The squares in a 3×3 square

Figure 6.2 A $k \times k$ square in an $n \times n$ square

In order to determine the number of different $k \times k$ squares in an $n \times n$ square S, we place S in the first quadrant of the coordinate plane so that the lower left corner of S is at the origin $(0, 0)$. (See Figure 6.2.) Then the upper right corner of S is at the point (n, n). Consequently, the lower left corner of a $k \times k$ square S^*, where $1 \leq k \leq n$, is at some point (x, y), while the upper right corner of S^* is at $(x + k, y + k)$. Necessarily, x and y are nonnegative integers with $x + k \leq n$ and $y + k \leq n$ (again, see Figure 6.2). Since $0 \leq x \leq n - k$ and $0 \leq y \leq n - k$, the number of choices for each of x and y is $n - k + 1$ and so the number of possibilities for (x, y) is $(n - k + 1)^2$. Because k is any of the integers $1, 2, \ldots, n$, the total number of different squares in S is

$$\sum_{k=1}^{n}(n - k + 1)^2 = n^2 + (n - 1)^2 + \cdots + 2^2 + 1^2$$

$$= 1^2 + 2^2 + \cdots + n^2 = \sum_{k=1}^{n} k^2.$$

Is there a compact formula for the expression

$$\sum_{k=1}^{n} k^2 = 1^2 + 2^2 + \cdots + n^2?$$

For the problem we are describing, it would be very helpful to know the answer to this question. Since we brought up this question, you might have already guessed that the answer is yes. A formula is given next, along with a proof by induction.

Result 6.5   *For every positive integer n,*

$$1^2 + 2^2 + \cdots + n^2 = \frac{n(n+1)(2n+1)}{6}.$$

Proof   We proceed by induction. Since $1^2 = (1 \cdot 2 \cdot 3)/6 = 1$, the statement is true when $n = 1$. Assume that

$$1^2 + 2^2 + \cdots + k^2 = \frac{k(k+1)(2k+1)}{6}$$

for an arbitrary positive integer k. We show that

$$1^2 + 2^2 + \cdots + (k+1)^2 = \frac{(k+1)(k+2)(2k+3)}{6}.$$

Observe that

$$
\begin{aligned}
1^2 + 2^2 + \cdots + (k+1)^2 &= [1^2 + 2^2 + \cdots + k^2] + (k+1)^2 \\
&= \frac{k(k+1)(2k+1)}{6} + (k+1)^2 \\
&= \frac{k(k+1)(2k+1)}{6} + \frac{6(k+1)^2}{6} \\
&= \frac{(k+1)[k(2k+1) + 6(k+1)]}{6} \\
&= \frac{(k+1)(2k^2 + 7k + 6)}{6} \\
&= \frac{(k+1)(k+2)(2k+3)}{6},
\end{aligned}
$$

as desired.

By the Principle of Mathematical Induction,

$$1^2 + 2^2 + \cdots + n^2 = \frac{n(n+1)(2n+1)}{6}$$

for every positive integer n.   ∎

Strictly speaking, the last sentence in the proof of Result 6.5 is typical of the last sentence of every proof using mathematical induction, for the idea is to show that the hypothesis of the Principle of Mathematical Induction is satisfied and so the conclusion follows. Some therefore omit this final sentence since it is understood that once properties (1) and (2) of Theorem 6.2 are satisfied, we have a proof. For emphasis, we will continue to include this concluding sentence, however.

There is another question that might have occurred to you. We explained why $1 + 2 + \cdots + n$ equals $n(n+1)/2$, but how did we know that $1^2 + 2^2 + \cdots + n^2$ equals $n(n+1)(2n+1)/6$? We can actually show that $1^2 + 2^2 + \cdots + n^2 =$

$n(n + 1)(2n + 1)/6$ by using the formula $1 + 2 + \cdots + n = n(n + 1)/2$. We begin by solving

$$(k + 1)^3 = k^3 + 3k^2 + 3k + 1$$

for k^2. Since $3k^2 = (k + 1)^3 - k^3 - 3k - 1$, it follows that

$$k^2 = \frac{1}{3}\left[(k + 1)^3 - k^3\right] - k - \frac{1}{3}$$

and so

$$\sum_{k=1}^{n} k^2 = \frac{1}{3}\left[\sum_{k=1}^{n}(k + 1)^3 - \sum_{k=1}^{n} k^3\right] - \sum_{k=1}^{n} k - \frac{1}{3}\sum_{k=1}^{n} 1.$$

Therefore,

$$\sum_{k=1}^{n} k^2 = \frac{1}{3}\left[(n + 1)^3 - 1^3\right] - \frac{1}{2}n(n + 1) - \frac{1}{3}n$$

$$= \frac{n^3 + 3n^2 + 3n}{3} - \frac{n^2 + n}{2} - \frac{n}{3}$$

$$= \frac{2n^3 + 6n^2 + 6n - 3n^2 - 3n - 2n}{6} = \frac{2n^3 + 3n^2 + n}{6}$$

$$= \frac{n(2n^2 + 3n + 1)}{6} = \frac{n(n + 1)(2n + 1)}{6}.$$

This is actually an alternative proof that

$$1^2 + 2^2 + \cdots + n^2 = \frac{n(n + 1)(2n + 1)}{6}$$

for every positive integer n but of course this proof depends on knowing that

$$1 + 2 + 3 + \cdots + n = \frac{n(n + 1)}{2}$$

for every positive integer n.

We have now used mathematical induction to establish the formulas

$$1 + 2 + \cdots + n = \frac{n(n + 1)}{2} \tag{6.5}$$

and

$$1^2 + 2^2 + \cdots + n^2 = \frac{n(n + 1)(2n + 1)}{6} \tag{6.6}$$

for every positive integer n. We saw that (6.6) gives the number of different squares in an $n \times n$ square composed of n^2 1×1 squares. Actually, (6.5) gives the number of intervals in an interval of length n composed of n intervals of length 1. You can probably guess what $1^3 + 2^3 + \cdots + n^3$ counts. Exercise 6.6 deals with this expression.

We now present a formula for

$$\frac{1}{2 \cdot 3} + \frac{1}{3 \cdot 4} + \cdots + \frac{1}{(n + 1)(n + 2)}$$

for every positive integer n.

Result 6.6   *For every positive integer n,*

$$\frac{1}{2\cdot 3} + \frac{1}{3\cdot 4} + \cdots + \frac{1}{(n+1)(n+2)} = \frac{n}{2n+4}.$$

Proof   We use induction. Since

$$\frac{1}{2\cdot 3} = \frac{1}{2\cdot 1+4} = \frac{1}{6},$$

the formula holds for $n = 1$. Assume that

$$\frac{1}{2\cdot 3} + \frac{1}{3\cdot 4} + \cdots + \frac{1}{(k+1)(k+2)} = \frac{k}{2k+4}$$

for a positive integer k. We show that

$$\frac{1}{2\cdot 3} + \frac{1}{3\cdot 4} + \cdots + \frac{1}{(k+2)(k+3)} = \frac{k+1}{2(k+1)+4} = \frac{k+1}{2k+6}.$$

Observe that

$$\frac{1}{2\cdot 3} + \frac{1}{3\cdot 4} + \cdots + \frac{1}{(k+2)(k+3)}$$

$$= \left[\frac{1}{2\cdot 3} + \frac{1}{3\cdot 4} + \cdots + \frac{1}{(k+1)(k+2)} \right] + \frac{1}{(k+2)(k+3)}$$

$$= \frac{k}{2k+4} + \frac{1}{(k+2)(k+3)} = \frac{k}{2(k+2)} + \frac{1}{(k+2)(k+3)}$$

$$= \frac{k(k+3)+2}{2(k+2)(k+3)} = \frac{k^2+3k+2}{2(k+2)(k+3)}$$

$$= \frac{(k+1)(k+2)}{2(k+2)(k+3)} = \frac{k+1}{2(k+3)} = \frac{k+1}{2k+6},$$

giving us the desired result. By the Principle of Mathematical Induction,

$$\frac{1}{2\cdot 3} + \frac{1}{3\cdot 4} + \cdots + \frac{1}{(n+1)(n+2)} = \frac{n}{2n+4}$$

for every positive integer n.   ∎

PROOF ANALYSIS   Each of the examples of mathematical induction proofs that we have seen involves a certain amount of algebra. We'll need to recall even more algebra soon. Many mistakes in these proofs are due to algebra errors. Therefore, care must be taken. For example, in the proof of Result 6.6, we encountered the sum

$$\frac{k}{2(k+2)} + \frac{1}{(k+2)(k+3)}.$$

To add these fractions, we needed to find a common denominator (actually a *least* common denominator), which is $2(k+2)(k+3)$. This was used to obtain the next fraction, that is,

$$\frac{k}{2(k+2)} + \frac{1}{(k+2)(k+3)} = \frac{k(k+3)}{2(k+2)(k+3)} + \frac{2}{2(k+2)(k+3)} = \frac{k(k+3)+2}{2(k+2)(k+3)}.$$

When we expanded and factored the numerator and then cancelled the term $k + 2$, this was actually expected since the final result we were looking for was

$$\frac{k+1}{2k+6} = \frac{k+1}{2(k+3)},$$

which does not contain $k + 2$ as a factor in the denominator.   ◆

6.2 A More General Principle of Mathematical Induction

The Principle of Mathematical Induction, described in the preceding section, gives us a technique for proving that a statement of the type

For every positive integer n, $P(n)$.

is true. There are situations, however, when the domain of $P(n)$ consists of those integers greater than or equal to some fixed integer m different from 1. We now describe an analogous technique to verify the truth of a statement of the following type where m denotes some fixed integer:

For every integer $n \geq m$, $P(n)$.

According to the Well-Ordering Principle, the set **N** of natural numbers is well-ordered; that is, every nonempty subset of **N** has a least element. As a consequence of the Well-Ordering Principle, other sets are also well-ordered.

Theorem 6.7   *For each integer m, the set*

$$S = \{i \in \mathbf{Z} : i \geq m\}$$

is well-ordered.

The proof of Theorem 6.7 is left as an exercise (see Exercise 6.17). The following is a consequence of Theorem 6.7. This is a slightly more general form of the Principle of Mathematical Induction. Consequently, it is commonly referred to by the same name.

Theorem 6.8   **(The Principle of Mathematical Induction)** *For a fixed integer m, let $S = \{i \in \mathbf{Z} : i \geq m\}$. For each integer $n \in S$, let $P(n)$ be a statement. If*

(1) *$P(m)$ is true and*
(2) *the implication*

If $P(k)$, then $P(k + 1)$.

is true for every integer $k \in S$,

then $P(n)$ is true for every integer $n \in S$.

The proof of Theorem 6.8 is similar to the proof of Theorem 6.2. We also state Theorem 6.8 symbolically.

The Principle of Mathematical Induction

For a fixed integer m, let $S = \{i \in \mathbf{Z} : i \geq m\}$. For each $n \in S$, let $P(n)$ be a statement. If

(1) $P(m)$ *is true and*
(2) $\forall k \in S, \; P(k) \Rightarrow P(k+1)$ *is true,*

then $\forall n \in S, \; P(n)$ is true.

This (more general) Principle of Mathematical Induction can be used to prove that certain quantified statements of the type $\forall n \in S, \; P(n)$ are true when $S = \{i \in \mathbf{Z} : i \geq m\}$ for a prescribed integer m. Of course, if $m = 1$, then $S = \mathbf{N}$. We now consider several examples.

Result 6.9   *For every nonnegative integer n,*

$$2^n > n.$$

Proof   We proceed by induction. The inequality holds for $n = 0$ since $2^0 > 0$. Assume that $2^k > k$, where k is a nonnegative integer. We show that $2^{k+1} > k + 1$. When $k = 0$, we have $2^{k+1} = 2 > 1 = k + 1$. We therefore assume that $k \geq 1$. Then

$$2^{k+1} = 2 \cdot 2^k > 2k = k + k \geq k + 1.$$

By the Principle of Mathematical Induction, $2^n > n$ for every nonnegative integer n.   ∎

PROOF ANALYSIS   Let's review the proof of Result 6.9. First, since Result 6.9 concerns nonnegative integers, we are applying Theorem 6.8 when $m = 0$. We began by observing that $2^n > n$ when $n = 0$. Next we assumed that $2^k > k$, where k is a nonnegative integer. Our goal was to show that $2^{k+1} > k + 1$. It seems logical to observe that $2^{k+1} = 2 \cdot 2^k$. Since we knew that $2^k > k$, we have $2^{k+1} = 2 \cdot 2^k > 2k$. If we could show that $2k \geq k + 1$, then we have a proof. However, when $k = 0$, the inequality $2k \geq k + 1$ doesn't hold. That's why we handled $k = 0$ separately in the proof. This allowed us to assume that $k \geq 1$ and then conclude that $2k \geq k + 1$.

We could have proved Result 6.9 a bit differently. We could have observed first that $2^n > n$ when $n = 0$ and then proved that $2^n > n$ for $n \geq 1$ by induction.   ◆

Our next example is to show that $2^n > n^2$ if n is a sufficiently large integer. We begin by trying a few values of n, as shown below. It appears that $2^n > n^2$ whenever $n \geq 5$.

n	2^n	n^2
0	1	0
1	2	1
2	4	4
3	8	9
4	16	16
5	32	25
6	64	36

Comparing 2^n and n^2

Result to Prove   For every integer $n \geq 5$,

$$2^n > n^2.$$

<u>PROOF STRATEGY</u>   Let's see what an induction proof of this result might look like. Of course, $2^n > n^2$ when $n = 5$. We assume that $2^k > k^2$ where $k \geq 5$ (and k is an integer) and we want to prove that $2^{k+1} > (k+1)^2$. We start with

$$2^{k+1} = 2 \cdot 2^k > 2k^2.$$

We would have a proof if we could show that $2k^2 \geq (k+1)^2$ or that $2k^2 \geq k^2 + 2k + 1$. There are several convincing ways to show that $2k^2 \geq k^2 + 2k + 1$ for integers $k \geq 5$. Here's one way:

Observe that $2k^2 = k^2 + k^2 = k^2 + k \cdot k \geq k^2 + 5k$ since $k \geq 5$. Also $k^2 + 5k = k^2 + 2k + 3k \geq k^2 + 2k + 3 \cdot 5 = k^2 + 2k + 15$, again since $k \geq 5$. Finally, $k^2 + 2k + 15 > k^2 + 2k + 1$. We now present a formal proof. (Here we are using the Principle of Mathematical Induction with $m = 5$.)   ◆

Result 6.10   *For every integer $n \geq 5$,*

$$2^n > n^2.$$

Proof   We proceed by induction. Since $2^5 > 5^2$, the inequality holds for $n = 5$. Assume that $2^k > k^2$ where $k \geq 5$. We show that $2^{k+1} > (k+1)^2$. Observe that

$$2^{k+1} = 2 \cdot 2^k > 2k^2 = k^2 + k^2 \geq k^2 + 5k$$
$$= k^2 + 2k + 3k \geq k^2 + 2k + 15$$
$$> k^2 + 2k + 1 = (k+1)^2.$$

Therefore, $2^{k+1} > (k+1)^2$. By the Principle of Mathematical Induction, $2^n > n^2$ for every integer $n \geq 5$.   ■

Result 6.11   *For every nonnegative integer n,*

$$3 \mid \left(2^{2n} - 1 \right).$$

Proof We proceed by induction. The result is true when $n = 0$ since in this case $2^{2n} - 1 = 0$ and $3 \mid 0$. Assume that $3 \mid (2^{2k} - 1)$, where k is a nonnegative integer. We show that $3 \mid (2^{2k+2} - 1)$. Since $3 \mid (2^{2k} - 1)$, there exists an integer x such that $2^{2k} - 1 = 3x$ and so $2^{2k} = 3x + 1$. Now

$$2^{2k+2} - 1 = 4 \cdot 2^{2k} - 1 = 4(3x + 1) - 1 = 12x + 3 = 3(4x + 1).$$

Since $4x + 1$ is an integer, $3 \mid (2^{2k+2} - 1)$.

By the Principle of Mathematical Induction, $3 \mid (2^{2n} - 1)$ for every nonnegative integer n. ∎

PROOF ANALYSIS   Let's review the preceding proof. As expected, to establish the inductive step, we assumed that $3 \mid (2^{2k} - 1)$ for an arbitrary nonnegative integer k and attempted to show that $3 \mid (2^{2k+2} - 1)$. To verify that $3 \mid (2^{2k+2} - 1)$, it was necessary to show that $2^{2k+2} - 1$ is a multiple of 3; that is, we needed to show that $2^{2k+2} - 1$ can be expressed as $3z$ for some integer z. Since our goal was to show that $2^{2k+2} - 1$ can be expressed in a certain form, it is natural to consider $2^{2k+2} - 1$ and see how we might write it. Since we knew that $2^{2k} - 1 = 3x$, where $x \in \mathbf{Z}$, it was logical to rewrite $2^{2k+2} - 1$ so that 2^{2k} appears. Actually, this is quite easy since

$$2^{2k+2} = 2^2 \cdot 2^{2k} = 4 \cdot 2^{2k}.$$

Therefore, $2^{2k+2} - 1 = 4 \cdot 2^{2k} - 1$. At this point, we need to be a bit careful because the expression we are currently considering is $4 \cdot 2^{2k} - 1$, not $4(2^{2k} - 1)$. That is, it would be incorrect to say that $4 \cdot 2^{2k} - 1 = 4(3x)$. Hence we need to substitute for 2^{2k} in this case, not for $2^{2k} - 1$. This is the reason that in the proof we rewrote $2^{2k} - 1 = 3x$ as $2^{2k} = 3x + 1$. ◆

We reinforce this kind of proof with another example.

Result 6.12   *For every nonnegative integer n,*

$$9 \mid (4^{3n} - 1).$$

Proof We proceed by induction. When $n = 0$, $4^{3n} - 1 = 0$. Since $9 \mid 0$, the statement is true when $n = 0$. Assume that $9 \mid (4^{3k} - 1)$, where k is a nonnegative integer. We now show that $9 \mid (4^{3k+3} - 1)$. Since $9 \mid (4^{3k} - 1)$, it follows that $4^{3k} - 1 = 9x$ for some integer x. Hence $4^{3k} = 9x + 1$. Now observe that

$$
\begin{aligned}
4^{3k+3} - 1 &= 4^3 \cdot 4^{3k} - 1 = 64(9x + 1) - 1 \\
&= 64 \cdot 9x + 64 - 1 = 64 \cdot 9x + 63 \\
&= 9(64x + 7).
\end{aligned}
$$

Since $64x + 7$ is an integer, $9 \mid (4^{3k+3} - 1)$.

By the Principle of Mathematical Induction, $9 \mid (4^{3n} - 1)$ for every nonnegative integer n. ∎

As a final comment regarding the preceding proof, notice that we did not multiply 64 and 9 since we were about to factor 9 from the expression in the next step in any case. For a positive integer n, the integer n **factorial**, denoted by $n!$, is defined as

$$n! = n(n-1)\cdots 3 \cdot 2 \cdot 1.$$

In particular, $1! = 1$, $2! = 2 \cdot 1 = 2$ and $3! = 3 \cdot 2 \cdot 1 = 6$. Also, $0!$ is defined as $0! = 1$. Among the many equalities and inequalities involving $n!$ is the following.

Result to Prove   For every positive integer n,

$$1 \cdot 3 \cdot 5 \cdots (2n-1) = \frac{(2n)!}{2^n \cdot n!}.$$

PROOF STRATEGY   First, observe that

$$(2n)! = 2n \cdot (2n-1) \cdot (2n-2) \cdots 3 \cdot 2 \cdot 1. \tag{6.7}$$

Of the $2n$ terms in the right side of expression (6.7), n of them are even, namely $2, 4, 6, \ldots, 2n$. If we were to factor 2 from each of these numbers, we obtain $2^n \cdot 1 \cdot 2 \cdot 3 \cdots n = 2^n \cdot n!$, while the remaining n integers, namely $1, 3, 5, \ldots, 2n-1$, are all odd. Thus

$$(2n)! = 2^n \cdot n! \cdot 1 \cdot 3 \cdot 5 \cdots (2n-1).$$

Therefore,

$$1 \cdot 3 \cdot 5 \cdots (2n-1) = \frac{(2n)!}{2^n \cdot n!}.$$

This equality can also be established by induction.   ◆

Result 6.13   *For every positive integer n,*

$$1 \cdot 3 \cdot 5 \cdots (2n-1) = \frac{(2n)!}{2^n \cdot n!}.$$

Proof   We proceed by induction. Since

$$1 = \frac{(2 \cdot 1)!}{2^1 \cdot 1!} = \frac{2}{2},$$

the statement is true for $n = 1$. Assume, for a positive integer k, that

$$1 \cdot 3 \cdot 5 \cdots (2k-1) = \frac{(2k)!}{2^k \cdot k!}.$$

We show that

$$1 \cdot 3 \cdot 5 \cdots (2k+1) = \frac{(2k+2)!}{2^{k+1} \cdot (k+1)!}.$$

Observe that

$$\frac{(2k+2)!}{2^{k+1} \cdot (k+1)!} = \frac{(2k+2)(2k+1)}{2 \cdot (k+1)} \cdot \frac{(2k)!}{2^k \cdot (k)!} = (2k+1)[1 \cdot 3 \cdot 5 \cdots (2k-1)]$$

$$= 1 \cdot 3 \cdot 5 \cdots (2k+1).$$

By the Principle of Mathematical Induction,

$$1 \cdot 3 \cdot 5 \cdots (2n - 1) = \frac{(2n)!}{2^n \cdot n!}$$

for every positive integer n.   ∎

We saw in Theorem 3.12 that for an integer x, its square x^2 is even if and only if x is even. This is actually a consequence of Theorem 3.17, which states that for integers a and b, their product ab is even if and only if a or b is even. We now present a generalization of Theorem 3.12.

Result 6.14   *Let $x \in \mathbf{Z}$. For every integer $n \geq 2$, x^n is even if and only if x is even.*

Proof   Assume, first, that x is even. Then $x = 2y$ for some integer y. Hence

$$x^n = x \cdot x^{n-1} = (2y)x^{n-1} = 2\left(yx^{n-1}\right).$$

Since yx^{n-1} is an integer, x^n is even.

We now verify the converse, namely if x^n is even, where $n \geq 2$, then x is even. We proceed by induction. If x^2 is even, then we have already seen that x is even. Hence the statement is true for $n = 2$. Assume that if x^k is even for some integer $k \geq 2$, then x is even. We show that if x^{k+1} is even, then x is even. Let x^{k+1} be an even integer. Then $x \cdot x^k$ is even. By Theorem 3.17, x is even or x^k is even. If x is even, then the result is proved. On the other hand, if x^k is even, then, by the induction hypothesis, x is even as well. By the Principle of Mathematical Induction, it follows, for every integer $n \geq 2$, that if x^n is even, then x is even.   ∎

Although it is impossible to illustrate every type of result where induction can be used, we give two examples that are considerably different than those we have seen.

One of De Morgan's laws (see Theorem 4.22) states that

$$\overline{A \cup B} = \overline{A} \cap \overline{B}$$

for every two sets A and B. It is possible to use this law to show that

$$\overline{A \cup B \cup C} = \overline{A} \cap \overline{B} \cap \overline{C}$$

for every three sets A, B and C. We show how induction can be used to prove De Morgan's law for any finite number of sets.

Theorem 6.15   *If $A_1, A_2, \ldots, A_n$ are $n \geq 2$ sets, then*

$$\overline{A_1 \cup A_2 \cup \cdots \cup A_n} = \overline{A_1} \cap \overline{A_2} \cap \cdots \cap \overline{A_n}.$$

Proof   We proceed by induction. For $n = 2$, the result *is* De Morgan's law and is therefore true. Assume that the result is true for any k sets, where $k \geq 2$; that is, assume that if $B_1, B_2, \ldots, B_k$ are any k sets, then

$$\overline{B_1 \cup B_2 \cup \cdots \cup B_k} = \overline{B_1} \cap \overline{B_2} \cap \cdots \cap \overline{B_k}.$$

We prove that the result is true for any $k + 1$ sets. Let $S_1, S_2, \ldots, S_{k+1}$ be $k + 1$ sets. We show that

$$\overline{S_1 \cup S_2 \cup \cdots \cup S_{k+1}} = \overline{S_1} \cap \overline{S_2} \cap \cdots \cap \overline{S_{k+1}}.$$

Let $T = S_1 \cup S_2 \cup \cdots \cup S_k$. Then

$$\overline{S_1 \cup S_2 \cup \cdots \cup S_{k+1}} = \overline{(S_1 \cup S_2 \cup \cdots \cup S_k) \cup S_{k+1}} = \overline{T \cup S_{k+1}}.$$

Now, by De Morgan's law,

$$\overline{T \cup S_{k+1}} = \overline{T} \cap \overline{S_{k+1}}.$$

By the definition of T and by the inductive hypothesis, we have

$$\overline{T} = \overline{S_1 \cup S_2 \cup \cdots \cup S_k} = \overline{S_1} \cap \overline{S_2} \cap \cdots \cap \overline{S_k}.$$

Therefore,

$$\overline{S_1 \cup S_2 \cup \cdots \cup S_{k+1}} = \overline{T \cup S_{k+1}} = \overline{T} \cap \overline{S_{k+1}}$$
$$= \overline{S_1} \cap \overline{S_2} \cap \cdots \cap \overline{S_k} \cap \overline{S_{k+1}}.$$

By the Principle of Mathematical Induction, for every $n \geq 2$ sets $A_1, A_2, \ldots, A_n$,

$$\overline{A_1 \cup A_2 \cup \cdots \cup A_n} = \overline{A_1} \cap \overline{A_2} \cap \cdots \cap \overline{A_n},$$

as desired.   ∎

PROOF ANALYSIS   A few comments may be useful concerning the notation used in the statement and the proof of Theorem 6.15. First, the sets $A_1, A_2, \ldots, A_n$ were used in the statement of Theorem 6.15 only as an aid to describe the result. Theorem 6.15 could have also been stated as:

> *For every integer $n \geq 2$, the complement of the union of any n sets equals the intersection of the complements of these sets.*

To verify the inductive step in the proof of Theorem 6.15, we assumed that the statement is true for any $k \geq 2$ sets, which we denoted by $B_1, B_2, \ldots, B_k$. The fact that we used $A_1, A_2, \ldots, A_n$ to describe the statement of Theorem 6.15 did not mean that we should use $A_1, A_2, \ldots, A_k$ for the k sets in the inductive hypothesis. In fact, it is probably better that we do not use this notation. In the inductive step, we now need to show that the result is true for any $k + 1$ sets. We used $S_1, S_2, \ldots, S_{k+1}$ for these sets. It would have been a bad idea to denote the $k + 1$ sets by $B_1, B_2, \ldots, B_{k+1}$ because that would have (improperly) suggested that k of the $k + 1$ sets must specifically be the sets mentioned in the inductive hypothesis.   ◆

We are now able to prove another well-known theorem concerning sets, to which we earlier referred.

Theorem 6.16   *If A is a finite set of cardinality $n \geq 0$, then the cardinality of its power set $\mathcal{P}(A)$ is 2^n.*

Proof   We proceed by induction. If A is a set with $|A| = 0$, then $A = \emptyset$. Thus $\mathcal{P}(A) = \{\emptyset\}$ and so $|\mathcal{P}(A)| = 1 = 2^0$. Therefore, the theorem is true for $n = 0$. Assume that if B is any set with $|B| = k$ for some nonnegative integer k, then $|\mathcal{P}(B)| = 2^k$. We show that if C is a set with $|C| = k + 1$, then $|\mathcal{P}(C)| = 2^{k+1}$. Let

$$C = \{c_1, c_2, \ldots, c_{k+1}\}.$$

By the inductive hypothesis, there are 2^k subsets of the set $\{c_1, c_2, \ldots, c_k\}$; that is, there are 2^k subsets of C not containing c_{k+1}. Any subset of C containing c_{k+1} can be expressed as $D \cup \{c_{k+1}\}$, where $D \subseteq \{c_1, c_2, \ldots, c_k\}$. Again, by the inductive hypothesis, there are 2^k such subsets D. Therefore, there are $2^k + 2^k = 2 \cdot 2^k = 2^{k+1}$ subsets of C.

By the Principle of Mathematical Induction, it follows for every nonnegative integer n that if $|A| = n$, then $|\mathcal{P}(A)| = 2^n$. ∎

Of course, Theorem 6.16 could also be stated as:

The number of subsets of a finite set with n elements is 2^n.

6.3 Proof by Minimum Counterexample

For each positive integer n, let $P(n)$ be a statement. We have seen that induction is a natural proof technique that can be used to verify the truth of the quantified statement

$$\forall n \in \mathbf{N}, \ P(n). \tag{6.8}$$

There are certainly such statements where induction does not work or does not work well. If we would attempt to prove (6.8) using a proof by contradiction, then we would begin such a proof by assuming that the statement $\forall n \in \mathbf{N}, P(n)$ is false. Consequently, there are positive integers n such that $P(n)$ is a false statement. By the Well-Ordering Principle, there exists a smallest positive integer n such that $P(n)$ is a false statement. Denote this integer by m. Therefore, $P(m)$ is a false statement, and for any integer k with $1 \leq k < m$, the statement $P(k)$ is true. The integer m is referred to as a **minimum counterexample** of the statement (6.8). If a proof (by contradiction) of $\forall n \in \mathbf{N}, P(n)$ can be given using the fact that m is a minimum counterexample, then such a proof is called a **proof by minimum counterexample**.

We now illustrate this proof technique. For the example we are about to describe, it is useful to recall from algebra that

$$(a + b)^3 = a^3 + 3a^2b + 3ab^2 + b^3.$$

Suppose that we wish to prove that $6 \mid (n^3 - n)$ for every positive integer n. An induction proof would probably start like this:

If $n = 1$, then $n^3 - n = 0$. Since $6 \mid 0$, the result is true for $n = 1$. Assume that $6 \mid (k^3 - k)$, where k is a positive integer. We wish to prove that $6 \mid [(k + 1)^3 - (k + 1)]$. Since $6 \mid (k^3 - k)$, it follows that $k^3 - k = 6x$ for some integer x. Then

$$\begin{aligned}
(k + 1)^3 - (k + 1) &= k^3 + 3k^2 + 3k + 1 - k - 1 \\
&= (k^3 - k) + 3k^2 + 3k \\
&= 6x + 3k(k + 1).
\end{aligned}$$

If we can show that $6 \mid 3k(k+1)$, we have a proof. Thus we need to show that $k(k+1)$ is even for every positive integer k. A lemma could be introduced to verify this. This lemma could be proved in two cases (k is even and k is odd) or induction could be used. Although such a lemma would not be difficult to prove, we give an alternative proof that avoids the need for a lemma.

Result 6.17   *For every positive integer n,*

$$6 \mid \left(n^3 - n\right).$$

Proof   Assume, to the contrary, that there are positive integers n such that $6 \nmid (n^3 - n)$. Then there is a smallest positive integer n such that $6 \nmid \left(n^3 - n\right)$. Let m be this integer. If $n = 1$, then $n^3 - n = 0$; while if $n = 2$, then $n^3 - n = 6$. Since $6 \mid 0$ and $6 \mid 6$, it follows that $6 \mid \left(n^3 - n\right)$ for $n = 1$ and $n = 2$. Therefore, $m \geq 3$. So we can write $m = k + 2$, where $1 \leq k < m$. Observe that

$$m^3 - m = (k+2)^3 - (k+2) = (k^3 + 6k^2 + 12k + 8) - (k+2)$$
$$= (k^3 - k) + (6k^2 + 12k + 6).$$

Since $k < m$, it follows that $6 \mid \left(k^3 - k\right)$. Hence $k^3 - k = 6x$ for some integer x. So we have

$$m^3 - m = 6x + 6(k^2 + 2k + 1) = 6(x + k^2 + 2k + 1).$$

Since $x + k^2 + 2k + 1$ is an integer, $6 \mid \left(m^3 - m\right)$, which produces a contradiction.   ∎

PROOF ANALYSIS   Let's see how this proof was constructed. In this proof, m is a positive integer such that $6 \nmid \left(m^3 - m\right)$; while for every positive integer n with $n < m$, we have $6 \mid \left(n^3 - n\right)$. We are trying to determine just how large m needs to be to obtain a contradiction. We saw that $6 \mid (1^3 - 1)$ and $6 \mid (2^3 - 2)$; so $m \geq 3$. Knowing that $m \geq 3$ allowed us to write m as $k + 2$, where $1 \leq k < m$. Because $1 \leq k < m$, we know that $6 \mid \left(k^3 - k\right)$ and so $k^3 - k = 6x$, where $x \in \mathbf{Z}$. So, in the proof, we wrote

$$m^3 - m = (k+2)^3 - (k+2) = (k^3 + 6k^2 + 12k + 8) - (k+2)$$
$$= (k^3 - k) + (6k^2 + 12k + 6) = 6x + 6k^2 + 12k + 6.$$

The fact that we can factor 6 from $6x + 6k^2 + 12k + 6$ is what allowed us to conclude that $6 \mid \left(m^3 - m\right)$ and obtain a contradiction. But how did we know that we wanted $m \geq 3$? If we had only observed that $6 \mid (1^3 - 1)$ and not that $6 \mid (2^3 - 2)$, then we would have known only that $m \geq 2$, which would have allowed us to write $m = k + 1$, where $1 \leq k < m$. Of course, we would still know that $6 \mid \left(k^3 - k\right)$ and so $k^3 - k = 6x$, where $x \in \mathbf{Z}$. However, when we consider $m^3 - m$, we would have

$$m^3 - m = (k+1)^3 - (k+1) = (k^3 + 3k^2 + 3k + 1) - (k+1)$$
$$= (k^3 - k) + 3k^2 + 3k = (k^3 - k) + 3k(k+1)$$
$$= 6x + 3k(k+1).$$

As it stands, we can factor 3 from $6x + 3k(k + 1)$ but cannot factor 6 unless we can prove that $k(k + 1)$ is even. This is the same difficulty we encountered when we were considering an induction proof. In any case, no contradiction is obtained.   ♦

If a result can be proved by induction, then it can also be proved by minimum counterexample. It is not difficult to use induction to prove that $3 \mid (2^{2n} - 1)$ for every nonnegative integer n. We also give a proof by minimum counterexample of this statement.

Result 6.18   *For every nonnegative integer n,*

$$3 \mid (2^{2n} - 1).$$

Proof   Assume, to the contrary, that there are nonnegative integers n for which $3 \nmid (2^{2n} - 1)$. By Theorem 6.7, there is a smallest nonnegative integer n such that $3 \nmid (2^{2n} - 1)$. Denote this integer by m. Thus $3 \nmid (2^{2m} - 1)$ and $3 \mid (2^{2n} - 1)$ for all nonnegative integers n for which $0 \le n < m$. Since $3 \mid (2^{2n} - 1)$ when $n = 0$, it follows that $m \ge 1$. Hence m can be expressed by $m = k + 1$, where $0 \le k < m$. Thus $3 \mid (2^{2k} - 1)$, which implies that $2^{2k} - 1 = 3x$ for some integer x. Consequently, $2^{2k} = 3x + 1$. Observe that

$$2^{2m} - 1 = 2^{2(k+1)} - 1 = 2^{2k+2} - 1 = 2^2 \cdot 2^{2k} - 1$$
$$= 4(3x + 1) - 1 = 12x + 3 = 3(4x + 1).$$

Since $4x + 1$ is an integer, $3 \mid (2^{2m} - 1)$, which produces a contradiction.   ■

We give one additional example using a proof by minimum counterexample.

Result 6.19   *For every positive integer n,*

$$1 + 2 + 3 + \cdots + n = \frac{n(n + 1)}{2}.$$

Proof   Assume, to the contrary, that

$$1 + 2 + 3 + \cdots + n \ne \frac{n(n + 1)}{2}$$

for some positive integers n. By the Well-Ordering Principle, there is a smallest positive integer n such that

$$1 + 2 + 3 + \cdots + n \ne \frac{n(n + 1)}{2}.$$

Denote this integer by m. Therefore,

$$1 + 2 + 3 + \cdots + m \ne \frac{m(m + 1)}{2},$$

while

$$1 + 2 + 3 + \cdots + n = \frac{n(n + 1)}{2}$$

for every integer n with $1 \leq n < m$. Since $1 = 1(1+1)/2$, it follows that $m \geq 2$. Hence we can write $m = k+1$, where $1 \leq k < m$. Consequently,

$$1 + 2 + 3 + \cdots + k = \frac{k(k+1)}{2}.$$

Observe that

$$
\begin{aligned}
1 + 2 + 3 + \cdots + m &= 1 + 2 + 3 + \cdots + (k+1) = (1 + 2 + 3 + \cdots + k) + (k+1) \\
&= \frac{k(k+1)}{2} + (k+1) = \frac{k(k+1) + 2(k+1)}{2} \\
&= \frac{(k+1)(k+2)}{2} = \frac{m(m+1)}{2},
\end{aligned}
$$

which produces a contradiction. ∎

6.4 The Strong Principle of Mathematical Induction

We close with one last form of mathematical induction. This principle goes by many names: the Strong Principle of Mathematical Induction, the Strong Form of Induction, the Alternate Form of Mathematical Induction and the Second Principle of Mathematical Induction are common names.

Theorem 6.20   (**The Strong Principle of Mathematical Induction**) *For each positive integer n, let $P(n)$ be a statement. If*

(a) *$P(1)$ is true and*
(b) *the implication*

> *If $P(i)$ for every integer i with $1 \leq i \leq k$, then $P(k+1)$.*

> *is true for every positive integer k,*

then $P(n)$ is true for every positive integer n.

As with the Principle of Mathematical Induction (Theorem 6.2), the Strong Principle of Mathematical Induction is also a consequence of the Well-Ordering Principle. The Strong Principle of Mathematical Induction is now stated more symbolically below.

The Strong Principle of Mathematical Induction   *For each positive integer n, let $P(n)$ be a statement. If*

(1) *$P(1)$ is true and*
(2) *$\forall k \in \mathbf{N},\ P(1) \wedge P(2) \wedge \cdots \wedge P(k) \Rightarrow P(k+1)$ is true,*

then $\forall n \in \mathbf{N},\ P(n)$ is true.

The difference in the statements of the Principle of Mathematical Induction and the Strong Principle of Mathematical Induction lies in the inductive step (condition 2).

To prove that $\forall n \in \mathbf{N}$, $P(n)$ is true by the Principle of Mathematical Induction, we are required to show that $P(1)$ is true and to verify the implication:

$$\text{If } P(k), \text{ then } P(k+1). \tag{6.9}$$

is true for every positive integer k. On the other hand, to prove $\forall n \in \mathbf{N}$, $P(n)$ is true by the Strong Principle of Mathematical Induction, we are required to show that $P(1)$ is true and to verify the implication:

$$\text{If } P(i) \text{ for every } i \text{ with } 1 \le i \le k, \text{ then } P(k+1). \tag{6.10}$$

is true for every positive integer k. If we were to give direct proofs of the implications (6.9) and (6.10), then we are permitted to assume more in the inductive step (6.10) of the Strong Principle of Mathematical Induction than in the induction step (6.9) of the Principle of Mathematical Induction and yet obtain the same conclusion. If the assumption that $P(k)$ is true is insufficient to verify the truth of $P(k+1)$ for an arbitrary positive integer k, but the assumption that all of the statements $P(1), P(2), \ldots, P(k)$ are true is sufficient to verify the truth of $P(k+1)$, then this suggests that we should use the Strong Principle of Mathematical Induction. Indeed, any result that can be proved by the Principle of Mathematical Induction can also be proved by the Strong Principle of Mathematical Induction.

Just as there is a more general version of the Principle of Mathematical Induction (namely, Theorem 6.8), there is a more general version of the Strong Principle of Mathematical Induction. We shall also refer to this as the Strong Principle of Mathematical Induction.

Theorem 6.21   **(The Strong Principle of Mathematical Induction)** *For a fixed integer m, let $S = \{i \in \mathbf{Z} : i \ge m\}$. For each $n \in S$, let $P(n)$ be a statement. If*

(1) *$P(m)$ is true and*
(2) *the implication*

$$\text{If } P(i) \text{ for every integer } i \text{ with } m \le i \le k, \text{ then } P(k+1).$$

is true for every integer $k \in S$,

then $P(n)$ is true for every integer $n \in S$.

We now consider a class of mathematical statements where the Strong Principle of Mathematical Induction is commonly the appropriate proof technique.

Suppose that we are considering a sequence $a_1, a_2, a_3, \ldots$ of numbers. One way of defining a sequence $\{a_n\}$ is to specify explicitly the nth term a_n (as a function of n). For example, we might have $a_n = \dfrac{1}{n}$, $a_n = \dfrac{(-1)^n}{n^2}$ or $a_n = n^3 + n$ for each $n \in \mathbf{N}$. A sequence can also be **defined recursively**. In a **recursively defined sequence** $\{a_n\}$, only the first term or perhaps the first few terms are defined specifically, say $a_1, a_2, \ldots, a_k$ for some fixed $k \in \mathbf{N}$. These are called the **initial values**. Then a_{k+1} is expressed in terms of $a_1, a_2, \ldots, a_k$; and more generally, for $n > k$, a_n is expressed in terms of $a_1, a_2, \ldots, a_{n-1}$. This is called the **recurrence relation**.

A specific example of this is the sequence $\{a_n\}$ defined by $a_1 = 1, a_2 = 3$ and $a_n = 2a_{n-1} - a_{n-2}$ for $n \geq 3$. In this case, there are two initial values, namely $a_1 = 1$ and $a_2 = 3$. The recurrence relation here is

$$a_n = 2a_{n-1} - a_{n-2} \text{ for } n \geq 3.$$

Letting $n = 3$, we find that $a_3 = 2a_2 - a_1 = 5$; while letting $n = 4$, we have $a_4 = 2a_3 - a_2 = 7$. Similarly, $a_5 = 9$ and $a_6 = 11$. From this information, one might well conjecture (guess) that $a_n = 2n - 1$ for every $n \in \mathbf{N}$. (Conjectures will be discussed in more detail in Section 7.1.) Using the Strong Principle of Mathematical Induction, we can, in fact, prove that this conjecture is true.

Result 6.22   *A sequence $\{a_n\}$ is defined recursively by*

$$a_1 = 1, a_2 = 3 \text{ and } a_n = 2a_{n-1} - a_{n-2} \text{ for } n \geq 3.$$

Then $a_n = 2n - 1$ for all $n \in \mathbf{N}$.

Proof   We proceed by induction. Since $a_1 = 2 \cdot 1 - 1 = 1$, the formula holds for $n = 1$. Assume for an arbitrary positive integer k that $a_i = 2i - 1$ for all integers i with $1 \leq i \leq k$. We show that $a_{k+1} = 2(k + 1) - 1 = 2k + 1$. If $k = 1$, then $a_{k+1} = a_2 = 2 \cdot 1 + 1 = 3$. Since $a_2 = 3$, it follows that $a_{k+1} = 2k + 1$ when $k = 1$. Hence we may assume that $k \geq 2$. Since $k + 1 \geq 3$, it follows that

$$a_{k+1} = 2a_k - a_{k-1} = 2(2k - 1) - (2k - 3) = 2k + 1,$$

which is the desired result. By the Strong Principle of Mathematical Induction, $a_n = 2n - 1$ for all $n \in \mathbf{N}$. ∎

<u>**PROOF ANALYSIS**</u>   A few comments about the proof of Result 6.22 are in order. At one point, we assumed for an arbitrary positive integer k that $a_i = 2i - 1$ for all integers i with $1 \leq i \leq k$. Our goal was to show that $a_{k+1} = 2k + 1$. Since k is a positive integer, it may occur that $k = 1$ or $k \geq 2$. If $k = 1$, then we need to show that $a_{k+1} = a_2 = 2 \cdot 1 + 1 = 3$. That $a_2 = 3$ is known because this is one of the initial values. If $k \geq 2$, then $k + 1 \geq 3$ and a_{k+1} can be expressed as $2a_k - a_{k-1}$ by the recurrence relation. In order to show that $a_{k+1} = 2k + 1$ when $k \geq 2$, it was necessary to know that $a_k = 2k - 1$ and that $a_{k-1} = 2(k - 1) - 1 = 2k - 3$. Because we were using the Strong Principle of Mathematical Induction, we knew both pieces of information. If we had used the Principle of Mathematical Induction, we would have assumed (and therefore knew) that $a_k = 2k - 1$ but we would not have known that $a_{k-1} = 2k - 3$, and so we would have been unable to establish the desired expression for a_{k+1}. ◆

Problem 6.23   *A sequence $\{a_n\}$ is defined recursively by*

$$a_1 = 1, a_2 = 4 \text{ and } a_n = 2a_{n-1} - a_{n-2} + 2 \text{ for } n \geq 3.$$

Conjecture a formula for a_n and verify that your conjecture is correct.

Solution   We begin by finding a few more terms of the sequence. Observe that $a_3 = 2a_2 - a_1 + 2 = 9$, while $a_4 = 2a_3 - a_2 + 2 = 16$ and $a_5 = 2a_4 - a_3 + 2 = 25$. The obvious conjecture is that $a_n = n^2$ for every positive integer n. We verify that this conjecture is correct in the next result.   ◆

Result 6.24   *A sequence $\{a_n\}$ is defined recursively by*

$$a_1 = 1, a_2 = 4 \text{ and } a_n = 2a_{n-1} - a_{n-2} + 2 \text{ for } n \geq 3.$$

Then $a_n = n^2$ for all $n \in \mathbf{N}$.

Proof   We proceed by induction. Since $a_1 = 1 = 1^2$, the formula holds for $n = 1$. Assume for an arbitrary positive integer k that $a_i = i^2$ for every integer i with $1 \leq i \leq k$. We show that $a_{k+1} = (k+1)^2$. Since $a_2 = 4$, it follows that $a_{k+1} = (k+1)^2$ when $k = 1$. Thus we may assume that $k \geq 2$. Hence $k + 1 \geq 3$ and so

$$a_{k+1} = 2a_k - a_{k-1} + 2 = 2k^2 - (k-1)^2 + 2$$
$$= 2k^2 - (k^2 - 2k + 1) + 2 = k^2 + 2k + 1 = (k+1)^2.$$

By the Strong Principle of Mathematical Induction, $a_n = n^2$ for all $n \in \mathbf{N}$.   ∎

Although we mentioned that problems involving recurrence relations are commonly solved with the aid of the Strong Principle of Mathematical Induction, it is by no means the only kind of problem where the Strong Principle of Mathematical Induction can be applied. Although the best examples of this require a knowledge of mathematics beyond what we have covered thus far, we do present another type of example.

Result 6.25   *For each integer $n \geq 8$, there are nonnegative integers a and b such that $n = 3a + 5b$.*

Proof   We proceed by induction. Since $8 = 3 \cdot 1 + 5 \cdot 1$, the statement is true for $n = 8$. Assume for each integer i with $8 \leq i \leq k$, where $k \geq 8$ is an arbitrary integer, that there are nonnegative integers s and t such that $i = 3s + 5t$. Consider the integer $k + 1$. We show that there are nonnegative integers x and y such that $k + 1 = 3x + 5y$. Since $9 = 3 \cdot 3 + 5 \cdot 0$ and $10 = 3 \cdot 0 + 5 \cdot 2$, this is true if $k + 1 = 9$ and $k + 1 = 10$. Hence we may assume that $k + 1 \geq 11$. Thus $8 \leq (k+1) - 3 < k$. By the induction hypothesis, there are nonnegative integers a and b such that

$$(k+1) - 3 = 3a + 5b \text{ and so } k + 1 = 3(a+1) + 5b.$$

Letting $x = a + 1$ and $y = b$, we have the desired conclusion.

By the Strong Principle of Mathematical Induction, for every integer $n \geq 8$, there are nonnegative integers a and b such that $n = 3a + 5b$.   ∎

EXERCISES FOR CHAPTER 6

Section 6.1: The Principle of Mathematical Induction

6.1. Which of the following sets are well-ordered?

    (a) $S = \{x \in \mathbf{Q} : x \geq -10\}$
    (b) $S = \{-2, -1, 0, 1, 2\}$
    (c) $S = \{x \in \mathbf{Q} : -1 \leq x \leq 1\}$
    (d) $S = \{p : p \text{ is a prime}\} = \{2, 3, 5, 7, 11, 13, 17, \ldots\}$.

6.2. Prove that if A is any well-ordered set of real numbers and B is a nonempty subset of A, then B is also well-ordered.

6.3. Prove that every nonempty set of negative integers has a largest element.

6.4. Prove that $1 + 3 + 5 + \cdots + (2n - 1) = n^2$ for every positive integer n

    (1) by mathematical induction and
    (2) by adding $1 + 3 + 5 + \cdots + (2n - 1)$ and $(2n - 1) + (2n - 3) + \cdots + 1$.

6.5. Use mathematical induction to prove that

$$1 + 5 + 9 + \cdots + (4n - 3) = 2n^2 - n$$

for every positive integer n.

6.6. (a) We have seen that $1^2 + 2^2 + \cdots + n^2$ is the number of squares in an $n \times n$ square composed of n^2 1×1 squares. What does $1^3 + 2^3 + 3^3 + \cdots + n^3$ represent geometrically?

    (b) Use mathematical induction to prove that $1^3 + 2^3 + 3^3 + \cdots + n^3 = \dfrac{n^2(n + 1)^2}{4}$ for every positive integer n.

6.7. Find another formula suggested by Exercises 6.4 and 6.5, and verify your formula by mathematical induction.

6.8. Find a formula for $1 + 4 + 7 + \cdots + (3n - 2)$ for positive integers n, and then verify your formula by mathematical induction.

6.9. Prove that $1 \cdot 3 + 2 \cdot 4 + 3 \cdot 5 + \cdots + n(n + 2) = \frac{n(n+1)(2n+7)}{6}$ for every positive integer n.

6.10. Let $r \neq 1$ be a real number. Use induction to prove that $a + ar + ar^2 + \cdots + ar^{n-1} = \frac{a(1-r^n)}{1-r}$ for every positive integer n.

6.11. Prove that $\frac{1}{3 \cdot 4} + \frac{1}{4 \cdot 5} + \cdots + \frac{1}{(n+2)(n+3)} = \frac{n}{3n+9}$ for every positive integer n.

6.12. Consider the open sentence $P(n): 9 + 13 + \cdots + (4n + 5) = \frac{4n^2 + 14n + 1}{2}$, where $n \in \mathbf{N}$.

    (a) Verify the implication $P(k) \Rightarrow P(k + 1)$ for an arbitrary positive integer k.
    (b) Is $\forall n \in \mathbf{N}, P(n)$ true?

6.13. Prove that $1 \cdot 1! + 2 \cdot 2! + \cdots + n \cdot n! = (n + 1)! - 1$ for every positive integer n.

6.14. Prove that $2! \cdot 4! \cdot 6! \cdot \cdots \cdot (2n)! \geq [(n + 1)!]^n$ for every positive integer n.

6.15. Prove that $\frac{1}{\sqrt{1}} + \frac{1}{\sqrt{2}} + \frac{1}{\sqrt{3}} + \cdots + \frac{1}{\sqrt{n}} \leq 2\sqrt{n} - 1$ for every positive integer n.

6.16. Prove that $7 \mid [3^{4n+1} - 5^{2n-1}]$ for every positive integer n.

Section 6.2: A More General Principle of Mathematical Induction

6.17. Prove Theorem 6.7: For each integer m, the set $S = \{i \in \mathbf{Z} : i \geq m\}$ is well-ordered.
[Hint: For every subset T of S, either $T \subseteq \mathbf{N}$ or $T - \mathbf{N}$ is a finite nonempty set.]

6.18. Prove that $2^n > n^3$ for every integer $n \geq 10$.

6.19. Prove the following implication for every integer $n \geq 2$: If $x_1, x_2, \ldots, x_n$ are any n real numbers such that $x_1 \cdot x_2 \cdots \cdot x_n = 0$, then at least one of the numbers $x_1, x_2, \ldots, x_n$ is 0. (Use the fact that if the product of two real numbers is 0, then at least one of the numbers is 0.)

6.20. (a) Use mathematical induction to prove that every finite nonempty set of real numbers has a largest element.
    (b) Use (a) to prove that every finite nonempty set of real numbers has a smallest element.

6.21. Prove that $4 \mid (5^n - 1)$ for every nonnegative integer n.

6.22. Prove that $3^n > n^2$ for every positive integer n.

6.23. Prove that $7 \mid (3^{2n} - 2^n)$ for every nonnegative integer n.

6.24. Prove Bernoulli's Identity: For every real number $x > -1$ and every positive integer n,

$$(1 + x)^n \geq 1 + nx.$$

6.25. Prove that $n! > 2^n$ for every integer $n \geq 4$.

6.26. Prove that $81 \mid (10^{n+1} - 9n - 10)$ for every nonnegative integer n.

6.27. Prove that $1 + \frac{1}{4} + \frac{1}{9} + \cdots + \frac{1}{n^2} \leq 2 - \frac{1}{n}$ for every positive integer n.

6.28. In Exercise 6 of Chapter 4, you were asked to prove that if $3 \mid 2a$, where $a \in \mathbf{Z}$, then $3 \mid a$. Assume that this result is true. Prove the following generalization: Let $a \in \mathbf{Z}$. For every positive integer n, if $3 \mid 2^n a$, then $3 \mid a$.

6.29. Prove that if $A_1, A_2, \ldots, A_n$ are any $n \geq 2$ sets, then

$$\overline{A_1 \cap A_2 \cap \cdots \cap A_n} = \overline{A_1} \cup \overline{A_2} \cup \cdots \cup \overline{A_n}.$$

6.30. Recall for integers $n \geq 2$, a, b, c, d, that if $a \equiv b \pmod{n}$ and $c \equiv d \pmod{n}$, then both $a + c \equiv b + d \pmod{n}$ and $ac \equiv bd \pmod{n}$. Use these results and mathematical induction to prove the following: For any $2m$ integers $a_1, a_2, \ldots, a_m$ and $b_1, b_2, \ldots, b_m$ for which $a_i \equiv b_i \pmod{n}$ for $1 \leq i \leq m$,
    (a) $a_1 + a_2 + \cdots + a_m \equiv b_1 + b_2 + \cdots + b_m \pmod{n}$ and
    (b) $a_1 a_2 \cdots a_m \equiv b_1 b_2 \cdots b_m \pmod{n}$.

6.31. Prove for every $n \geq 1$ positive real numbers $a_1, a_2, \ldots, a_n$ that

$$\left(\sum_{i=1}^{n} a_i \right) \left(\sum_{i=1}^{n} \frac{1}{a_i} \right) \geq n^2.$$

6.32. Prove for every $n \geq 2$ positive real numbers $a_1, a_2, \ldots, a_n$ that

$$(n - 1) \sum_{i=1}^{n} a_i^2 \geq 2 \sum_{1 \leq i < j \leq n} a_i a_j. \tag{6.11}$$

[Note: When $n = 4$, for example, (6.11) states that
$3(a_1^2 + a_2^2 + a_3^2 + a_4^2) \geq 2(a_1 a_2 + a_1 a_3 + a_1 a_4 + a_2 a_3 + a_2 a_4 + a_3 a_4).$]

Section 6.3: Proof by Minimum Counterexample

6.33. Use proof by minimum counterexample to prove that $6 \mid 7n \left(n^2 - 1\right)$ for every positive integer n.

6.34. Use the method of minimum counterexample to prove that $3 \mid \left(2^{2n} - 1\right)$ for every positive integer n.

6.35. Give a proof by minimum counterexample that $1 + 3 + 5 + \cdots + (2n - 1) = n^2$ for every positive integer n.

6.36. Prove that $5 \mid (n^5 - n)$ for every integer n.

6.37. Use proof by minimum counterexample to prove that $3 \mid (2^n + 2^{n+1})$ for every nonnegative integer n.

6.38. Give a proof by minimum counterexample that $2^n > n^2$ for every integer $n \geq 5$.

6.39. Prove that $12 \mid (n^4 - n^2)$ for every positive integer n.

6.40. Let $S = \{2^r : r \in \mathbf{Z}, r \geq 0\}$. Use proof by minimum counterexample to prove that for every $n \in \mathbf{N}$, there exists a subset S_n of S such that $\sum_{i \in S_n} i = n$.

Section 6.4: The Strong Principle of Mathematical Induction

6.41. A sequence $\{a_n\}$ is defined recursively by $a_1 = 1$ and $a_n = 2a_{n-1}$ for $n \geq 2$. Conjecture a formula for a_n and verify that your conjecture is correct.

6.42. A sequence $\{a_n\}$ is defined recursively by $a_1 = 1$, $a_2 = 2$ and $a_n = a_{n-1} + 2a_{n-2}$ for $n \geq 3$. Conjecture a formula for a_n and verify that your conjecture is correct.

6.43. A sequence $\{a_n\}$ is defined recursively by $a_1 = 1$, $a_2 = 4$, $a_3 = 9$ and

$$a_n = a_{n-1} - a_{n-2} + a_{n-3} + 2(2n - 3)$$

for $n \geq 4$. Conjecture a formula for a_n and prove that your conjecture is correct.

6.44. Consider the sequence $F_1, F_2, F_3, \ldots$, where

$$F_1 = 1, \ F_2 = 1, \ F_3 = 2, \ F_4 = 3, \ F_5 = 5 \text{ and } F_6 = 8.$$

The terms of this sequence are called **Fibonacci numbers**.

(a) Define the sequence of Fibonacci numbers by means of a recurrence relation.
(b) Prove that $2 \mid F_n$ if and only if $3 \mid n$.

6.45. Use the Strong Principle of Mathematical Induction to prove that for each integer $n \geq 12$, there are nonnegative integers a and b such that $n = 3a + 7b$.

6.46. Use the Strong Principle of Mathematical Induction to prove the following. Let $S = \{i \in \mathbf{Z} : i \geq 2\}$ and let P be a subset of S with the properties that $2, 3 \in P$ and if $n \in S$, then either $n \in P$ or $n = ab$, where $a, b \in S$. Then every element of S either belongs to P or can be expressed as a product of elements of P. [Note: You might recognize the set P of primes. This is an important theorem in mathematics, which appears as Theorem 11.17 in Chapter 11.]

6.47. Prove that there exists an odd integer m such that every odd integer n with $n \geq m$ can be expressed either as $3a + 11b$ or as $5c + 7d$ for nonnegative integers a, b, c and d.

ADDITIONAL EXERCISES FOR CHAPTER 6

6.48. Prove that $1 \cdot 2 + 2 \cdot 3 + 3 \cdot 4 + \cdots + n(n + 1) = \frac{n(n+1)(n+2)}{3}$ for every positive integer n.

6.49. Prove that $4^n > n^3$ for every positive integer n.

6.50. Prove that $24 \mid (5^{2n} - 1)$ for every positive integer n.

6.51. By Result 6.5,

$$1^2 + 2^2 + 3^2 + \cdots + n^2 = \frac{n(n + 1)(2n + 1)}{6} \tag{6.12}$$

for every positive integer n.

(a) Use (6.12) to determine a formula for $2^2 + 4^2 + 6^2 + \cdots + (2n)^2$ for every positive integer n.
(b) Use (6.12) and (a) to determine a formula for $1^2 + 3^2 + 5^2 + \cdots + (2n-1)^2$ for every positive integer n.
(c) Use (a) and (b) to determine a formula for

$$1^2 - 2^2 + 3^2 - 4^2 + \cdots + (-1)^{n+1} n^2$$

for every positive integer n.
(d) Use mathematical induction to verify the formulas in (b) and (c).

6.52. Use the Strong Principle of Mathematical Induction to prove that for each integer $n \geq 28$, there are nonnegative integers x and y such that $n = 5x + 8y$.

6.53. Find a positive integer m such that for each integer $n \geq m$, there are positive integers x and y such that $n = 3x + 5y$. Use the Principle of Mathematical Induction to prove this.

6.54. Find a positive integer m such that for each integer $n \geq m$, there are integers $x, y \geq 2$ such that $n = 2x + 3y$. Use the Principle of Mathematical Induction to prove this.

6.55. A sequence $\{a_n\}$ of real numbers is defined recursively by $a_1 = 1$, $a_2 = 2$ and $a_n = \sum_{i=1}^{n-1}(i-1)a_i$ for $n \geq 3$. Prove that $a_n = (n-1)!$ for every integer $n \geq 3$.

6.56. Consider the sequence $a_1 = 2$, $a_2 = 5$, $a_3 = 9$, $a_4 = 14$, etc.

(a) Find a recurrence relation that expresses a_n in terms of a_{n-1} for every integer $n \geq 2$.
(b) Conjecture an explicit formula for a_n, and then prove that your conjecture is correct.

6.57. The following theorem allows one to prove certain quantified statements over some finite sets.
The Principle of Finite Induction   *For a fixed positive integer m, let $S = \{1, 2, \ldots, m\}$. For each $n \in S$, let $P(n)$ be a statement. If*

(a) *$P(1)$ is true and*
(b) *the implication*

$$\text{If } P(k), \text{ then } P(k+1).$$

   is true for every integer k with $1 \leq k < m$,

then $P(n)$ is true for every integer $n \in S$.
Use the Principle of Finite Induction to prove the following result.
Let $S = \{1, 2, \ldots, 24\}$. For every integer t with $1 \leq t \leq 300$, there exists a subset $S_t \subseteq S$ such that $\sum_{i \in S_t} i = t$.

6.58. Evaluate the proposed proof of the following result.
Result   For every positive integer n,

$$1 + 3 + 5 + \cdots + (2n-1) = n^2.$$

Proof   We proceed by induction. Since $2 \cdot 1 - 1 = 1^2$, the formula holds for $n = 1$. Assume that $1 + 3 + 5 + \cdots + (2k-1) = k^2$ for a positive integer k. We prove that $1 + 3 + 5 + \cdots + (2k+1) = (k+1)^2$. Observe that

$$1 + 3 + 5 + \cdots + (2k+1) = (k+1)^2$$
$$1 + 3 + 5 + \cdots + (2k-1) + (2k+1) = (k+1)^2$$
$$k^2 + (2k+1) = (k+1)^2$$
$$(k+1)^2 = (k+1)^2.$$

    ∎

6.59. Below is given a proof of a result. Which result is being proved and which proof technique is being used?

Proof   Assume, to the contrary, that there is some positive integer n such that $8 \nmid (3^{2n} - 1)$. Let m be the smallest positive integer such that $8 \nmid (3^{2m} - 1)$. For $n = 1$, $3^{2n} - 1 = 8$. Since $8 \mid 8$, it follows that $m \geq 2$. Let $m = k + 1$. Since $1 \leq k < m$, it follows that $8 \mid (3^{2k} - 1)$. Therefore, $3^{2k} - 1 = 8x$ for some integer x and so $3^{2k} = 8x + 1$. Hence

$$3^{2m} - 1 = 3^{2(k+1)} - 1 = 3^{2k+2} - 1 = 9 \cdot 3^{2k} - 1$$
$$= 9(8x + 1) - 1 = 72x + 8 = 8(9x + 1).$$

Since $9x + 1$ is an integer, $8 \mid (3^{2m} - 1)$, which produces a contradiction. ∎

6.60. Below is given a proof of a result. Which result is being proved and which proof technique is being used?

Proof   First observe that $a_1 = 8 = 3 \cdot 1 + 5$ and $a_2 = 11 = 3 \cdot 2 + 5$. Thus $a_n = 3n + 5$ for $n = 1$ and $n = 2$. Assume that $a_i = 3i + 5$ for all integers i with $1 \leq i \leq k$, where $k \geq 2$. Since $k + 1 \geq 3$, it follows that

$$a_{k+1} = 5a_k - 4a_{k-1} - 9 = 5(3k + 5) - 4(3k + 2) - 9$$
$$= 15k + 25 - 12k - 8 - 9 = 3k + 8 = 3(k + 1) + 5.$$ ∎

6.61. By an n-gon, we mean an n-sided polygon. So a 3-gon is a triangle and a 4-gon is a quadrilateral. It is well known that the sum of the interior angles of a triangle is $180°$. Use induction to prove that for every integer $n \geq 3$, the sum of the interior angles of an n-gon is $(n - 2) \cdot 180°$.

6.62. Suppose that $\{a_n\}$ is a sequence of real numbers defined recursively by $a_1 = 1$, $a_2 = 2$, $a_3 = 3$ and $a_n = 2a_{n-1} - a_{n-3}$ for $n \geq 4$. Prove that $a_n = a_{n-1} + a_{n-2}$ for every integer $n \geq 3$.

6.63. Suppose that $\{a_n\}$ is a sequence of real numbers defined recursively by $a_1 = 1$, $a_2 = 2$ and $a_n = a_{n-1}/a_{n-2}$ for $n \geq 3$.

(a) Prove that

$$a_n = \begin{cases} 1 & \text{if } n \equiv 1, 4 \ (\text{mod } 6) \\ 2 & \text{if } n \equiv 2, 3 \ (\text{mod } 6) \\ 1/2 & \text{if } n \equiv 0, 5 \ (\text{mod } 6) \end{cases}$$

for every positive integer n.

(b) Prove, for each nonnegative integer j, that $\sum_{i=1}^{6} a_{j+i} = 7$.

6.64. Let $x \in \mathbf{R}$ where $x \geq 3$. Prove that $(1 + x)^n \geq \left[\frac{n(n-1)(n-2)}{6} \right] x^3$ for every integer $n \geq 3$.

6.65. Prove that $\sum_{j=1}^{n} \left(\sum_{i=1}^{j} i \right) = \frac{n(n+1)(n+2)}{6}$ for every positive integer n.

6.66. Prove that $\sum_{j=1}^{n} \left(\sum_{i=1}^{j} (2i - 1) \right) = \frac{n(n+1)(2n+1)}{6}$ for every positive integer n.

6.67. Prove that there exists an odd integer m such that every odd integer n with $n \geq m$ can be expressed as $3a + 5b + 7c$ for positive integers a, b and c.

7

Prove or Disprove

In every mathematical statement that you have seen so far, you have been informed of its truth value. If the statement was true, then we have either provided a proof or asked you to provide one of your own. What you didn't know (perhaps) was how we or you were to verify its truth. If the statement was false, then here too we either verified this or asked you to verify that it was false. As you proceed further into the world of mathematics, you will more and more often encounter statements whose truth is in question. Consequently, each such statement presents two problems for you: (1) Determine the truth or falseness of the statement. (2) Verify the correctness of your belief.

7.1 Conjectures in Mathematics

In mathematics, when we don't know whether a certain statement is true but there is good reason to believe that it is, then we refer to the statement as a **conjecture**. So the word *conjecture* is used in mathematics as a sophisticated synonym for an intelligent guess (or perhaps just a guess). Once a conjecture is proved, then the conjecture becomes a theorem. If, on the other hand, the conjecture is shown to be false, then we made an incorrect guess. This is the way mathematics develops—by guessing and showing that our guess is correct or wrong and then possibly making a new conjecture and then repeating the process (possibly often). As we learn what's true and what's false about the mathematics we're studying, this influences the questions we ask and the conjectures we make.

Let's consider an example of a conjecture (although there is always the possibility that someone has settled the conjecture between the time it was written here and the moment you read it). A word is called a **palindrome** if it reads the same forward and backward (such as *deed, noon* and *radar*). Indeed, a sentence is a palindrome if it reads the same forward and backward, ignoring spaces (*Name no one man*). A positive integer is called a **palindrome** if it is the same number when its digits are reversed. (It is *considerably easier* to give an example of a number that is a palindrome than a word that is a palindrome.) For example, 1221 and 47374 are palindromes. Consider the integer 27.

It is not a palindrome. Reverse its digits and we obtain 72. Needless to say, 72 is not a palindrome either. Adding 27 and 72, we have:

$$
\begin{array}{r}
27 \\
+\ 72 \\
\hline
99
\end{array}
$$

A palindrome results. Consider another positive integer, say 59. It is not a palindrome. Reverse its digits and add:

$$
\begin{array}{r}
59 \\
+\ 95 \\
\hline
154
\end{array}
$$

The result is not a palindrome either. Reverse its digits and add:

$$
\begin{array}{r}
154 \\
+\ 451 \\
\hline
605
\end{array}
$$

Once again we arrive at a number that is not a palindrome. But reverse its digits and add:

$$
\begin{array}{r}
605 \\
+\ 506 \\
\hline
1111
\end{array}
$$

This time the result *is* a palindrome. It has been conjectured that if we begin with any positive integer and apply the technique described above to it, then we will eventually arrive at a palindrome. However, no one knows if this is true. (It is known to be true for all two-digit numbers.)

Some conjectures have become famous because it has taken years, decades or *even centuries* to establish their truth or falseness. Other conjectures remain undecided still today. We now consider four conjectures in mathematics, each of which has a long history.

In 1852, a question occurred to the British student Francis Guthrie when he was coloring a map of the counties of England. Suppose that some country (real or imaginary) has been divided into counties in some manner. Is it possible to color the counties in this map with four or fewer colors such that one color is used for each county and two counties that share a common boundary (not simply a single point) are colored differently? For example, the map of the "country" shown in Figure 7.1 has eight "counties," which are colored with the four colors red (R), blue (B), green (G) and yellow (Y), according to the rules described above. This map can also be colored with more than four colors but not less than four.

Within a few years, some of the best known mathematicians of the time had become aware of Francis Guthrie's question, and eventually a famous conjecture developed from this.

The Four Color Conjecture Every map can be colored with four or fewer colors.

Many attempted to settle this conjecture. In fact, in 1879 an article was published containing a reported proof of the conjecture. However, in 1890, an error was discovered in the proof and the "theorem" returned to its conjecture status. It was not until 1976 when

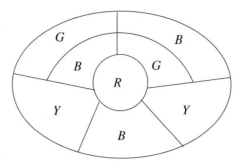

Figure 7.1   Coloring the counties in a country with four colors

an actual proof by Kenneth Appel and Wolfgang Haken, combining both mathematics and computers, was presented. The period between the origin of the problem and its solution covered some 124 years. This is now a theorem.

The Four   Every map can be colored with four or fewer colors.
Color Theorem

We now describe a conjecture with an even longer history. One of the famous mathematicians of the 17th century was Pierre Fermat. He is undoubtedly best known for one particular assertion he made. He wrote that for each integer $n \geq 3$, there are no nonzero integers x, y and z such that $x^n + y^n = z^n$. Of course, there are many nonzero integer solutions to the equation $x^2 + y^2 = z^2$. For example, $3^2 + 4^2 = 5^2$, $5^2 + 12^2 = 13^2$ and $8^2 + 15^2 = 17^2$. A triple (x, y, z) of positive integers such that $x^2 + y^2 = z^2$ is often called a **Pythagorean triple**. Therefore, $(3, 4, 5)$, $(5, 12, 13)$ and $(8, 15, 17)$ are Pythagorean triples. Indeed, if (a, b, c) is a Pythagorean triple and $k \in \mathbf{N}$, then (ka, kb, kc) is also a Pythagorean triple. Fermat's assertion was discovered, unproved, in a margin of a book of Fermat's after his death. In the margin it was written that there was insufficient space to contain his "truly remarkable demonstration." Consequently, this statement became known as Fermat's Last Theorem. It would have been more appropriate, however, to have referred to this statement as Fermat's Last Conjecture, as the truth or falseness of this statement remained in question for approximately 350 years. However, in 1993, the British mathematician Andrew Wiles settled the conjecture by giving a truly remarkable proof of it. Hence Fermat's Last Theorem is at last a theorem.

Fermat's Last   For each integer $n \geq 3$, there are no nonzero integers x, y and z such that $x^n + y^n = z^n$.
Theorem

The final two conjectures we mention concern primes. Although we have mentioned primes from time to time, we have not yet presented a formal definition. We do this now. An integer $p \geq 2$ is a **prime** if its only positive integer divisors are 1 and p. A **Fermat number** (Yes, the same Fermat!) is an integer of the form $F_t = 2^{2^t} + 1$, where t is a nonnegative integer. The first five Fermat numbers are

$$F_0 = 3, \; F_1 = 5, \; F_2 = 17, \; F_3 = 257, \; F_4 = 65,537,$$

all of which happen to be primes.

In 1640, Fermat wrote to many (including to the famous mathematician Blaise Pascal) that he believed *every* such number (he didn't call them Fermat numbers) was a prime, but he was unable to prove this. Hence we have the following.

Fermat's Conjecture   Every Fermat number is a prime.

Nearly one century later (in 1739), the famous mathematician Leonhard Euler proved that $F_5 = 4,294,967,297$ is divisible by 641, thereby disproving Fermat's Conjecture. More specifically, Euler proved the following.

Euler's Theorem   If p is a prime factor of F_t, then $p = 2^{t+1}k + 1$ for some positive integer k.

Letting $t = 5$ in Euler's Theorem, we see that each prime factor of F_5 is of the form $64k + 1$. The first five primes of this form are 193, 257, 449, 577 and 641, the last of which divides F_5.

In recent decades, other Fermat numbers have been studied and have been shown not to be prime. Indeed, many students of this topic now lean toward the opposing viewpoint (and conjecture): Except for the Fermat numbers $F_0, F_1, \cdots, F_4$ (all of which were observed to be prime by Fermat), *no* Fermat number is prime.

The last conjecture we describe here has its origins around 1742. The German mathematician Christian Goldbach conjectured that every even integer exceeding 2 is the sum of two primes. Of course, this is easy to see for small even integers. For example, $4 = 2 + 2$, $6 = 3 + 3$, $8 = 5 + 3$ and $10 = 7 + 3 = 5 + 5$. The major difference between this conjecture and the three preceding conjectures is that this conjecture has never been resolved. Hence we conclude with the following.

Goldbach's Conjecture   Every even integer at least 4 is the sum of two primes.

7.2 Revisiting Quantified Statements

Many (in fact, most) of the statements we have encountered are quantified statements. Indeed, for an open sentence $P(x)$ over a domain S, we have often considered a quantified statement with a universal quantifier, namely

$$\forall x \in S, P(x): \text{ For every } x \in S, P(x). \text{ or If } x \in S, \text{ then } P(x).$$

or a quantified statement with an existential quantifier, namely

$$\exists x \in S, P(x): \text{ There exists } x \in S \text{ such that } P(x).$$

Recall that $\forall x \in S, P(x)$ is a true statement if $P(x)$ is true for every $x \in S$; while $\exists x \in S, P(x)$ is a true statement if $P(x)$ is true for at least one $x \in S$. Let's review these again.

Example 7.1   *Let $S = \{1, 3, 5, 7\}$ and consider*

$$P(n): n^2 + n + 1 \text{ is prime.}$$

for each n ∈ S. Then both

$$\forall n \in S, \, P(n): \text{ For every } n \in S, \, n^2 + n + 1 \text{ is prime.}$$

and

$$\exists n \in S, \, P(n): \text{ There exists } n \in S \text{ such that } n^2 + n + 1 \text{ is prime.}$$

are quantified statements. Since

$$
\begin{array}{lll}
P(1): & 1^2 + 1 + 1 = 3 \text{ is prime.} & \text{is true,} \\
P(3): & 3^2 + 3 + 1 = 13 \text{ is prime.} & \text{is true,} \\
P(5): & 5^2 + 5 + 1 = 31 \text{ is prime.} & \text{is true,} \\
P(7): & 7^2 + 7 + 1 = 57 \text{ is prime.} & \text{is false,}
\end{array}
$$

it follows that ∀n ∈ S, P(n) is false and ∃n ∈ S, P(n) is true. On the other hand, the statement

$$Q: \ 323 \text{ is prime.}$$

is not a quantified statement but Q is false (as 323 = 17 · 19 is not prime). ♦

Let $P(x)$ be a statement for each x in some domain S. Recall that the negation of $\forall x \in S, \, P(x)$ is

$$\sim (\forall x \in S, \, P(x)) \equiv \exists x \in S, \sim P(x).$$

and the negation of $\exists x \in S, \, P(x)$ is

$$\sim (\exists x \in S, \, P(x)) \equiv \forall x \in S, \sim P(x).$$

Again, consider

$$P(n): \ n^2 + n + 1 \text{ is prime.}$$

from Example 7.1, which is a statement for each n in $S = \{1, 3, 5, 7\}$. The negation of $\forall n \in S, \, P(n)$ is

$$\exists n \in S, \sim P(n): \text{ There exists } n \in S \text{ such that } n^2 + n + 1 \text{ is not prime.}$$

is true as $7 \in S$ but $7^2 + 7 + 1 = 57$ is not prime. On the other hand, the negation of $\exists n \in S, \, P(n)$ is

$$\forall n \in S, \sim P(n): \text{ If } n \in S, \text{ then } n^2 + n + 1 \text{ is not prime.}$$

is false since, for example, $1 \in S$ and $1^2 + 1 + 1 = 3$ *is* prime.

In Section 2.10 we began a discussion of quantified statements containing two quantifiers. We continue this discussion here.

Example 7.2   *Consider*

$$P(s, t): \ 2^s + 3^t \text{ is prime.}$$

where s is a positive even integer and t is a positive odd integer. If we let S denote the set of positive even integers and T the set of positive odd integers, then the quantified statement

$$\exists s \in S, \exists t \in T, P(s, t).$$

can be expressed in words as

There exist a positive even integer s and a
positive odd integer t such that $2^s + 3^t$ is prime. **(7.1)**

The statement (7.1) is true since

$$P(2, 1): 2^2 + 3^1 = 7 \text{ is prime.}$$

is true. On the other hand, the quantified statement

$$\forall s \in S, \forall t \in T, P(s, t).$$

can be expressed in words as

For every positive even integer s and every positive
odd integer t, $2^s + 3^t$ is prime. **(7.2)**

The statement (7.2) is false since

$$P(6, 3): 2^6 + 3^3 = 91 \text{ is a prime.}$$

is false, as $91 = 7 \cdot 13$ is not a prime. ◆

Let $P(s, t)$ be an open sentence, where the domain of the variable s is S and the domain of the variable t is T. Recall that the negations of the quantified statements $\exists s \in S, \exists t \in T, P(s, t)$ and $\forall s \in S, \forall t \in T, P(s, t)$ are

$$\sim (\exists s \in S, \exists t \in T, P(s, t)) \equiv \forall s \in S, \forall t \in T, \sim P(s, t).$$

and

$$\sim (\forall s \in S, \forall t \in T, P(s, t)) \equiv \exists s \in S, \exists t \in T, \sim P(s, t).$$

Therefore, the negation of the statement (7.1) is

For every positive even integer s and every positive odd integer t, $2^s + 3^t$ is not prime.

which is a false statement. On the other hand, the negation of the statement (7.2) is

There exist a positive even integer s and a
positive odd integer t such that $2^s + 3^t$ is not prime.

which is a true statement.

We have seen that quantified statements may also contain different kinds of quantifiers. For example, it follows by the definition of an even integer that for every even integer n, there exists an integer k such that $n = 2k$. There is another mathematical symbol with which you should be familiar. The symbol $\ni$ denotes the phrase *such that* (although some mathematicians simply write s.t. for "such that"). For example, let S denote the set of even integers again. Then

$$\forall n \in S, \ \exists k \in \mathbf{Z} \ni n = 2k. \tag{7.3}$$

states:

> *For every even integer n, there exists an integer k such that n = 2k.*

This statement can be reworded as:

> *If n is an even integer, then n = 2k for some integer k.*

If we interchange the two quantifiers in (7.3), we obtain, in words:

> *There exists an even integer n such that for every integer k, n = 2k.*

This statement can also be reworded as

> *There exists an even integer n such that n = 2k for every integer k.*

This statement can be expressed in symbols as

$$\exists n \in S, \ \forall k \in \mathbf{Z}, n = 2k. \tag{7.4}$$

Certainly, the statements (7.3) and (7.4) say something totally different. Indeed, (7.3) is true and (7.4) is false.

Another such example of this is

> For every real number x, there exists an integer n such that $|x - n| < 1$. $\qquad$ **(7.5)**

This statement can also be expressed as

> *If x is a real number, then there exists an integer n such that $|x - n| < 1$.*

In order to state (7.5) in symbols, let

$$P(x, n)\colon \ |x - n| < 1.$$

where the domain of the variable x is $\mathbf{R}$ and the domain of the variable n is $\mathbf{Z}$. Thus, (7.5) can be expressed in symbols as

$$\forall x \in \mathbf{R}, \exists n \in \mathbf{Z}, P(x, n).$$

The statement (7.5) is true, as we now verify. In the proof of the following result, we will refer to the **ceiling** $\lceil x \rceil$ of a real number, which is the smallest integer greater than or equal to x.

Result 7.3   *For every real number x, there exists an integer n such that $|x - n| < 1$.*

Proof   Let x be a real number. If we let $n = \lceil x \rceil$, then $|x - n| = |x - \lceil x \rceil| = \lceil x \rceil - x < 1$. $\blacksquare$

Another example of a quantified statement containing two different quantifiers is

> There exists a positive even integer m such that for every positive integer n, $\left|\frac{1}{m} - \frac{1}{n}\right| \leq \frac{1}{2}$. $\qquad$ **(7.6)**

Let S denote the set of positive even integers and let

$$P(m, n): \ \left|\frac{1}{m} - \frac{1}{n}\right| \leq \frac{1}{2}.$$

where the domain of the variable m is S and the domain of the variable n is **N**. Thus, (7.6) can be expressed in symbols as

$$\exists m \in S, \forall n \in \mathbf{N}, \ P(m, n).$$

The truth of the statement (7.6) is now verified.

Result 7.4 *There exists a positive even integer m such that for every positive integer n,*

$$\left|\frac{1}{m} - \frac{1}{n}\right| \leq \frac{1}{2}.$$

Proof Consider $m = 2$. Let n be a positive integer. We consider three cases.

Case 1. $n = 1$. Then $\left|\frac{1}{m} - \frac{1}{n}\right| = \left|\frac{1}{2} - \frac{1}{1}\right| = \frac{1}{2}$.

Case 2. $n = 2$. Then $\left|\frac{1}{m} - \frac{1}{n}\right| = \left|\frac{1}{2} - \frac{1}{2}\right| = 0 < \frac{1}{2}$.

Case 3. $n \geq 3$. Then $\left|\frac{1}{m} - \frac{1}{n}\right| = \left|\frac{1}{2} - \frac{1}{n}\right| = \frac{1}{2} - \frac{1}{n} < \frac{1}{2}$.

Thus $\left|\frac{1}{2} - \frac{1}{n}\right| \leq \frac{1}{2}$ for every $n \in \mathbf{N}$. $\blacksquare$

Let $P(s, t)$ be an open sentence, where the domain of the variable s is S and the domain of the variable t is T. The negation of the quantified statement $\forall s \in S, \exists t \in T, P(s, t)$ is

$$\sim (\forall s \in S, \exists t \in T, P(s, t)) \equiv \exists s \in S, \sim (\exists t \in T, P(s, t))$$
$$\equiv \exists s \in S, \forall t \in T, \sim P(s, t);$$

while the negation of the quantified statement $\exists s \in S, \forall t \in T, P(s, t)$ is

$$\sim (\exists s \in S, \forall t \in T, P(s, t)) \equiv \forall s \in S, \sim (\forall t \in T, P(s, t))$$
$$\equiv \forall s \in S, \exists t \in T, \sim P(s, t).$$

Consequently, the negation of the statement (7.5) is

> There exists a real number x such that for every integer n, $|x - n| \geq 1$.

This statement is therefore false. The negation of the statement (7.6) is

> For every positive even integer m, there exists a positive integer n such that $\left|\frac{1}{m} - \frac{1}{n}\right| > \frac{1}{2}$.

This too is false.

Let's consider the following statement, which has more than two quantifiers.

For every positive real number e, there exists a positive real number d

such that for every real number x, $|x| < d$ implies that $|2x| < e$.   **(7.7)**

If we let

$$P(x, d): \ |x| < d. \ \text{ and } \ Q(x, e): \ |2x| < e.$$

where the domain of the variables e and d is $\mathbf{R}^+$ and the domain of the variable x is $\mathbf{R}$, then (7.7) can be expressed in symbols as

$$\forall e \in \mathbf{R}^+, \exists d \in \mathbf{R}^+, \forall x \in \mathbf{R}, \ P(x, d) \Rightarrow Q(x, e).$$

The statement (7.7) is in fact true, which we now verify.

Result 7.5   *For every positive real number e, there exists a positive real number d such that if x is a real number with $|x| < d$, then $|2x| < e$.*

Proof   Let e be a positive real number. Now choose $d = e/2$. Let x be a real number with $|x| < d = e/2$. Then

$$|2x| = 2|x| < 2\left(\frac{e}{2}\right) = e,$$

as desired.   ■

7.3 Testing Statements

We now turn our attention to the main topic of this chapter. For a given statement whose truth value is not provided to us, our task is to determine the truth or falseness of the statement and, in addition, show that our conclusion is correct by proving or disproving the statement, as appropriate.

Example 7.6   *Prove or disprove: There is a real number solution of the equation*

$$x^6 + 2x^2 + 1 = 0.$$

Strategy   Observe that x^6 and x^2 are even powers of x. Thus if x is any real number, then $x^6 \geq 0$ and $x^2 \geq 0$, so $2x^2 \geq 0$. Adding 1 to $x^6 + 2x^2$ shows that $x^6 + 2x^2 + 1 \geq 1$. Hence it is impossible for $x^6 + 2x^2 + 1$ to be 0. These thoughts lead us to our solution. We begin by informing the reader that the statement is false, so the reader knows what we will be trying to do.   ◆

Solution of   The statement is false. Let $x \in \mathbf{R}$. Since $x^6 \geq 0$ and $x^2 \geq 0$, it follows that $x^6 + 2x^2 +$
Example 7.6 $1 \geq 1$ and so $x^6 + 2x^2 + 1 \neq 0$.   ◆

For the preceding example, we wrote "Strategy" rather than "Proof Strategy" for two reasons: (1) Since the statement may be false, there may be no proof in this case. (2) We are essentially "thinking out loud," trying to convince ourselves whether the statement is true or false. Of course, if the statement turns out to be true, then our strategy may very well turn into a proof strategy.

Example 7.7   *Prove or disprove: Let $x, y, z \in \mathbf{Z}$. Then two of the integers x, y and z are of the same parity.*

Strategy   For any three given integers, either two are even or two are odd. So it certainly seems as if the statement is true. The only question appears to be whether what we said in the preceding sentence is convincing enough to all readers. We try another approach.   ◆

Solution of Example 7.7   The statement is true.

Proof   Consider x and y. If x and y are of the same parity, then the proof is complete. Thus we may assume that x and y are of opposite parity, say x is even and y is odd. If z is even, then x and z are of the same parity; while if z is odd, then y and z are of the same parity.   ∎

Of course, the preceding proof could have been done by cases as well.

Example 7.8   *Prove or disprove: Let A, B and C be sets. If $A \times C = B \times C$, then $A = B$.*

Strategy   The elements of the set $A \times C$ are ordered pairs of elements, namely they are of the form (x, y), where $x \in A$ and $y \in C$. Let $(x, y) \in A \times C$. If $A \times C = B \times C$, then it follows that (x, y) must be an element of $B \times C$ as well. This says that $x \in B$ and $y \in C$. Conversely, if $(x, y) \in B \times C$, then $(x, y) \in A \times C$, which implies that $x \in A$ as well. These observations certainly seem to suggest that it should be possible to show that $A = B$ under these conditions. However, this argument depends on $A \times C$ containing an element (x, y). Could it happen that $A \times C$ contains no elements? If A or C is empty, this would happen. However, if $C \neq \emptyset$ and $A \times C = \emptyset$, then A must be empty. But $B \times C = A \times C = \emptyset$ would mean that B must also be empty and so $A = B$. This suggests a different response.   ◆

Solution of Example 7.8   The statement is false. Let $A = \{1\}$, $B = \{2\}$ and $C = \emptyset$. Then $A \times C = B \times C = \emptyset$, but $A \neq B$. Hence these sets A, B and C form a counterexample.   ◆

In some instances, we might consider modifying a false statement so that the revised statement is true. Our preceding discussion seems to suggest that if the set C were required to be nonempty, then the statement would have been true.

Result 7.9   *Let A, B and C be sets, where $C \neq \emptyset$. If $A \times C = B \times C$, then $A = B$.*

Proof   Assume that $A \times C = B \times C$. Since $C \neq \emptyset$, the set C contains some element c. Let $x \in A$. Then $(x, c) \in A \times C$. Since $A \times C = B \times C$, it follows that $(x, c) \in B \times C$. Hence $x \in B$ and so $A \subseteq B$. By a similar argument, it follows that $B \subseteq A$. Thus $A = B$.   ∎

Example 7.10   *Prove or disprove: There exists a real number x such that $x^3 < x < x^2$.*

Strategy   If there is a real number x such that $x^3 < x < x^2$, then this number is certainly not 0. Consequently, any real number x with this property is either positive or negative. If $x > 0$, then we can divide $x^3 < x < x^2$ by x, obtaining $x^2 < 1 < x$. However, if $x > 1$, then $x^2 > 1$. Therefore, there is no positive real number x for which $x^3 < x < x^2$. Hence any real number x satisfying $x^3 < x < x^2$ must be negative. Dividing $x^3 < x < x^2$ by x gives us $x^2 > 1 > x$ or $x < 1 < x^2$. Experimenting with some negative numbers tells us that any number less than -1 has the desired property.   ◆

Solution of Example 7.10   The statement is true.

Proof   Consider $x = -2$. Then $x^3 = -8$ and $x^2 = 4$. Thus $x^3 < x < x^2$.   ∎

Example 7.11   *Prove or disprove: For every positive irrational number b, there exists an irrational number a such that $0 < a < b$.*

Strategy   We begin with a positive irrational number b. If this statement is true, then we must show that there is an irrational number a such that $0 < a < b$. If we let $a = b/2$, then certainly $0 < a < b$. The only question is whether $b/2$ is necessarily irrational. We have seen, however, that $b/2$ is irrational (in Exercise 5.17 in Section 5.2).   ◆

Solution of Example 7.11   The statement is true.

Proof   Let b be a positive irrational number. Now let $a = b/2$. Then $0 < a < b$ and a is irrational by Exercise 5.17 in Section 5.2.   ∎

Example 7.12   *Prove or disprove: Every even integer is the sum of three distinct even integers.*

Strategy   This statement can be reworded in a variety of ways. One rewording of this statement is: If n is an even integer, then there exist three distinct even integers a, b and c such that $n = a + b + c$. What this statement does *not* say is that the sum of three distinct even integers is even; that is, we do not begin with three distinct even integers and show that their sum is even. We begin with an even integer n and ask whether we can find three distinct even integers a, b and c such that $n = a + b + c$. This is certainly true for $n = 0$ since $0 = (-2) + 0 + 2$. It is also true for $n = 2$ since $2 = (-2) + 0 + 4$. If $n = 4$, then $4 = (-2) + 2 + 4$. This last example may suggest a proof in general. For every even integer n, we can write $n = 2 + (-2) + n$. Certainly, n, 2 and -2 are even. But are they distinct? They are not distinct if $n = 2$ or $n = -2$. This provides a plan for a proof.   ◆

Solution of Example 7.12   The statement is true.

Proof   Let n be an even integer. We show that n is the sum of three distinct even integers by considering the following three cases.

Case 1. $n = 2$. Observe that $2 = (-2) + 0 + 4$.

Case 2. $n = -2$. Observe that $-2 = (-4) + 0 + 2$.

Case 3. $n \neq 2, -2$. Then $n = 2 + (-2) + n$. ∎

Example 7.13 *Prove or disprove: Let $k \in \mathbf{N}$. If $k^2 + 5k$ is odd, then $(k + 1)^2 + 5(k + 1)$ is odd.*

Strategy One idea that might occur to us is to assume that $k^2 + 5k$ is an odd integer, where $k \in \mathbf{N}$, and see if we can show that $(k + 1)^2 + 5(k + 1)$ is also odd. If $k^2 + 5k$ is odd, then we can write $k^2 + 5k = 2\ell + 1$ for some integer ℓ. Then

$$(k + 1)^2 + 5(k + 1) = k^2 + 2k + 1 + 5k + 5 = (k^2 + 5k) + (2k + 6)$$
$$= (2\ell + 1) + (2k + 6) = (2\ell + 2k + 6) + 1$$
$$= 2(\ell + k + 3) + 1,$$

which is an odd integer and we have a proof. ◆

Solution of Example 7.13 The statement is true.

Proof Assume that $k^2 + 5k$ is an odd integer, where $k \in \mathbf{N}$. Then $k^2 + 5k = 2\ell + 1$ for some integer ℓ. Hence

$$(k + 1)^2 + 5(k + 1) = k^2 + 2k + 1 + 5k + 5 = (k^2 + 5k) + (2k + 6)$$
$$= (2\ell + 1) + (2k + 6) = (2\ell + 2k + 6) + 1$$
$$= 2(\ell + k + 3) + 1.$$

Since $\ell + k + 3$ is an integer, $(k + 1)^2 + 5(k + 1)$ is an odd integer. ∎

Example 7.14 *Prove or disprove: For every positive integer n, $n^2 + 5n$ is an odd integer.*

Strategy It seems like the reasonable thing to do is to investigate $n^2 + 5n$ for a few values of n. For $n = 1$, we have $n^2 + 5n = 1 + 5 \cdot 1 = 6$. We have already solved the problem! For $n = 1$, $n^2 + 5n$ is not an odd integer. We have discovered a counterexample. ◆

Solution of Example 7.14 The statement is false. For $n = 1$, $n^2 + 5n = 1 + 5 \cdot 1 = 6$, which is even. Thus $n = 1$ is a counterexample. ◆

Looking at Examples 7.13 and 7.14 again, we might be wondering what exactly is going on. Certainly, these two examples seem to be related. Perhaps the following thought may occur to us. For each positive integer n, let

$$P(n): \text{ The integer } n^2 + 5n \text{ is odd.}$$

and consider the (quantified) statement

For every positive integer n, $n^2 + 5n$ is odd.

or in symbols,

$$\forall n \in \mathbf{N}, \; P(n). \tag{7.8}$$

We might ask whether (7.8) is true. Because of the domain, a proof by induction seems appropriate. In fact, the statement in Example 7.13 is the inductive step in an induction proof of (7.8). By Example 7.13, the inductive step is true. On the other hand, the statement (7.8) is false as $n = 1$ is a counterexample. This emphasizes the importance of verifying both the basis step and the inductive step in an induction proof. Returning to Example 7.13 once again, we can show (using a proof by cases) that $k^2 + 5k$ is even for every $k \in \mathbf{N}$, which would provide a vacuous proof of the statement in Example 7.13.

In this chapter, we have discussed analyzing statements, particularly understanding statements, determining whether they are true or false and proving or disproving them. All of the statements that we have analyzed were provided to us. But how do we obtain statements to analyze for ourselves? This is an important question and concerns the creative aspect of mathematics—how new mathematics is discovered. Obviously, there is no rule or formula for creativity, but creating new statements often comes from studying old statements.

Let's illustrate how we might create statements to analyze. In Exercise 4.6, you were asked to prove the following:

$$\text{Let } a \in \mathbf{Z}. \text{ If } 3 \mid 2a, \text{ then } 3 \mid a. \tag{7.9}$$

What other statements does this suggest? For example, is its converse true? (The answer is yes, but the converse is not very interesting.) Is (7.9) true if 3 and 2 are interchanged? What integers can we replace 2 by in (7.9) and obtain a true statement? That is, for which positive integers k is it true that if $3 \mid ka$, then $3 \mid a$? Of course, this is true for $k = 1$. And we know that it's true for $k = 2$. It is not true for $k = 3$; that is, it is not true that if $3 \mid 3a$, then $3 \mid a$. The integer $a = 1$ is a counterexample. On the other hand, it is possible to prove that if $3 \mid 4a$, where $a \in \mathbf{Z}$, then $3 \mid a$. What we are attempting to do is to **extend** the result in (7.9) so that we have a result of the type:

$$\text{Let } a \in \mathbf{Z}. \text{ If } 3 \mid ka, \text{ then } 3 \mid a. \tag{7.10}$$

for a fixed integer k greater than 2. We would like to find a set S of positive integers such that the following is true:

$$\text{Let } a \in \mathbf{Z}. \text{ If } 3 \mid ka, \text{ where } k \in S, \text{ then } 3 \mid a. \tag{7.11}$$

Surely $2 \in S$. Result (7.9) then becomes a special case and a corollary of (7.11). For this reason, (7.11) is called a **generalization** of (7.9). Ideally, we would like S to have the added property that (7.11) is true if $k \in S$, while (7.11) is false if $k \notin S$. We are thus seeking a set S of integers such that the following is true:

$$\text{Let } a \in \mathbf{Z}. \text{ Then } 3 \mid ka \text{ implies that } 3 \mid a \text{ if and only if } k \in S.$$

If we were successful in finding this set S, then we might start all over again by replacing 3 in (7.9) by some other positive integer.

In mathematics it is often the case that a new result is obtained by looking at an old result in a new way and extending it to obtain a generalization of the old result. Hence it frequently happens that: *Today's theorem becomes tomorrow's corollary.*

We conclude this chapter with a quiz. Solutions are given following the quiz.

Quiz

Prove or disprove each of the following statements.

1. If n is a positive integer and s is an irrational number, then n/s is an irrational number.
2. For every integer b, there exists a positive integer a such that $|a - |b|| \le 1$.
3. If x and y are integers of the same parity, then xy and $(x + y)^2$ are of the same parity.
4. Let $a, b \in \mathbf{Z}$. If $6 \nmid ab$, then either (1) $2 \nmid a$ and $3 \nmid b$ or (2) $3 \nmid a$ and $2 \nmid b$.
5. For every positive integer n, $2^{2^n} \ge 4^{n!}$.
6. If A, B and C are sets, then $(A - B) \cup (A - C) = A - (B \cup C)$.
7. Let $n \in \mathbf{N}$. If $(n + 1)(n + 4)$ is odd, then $(n + 1)(n + 4) + 3^n$ is odd.
8. (a) There exist distinct rational numbers a and b such that $(a - 1)(b - 1) = 1$.
   (b) There exist distinct rational numbers a and b such that $\frac{1}{a} + \frac{1}{b} = 1$.
9. Let $a, b, c \in \mathbf{Z}$. If every two of a, b and c are of the same parity, then $a + b + c$ is even.
10. If n is a nonnegative integer, then 5 divides $2 \cdot 4^n + 3 \cdot 9^n$.

Solutions for Quiz

1. The statement is true.

   Proof   Assume, to the contrary, that there exist a positive integer n and an irrational number s such that n/s is a rational number. Then $n/s = a/b$, where $a, b \in \mathbf{Z}$ and $a, b \ne 0$. Therefore, $s = nb/a$, where $nb, a \in \mathbf{Z}$ and $a \ne 0$. Thus s is rational, producing a contradiction.   ∎

2. The statement is true.

   Proof   Let $b \in \mathbf{Z}$. Now let $a = |b| + 1$. Thus $a \in \mathbf{N}$ and $|a - |b|| = |(|b| + 1) - |b|| = 1$.   ∎

3. The statement is false. Observe that $x = 1$ and $y = 3$ are of the same parity. Then $xy = 3$ and $(x + y)^2 = 16$ are of opposite parity. Hence $x = 1$ and $y = 3$ produce a counterexample.   ◆

4. The statement is false. Let $a = b = 2$. So $ab = 4$. Hence $6 \nmid ab$. Since $2 \mid a$ and $2 \mid b$, both (1) and (2) are false. Thus $a = b = 2$ constitute a counterexample.   ◆

5. The statement is false. For $n = 3$, $2^{2^n} = 2^8 = 256$ while $4^{n!} = 4^{3!} = 4^6 = 4096$. Thus $2^{2^3} < 4^{3!}$ and so $n = 3$ is a counterexample.   ◆

6. The statement is false. Let $A = \{1, 2, 3\}$, $B = \{2\}$ and $C = \{3\}$. Thus $B \cup C = \{2, 3\}$. Hence $A - B = \{1, 3\}$, $A - C = \{1, 2\}$ and $A - (B \cup C) = \{1\}$. Therefore, $(A - B) \cup (A - C) = \{1, 2, 3\} \ne A - (B \cup C)$. So $A = \{1, 2, 3\}$, $B = \{2\}$ and $C = \{3\}$ constitute a counterexample.   ◆

7. The statement is true.

   Proof   Let $n \in \mathbf{N}$ and consider $(n + 1)(n + 4)$. We show that $(n + 1)(n + 4)$ is even, thereby giving a vacuous proof. There are two cases.

   Case 1. n is even. Then $n = 2k$ for some integer k. Thus

   $$(n + 1)(n + 4) = (2k + 1)(2k + 4) = 2(2k + 1)(k + 2).$$

   Since $(2k + 1)(k + 2) \in \mathbf{Z}$, it follows that $(n + 1)(n + 4)$ is even.

Case 2. n is odd. Then $n = 2\ell + 1$ for some integer ℓ. Thus

$$(n + 1)(n + 4) = (2\ell + 2)(2\ell + 5) = 2(\ell + 1)(2\ell + 5).$$

Since $(\ell + 1)(2\ell + 5) \in \mathbf{Z}$, it follows that $(n + 1)(n + 4)$ is even. ∎

8. (a) The statement is true.

   Proof Let $a = 3$ and $b = \frac{3}{2}$. Then $(a - 1)(b - 1) = 2\left(\frac{1}{2}\right) = 1$. ∎

   (b) The statement is true.

   Proof Let $a = \frac{1}{2}$ and $b = -1$. Then

   $$\frac{1}{a} + \frac{1}{b} = \frac{1}{\frac{1}{2}} + \frac{1}{-1} = 2 - 1 = 1.$$ ∎

   Proof Analysis Observe that if a and b are two (distinct) rational numbers that satisfy $\frac{1}{a} + \frac{1}{b} = 1$, then $\frac{a+b}{ab} = 1$ and so $a + b = ab$. Thus $ab - a - b = 0$, which is equivalent to $ab - a - b + 1 = 1$ and so $(a - 1)(b - 1) = 1$. Therefore, two distinct rational numbers a and b satisfy

   $$(a - 1)(b - 1) = 1$$

   if and only if a and b satisfy

   $$\frac{1}{a} + \frac{1}{b} = 1$$

   if and only if a and b satisfy

   $$a + b = ab.$$ ◆

9. The statement is false. Let $a = 1, b = 3$ and $c = 5$. Then every two of a, b and c are of the same parity; yet $a + b + c$ is odd. Hence $a = 1, b = 3$ and $c = 5$ produce a counterexample. ◆

10. The statement is true.

    Proof We proceed by induction. For $n = 0, 2 \cdot 4^n + 3 \cdot 9^n = 2 \cdot 1 + 3 \cdot 1 = 5$. Thus $5 \mid (2 \cdot 4^0 + 3 \cdot 9^0)$ and the statement is true for $n = 0$.
    Assume that $5 \mid (2 \cdot 4^k + 3 \cdot 9^k)$ for a nonnegative integer k. We show that $5 \mid (2 \cdot 4^{k+1} + 3 \cdot 9^{k+1})$. Since $5 \mid (2 \cdot 4^k + 3 \cdot 9^k)$, it follows that $2 \cdot 4^k + 3 \cdot 9^k = 5x$ for some integer x. Thus $2 \cdot 4^k = 5x - 3 \cdot 9^k$. Hence

    $$\begin{aligned}
    2 \cdot 4^{k+1} + 3 \cdot 9^{k+1} &= 4(2 \cdot 4^k) + 3 \cdot 9^{k+1} \\
    &= 4(5x - 3 \cdot 9^k) + 3 \cdot 9^{k+1} \\
    &= 20x - 12 \cdot 9^k + 27 \cdot 9^k \\
    &= 20x + 15 \cdot 9^k = 5(4x + 3 \cdot 9^k).
    \end{aligned}$$

    Since $4x + 3 \cdot 9^k \in \mathbf{Z}$, it follows that $5 \mid (2 \cdot 4^{k+1} + 3 \cdot 9^{k+1})$. By the Principle of Mathematical Induction, 5 divides $2 \cdot 4^n + 3 \cdot 9^n$ for every nonnegative integer n. ∎

EXERCISES FOR CHAPTER 7

Section 7.1: Conjectures in Mathematics

7.1. Consider the following sequence of equalities:

$1 = 0 + 1$
$2 + 3 + 4 = 1 + 8$
$5 + 6 + 7 + 8 + 9 = 8 + 27$
$10 + 11 + 12 + 13 + 14 + 15 + 16 = 27 + 64$

(a) What is the next equality in this sequence?
(b) What conjecture is suggested by these equalities?
(c) Prove the conjecture in (b) by induction.

7.2. Consider the following statements:

$(1 + 2)^2 - 1^2 = 2^3$
$(1 + 2 + 3)^2 - (1 + 2)^2 = 3^3$
$(1 + 2 + 3 + 4)^2 - (1 + 2 + 3)^2 = 4^3$

(a) Based on the three statements given above, what is the next statement suggested by these?
(b) What conjecture is suggested by these statements?
(c) Verify the conjecture in (b).

7.3. A sequence $\{a_n\}$ of real numbers is defined recursively by $a_1 = 2$ and for $n \geq 2$,

$$a_n = \frac{2 + 1 \cdot a_1^2 + 2 \cdot a_2^2 + \cdots + (n-1)a_{n-1}^2}{n}.$$

(a) Determine a_2, a_3 and a_4.
(b) Clearly, a_n is a rational number for each $n \in \mathbf{N}$. Based on the information in (a), however, what conjecture does this suggest?

7.4. It has been stated that the German mathematician Christian Goldbach is known for a conjecture he made concerning primes. We refer to this conjecture as Conjecture A.
Conjecture A *Every even integer at least 4 is the sum of two primes.*
Goldbach made two other conjectures concerning primes.
Conjecture B *Every integer at least 6 is the sum of three primes.*
Conjecture C *Every odd integer at least 9 is the sum of three odd primes.*
Prove that the truth of one or more of these three conjectures implies the truth of one or two of the other conjectures.

7.5. By an ordered partition of an integer $n \geq 2$ is meant a sequence of positive integers whose sum is n. For example, the ordered partitions of 3 are $3, 1 + 2, 2 + 1, 1 + 1 + 1$.

(a) Determine the ordered partitions of 4.
(b) Make a conjecture concerning the number of ordered partitions of an integer $n \geq 2$.

7.6. Two recursively defined sequences $\{a_n\}$ and $\{b_n\}$ of positive integers have the same recurrence relation, namely $a_n = 2a_{n-1} + a_{n-2}$ and $b_n = 2b_{n-1} + b_{n-2}$ for $n \geq 3$. The initial values for $\{a_n\}$ are $a_1 = 1$ and $a_2 = 3$, while the initial values for $\{b_n\}$ are $b_1 = 1$ and $b_2 = 2$.

(a) Determine a_3 and a_4.
(b) Determine whether the following is true or false:
    Conjecture: $a_n = 2^{n-2} \cdot n + 1$ for every integer $n \geq 2$.

(c) Determine b_3 and b_4.
(d) Determine whether the following is true or false:
   Conjecture: $b_n = \frac{(1+\sqrt{2})^n - (1-\sqrt{2})^n}{2\sqrt{2}}$ for every integer $n \geq 2$.

7.7. We know that $1 + 2 + 3 = 1 \cdot 2 \cdot 3$; that is, there exist three positive integers whose sum equals their product. Prove or disprove (a) and (b).

   (a) There exist four positive integers whose sum equals their product.
   (b) There exist five positive integers whose sum equals their product.
   (c) What conjecture does this suggest to you?

7.8. Observe that $3 = 1 + 2$, $6 = 1 + 2 + 3$, $9 = 4 + 5$ and $12 = 3 + 4 + 5$.

   (a) Show that 13 and 14 can be expressed as the sum of two or more consecutive positive integers.
   (b) Make a conjecture as for which integers $n \geq 3$ can be expressed as the sum of two or more consecutive positive integers.
   (c) Prove your conjecture in (b). [Note that every positive integer n can be expressed as $n = 2^r s$, where r is a nonnegative integer and s is a positive odd integer.]

7.9. An $m \times n$ checkerboard has $m \geq 2$ rows, $n \geq 2$ columns and mn squares in all. Two squares are adjacent if they belong to the same row or column and there is no square strictly between them. Conjecture for which integers m and n it is possible to number the squares from 1 to mn such that consecutively numbered squares are adjacent as are the squares numbered 1 and mn.

Section 7.2: Revisiting Quantified Statements

7.10. (a) Express the following quantified statement in symbols:
      For every odd integer n, the integer $3n + 1$ is even.
   (b) Prove that the statement in (a) is true.

7.11. (a) Express the following quantified statement in symbols:
      There exists a positive even integer n such that $3n + 2^{n-2}$ is odd.
   (b) Prove that the statement in (a) is true.

7.12. (a) Express the following quantified statement in symbols:
      For every positive integer n, the integer n^{n-1} is even.
   (b) Show that the statement in (a) is false.

7.13. (a) Express the following quantified statement in symbols:
      There exists an integer n such that $3n^2 - 5n + 1$ is an even integer.
   (b) Show that the statement in (a) is false.

7.14. (a) Express the following quantified statement in symbols:
      For every integer $n \geq 2$, there exists an integer m such that $n < m < 2n$.
   (b) Prove that the statement in (a) is true.

7.15. (a) Express the following quantified statement in symbols:
      There exists an integer n such that $m(n - 3) < 1$ for every integer m.
   (b) Prove that the statement in (a) is true.

7.16. (a) Express the following quantified statement in symbols:
      For every integer n, there exists an integer m such that $(n - 2)(m - 2) > 0$.
   (b) Express in symbols the negation of the statement in (a).
   (c) Show that the statement in (a) is false.

7.17. (a) Express the following quantified statement in symbols:
      There exists a positive integer n such that $-nm < 0$ for every integer m.

(b) Express in symbols the negation of the statement in (a).

(c) Show that the statement in (a) is false.

7.18. (a) Express the following quantified statement in symbols:

For every positive integer a, there exists an integer b with $|b| < a$ such that $|bx| < a$ for every real number x.

(b) Prove that the statement in (a) is true.

7.19. (a) Express the following quantified statement in symbols:

For every real number x, there exist integers a and b such that $a \le x \le b$ and $b - a = 1$.

(b) Prove that the statement in (a) is true.

7.20. (a) Express the following quantified statement in symbols:

There exists an integer n such that for every two real numbers x and y, $x^2 + y^2 \ge n$.

(b) Prove that the statement in (a) is true.

7.21. (a) Express the following quantified statement in symbols:

For every even integer a and odd integer b, there exists a rational number c such that either $a < c < b$ or $b < c < a$.

(b) Prove that the statement in (a) is true.

7.22. (a) Express the following quantified statement in symbols:

There exist two integers a and b such that for every positive integer n, $a < \frac{1}{n} < b$.

(b) Prove that the statement in (a) is true.

7.23. (a) Express the following quantified statement in symbols:

There exist odd integers a, b and c such that such that $a + b + c = 1$.

(b) Prove that the statement in (a) is true.

7.24. (a) Express the following quantified statement in symbols:

For every three odd integers a, b and c, their product abc is odd.

(b) Prove that the statement in (a) is true.

7.25. Consider the following statement.

R : There exists a real number L such that for every positive real

number e, there exists a positive real number d such that

if x is a real number with $|x| < d$, then $|3x - L| < e$.

(a) Use $P(x, d)$: $|x| < d$ and $Q(x, L, e)$: $|3x - L| < e$ to express the statement R in symbols.

(b) Prove that the statement R is true.

7.26. Prove the following statement. For every positive real number a and positive rational number b, there exist a real number c and irrational number d such that $ac + bd = 1$.

7.27. Prove the following statement. For every integer a, there exist integers b and c such that $|a - b| > cd$ for every integer d.

Section 7.3: Testing Statements

7.28. For the set $S = \{1, 2, 3, 4\}$, let

$P(n)$: $2^{n+1} + (-1)^{n+1} \left(2^n + 2^{n-1}\right)$ is prime. and $Q(n)$: $2n + 3$ is prime.

Prove or disprove: $\forall n \in S, P(n) \Rightarrow Q(n)$.

7.29. Let $P(n)$: $n^2 + 3n + 1$ is even. Prove or disprove:

(a) $\forall k \in \mathbf{N}, P(k) \Rightarrow P(k + 1)$.

(b) $\forall n \in \mathbf{N}, P(n)$.

For each of the exercises 7.30–7.81: Prove or disprove.

7.30. Let $x \in \mathbf{Z}$. If $4x + 7$ is odd, then x is even.

7.31. For every nonnegative integer n, there exists a nonnegative integer k such that $k < n$.

7.32. Every even integer can be expressed as the sum of two odd integers.

7.33. If $x, y, z \in \mathbf{Z}$ such that $x + y + z = 101$, then two of the integers x, y and z are of opposite parity.

7.34. For every two sets A and B, $(A \cup B) - B = A$.

7.35. Let A be a set. If $A \cap B = \emptyset$ for every set B, then $A = \emptyset$.

7.36. There exists an odd integer, the sum of whose digits is even and the product of whose digits is odd.

7.37. For every nonempty set A, there exists a set B such that $A \cup B = \emptyset$.

7.38. If x and y are real numbers, then $|x + y| = |x| + |y|$.

7.39. Let S be a nonempty set. For every proper subset A of S, there exists a nonempty subset B of S such that $A \cup B = S$ and $A \cap B = \emptyset$.

7.40. There is a real number solution of the equation $x^4 + x^2 + 1 = 0$.

7.41. There exists an integer a such that $a \cdot c \geq 0$ for every integer c.

7.42. There exist real numbers a, b and c such that $\frac{a+b}{a+c} = \frac{b}{c}$.

7.43. If $x, y \in \mathbf{R}$ and $x^2 < y^2$, then $x < y$.

7.44. Let $x, y, z \in \mathbf{Z}$. If $z = x - y$ and z is even, then x and y are odd.

7.45. Every odd integer can be expressed as the sum of three odd integers.

7.46. Let $x, y, z \in \mathbf{Z}$. If $z = x + y$ and x is odd, then y is even and z is odd.

7.47. For every two integers a and c, there exists an integer b such that $a + b = c$.

7.48. Every even integer can be expressed as the sum of two even integers.

7.49. For every two rational numbers a and b with $a < b$, there exists a rational number r such that $a < r < b$.

7.50. Let A, B, C and D be sets with $A \subseteq C$ and $B \subseteq D$. If A and B are disjoint, then C and D are disjoint.

7.51. Let A and B be sets. If $A \cup B \neq \emptyset$, then both A and B are nonempty.

7.52. For every two *positive* integers a and c, there exists a *positive* integer b such that $a + b = c$.

7.53. For every odd integer a, there exist integers b and c of opposite parity such that $a + b = c$.

7.54. For every rational number a/b, where $a, b \in \mathbf{N}$, there exists a rational number c/d, where c and d are positive odd integers, such that $0 < \frac{c}{d} < \frac{a}{b}$.

7.55. The equation $x^3 + x^2 - 1 = 0$ has a real number solution between $x = 0$ and $x = 1$.

7.56. There exists a real number x such that $x^2 < x < x^3$.

7.57. Let A and B be sets. If $A - B = B - A$, then $A - B = \emptyset$.

7.58. If $x \in \mathbf{Z}$, then $\frac{x^3+x}{x^4-1} = \frac{x}{x^2-1}$.

7.59. For every positive rational number b, there exists an irrational number a with $0 < a < b$.

7.60. Let A be a set. If $A - B = \emptyset$ for every set B, then $A = \emptyset$.

7.61. Let A, B and C be sets. If $A \cap B = A \cap C$, then $B = C$.

7.62. For every nonempty set A, there exists a set B such that $|A - B| = |B - A|$.

7.63. Let A be a set. If $A \cup B \neq \emptyset$ for every set B, then $A \neq \emptyset$.

7.64. There exist an irrational number a and a rational number b such that a^b is irrational.

7.65. There exists a real number solution of the equation $x^2 + x + 1 = 0$.

7.66. For every two sets A and B, $\mathcal{P}(A \cup B) = \mathcal{P}(A) \cup \mathcal{P}(B)$.

7.67. Every nonzero rational number can be expressed as the product of two irrational numbers.

7.68. If A and B are disjoint sets, then $\mathcal{P}(A)$ and $\mathcal{P}(B)$ are disjoint.

7.69. Let S be a nonempty set and let T be a collection of subsets of S. If $A \cap B \neq \emptyset$ for all pairs A, B of elements of T, then there exists an element $x \in S$ such that $x \in C$ for all $C \in T$.

7.70. Let A, B and C be sets. Then $A \cup (B - C) = (A \cup B) - (A \cup C)$.

7.71. Let $a, b, c \in \mathbf{Z}$. If ab, ac and bc are even, then a, b and c are even.

7.72. Let $n \in \mathbf{Z}$. If $n^3 + n$ is even, then n is even.

7.73. There exist three distinct integers a, b and c such that $a^b = b^c$.

7.74. Let $a, b, c \in \mathbf{Z}$. Then at least one of the numbers $a + b$, $a + c$ and $b + c$ is even.

7.75. Every integer can be expressed as the sum of two unequal integers.

7.76. There exist positive integers x and y such that $x^2 - y^2 = 101$.

7.77. For every positive integer n, $n^2 - n + 11$ is a prime.

7.78. For every odd prime p, there exist positive integers a and b such that $a^2 - b^2 = p$.

7.79. If the product of two consecutive integers is not divisible by 3, then their sum is.

7.80. The sum of every five consecutive integers is divisible by 5 and the sum of no six consecutive integers is divisible by 6.

7.81. There exist three distinct positive real numbers a, b, c, none of which are integers, such that all of ab, ac, bc and abc are integers.

ADDITIONAL EXERCISES FOR CHAPTER 7

7.82. (a) Show that the following statement is false: For every natural number x, there exists a natural number y such that $x < y < x^2$.
  (b) Make a small addition to the statement in (a) so that the new statement is true. Prove the new statement.

7.83. (a) Show that the following statement is false: Every positive integer is the sum of two distinct positive odd integers.
  (b) Make a small addition to the statement in (a) so that the new statement is true. Prove the new statement.

7.84. (a) Prove or disprove: There exist two distinct positive integers whose sum exceeds their product.
  (b) Your solution to (a) should suggest another problem to you. State and solve this new problem.

7.85. (a) Prove or disprove: If a and b are positive integers, then $\sqrt{a + b} = \sqrt{a} + \sqrt{b}$.
  (b) Prove or disprove: There exist positive real numbers a and b such that $\sqrt{a + b} = \sqrt{a} + \sqrt{b}$.
  (c) Complete the following statement so that it's true and provide a proof:
    Let $a, b \in \mathbf{R}^+ \cup \{0\}$. Then $\sqrt{a + b} = \sqrt{a} + \sqrt{b}$ if and only if _____.

7.86. Consider the open sentence $P(n): n! > \left(\frac{n}{2}\right)^n$ for $n \in \mathbf{N}$. Prove or disprove: $\forall n \in \mathbf{N}, P(n)$.

7.87. Evaluate the proof of the following statement.
  Result   Every even integer can be expressed as the sum of three distinct even integers.
  Proof   Let n be an even integer. Since $n + 2$, $n - 2$ and $-n$ are distinct even integers and

$$n = (n + 2) + (n - 2) + (-n),$$

the desired result follows.   ∎

7.88. It is known (although challenging to prove) that for every nonnegative integer m, the integer $8m + 3$ can be expressed as $a^2 + b^2 + c^2$ for positive integers a, b and c.

   (a) For every integer m with $0 \le m \le 10$, find positive integers a, b and c such that $8m + 3 = a^2 + b^2 + c^2$.

   (b) Prove or disprove: If a, b and c are positive integers such that $a^2 + b^2 + c^2 = 8m + 3$ for some integer m, then all of a, b and c are odd.

7.89. In Exercise 6 in Chapter 4, you were asked to prove the statement

$$P: \text{ Let } a \in \mathbf{Z}. \text{ If } 3 \mid 2a, \text{ then } 3 \mid a.$$

   (a) Prove that the converse of P is true. Now state P and its converse in a more familiar manner.

   (b) Is the statement obtained by interchanging 2 and 3 in P true?

   (c) Find a set S of positive integers with $2 \in S$ and $|S| \ge 3$ such that the following is true:

$$\text{Let } a \in \mathbf{Z}. \text{ If } 3 \mid ka, \text{ where } k \in S, \text{ then } 3 \mid a.$$

     Prove this generalization of the statement P.

   (d) In Exercise 72 in Chapter 4, it was shown for integers a and b that $3 \mid ab$ if and only if $3 \mid a$ or $3 \mid b$. How can this be used to answer (c)?

7.90. In Exercise 20 in Chapter 5, you were asked to prove that $\sqrt{2} + \sqrt{3}$ is irrational.

   (a) Prove that $\sqrt{2} + \sqrt{5}$ is irrational.

   (b) Determine, with proof, another positive integer a such that $\sqrt{2} + \sqrt{a}$ is irrational.

   (c) State and prove a generalization of the result in (a).

7.91. In Exercise 27 in Chapter 3, you were asked to prove the statement:

$$P: \text{ If } n \in \mathbf{Z}, \text{ then } n^3 - n \text{ is even.}$$

This can be restated as:

$$P: \text{ If } n \in \mathbf{Z}, \text{ then } 2 \mid (n^3 - n).$$

   (a) Find a positive integer $a \ne 2$ such that

$$\text{If } n \in \mathbf{Z}, \text{ then } a \mid (n^3 - n).$$

     is true and prove this statement.

   (b) Find a positive integer $k \ne 3$ such that

$$\text{If } n \in \mathbf{Z}, \text{ then } 2 \mid (n^k - n).$$

     is true and prove this statement.

   (c) Ask a question of your own dealing with P and provide an answer.

7.92. Let A denote the set of odd integers. Investigate the truth (or falseness) of the following statements.

   (a) For all $x, y \in A, 2 \mid (x^2 + 3y^2)$.

   (b) There exist $x, y \in A$ such that $4 \mid (x^2 + 3y^2)$.

   (c) For all $x, y \in A, 4 \mid (x^2 + 3y^2)$.

   (d) There exist $x, y \in A$ such that $8 \mid (x^2 + 3y^2)$.

   (e) There exist $x, y \in A$ such that $6 \mid (x^2 + 3y^2)$.

   (f) Provide a related statement of your own and determine whether it is true or false.

7.93. (a) Prove or disprove the following: There exist four positive integers a, b, c and d such that
$a^2 + b^2 + c^2 = d^2$.

(b) Prove or disprove the following: There exist four *distinct* positive integers a, b, c and d such that
$a^2 + b^2 + c^2 = d^2$.

(c) The problems in (a) or (b) above should suggest another problem that you can solve. State and solve such a problem.

(d) The problems in (a) or (b) above should suggest a conjecture to you (that you probably cannot solve). State such a conjecture.

8

Equivalence Relations

There are many common examples of relations in mathematics. For instance, three different ways that a real number x can be related to a real number y are:

$$(1)\ x < y,\ (2)\ y = x^2 + 1 \text{ or } (3)\ x = y.$$

Three different ways that an integer a can be related to an integer b are:

$$(1)\ a \mid b,\ (2)\ a \text{ and } b \text{ are of opposite parity or } (3)\ a \equiv b\ (\text{mod } 3).$$

In the area of geometry, three different ways that a line ℓ in 3-space can be related to a plane Π in 3-space are:

$$(1)\ \ell \text{ lies on } \Pi,\ (2)\ \ell \text{ is parallel to } \Pi \text{ or } (3)\ \ell \text{ intersects } \Pi \text{ in exactly one point.}$$

Three different ways that a triangle T can be related to a triangle T' are:

$$(1)\ T \text{ is congruent to } T',\ (2)\ T \text{ is similar to } T' \text{ or } (3)\ T \text{ has the same area as } T'.$$

All of the preceding examples concern two sets A and B (where possibly $A = B$) such that elements of A are related to elements of B in some manner. We now study this idea in a more general setting.

8.1 Relations

Let A and B be two sets. By a **relation R from A to B** we mean a subset of $A \times B$. That is, R is a set of ordered pairs, where the first coordinate of the pair belongs to A and the second coordinate belongs to B. If $(a, b) \in R$, then we say that a is **related** to b by R and write $a\,R\,b$. If $(a, b) \notin R$, then a is *not* related to b by R and we write $a\,\not{R}\,b$. For the sets $A = \{x, y, z\}$ and $B = \{1, 2\}$, the set

$$R = \{(x, 2), (y, 1), (y, 2)\} \tag{8.1}$$

is a subset of $A \times B$ and is therefore a relation from A to B. Thus, $x\,R\,2$ (x is related to 2) and $x\,\not{R}\,1$ (x is not related to 1). For two given sets A and B, it is always the case that $\emptyset$ and $A \times B$ are subsets of $A \times B$. Therefore, $\emptyset$ and $A \times B$ are both examples of relations from A to B. (Indeed, these are the extreme examples.) For the relation $\emptyset$, no element of A is related to any element of B; while for the relation $A \times B$, each element of A is related to every element of B. Simply said then, a relation from a set A to a set

B tells us which elements of A are related to which elements of B. Although this may seem like a fairly simple idea, it is very important that we have a thorough understanding of it.

Let R be a relation from A to B. The **domain** of R, denoted by dom(R), is the subset of A defined by

$$\text{dom}(R) = \{a \in A : (a, b) \in R \text{ for some } b \in B\};$$

while the **range** of R, denoted by range(R), is the subset of B defined by

$$\text{range}(R) = \{b \in B : (a, b) \in R \text{ for some } a \in A\}.$$

Hence dom(R) is that set of elements of A that occur as first coordinates among the ordered pairs in R and range(R) is the set of elements of B that occur as second coordinates among the ordered pairs in R. The domain and range of the relation R given in (8.1) are dom(R) = $\{x, y\}$ and range(R) = $\{1, 2\}$. The reason that $z \notin$ dom(R) is because there is no ordered pair in R whose first coordinate is z.

Let R be a relation from A to B. By the **inverse relation** of R is meant the relation R^{-1} from B to A defined by

$$R^{-1} = \{(b, a) : (a, b) \in R\}.$$

For example, the inverse relation of the relation $R = \{(x, 2), (y, 1), (y, 2)\}$ from $A = \{x, y, z\}$ to $B = \{1, 2\}$ is the relation

$$R^{-1} = \{(1, y), (2, x), (2, y)\}$$

from B to A.

By a **relation on a set** A, we mean a relation from A to A. That is, a relation on a single set A is a collection of ordered pairs whose first *and* second coordinates belong to A. Therefore, $\{(1, 2), (1, 3), (2, 2), (2, 3)\}$ is an example of a relation on the set $A = \{1, 2, 3, 4\}$.

If $A = \{1, 2\}$, then

$$A \times A = \{(1, 1), (1, 2), (2, 1), (2, 2)\}.$$

Since $|A \times A| = 4$, the number of subsets of $A \times A$ is $2^4 = 16$. However, a relation on A is, by definition, a subset of $A \times A$. Consequently, there are 16 relations on A. Six of these 16 relations are

$$\emptyset, \{(1, 2)\}, \{(1, 1), (1, 2)\}, \{(1, 2), (2, 1)\}, \{(1, 1), (1, 2), (2, 2)\}, A \times A.$$

8.2 Properties of Relations

For a relation defined on a single set, there are three properties that a relation may possess and which will be of special interest to us. A relation R defined on a set A is called **reflexive** if $x \, R \, x$ for every $x \in A$. That is, R is reflexive if $(x, x) \in R$ for every $x \in A$. Let $S = \{a, b, c\}$ and consider the following six relations defined on S:

$$R_1 = \{(a, b), (b, a), (c, a)\}$$
$$R_2 = \{(a, b), (b, b), (b, c), (c, b), (c, c)\}$$
$$R_3 = \{(a, a), (a, c), (b, b), (c, a), (c, c)\}$$

$$R_4 = \{(a, a), (a, b), (b, b), (b, c), (a, c)\}$$
$$R_5 = \{(a, a), (a, b)\}$$
$$R_6 = \{(a, b), (a, c)\}.$$

The relation R_1 is not reflexive since $(a, a) \notin R_1$, for example. Since $(a, a) \notin R_2$, it follows that R_2 is not reflexive either. Because $(a, a), (b, b), (c, c) \in R_3$, the relation R_3 is reflexive. None of the relations R_4, R_5, R_6 are reflexive.

A relation R defined on a set A is called **symmetric** if whenever $x \ R \ y$, then $y \ R \ x$ for all $x, y \in A$. Hence for a relation R on A to be "not symmetric," there must be some ordered pair (w, z) in R for which $(z, w) \notin R$. Certainly, if such an ordered pair (w, z) exists, then $w \neq z$. The relation R_1 is not symmetric since $(c, a) \in R_1$ but $(a, c) \notin R_1$. Notice that $(a, b) \in R_1$ and $(b, a) \in R_1$, but this does not mean that R_1 is symmetric. Recall that the definition of a symmetric relation R on a set A says that whenever $x \ R \ y$, then $y \ R \ x$ *for all* $x, y \in A$.

The relation R_3 *is* symmetric, however, since both (a, c) and (c, a) belong to R_3. None of the ordered pairs $(a, a), (b, b), (c, c)$ in R_3 are relevant as to whether R_3 is symmetric. None of the relations R_2, R_4, R_5, R_6 are symmetric.

A relation R defined on a set A is called **transitive** if whenever $x \ R \ y$ and $y \ R \ z$, then $x \ R \ z$ for all $x, y, z \in A$. Notice that in this definition, it is *not* required that x, y and z be distinct. Hence for a relation R on A to be "not transitive," there must exist two ordered pairs (u, v) and (v, w) in R such that $(u, w) \notin R$. If this should occur, then necessarily $u \neq v$ and $v \neq w$ (although perhaps $u = w$). For example, the relation R_2 is not transitive since $(a, b), (b, c) \in R_2$ but $(a, c) \notin R_2$. Actually, R_1 is not transitive either because $(a, b), (b, a) \in R_1$ but $(a, a) \notin R_1$. The example (counterexample) that shows that R_1 is not transitive illustrates the fact that showing a relation is not transitive may not be easy. All of the relations R_3, R_4, R_5, R_6 are transitive. It is not always easy to convince oneself that a relation *is* transitive either. Let's give a careful argument as to why the relations R_5 and R_6 are transitive.

For R_5 to be transitive, it is required that (x, z) belongs to R_5 whenever (x, y) and (y, z) belong to R_5 for all $x, y, z \in A$. To verify that the transitive property holds in R_5, we must consider *all* possible pairs of ordered pairs of the type (x, y) and (y, z). We have two choices for (x, y) in R_5, namely (a, a) and (a, b), that is, $x = a$ and $y = a$, or $x = a$ and $y = b$. If $(x, y) = (a, a)$, then $y = a$ and so either $(y, z) = (a, a)$ or $(y, z) = (a, b)$. In the first case, we have

$$(a, a) \in R_5 \text{ and } (a, a) \in R_5,$$

and $(x, z) = (a, a)$ belongs to R_5. In the second case,

$$(a, a) \in R_5 \text{ and } (a, b) \in R_5,$$

and $(x, z) = (a, b)$ belongs to R_5. This example suggests (correctly!) that if (x, y) and (y, z) belong to some relation R and $x = y$, then certainly $(x, z) \in R$. The same could be said if $y = z$. Thus, when checking transitivity, we need only consider ordered pairs (x, y) and (y, z) for which $x \neq y$ and $y \neq z$. Suppose next that $(x, y) = (a, b)$, so that $y = b$. Here there is no ordered pair of the type (y, z); that is, there is no ordered pair of R_5 whose first coordinate is b. Thus, there is nothing to check when $(x, y) = (a, b)$. For R_5, there are only two possibilities for two ordered pairs of the type $(x, y), (y, z)$ and in each case, $(x, z) \in R_5$. Thus R_5 is transitive.

Let's turn to R_6 now. The relation R_6 does not contain two ordered pairs of the type (x, y), (y, z) since if $(x, y) = (a, b)$, no ordered pair has b as its first coordinate; while if $(x, y) = (a, c)$, no ordered pair has c as its first coordinate. Consequently, the hypothesis of the transitive property is false and the implication "If $(x, y) \in R_6$ and $(y, z) \in R_6$, then $(x, z) \in R_6$." is satisfied vacuously. Hence, R_6 is transitive. Another way to convince yourself that R_6 is transitive is to think of what must happen if R_6 is not transitive; namely, there must be two ordered pairs (x, y), (y, z) in R_6 such that $(x, z) \notin R_6$. But there are no such ordered pairs (x, y) and (y, z)!

In the preceding discussions, we have made use of an important point when testing a relation for transitivity. It bears repeating here.

When we are attempting to determine whether a relation R is transitive and, consequently, checking all pairs of the type (x, y) and (y, z), we need not consider the situation where $x = y$ or $y = z$.

In this case, the ordered pair (x, z) will always be present in R. If a relation R is not transitive, then there must exist ordered pairs (x, y) and (y, z) in R, where $x \neq y$ and $y \neq z$, such that (x, z) is not in R. That is, (x, y) and (y, z) constitute a counterexample to the implication "If $(x, y) \in R$ and $(y, z) \in R$, then $(x, z) \in R$," which is a requirement of R being transitive.

We already mentioned that relations occur frequently in mathematics. Let R be the relation defined on the set $\mathbf{Z}$ of integers by $a\ R\ b$ if $a \leq b$; that is, R is the relation $\leq$. Since $x \leq x$ for every integer x, it follows that $x\ R\ x$ for every $x \in \mathbf{Z}$; that is, R is reflexive. Certainly, $2\ R\ 3$ since $2 \leq 3$. However, $3 > 2$; so $3\ \not R\ 2$. Therefore, R is not symmetric. On the other hand, it is a well-known property of integers that if $a \leq b$ and $b \leq c$, then $a \leq c$. Therefore, if $a\ R\ b$ and $b\ R\ c$, then $a\ R\ c$. So R is transitive.

Another relation R we could consider on the set $\mathbf{Z}$ is defined by $a\ R\ b$ if $a \neq b$. However, then $1\ \not R\ 1$ since $1 = 1$. Consequently, this relation is not reflexive. If a and b are integers such that $a \neq b$, then we also have $b \neq a$. So if $a\ R\ b$, then $b\ R\ a$. This says that this relation is symmetric. Notice that $2 \neq 3$ and $3 \neq 2$ but $2 = 2$. That is, $2\ R\ 3$ and $3\ R\ 2$ but $2\ \not R\ 2$. Therefore, R is not transitive.

The **distance** between two real numbers a and b is $|a - b|$. So the distance between 2 and 4.5 is $|2 - 4.5| = |-2.5| = 2.5$. Thus if the real numbers (points) a and b are plotted on the real number line (x-axis), then the length of the segment between them is the distance. This is illustrated in Figure 8.1 for the real numbers $a = 3$ and $b = -2$, where the distance between them is thus $|a - b| = |3 - (-2)| = 5 = |(-2) - 3| = |b - a|$.

Define a relation R on the set $\mathbf{R}$ of real numbers by $a\ R\ b$ if $|a - b| \leq 1$; that is, a is related to b if the distance between a and b is at most 1. Certainly, the distance from a real number to itself is 0; that is, $|a - a| = 0 \leq 1$ for every $x \in \mathbf{R}$. So $a\ R\ a$ and R is reflexive. If the distance between two real numbers a and b is at most 1, then the distance

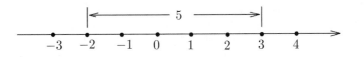

Figure 8.1   The distance between 3 and -2

between b and a is at most 1. In symbols, if $|a - b| \leq 1$, then $|b - a| = |a - b| \leq 1$; that is, if $a \, R \, b$, then $b \, R \, a$. Therefore, R is symmetric. Now to the transitive property. If $a \, R \, b$ and $b \, R \, c$, is $a \, R \, c$? That is, if the distance between a and b is at most 1 and the distance between b and c is at most 1, does it follow that the distance between a and c is at most 1? The answer is no. For example, $3 \, R \, 2$ and $2 \, R \, 1$ since $|3 - 2| \leq 1$ and $|2 - 1| \leq 1$. However, $|3 - 1| = 2$. So $3 \, \not{R} \, 1$ and R is not transitive.

8.3 Equivalence Relations

Perhaps the most familiar relation in mathematics is the equals relation. For example, let R be the relation defined on $\mathbf{Z}$ by $a \, R \, b$ if $a = b$. For every integer a, we have $a = a$ and so $a \, R \, a$. If $a = b$, then $b = a$. Hence if $a \, R \, b$, then $b \, R \, a$. Also, if $a = b$ and $b = c$, then $a = c$. So if $a \, R \, b$ and $b \, R \, c$, then $a \, R \, c$. These observations tell us that the equals relation on the set of integers possesses all three of the properties reflexive, symmetric and transitive. This suggests the question of asking what other relations (on the set $\mathbf{Z}$ or indeed on any set) have these same three properties possessed by the equals relation. These are the relations that will be our primary focus in this chapter.

A relation R on a set A is called an **equivalence relation** if R is reflexive, symmetric and transitive. Of course then, the equals relation R defined on $\mathbf{Z}$ by $a \, R \, b$ if $a = b$ is an equivalence relation on $\mathbf{Z}$. For another example, consider the set $A = \{1, 2, 3, 4, 5, 6\}$ and the relation

$$R = \{(1, 1), (2, 2), (3, 3), (4, 4), (5, 5), (6, 6), (1, 3), (1, 6), (6, 1), (6, 3),$$

$$(3, 1), (3, 6), (2, 4), (4, 2)\} \tag{8.2}$$

defined on A. This relation has all three of the properties reflexive, symmetric and transitive and is consequently an equivalence relation.

Suppose that R is an equivalence relation on some set A. If $a \in A$, then a is related to a since R is reflexive. Quite possibly, other elements of A are related to a as well. The set of all elements that are related to a given element of A will turn out to be important and, for this reason, these sets are given special names. For an equivalence relation R defined on a set A and for $a \in A$, the set

$$[a] = \{x \in A : x \, R \, a\}$$

consisting of all elements in A that are related to a, is called an **equivalence class**, in fact, the equivalence class containing a since $a \in [a]$ (because R is reflexive). Loosely speaking, then, $[a]$ consists of the "relatives" of a. For the equivalence relation R defined in (8.2), the resulting equivalence classes are

$$[1] = \{1, 3, 6\}, \quad [2] = \{2, 4\}, \quad [3] = \{1, 3, 6\},$$

$$[4] = \{2, 4\}, \quad [5] = \{5\}, \quad [6] = \{1, 3, 6\}. \tag{8.3}$$

Since $[1] = [3] = [6]$ and $[2] = [4]$, there are only three distinct equivalence classes in this case, namely $[1]$, $[2]$ and $[5]$.

Let's return to the equals relation R defined on $\mathbf{Z}$ by $a \, R \, b$ if $a = b$ and determine the equivalence classes for this equivalence relation. For $a \in \mathbf{Z}$,

$$[a] = \{x \in \mathbf{Z} : x \, R \, a\} = \{x \in \mathbf{Z} : x = a\} = \{a\};$$

that is, every integer is in an equivalence class by itself.

As another illustration, define a relation R on the set L of straight lines in a plane by $\ell_1\ R\ \ell_2$ if either $\ell_1 = \ell_2$ (the lines coincide) or ℓ_1 is parallel to ℓ_2. Since every line coincides with itself, R is reflexive. If a line ℓ_1 is parallel to a line ℓ_2 (or they coincide), then ℓ_2 is parallel to ℓ_1 (or they coincide). Thus R is symmetric. Finally, if ℓ_1 is parallel to ℓ_2 and ℓ_2 is parallel to ℓ_3 (including the possibility that such pairs of lines may coincide), then either ℓ_1 is parallel to ℓ_3 or they coincide. Indeed, it may very well occur that ℓ_1 and ℓ_2 are distinct parallel lines, as are ℓ_2 and ℓ_3 but ℓ_1 and ℓ_3 coincide. In any case, though, this relation is transitive. Therefore, R is an equivalence relation. Hence for $\ell \in L$, the equivalence class

$$[\ell] = \{x \in L : x\ R\ \ell\} = \{x \in L : x = \ell \text{ or } x \text{ is parallel to } \ell\};$$

that is, the equivalence class $[\ell]$ consists of ℓ and all lines in the plane parallel to ℓ.

To describe additional examples of relations from geometry, let $\mathcal{T}$ be the set of all triangles in a plane. For two triangles T and T' in $\mathcal{T}$, define relations R_1 and R_2 on $\mathcal{T}$ by $T\ R_1\ T'$ if T is congruent to T' and $T\ R_2\ T'$ if T is similar to T'. Then both R_1 and R_2 are equivalence relations. For a triangle T and the relation R_1, $[T]$ is the set of triangles in $\mathcal{T}$ that are congruent to T; while for R_2, $[T]$ is the set of triangles in $\mathcal{T}$ that are similar to T.

The relation R defined on $\mathbf{Z}$ by $x\ R\ y$ if $|x| = |y|$ is also an equivalence relation. In this case, for $a \in \mathbf{Z}$, the equivalence class $[a]$ consists of the two integers a and $-a$, unless $a = 0$, in which case $[0] = \{0\}$. We now consider an example that requires more thought and explanation.

Result 8.1   *A relation R is defined on $\mathbf{Z}$ by $x\ R\ y$ if $x + 3y$ is even. Then R is an equivalence relation.*

Before proving this result, let's be certain that we understand this relation. First, notice that $5\ R\ 7$ since $5 + 3 \cdot 7 = 26$ is even. However, $8\ \cancel{R}\ 9$ since $8 + 3 \cdot 9 = 35$ is not even. On the other hand, $4\ R\ 4$ because $4 + 3 \cdot 4 = 16$ is even.

Proof of Result 8.1   First we show that R is reflexive. Let $a \in \mathbf{Z}$. Then $a + 3a = 4a = 2(2a)$ is even since $2a \in \mathbf{Z}$. Therefore $a\ R\ a$ and R is reflexive.

Next we show that R is symmetric. Assume that $a\ R\ b$. Thus $a + 3b$ is even. Hence $a + 3b = 2k$ for some integer k. So $a = 2k - 3b$. Therefore,

$$b + 3a = b + 3(2k - 3b) = b + 6k - 9b = 6k - 8b = 2(3k - 4b).$$

Since $3k - 4b$ is an integer, $b + 3a$ is even. Therefore, $b\ R\ a$ and R is symmetric.

Finally, we show that R is transitive. Assume that $a\ R\ b$ and $b\ R\ c$. Hence $a + 3b$ and $b + 3c$ are even; so $a + 3b = 2k$ and $b + 3c = 2\ell$ for some integers k and ℓ. Adding these two equations, we obtain $(a + 3b) + (b + 3c) = 2k + 2\ell$. So $a + 4b + 3c = 2k + 2\ell$ and $a + 3c = 2k + 2\ell - 4b = 2(k + \ell - 2b)$. Since $k + \ell - 2b$ is an integer, $a + 3c$ is even. Hence $a\ R\ c$ and so R is transitive. Therefore, R is an equivalence relation.   ∎

PROOF ANALYSIS   A few remarks concerning the preceding proof are in order. Recall that a relation R defined on a set A is reflexive if $x\ R\ x$ for every $x \in A$. The reflexive property may also be reworded to read: For every $x \in A$, $x\ R\ x$ or: If $x \in A$, then $x\ R\ x$. Hence when we proved that R is reflexive in Result 8.1, we began by assuming that a was an arbitrary element of **Z**. (We're giving a direct proof.) We were then required to show that $a + 3a$ is even, which we did. It would be incorrect, however, to assume that $a + 3a$ is even or that $a\ R\ a$. This, in fact, is what we want to prove. ◆

Since the relation defined in Result 8.1 is an equivalence relation, there are equivalence classes, namely an equivalence class $[a]$ for each $a \in$ **Z**. Let's start with 0, say. The equivalence class $[0]$ is the set of all integers related to 0. In symbols, this equivalence class is

$$[0] = \{x \in \mathbf{Z}:\ x\ R\ 0\} = \{x \in \mathbf{Z}:\ x + 3 \cdot 0 \text{ is even}\}$$
$$= \{x \in \mathbf{Z}:\ x \text{ is even}\} = \{0, \pm 2, \pm 4, \ldots\};$$

that is, $[0]$ is the set of even integers. It shouldn't be difficult to see that if a is an even integer, say $a = 2k$, where $k \in$ **Z**, then

$$[a] = \{x \in \mathbf{Z}:\ x\ R\ a\} = \{x \in \mathbf{Z}:\ x + 3a \text{ is even}\}$$
$$= \{x \in \mathbf{Z}:\ x + 3(2k) \text{ is even}\} = \{x \in \mathbf{Z}:\ x + 6k \text{ is even}\}$$

is also the set of even integers. On the other hand, the equivalence class consisting of those integers related to 1 is

$$[1] = \{x \in \mathbf{Z}:\ x\ R\ 1\} = \{x \in \mathbf{Z}:\ x + 3 \cdot 1 \text{ is even}\}$$
$$= \{x \in \mathbf{Z}:\ x + 3 \text{ is even}\} = \{\pm 1, \pm 3, \pm 5, \ldots\},$$

which is the set of odd integers. In fact, if a is an odd integer, then $a = 2\ell + 1$ for some integer ℓ and

$$[a] = \{x \in \mathbf{Z}:\ x + 3a \text{ is even}\} = \{x \in \mathbf{Z}:\ x + 3(2\ell + 1) \text{ is even}\}$$
$$= \{x \in \mathbf{Z}:\ x + 6\ell + 3 \text{ is even}\}$$

is the set of odd integers. Therefore, if a and b are even, then $[a] = [b]$ is the set of even integers; while if a and b are odd, then $[a] = [b]$ is the set of odd integers. Hence there are only two distinct equivalence classes, namely $[0]$ and $[1]$, the sets of even and odd integers, respectively. We will soon see that there is a good reason for this observation.

8.4 Properties of Equivalence Classes

You may have noticed that in the preceding examples of equivalence relations, we have seen several situations where two equivalence classes are equal. It is possible to determine exactly when this happens.

Theorem 8.2 *Let R be an equivalence relation on a nonempty set A and let a and b be elements of A. Then $[a] = [b]$ if and only if $a\ R\ b$.*

Proof Assume that $a\ R\ b$. We show that the sets $[a]$ and $[b]$ are equal by verifying that $[a] \subseteq [b]$ and $[b] \subseteq [a]$. First, we show that $[a] \subseteq [b]$. Let $x \in [a]$. Then $x\ R\ a$. Since $a\ R\ b$ and R is transitive, $x\ R\ b$. Therefore, $x \in [b]$ and so $[a] \subseteq [b]$. Next, let $y \in [b]$. Thus, $y\ R\ b$. Since $a\ R\ b$ and R is symmetric, $b\ R\ a$. Again, by the transitivity of R, we have $y\ R\ a$. Therefore, $y \in [a]$ and so $[b] \subseteq [a]$. Hence $[a] = [b]$.

For the converse, assume that $[a] = [b]$. Because R is reflexive, $a \in [a]$. But, since $[a] = [b]$, it follows that $a \in [b]$. Consequently, $a \, R \, b$. ∎

According to Theorem 8.2 then, if R is an equivalence relation on a set A and a is related to b, then the set $[a]$ of elements of A related to a and the set $[b]$ of elements of A related to b are equal, that is, $[a] = [b]$. Because the theorem characterizes when $[a] = [b]$, we know that if $a \, \cancel{R} \, b$, then $[a] \neq [b]$.

Let's return once more to the equivalence relation defined in (8.2) on the set $A = \{1, 2, 3, 4, 5, 6\}$ and the equivalence classes given in (8.3). We observed earlier that $[1] = [3] = [6]$. Since every two of the integers 1, 3, 6 are related to each other (according to the definition of R), Theorem 8.2 tells us that the equality of $[1]$, $[3]$ and $[6]$ is expected. The same can be said of $[2]$ and $[4]$. However, since $(5, 6) \notin R$, for example, Theorem 8.2 tells us that $[5] \neq [6]$, which is the case. Therefore, as we also observed earlier, there are only three distinct equivalence classes, namely,

$$[1] = [3] = [6] = \{1, 3, 6\}, \quad [2] = [4] = \{2, 4\}, \quad [5] = \{5\}. \tag{8.4}$$

Now, you might have noticed one other thing. Every element of A belongs to exactly one equivalence class. This observation might remind you of a concept we discussed earlier.

Recall that a **partition** P of a nonempty set S is a collection of nonempty subsets of S with the property that every element of S belongs to exactly one of these subsets; that is, P is a collection of pairwise disjoint, nonempty subsets of S whose union is S. Hence the set of the distinct equivalence classes in (8.4) is a partition of the set $A = \{1, 2, 3, 4, 5, 6\}$. We now show that this too is expected.

Theorem 8.3 *Let R be an equivalence relation defined on a nonempty set A. Then the set*

$$P = \{[a] \, : \, a \in A\}$$

of equivalence classes resulting from R is a partition of A.

Proof Certainly, each equivalence class $[a]$ is nonempty since $a \in [a]$ and so each element of A belongs to at least one equivalence class. We show that every element of A belongs to exactly one equivalence class. Assume that some element x of A belongs to two equivalence classes, say $[a]$ and $[b]$. Since $x \in [a]$ and $x \in [b]$, it follows that $x \, R \, a$ and $x \, R \, b$. Because R is symmetric, $a \, R \, x$. Thus $a \, R \, x$ and $x \, R \, b$. Since R is transitive, $a \, R \, b$. By Theorem 8.2, it follows that $[a] = [b]$. So any two equivalence classes to which x belongs are equal. Hence x belongs to a unique equivalence class. ∎

In the proof of Theorem 8.3, we were required to show that each element $x \in A$ belongs to a unique equivalence class. During this proof, we assumed that x belongs to two equivalence classes $[a]$ and $[b]$. Observe that we made no assumption whether $[a]$ and $[b]$ are distinct. Later we learned that $[a] = [b]$; so x can only belong to one equivalence class. With a very small change, we could have reached the same conclusion by a different proof technique. We could have said: Assume, to the contrary, that x belongs to two *distinct* equivalence classes $[a]$ and $[b]$. By the same argument as above, we can show that $[a] = [b]$. However, *now*, this produces a contradiction, and we have just given a proof by contradiction.

According to Theorem 8.3 then, whenever we have an equivalence relation R defined on a nonempty set A, a partition of A into the associated equivalence classes of R results. Perhaps unexpectedly, the converse is also true. That is, if we are given a partition of A, then there is a corresponding equivalence relation that can be defined on A, whose resulting equivalence classes are the elements of the given partition. For example, let

$$P = \{\{1, 3, 4\}, \{2, 7\}, \{5, 6\}\}$$

be a given partition of the set $A = \{1, 2, 3, 4, 5, 6, 7\}$. (Notice that every element of A belongs to exactly one subset in P.) Then

$$R = \{(1, 1), (1, 3), (1, 4), (2, 2), (2, 7), (3, 1), (3, 3), (3, 4), (4, 1),$$
$$(4, 3), (4, 4), (5, 5), (5, 6), (6, 5), (6, 6), (7, 2), (7, 7)\}$$

is an equivalence relation on A, whose distinct equivalence classes are

$$[1] = \{1, 3, 4\}, \quad [2] = \{2, 7\} \quad \text{and} \quad [5] = \{5, 6\},$$

and so $P = \{[1], [2], [5]\}$. We now establish this result in general; that is, if we have a nonempty set A and a partition P of A, then it is possible to create an equivalence relation R on A such that the distinct equivalence classes of R are precisely the subsets in P. Since we are trying to verify this in general (and not for a specific example), we need to describe the subsets in P with the aid of an index set. Since we will want every subset in P to be an equivalence class, we will need every two elements in the same subset to be related. On the other hand, since we will want two different subsets in P to be different equivalence classes, we will need elements in distinct subsets not to be related.

Theorem 8.4   *Let $P = \{A_\alpha : \alpha \in I\}$ be a partition of a nonempty set A. Then there exists an equivalence relation R on A such that P is the set of equivalence classes determined by R, that is, $P = \{[a] : a \in A\}$.*

Proof   Define a relation R on A by $x \ R \ y$ if x and y belong to the same subset in P; that is, $x \ R \ y$ if $x, y \in A_\alpha$ for some $\alpha \in I$. We now show that R is an equivalence relation. Let $a \in A$. Since P is a partition of A, it follows that $a \in A_\beta$ for some $\beta \in I$. Trivially, a and a belong to A_β; so $a \ R \ a$ and R is reflexive.

Next, let $a, b \in A$ and assume that $a \ R \ b$. Then a and b belong to A_γ for some $\gamma \in I$. Hence b and a belong to A_γ; so $b \ R \ a$ and R is symmetric.

Finally, let a, b and c be elements of A such that $a \ R \ b$ and $b \ R \ c$. Therefore, $a, b \in A_\beta$ and $b, c \in A_\gamma$ for some $\beta, \gamma \in I$. Since P is a partition of A, the element b belongs to only one set in P. Hence $A_\beta = A_\gamma$ and so $a, c \in A_\beta$. Thus $a \ R \ c$ and R is transitive. Therefore, R is an equivalence relation on A.

We now consider the equivalence classes resulting from R. Let $a \in A$. Then $a \in A_\alpha$ for some $\alpha \in I$. The equivalence class $[a]$ consists of all elements of A related to a. From the way that R is defined, however, the only elements related to a are those elements belonging to the same subset in P to which a belongs, that is, $[a] = A_\alpha$. Hence

$$\{[a] : a \in A\} = \{A_\alpha : \alpha \in I\} = P. \qquad \blacksquare$$

We now give an additional example of an equivalence relation. Although this example is similar to the one described in Result 8.1, it is different enough to require some thought.

Result to Prove   A relation R is defined on $\mathbf{Z}$ by $x \ R \ y$ if $11x - 5y$ is even. Then R is an equivalence relation.

PROOF STRATEGY   Since we want to verify that R is an equivalence relation, we need to show that R is reflexive, symmetric and transitive. Let's start with the first of these. We begin with an integer a. To show that $a \ R \ a$, we need to show that $11a - 5a$ is even. However, $11a - 5a = 6a = 2(3a)$, so this shouldn't cause any difficulties.

To verify that R is symmetric, we begin with $a \ R \ b$ (where $a, b \in \mathbf{Z}$, of course) and attempt to show that $b \ R \ a$. Since $a \ R \ b$, it follows that $11a - 5b$ is even. To show that $b \ R \ a$, we need to show that $11b - 5a$ is even. Since $11a - 5b$ is even, we can write $11a - 5b = 2k$ for some integer k. At first though, it might seem like a good idea to solve for a in terms of b or solve for b in terms of a. However, because neither the coefficient of a nor the coefficient of b is 1 or -1 in $11a - 5b = 2k$, fractions would be introduced. We need another approach. Notice that if we write

$$11b - 5a = (11a - 5b) + (?\ a + ?\ b),$$

then we have

$$11b - 5a = (11a - 5b) + (-16a + 16b)$$
$$= 2k - 16a + 16b = 2(k - 8a + 8b).$$

This will work.

To verify that R is transitive, We begin by assuming that $a \ R \ b$ and $b \ R \ c$ (and attempt to show that $a \ R \ c$). Thus $11a - 5b$ and $11b - 5c$ are even and so

$$11a - 5b = 2k \quad \text{and} \quad 11b - 5c = 2\ell \tag{8.5}$$

for integers k and ℓ. To show that $a \ R \ c$, we must verify that $11a - 5c$ is even. We need to work the expression $11a - 5c$ into the discussion. However, this can be done by adding the expressions in (8.5). We're ready to give a proof now. ◆

Result 8.5   *A relation R is defined on $\mathbf{Z}$ by $x \ R \ y$ if $11x - 5y$ is even. Then R is an equivalence relation.*

Proof   First, we show that R is reflexive. Let $a \in \mathbf{Z}$. Then $11a - 5a = 6a = 2(3a)$. Since $3a$ is an integer, $11a - 5a$ is even. Thus $a \ R \ a$ and R is reflexive.

Next we show that R is symmetric. Assume that $a \ R \ b$, where $a, b \in \mathbf{Z}$. Thus $11a - 5b$ is even. Therefore, $11a - 5b = 2k$, where $k \in \mathbf{Z}$. Observe that

$$11b - 5a = (11a - 5b) + (-16a + 16b)$$
$$= 2k - 16a + 16b = 2(k - 8a + 8b).$$

Since $k - 8a + 8b$ is an integer, $11b - 5a$ is even. Hence $b \ R \ a$ and R is symmetric.

Finally, we show that R is transitive. Assume that $a \ R \ b$ and $b \ R \ c$. Hence $11a - 5b$ and $11b - 5c$ are even. Therefore, $11a - 5b = 2k$ and $11b - 5c = 2\ell$, where $k, \ell \in \mathbf{Z}$.

Adding these equations, we obtain $(11a - 5b) + (11b - 5c) = 2k + 2\ell$. Solving for $11a - 5c$, we have

$$11a - 5c = 2k + 2\ell - 6b = 2(k + \ell - 3b).$$

Since $k + \ell - 3b$ is an integer, $11a - 5c$ is even. Hence $a \; R \; c$ and R is transitive. Therefore, R is an equivalence relation. ∎

We now determine the equivalence classes for the equivalence relation just discussed. Let's begin with the equivalence class containing 0, say. Then

$$[0] = \{x \in \mathbf{Z} : \; x \; R \; 0\} = \{x \in \mathbf{Z} : \; 11x \text{ is even}\}$$
$$= \{x \in \mathbf{Z} : \; x \text{ is even}\} = \{0, \pm 2, \pm 4, \ldots\}.$$

Recall that the distinct equivalence classes always produce a partition of the set involved (in this case $\mathbf{Z}$). Since the class $[0]$ does not consist of all integers, there is at least one other equivalence class. To determine another equivalence class, we look for an element that does not belong to $[0]$. Since $1 \notin [0]$, the equivalence class $[1]$ is distinct (and disjoint) from $[0]$. Thus

$$[1] = \{x \in \mathbf{Z} : \; x \; R \; 1\} = \{x \in \mathbf{Z} : \; 11x - 5 \text{ is even}\}$$
$$= \{x \in \mathbf{Z} : \; x \text{ is odd}\} = \{\pm 1, \pm 3, \pm 5, \ldots\}.$$

Since $[0]$ and $[1]$ produce a partition of $\mathbf{Z}$ (that is, every integer belongs to exactly one of $[0]$ and $[1]$), these are the only equivalence classes in this case.

8.5 Congruence Modulo n

Next we describe one of the most important equivalence relations. If you have more mathematics in your future, it is likely that you will see the equivalence relation we are about to describe again—indeed often. Recall again that for integers a and b, where $a \neq 0$, the integer a is said to **divide** b, written as $a \mid b$, if there exists an integer c such that $b = ac$. Also, for integers a, b and $n \geq 2$, a is said to be **congruent to** b **modulo** n, written $a \equiv b \pmod{n}$, if $n \mid (a - b)$. For example, $24 \equiv 6 \pmod{9}$ since $9 \mid (24 - 6)$; while $1 \equiv 5 \pmod{2}$ since $2 \mid (1 - 5)$. Also, $4 \equiv 4 \pmod{5}$ since $5 \mid (4 - 4)$. However, $8 \not\equiv 2 \pmod{4}$ since $4 \nmid (8 - 2)$. These concepts were introduced in Chapter 4.

Let's consider a few examples of pairs a, b of integers such that $a \equiv b \pmod{5}$. Notice that $7 \equiv 7 \pmod{5}$, $-1 \equiv -1 \pmod{5}$ and $0 \equiv 0 \pmod{5}$. Also, $2 \equiv -8 \pmod{5}$ and $-8 \equiv 2 \pmod{5}$. Notice also that $2 \equiv 17 \pmod{5}$. Therefore, both $-8 \equiv 2 \pmod{5}$ and $2 \equiv 17 \pmod{5}$. Furthermore, $-8 \equiv 17 \pmod{5}$. These examples might suggest that the reflexive, symmetric and transitive properties are satisfied here, a fact which we are about to verify. This is the important equivalence relation we referred to at the beginning of this section, not just for $n = 5$ but for any integer $n \geq 2$.

Theorem 8.6   *Let $n \in \mathbf{Z}$, where $n \geq 2$. Then congruence modulo n (that is, the relation R defined on $\mathbf{Z}$ by $a \; R \; b$ if $a \equiv b \pmod{n}$) is an equivalence relation on $\mathbf{Z}$.*

Proof   Let $a \in \mathbf{Z}$. Since $n \mid 0$, it follows that $n \mid (a - a)$ and so $a \equiv a \pmod{n}$. Thus, $a \; R \; a$, implying that R is reflexive.

Next, we show that R is symmetric. Assume that $a \, R \, b$, where $a, b \in \mathbf{Z}$. Since $a \, R \, b$, it follows that $a \equiv b \pmod{n}$ and so $n \mid (a - b)$. Hence, there exists $k \in \mathbf{Z}$ such that $a - b = nk$. Thus,

$$b - a = -(a - b) = -(nk) = n(-k).$$

Since $-k \in \mathbf{Z}$, it follows that $n \mid (b - a)$ and so $b \equiv a \pmod{n}$. Therefore, $b \, R \, a$ and R is symmetric.

Finally, we show that R is transitive. Assume that $a \, R \, b$ and $b \, R \, c$, where $a, b, c \in \mathbf{Z}$. We show that $a \, R \, c$. Since $a \, R \, b$ and $b \, R \, c$, we know that $a \equiv b \pmod{n}$ and $b \equiv c \pmod{n}$. Thus, $n \mid (a - b)$ and $n \mid (b - c)$. Consequently,

$$a - b = nk \quad \text{and} \quad b - c = n\ell \tag{8.6}$$

for some integers k and ℓ. Adding the equations in (8.6), we obtain

$$(a - b) + (b - c) = nk + n\ell = n(k + \ell);$$

so $a - c = n(k + \ell)$. Since $k + \ell \in \mathbf{Z}$, we have $n \mid (a - c)$ and so $a \equiv c \pmod{n}$. Therefore, $a \, R \, c$ and R is transitive. $\blacksquare$

PROOF ANALYSIS   Theorem 8.6 describes a well-known equivalence relation. Let's review how we verified this. The proof we gave to show that congruence modulo n is an equivalence relation is a common proof technique for this kind of result and we need to be familiar with it. To prove that R is reflexive, we began with an arbitrary element of $\mathbf{Z}$. We called this element a. Our goal was to show that $a \, R \, a$. By definition, $a \, R \, a$ if and only if $a \equiv a \pmod{n}$. However, $a \equiv a \pmod{n}$ if and only if $n \mid (a - a)$, which is the same as the statement $n \mid 0$. Clearly, $n \mid 0$ and this is where we decided to start.

To prove that R is symmetric, we started (as always) by assuming that $a \, R \, b$. Our goal was to show that $b \, R \, a$. Since $a \, R \, b$, the definition of the relation R tells us that $a \equiv b \pmod{n}$. From this, we knew that $n \mid (a - b)$ and $a - b = nk$ for some integer k. However, to show that $b \, R \, a$, we needed to verify that $b \equiv a \pmod{n}$. But this can only be done if we can show that $n \mid (b - a)$ or, equivalently, that $b - a = n\ell$ for some integer ℓ. Hence we needed to verify that $b - a$ can be expressed as the product of n and some other integer. Since $b - a$ is the negative of $a - b$ and we have a convenient expression for $a - b$, this provided us with a key step.

Finally, to prove that R is transitive, we began by assuming that $a \, R \, b$ and $b \, R \, c$, which led us to the expressions $a - b = nk$ and $b - c = n\ell$, where $k, \ell \in \mathbf{Z}$. Since our goal was to show that $a \, R \, c$, we were required to show that $a - c$ is a multiple of n. Somehow then, we needed to work the term $a - c$ into the problem, knowing that $a - b = nk$ and $b - c = n\ell$. The key step here was to observe that $a - c = (a - b) + (b - c)$. $\blacklozenge$

According to Theorem 8.6 then, congruence modulo 3 is an equivalence relation. In other words, if we define a relation R on $\mathbf{Z}$ by $a \, R \, b$ if $a \equiv b \pmod{3}$, then it follows that R is an equivalence relation. Let's determine the distinct equivalence classes in this case. First, select an integer, say 0. Then [0] is an equivalence class. Indeed,

$$[0] = \{x \in \mathbf{Z} : x \, R \, 0\} = \{x \in \mathbf{Z} : x \equiv 0 \pmod{3}\}$$
$$= \{x \in \mathbf{Z} : 3 \mid x\} = \{0, \pm 3, \pm 6, \pm 9, \ldots\}.$$

Hence the class [0] consists of the multiples of 3. This class could be denoted by [3] or [6] or even [−300]. Since there is an integer that is not in [0], there must be at least one equivalence class distinct from [0]. In particular, since $1 \notin [0]$, it follows that $[1] \neq [0]$; in fact, necessarily, $[1] \cap [0] = \emptyset$. The equivalence class

$$[1] = \{x \in \mathbf{Z} : x \ R \ 1\} = \{x \in \mathbf{Z} : x \equiv 1 \ (\text{mod } 3)\}$$
$$= \{x \in \mathbf{Z} : 3 \mid (x - 1)\} = \{1, -2, 4, -5, 7, -8, \ldots\}.$$

Since $2 \notin [0]$ and $2 \notin [1]$, the equivalence class [2] is different from both [0] and [1]. By definition,

$$[2] = \{x \in \mathbf{Z} : x \ R \ 2\} = \{x \in \mathbf{Z} : x \equiv 2 \ (\text{mod } 3)\}$$
$$= \{x \in \mathbf{Z} : 3 \mid (x - 2)\} = \{2, -1, 5, -4, 8, -7, \ldots\}.$$

Since every integer belongs to (exactly) one of these classes, we have exactly three distinct equivalence classes in this case, namely:

$$[0] = \{0, \pm 3, \pm 6, \pm 9, \ldots\},$$
$$[1] = \{1, -2, 4, -5, 7, -8, \ldots\},$$
$$[2] = \{2, -1, 5, -4, 8, -7, \ldots\}.$$

These equivalence classes have a connection with some very familiar mathematical concepts, division and remainders, which we encountered in Section 4.1 and which is useful to review here. If m and $n \geq 2$ are integers and m is divided by n, then we can express this division as $m = nq + r$, where q is the quotient and r is the remainder. The remainder r has the requirement that $0 \leq r < n$. With this requirement, q and r are unique and the result that we have just referred to is the **Division Algorithm**. (The Division Algorithm will be studied in considerably more detail in Chapter 11.) Consequently, every integer m can be expressed as $3q + r$, where $0 \leq r < 3$; that is, r has one of the values 0, 1 or 2. Hence, every integer can be expressed as $3q$, $3q + 1$ or $3q + 2$ for some integer q. In this case, the equivalence class [0] consists of the multiples of 3, and so every integer having a remainder of 0 when divided by 3 belongs to [0]. Furthermore, every integer having a remainder of 1 when divided by 3 belongs to [1], while every integer having remainder of 2 when divided by 3 belongs to [2]. Since

$$73 = 24 \cdot 3 + 1 \quad \text{and} \quad -22 = (-8) \cdot 3 + 2,$$

for example, it follows that $73 \in [1]$ and $-22 \in [2]$. In fact, $[73] = [1]$ and $[-22] = [2]$.

In general, for $n \geq 2$, the equivalence relation congruence modulo n results in n distinct equivalence classes. In other words, if we define $a \ R \ b$ by $a \equiv b \ (\text{mod } n)$, then there are n distinct equivalence classes: $[0], [1], \ldots, [n-1]$. In fact, for an integer r with $0 \leq r < n$, an integer m belongs to the set $[r]$ if and only if there is an integer q (the quotient) such that $m = nq + r$. In fact, the equivalence class $[r]$ consists of all integers having a remainder of r when divided by n.

Let's consider another equivalence relation defined on $\mathbf{Z}$ involving congruence but which is seemingly different from the class of examples we have just described.

Result to Prove   Let R be the relation defined on $\mathbf{Z}$ by $a \ R \ b$ if $2a + b \equiv 0 \ (\text{mod } 3)$. Then R is an equivalence relation.

PROOF STRATEGY   To prove that R is reflexive, we must show that $x \, R \, x$ for every $x \in \mathbf{Z}$. This means that we must show that $2x + x \equiv 0 \pmod{3}$ or that $3x \equiv 0 \pmod{3}$. This is equivalent to showing that $3 \mid 3x$, which is obvious. This tells us where to begin the proof of the reflexive property.

Proving that R is symmetric is somewhat more subtle. Of course, we know where to begin. We assume that $x \, R \, y$. From this, we have $2x + y \equiv 0 \pmod{3}$. So $3 \mid (2x + y)$ or $2x + y = 3r$ for some integer r. Our goal is to show that $y \, R \, x$ or, equivalently, that $2y + x \equiv 0 \pmod{3}$. Eventually then, we must show that $2y + x = 3s$ for some integer s. We cannot assume this of course. Since $2x + y = 3r$, it follows that $y = 3r - 2x$. So

$$2y + x = 2(3r - 2x) + x = 6r - 3x = 3(2r - x).$$

Since $2r - x \in \mathbf{Z}$, we have $3 \mid (2y + x)$ and the verification of symmetry is nearly complete.

Proving that R is transitive should be as expected.   ♦

Result 8.7   *Let R be the relation defined on $\mathbf{Z}$ by a R b if $2a + b \equiv 0 \pmod{3}$. Then R is an equivalence relation.*

Proof   Let $x \in \mathbf{Z}$. Since $3 \mid 3x$, it follows that $3x \equiv 0 \pmod{3}$. So $2x + x \equiv 0 \pmod{3}$. Thus, $x \, R \, x$ and R is reflexive.

Next we verify that R is symmetric. Assume that $x \, R \, y$, where $x, y \in \mathbf{Z}$. Thus $2x + y \equiv 0 \pmod{3}$ and so $3 \mid (2x + y)$. Therefore, $2x + y = 3r$ for some integer r. Hence $y = 3r - 2x$. So

$$2y + x = 2(3r - 2x) + x = 6r - 3x = 3(2r - x).$$

Since $2r - x$ is an integer, $3 \mid (2y + x)$. So $2y + x \equiv 0 \pmod{3}$. Therefore, $y \, R \, x$ and R is symmetric.

Finally, we show that R is transitive. Assume that $x \, R \, y$ and $y \, R \, z$, where $x, y, z \in \mathbf{Z}$. Then $2x + y \equiv 0 \pmod{3}$ and $2y + z \equiv 0 \pmod{3}$. Thus, $3 \mid (2x + y)$ and $3 \mid (2y + z)$. From this, it follows that $2x + y = 3r$ and $2y + z = 3s$ for some integers r and s. Adding these two equations, we obtain

$$2x + 3y + z = 3r + 3s;$$

so

$$2x + z = 3r + 3s - 3y = 3(r + s - y).$$

Since $r + s - y$ is an integer, $3 \mid (2x + z)$; so $2x + z \equiv 0 \pmod{3}$. Hence $x \, R \, z$ and R is transitive.   ■

PROOF ANALYSIS   A few additional comments about the proof of the symmetric property in Result 8.7 might be helpful. At one point in the proof we knew that $2x + y = 3r$ for some integer r and we wanted to show that $2y + x = 3s$ for some integer s. If we added these two equations, then we would obtain $3x + 3y = 3r + 3s$. Of course, we can't add these because we don't know that $2y + x = 3s$. But this does suggest another idea.

Assume that $x \mathrel{R} y$. Thus $2x + y \equiv 0 \pmod 3$. Hence $3 \mid (2x + y)$; so $2x + y = 3r$ for some integer r. Observe that

$$3x + 3y = (2x + y) + (2y + x) = 3r + (2y + x).$$

Therefore,

$$2y + x = 3x + 3y - 3r = 3(x + y - r).$$

Because $x + y - r \in \mathbf{Z}$, it follows that $3 \mid (2y + x)$. Consequently, $2y + x \equiv 0 \pmod 3$, $y \mathrel{R} x$ and R is symmetric.   ♦

The distinct equivalence classes for the equivalence relation described in Result 8.7 are

$$[0] = \{x \in \mathbf{Z} : \; x \mathrel{R} 0\} = \{x \in \mathbf{Z} : \; 2x \equiv 0 \pmod 3\}$$
$$= \{x \in \mathbf{Z} : \; 3 \mid 2x\} = \{0, \pm 3, \pm 6, \pm 9, \ldots\},$$
$$[1] = \{x \in \mathbf{Z} : \; x \mathrel{R} 1\} = \{x \in \mathbf{Z} : \; 2x + 1 \equiv 0 \pmod 3\}$$
$$= \{x \in \mathbf{Z} : \; 3 \mid (2x + 1)\} = \{1, -2, 4, -5, 7, -8, \ldots\},$$
$$[2] = \{x \in \mathbf{Z} : \; x \mathrel{R} 2\} = \{x \in \mathbf{Z} : \; 2x + 2 \equiv 0 \pmod 3\}$$
$$= \{x \in \mathbf{Z} : \; 3 \mid (2x + 2)\} = \{2, -1, 5, -4, 8, -7, \ldots\}.$$

Let's discuss how we obtained these equivalence classes. We started with the integer 0 and saw that $[0] = \{x \in \mathbf{Z} : \; 3 \mid 2x\}$. By trying various values of x (namely, 0, 1, 2, 3, 4, 5, etc. and $-1, -2, -3, -4$, etc.), we see that we are obtaining the multiples of 3. (Exercise 6 of Chapter 4 asks you to show that if $3 \mid 2x$, then x is a multiple of 3.) The contents of $[1]$ and $[2]$ can be justified, if necessary, in a similar manner.

We have seen that if we define a relation R_1 on $\mathbf{Z}$ by $a \mathrel{R_1} b$ if $a \equiv b \pmod 3$, then we have three distinct equivalence classes; while if we define a relation R_2 on $\mathbf{Z}$ by $a \mathrel{R_2} b$ if $2a + b \equiv 0 \pmod 3$, then we also have three distinct classes—in fact, the *same* equivalence classes. Let's see why this is true.

Result 8.8   *Let $a, b \in \mathbf{Z}$. Then $a \equiv b \pmod 3$ if and only if $2a + b \equiv 0 \pmod 3$.*

Proof   First, assume that $a \equiv b \pmod 3$. Then $3 \mid (a - b)$ and so $a - b = 3x$ for some integer x. Thus $a = 3x + b$. Now

$$2a + b = 2(3x + b) + b = 6x + 3b = 3(2x + b).$$

Since $2x + b$ is an integer, $3 \mid (2a + b)$ and so $2a + b \equiv 0 \pmod 3$.

For the converse, assume that $2a + b \equiv 0 \pmod 3$. Hence $3 \mid (2a + b)$, which implies that $2a + b = 3y$ for some integer y. Thus $b = 3y - 2a$. Observe that

$$a - b = a - (3y - 2a) = 3a - 3y = 3(a - y).$$

Since $a - y$ is an integer, $3 \mid (a - b)$ and so $a \equiv b \pmod 3$.   ■

We shouldn't jump to the conclusion that just because we are dealing with an equivalence relation defined in terms of the integers modulo 3, we will necessarily have three distinct equivalence classes. For example, suppose that we define a relation R on

$\mathbf{Z}$ by $a \, R \, b$ if $a^2 \equiv b^2 \pmod 3$. Then, here too, R is an equivalence relation. In this case, however, there are only *two* distinct equivalence classes, namely:

$$[0] = \{0, \pm 3, \pm 6, \pm 9, \ldots\} \text{ and } [1] = \{\pm 1, \pm 2, \pm 4, \pm 5, \ldots\},$$

since whenever an integer n has a remainder 1 or 2 when it is divided by 3, then n^2 has a remainder of 1 when it is divided by 3.

8.6 The Integers Modulo n

We have already seen that for each integer $n \geq 2$, the relation R defined on $\mathbf{Z}$ by $a \, R \, b$ if $a \equiv b \pmod n$ is an equivalence relation. Furthermore, this equivalence relation results in the n distinct equivalence classes $[0], [1], \ldots, [n-1]$. We denote the set of these equivalence classes by $\mathbf{Z}_n$ and refer to this set as the **integers modulo** n. Thus, $\mathbf{Z}_3 = \{[0], [1], [2]\}$ and, in general,

$$\mathbf{Z}_n = \{[0], [1], \ldots, [n-1]\}\,.$$

Hence each element $[r]$ of $\mathbf{Z}_n$, where $0 \leq r < n$, is a set that contains infinitely many integers; indeed, as we have noted, $[r]$ consists of all those integers having the remainder r when divided by n. For this reason, the elements of $\mathbf{Z}_n$ are sometimes called **residue classes**.

Although it makes perfectly good sense to take the union and intersection of two elements of $\mathbf{Z}_n$ since these elements are sets (in fact, subsets of $\mathbf{Z}$), it doesn't make sense at this point to add or multiply two elements of $\mathbf{Z}_n$. However, since the elements of $\mathbf{Z}_n$ have the appearance of integers, say $[a]$ and $[b]$, where $a, b \in \mathbf{Z}$, it does suggest the possibility of defining addition and multiplication in $\mathbf{Z}_n$. We now discuss how these operations can be defined on the set $\mathbf{Z}_n$.

Of course, we have seen addition and multiplication defined many times before. When we speak of addition and multiplication being *operations* on a set S, we mean that for $x, y \in S$, the sum $x + y$ and the product xy should both belong to S. For example, in the set $\mathbf{Q}$ of rational numbers, the sum and product of two rational numbers a/b and c/d (so $a, b, c, d \in \mathbf{Z}$ and $b, d \neq 0$) are defined by

$$\frac{a}{b} + \frac{c}{d} = \frac{ad + bc}{bd} \quad \text{and} \quad \frac{a}{b} \cdot \frac{c}{d} = \frac{ac}{bd},$$

both of which are rational numbers and so belong to $\mathbf{Q}$.

As we mentioned, if addition and multiplication are operations on a set S, then $x + y \in S$ and $xy \in S$ for all $x, y \in S$. Therefore, if T is a nonempty subset of S and $x, y \in T$, then $x + y \in S$ and $xy \in S$. The set T is **closed under addition** if $x + y \in T$ whenever $x, y \in T$. Similarly, T is **closed under multiplication** if $xy \in T$ whenever $x, y \in T$. Necessarily, if addition and multiplication are operations on a set S, then S is closed under addition and multiplication.

For example, addition and multiplication are operations on $\mathbf{Z}$. If A and B denote the sets of even integers and odd integers, respectively, then A is closed under both addition and multiplication but B is closed under multiplication only.

However addition and multiplication might be defined in $\mathbf{Z}_n$, we would certainly expect that the sum and product of two elements of $\mathbf{Z}_n$ to be an element of $\mathbf{Z}_n$. There appears to be a natural definition of addition and multiplication in $\mathbf{Z}_n$; namely, for two equivalence classes $[a]$ and $[b]$ in $\mathbf{Z}_n$, we define

$$[a] + [b] = [a + b] \quad \text{and} \quad [a] \cdot [b] = [ab]. \tag{8.7}$$

Let's suppose that we are considering $\mathbf{Z}_6$, for example, where then $\mathbf{Z}_6 = \{[0], [1], \ldots, [5]\}$. From the definitions of addition and multiplication that we just gave, $[1] + [3] = [1 + 3] = [4]$ and $[1] \cdot [3] = [1 \cdot 3] = [3]$. This certainly seems harmless enough, but let's consider adding and multiplying two other equivalence classes, say $[2]$ and $[3]$. Again, according to the definitions in (8.7), $[2] + [3] = [2 + 3] = [5]$ and $[2] \cdot [3] = [2 \cdot 3] = [6]$. However, we have been expressing the elements of $\mathbf{Z}_6$ by $[0], [1], [2], [3], [4]$ and $[5]$ and we don't explicitly see $[2 \cdot 3] = [6]$ among these elements. Since $6 \equiv 0 \pmod{6}$, it follows that $6 \in [0]$, that is, $[6] = [0]$. Also, the remainder is 0 when 6 is divided by 6 and so $[6] = [0]$. Therefore, $[2] \cdot [3] = [0]$. By similar reasoning, $[3] + [5] = [2]$ and $[3] \cdot [5] = [3]$. In fact, the complete addition and multiplication tables for $\mathbf{Z}_6$ are given in Figure 8.2.

If we add $[1]$ to $[0]$, add $[1]$ to $[1]$ and continue in this manner, then we obtain $[0] + [1] = [1], [1] + [1] = [2], [2] + [1] = [3], \ldots, [5] + [1] = [6] = [0], [6] + [1] = [0] + [1] = [1]$, etc.; that is, we return to $[0]$ and cycle through all the classes of $\mathbf{Z}_6$ again (and again). If, instead of $\mathbf{Z}_6$, we were dealing with $\mathbf{Z}_{12}$, we would have $[0] + [1] = [1], [1] + [1] = [2], [2] + [1] = [3], \ldots, [11] + [1] = [12] = [0], [12] + [1] = [0] + [1] = [1]$, etc. and this should remind you of what occurs when a certain number of hours is added to a time (in hours), where, of course, 12 o'clock is represented here as 0 o'clock. (For example, if it is 11 o'clock now, what time will it be 45 hours from now?)

Although the definitions of addition and multiplication in $\mathbf{Z}_n$ that we gave in (8.7) should seem quite reasonable and expected, there is a possible point of concern here that needs to be addressed. According to the definition of addition in $\mathbf{Z}_6$, $[4] + [5] = [3]$. However, the class $[4]$, which consists of all integers x such that $x \equiv 4 \pmod{6}$, need not be represented this way. Since $10 \in [4]$, it follows that $[10] = [4]$. Also, $[16] = [4]$ and $[-2] = [4]$, for example. Moreover, $[11] = [5]$, $[17] = [5]$ and $[-25] = [5]$. Hence adding the equivalence classes $[4]$ and $[5]$ is the same as adding $[10]$ and $[-25]$, say, since $[10] = [4]$ and $[-25] = [5]$. But, according to the definition we have given, $[10] + [-25] = [-15]$. Luckily, $[-15] = [3]$ and so we obtain the same sum as before. But will this happen every time? That is, does the definition of the sum of the equivalence classes $[a]$ and $[b]$ that we gave in (8.7) depend on the representatives a and b of these classes? If the sum (or product) of two equivalence classes does not depend on the representatives, then we say that this sum (or product) is **well-defined**. We certainly

+	[0]	[1]	[2]	[3]	[4]	[5]
[0]	[0]	[1]	[2]	[3]	[4]	[5]
[1]	[1]	[2]	[3]	[4]	[5]	[0]
[2]	[2]	[3]	[4]	[5]	[0]	[1]
[3]	[3]	[4]	[5]	[0]	[1]	[2]
[4]	[4]	[5]	[0]	[1]	[2]	[3]
[5]	[5]	[0]	[1]	[2]	[3]	[4]

·	[0]	[1]	[2]	[3]	[4]	[5]
[0]	[0]	[0]	[0]	[0]	[0]	[0]
[1]	[0]	[1]	[2]	[3]	[4]	[5]
[2]	[0]	[2]	[4]	[0]	[2]	[4]
[3]	[0]	[3]	[0]	[3]	[0]	[3]
[4]	[0]	[4]	[2]	[0]	[4]	[2]
[5]	[0]	[5]	[4]	[3]	[2]	[1]

Figure 8.2   The Addition and Multiplication Tables for $\mathbf{Z}_6$

would want this to be the case, which, fortunately, it is. More precisely, addition and multiplication in $\mathbf{Z}_n$ are **well-defined** if whenever $[a] = [b]$ and $[c] = [d]$ in $\mathbf{Z}_n$, then $[a + c] = [b + d]$ and $[ac] = [bd]$.

Theorem 8.9   *Addition in $\mathbf{Z}_n$, $n \geq 2$, is well-defined.*

Proof   The set $\mathbf{Z}_n$ is the set of equivalence classes resulting from the equivalence relation R defined on $\mathbf{Z}$ by $a \, R \, b$ if $a \equiv b \pmod{n}$. Let $[a], [b], [c], [d] \in \mathbf{Z}_n$, where $[a] = [b]$ and $[c] = [d]$. We prove that $[a + c] = [b + d]$. Since $[a] = [b]$, it follows by Theorem 8.2 that $a \, R \, b$. Similarly, $c \, R \, d$. Therefore, $a \equiv b \pmod{n}$ and $c \equiv d \pmod{n}$. Thus, $n \mid (a - b)$ and $n \mid (c - d)$. Hence, there exist integers x and y so that

$$a - b = nx \text{ and } c - d = ny. \tag{8.8}$$

Adding the equations in (8.8), we obtain

$$(a - b) + (c - d) = nx + ny = n(x + y);$$

so $(a + c) - (b + d) = n(x + y)$. This implies that $n \mid [(a + c) - (b + d)]$. Thus, $(a + c) \equiv (b + d) \pmod{n}$. From this, we conclude that $(a + c) \, R \, (b + d)$, which implies that $[a + c] = [b + d]$.   ∎

If the proof of Theorem 8.9 looks a bit familiar, review Result 4.10 and its proof. As an example, in $\mathbf{Z}_7$, $[118] + [26] = [144]$. Since the remainder is 4 when 144 is divided by 7, it follows that $[118] + [26] = [4]$. Furthermore, $[118] = [6]$ and $[26] = [5]$; so $[118] + [26] = [6] + [5] = [11] = [4]$.

As we have mentioned, the multiplication in $\mathbf{Z}_n$ that we described in (8.7) is also well-defined. The verification of this fact has been left as an exercise (Exercise 8.58).

Addition and multiplication in $\mathbf{Z}_n$ satisfy many familiar properties. Among these are:

Commutative Properties

$$[a] + [b] = [b] + [a] \text{ and } [a] \cdot [b] = [b] \cdot [a] \quad \text{for all } a, b \in \mathbf{Z};$$

Associative Properties

$$([a] + [b]) + [c] = [a] + ([b] + [c]) \text{ and}$$
$$([a] \cdot [b]) \cdot [c] = [a] \cdot ([b] \cdot [c]) \quad \text{for all } a, b, c \in \mathbf{Z};$$

Distributive Property

$$[a] \cdot ([b] + [c]) = [a] \cdot [b] + [a] \cdot [c] \quad \text{for all } a, b, c \in \mathbf{Z}.$$

Although we defined multiplication in $\mathbf{Z}_n$ in a manner that was probably expected, this is not the only way it could have been defined. For example, suppose that we are considering the set $\mathbf{Z}_3$ of integers modulo 3. For equivalence classes $[a]$ and $[b]$ in $\mathbf{Z}_3$, define the "product" $[a] \cdot [b]$ to equal $[q]$, where $[q]$ is the quotient when ab is divided by 3. Since the "product" of every two elements of $\mathbf{Z}_3$ is an element of $\mathbf{Z}_3$, this operation is closed. In particular, $[2] \cdot [2] = [1]$ since the quotient is 1 when $2 \cdot 2 = 4$ is divided by 3. However, $[2] = [5]$ but $[5] \cdot [5] = [8] = [2]$. Notice also that $[5] \cdot [2] = [3] = [0]$. Hence *this* multiplication is not well-defined.

EXERCISES FOR CHAPTER 8

Section 8.1: Relations

8.1. Let $A = \{a, b, c\}$ and $B = \{r, s, t, u\}$. Furthermore, let $R = \{(a, s), (a, t), (b, t)\}$ be a relation from A to B. Determine dom(R) and range(R).

8.2. Let A be a nonempty set and $B \subseteq P(A)$. Define a relation R from A to B by $x \, R \, Y$ if $x \in Y$. Give an example of two sets A and B that illustrate this. What is R for these two sets?

8.3. Let $A = \{0, 1\}$. Determine all the relations on A.

8.4. Let $A = \{a, b, c\}$ and $B = \{1, 2, 3, 4\}$. Then $R_1 = \{(a, 2), (a, 3), (b, 1), (b, 3), (c, 4)\}$ is a relation from A to B, while $R_2 = \{(1, b), (1, c), (2, a), (2, b), (3, c), (4, a), (4, c)\}$ is a relation from B to A. A relation R is defined on A by $x \, R \, y$ if there exists $z \in B$ such that $x \, R_1 \, z$ and $z \, R_2 \, y$. Express R by listing its elements.

8.5. For the relation $R = \{(1, 1), (1, 2), (1, 3), (2, 2), (2, 3), (3, 3)\}$ defined on the set $\{1, 2, 3\}$, what is R^{-1}?

8.6. A relation R is defined on **N** by $a \, R \, b$ if $a/b \in \mathbf{N}$. For $c, d \in \mathbf{N}$, under what conditions is $c \, R^{-1} \, d$?

8.7. For the relation $R = \{(x, y) : \ x + 4y \text{ is odd}\}$ defined on **N**, what is R^{-1}?

8.8. For the relation $R = \{(x, y) : \ x \le y\}$ defined on **N**, what is R^{-1}?

8.9. Let A and B be sets with $|A| = |B| = 4$.

   (a) Prove or disprove: If R is a relation from A to B where $|R| = 9$ and $R = R^{-1}$, then $A = B$.
   (b) Show that by making a small change in the statement in (a), a different response to the resulting statement can be obtained.

8.10. Let A be a set with $|A| = 4$. What is the maximum number of elements that a relation R on A can contain so that $R \cap R^{-1} = \emptyset$?

Section 8.2: Properties of Relations

8.11. Let $A = \{a, b, c, d\}$ and let $R = \{(a, a), (a, b), (a, c), (a, d), (b, b), (b, c), (b, d), (c, c), (c, d), (d, d)\}$ be a relation on A. Which of the properties reflexive, symmetric and transitive does the relation R possess? Justify your answers.

8.12. Let $S = \{a, b, c\}$. Then $R = \{(a, a), (a, b), (a, c)\}$ is a relation on S. Which of the properties reflexive, symmetric and transitive does the relation R possess? Justify your answers.

8.13. Let $S = \{a, b, c\}$. Then $R = \{(a, b)\}$ is a relation on S. Which of the properties reflexive, symmetric and transitive does the relation R possess? Justify your answers.

8.14. Let $A = \{a, b, c, d\}$. Give an example (with justification) of a relation R on A that has none of the following properties: reflexive, symmetric, transitive.

8.15. A relation R is defined on **Z** by $a \, R \, b$ if $|a - b| \le 2$. Which of the properties reflexive, symmetric and transitive does the relation R possess? Justify your answers.

8.16. Let $A = \{a, b, c, d\}$. How many relations defined on A are reflexive, symmetric and transitive and contain the ordered pairs $(a, b), (b, c), (c, d)$?

8.17. Let $R = \emptyset$ be the empty relation on a nonempty set A. Which of the properties reflexive, symmetric and transitive does R possess?

8.18. Let $A = \{1, 2, 3, 4\}$. Give an example of a relation on A that is:

   (a) reflexive and symmetric but not transitive.
   (b) reflexive and transitive but not symmetric.
   (c) symmetric and transitive but not reflexive.
   (d) reflexive but neither symmetric nor transitive.
   (e) symmetric but neither reflexive nor transitive.
   (f) transitive but neither reflexive nor symmetric.

8.19. A relation R is defined on $\mathbf{Z}$ by $x\ R\ y$ if $x \cdot y \geq 0$. Prove or disprove the following: (a) R is reflexive, (b) R is symmetric, (c) R is transitive.

8.20. Determine the maximum number of elements in a relation R on a 3-element set such that R has none of the properties reflexive, symmetric and transitive.

8.21. Prove or disprove: If there exists a relation R_1 on the set $\{a_1, a_2\}$ that is not reflexive, not symmetric and not transitive, then there exists a relation R_2 on the set $\{b_1, b_2, b_3\}$ that is not reflexive, not symmetric and not transitive.

8.22. Let S be the set of all polynomials of degree at most 3. An element $s(x)$ of S can then be expressed as $s(x) = ax^3 + bx^2 + cx + d$, where $a, b, c, d \in \mathbf{R}$. A relation R is defined on S by $p(x)\ R\ q(x)$ if $p(x)$ and $q(x)$ have a real root in common. (For example, $p = (x - 1)^2$ and $q = x^2 - 1$ have the root 1 in common so that $p\ R\ q$.) Determine which of the properties reflexive, symmetric, and transitive are possessed by R.

8.23. A relation R is defined on $\mathbf{N}$ by $a\ R\ b$ if either $a \mid b$ or $b \mid a$. Determine which of the properties reflexive, symmetric and transitive are possessed by R.

Section 8.3: Equivalence Relations

8.24. Let R be an equivalence relation on $A = \{a, b, c, d, e, f, g\}$ such that $a\ R\ c,\ c\ R\ d,\ d\ R\ g$ and $b\ R\ f$. If there are three distinct equivalence classes resulting from R, then determine these equivalence classes and determine all elements of R.

8.25. Let $A = \{1, 2, 3, 4, 5, 6\}$. The relation

$$R = \{(1, 1), (1, 5), (2, 2), (2, 3), (2, 6), (3, 2), (3, 3), (3, 6), (4, 4),$$
$$(5, 1), (5, 5), (6, 2), (6, 3), (6, 6)\}$$

is an equivalence relation on A. Determine the distinct equivalence classes.

8.26. Let $A = \{1, 2, 3, 4, 5, 6\}$. The distinct equivalence classes resulting from an equivalence relation R on A are $\{1, 4, 5\}, \{2, 6\}$ and $\{3\}$. What is R?

8.27. Let R be a relation defined on $\mathbf{Z}$ by $a\ R\ b$ if $a^3 = b^3$. Show that R is an equivalence relation on $\mathbf{Z}$ and determine the distinct equivalence classes.

8.28. (a) Let R be the relation defined on $\mathbf{Z}$ by $a\ R\ b$ if $a + b$ is even. Show that R is an equivalence relation and determine the distinct equivalence classes.

   (b) Suppose that "even" is replaced by "odd" in (a). Which of the properties reflexive, symmetric and transitive does R possess?

8.29. Let R be an equivalence relation defined on a set A containing the elements a, b, c and d. Prove that if $a\ R\ b, c\ R\ d$ and $a\ R\ d$, then $b\ R\ c$.

8.30. Let $H = \{2^m : m \in \mathbf{Z}\}$. A relation R is defined on the set $\mathbf{Q}^+$ of positive rational numbers by $a\ R\ b$ if $a/b \in H$.

   (a) Show that R is an equivalence relation.
   (b) Describe the elements in the equivalence class [3].

8.31. A relation R on a nonempty set A is defined to be **circular** if whenever $x \ R \ y$ and $y \ R \ z$, then $z \ R \ x$ for all $x, y, z \in A$. Prove that a relation R on A is an equivalence relation if and only if R is circular and reflexive.

8.32. A relation R is defined on the set $A = \{a + b\sqrt{2} : a, b \in \mathbf{Q}, a + b\sqrt{2} \neq 0\}$ by $x \ R \ y$ if $x/y \in \mathbf{Q}$. Show that R is an equivalence relation and determine the distinct equivalence classes.

8.33. Let $H = \{4k : k \in \mathbf{Z}\}$. A relation R is defined on $\mathbf{Z}$ by $a \ R \ b$ if $a - b \in H$.

   (a) Show that R is an equivalence relation.
   (b) Determine the distinct equivalence classes.

8.34. Let H be a nonempty subset of $\mathbf{Z}$. Suppose that the relation R defined on $\mathbf{Z}$ by $a \ R \ b$ if $a - b \in H$ is an equivalence relation. Verify the following

   (a) $0 \in H$.
   (b) If $a \in H$, then $-a \in H$.
   (c) If $a, b \in H$, then $a + b \in H$.

8.35. Prove or disprove: There exist equivalence relations R_1 and R_2 on the set $S = \{a, b, c\}$ such that $R_1 \nsubseteq R_2$, $R_2 \nsubseteq R_1$ and $R_1 \cup R_2 = S \times S$.

Section 8.4: Properties of Equivalence Classes

8.36. Give an example of an equivalence relation R on the set $A = \{v, w, x, y, z\}$ such that there are exactly three distinct equivalence classes. What are the equivalence classes for your example?

8.37. A relation R is defined on $\mathbf{N}$ by $a \ R \ b$ if $a^2 + b^2$ is even. Prove that R is an equivalence relation. Determine the distinct equivalence classes.

8.38. Let R be a relation defined on the set $\mathbf{N}$ by $a \ R \ b$ if either $a \mid 2b$ or $b \mid 2a$. Prove or disprove: R is an equivalence relation.

8.39. Let S be a nonempty subset of $\mathbf{Z}$ and let R be a relation defined on S by $x \ R \ y$ if $3 \mid (x + 2y)$.

   (a) Prove that R is an equivalence relation.
   (b) If $S = \{-7, -6, -2, 0, 1, 4, 5, 7\}$, then what are the distinct equivalence classes in this case?

8.40. A relation R is defined on $\mathbf{Z}$ by $x \ R \ y$ if $3x - 7y$ is even. Prove that R is an equivalence relation. Determine the distinct equivalence classes.

8.41. (a) Prove that the intersection of two equivalence relations on a nonempty set is an equivalence relation.
   (b) Consider the equivalence relations R_2 and R_3 defined on $\mathbf{Z}$ by $a \ R_2 \ b$ if $a \equiv b \pmod 2$ and $a \ R_3 \ b$ if $a \equiv b \pmod 3$. By (a), $R_1 = R_2 \cap R_3$ is an equivalence relation on $\mathbf{Z}$. Determine the distinct equivalence classes in R_1.

8.42. Prove or disprove: The union of two equivalence relations on a nonempty set is an equivalence relation.

8.43. Let $A = \{u, v, w, x, y, z\}$. The relation

$$R = \{(u, u), (u, v), (u, w), (v, u), (v, v), (v, w), (w, u), (w, v),$$
$$(w, w), (x, x), (x, y), (y, x), (y, y), (z, z)\}$$

defined on A is an equivalence relation. In particular, $[u] = [v] = [w] = \{u, v, w\}$, $[x] = [y] = \{x, y\}$ and $[z] = \{z\}$; so $|[u]| = |[v]| = |[w]| = 3$ and $|[x]| = |[y]| = 2$, while $|[z]| = 1$. Therefore, $|[u]| + |[v]| + |[w]| + |[x]| + |[y]| + |[z]| = 14$. Let $A = \{a_1, a_2, \ldots, a_n\}$ be an n-element set and let R be an equivalence relation defined on A. Prove that $\sum_{i=1}^{n} |[a_i]|$ is even if and only if n is even.

Section 8.5: Congruence Modulo n

8.44. Classify each of the following statements as true or false.
(a) $25 \equiv 9 \pmod 8$, (b) $-17 \equiv 9 \pmod 8$, (c) $-14 \equiv -14 \pmod 4$, (d) $25 \equiv -3 \pmod{11}$.

8.45. A relation R is defined on $\mathbf{Z}$ by $a \ R \ b$ if $3a + 5b \equiv 0 \pmod 8$. Prove that R is an equivalence relation.

8.46. Let R be the relation defined on $\mathbf{Z}$ by $a \ R \ b$ if $a + b \equiv 0 \pmod 3$. Show that R is not an equivalence relation.

8.47. The relation R on $\mathbf{Z}$ defined by $a \ R \ b$ if $a^2 \equiv b^2 \pmod 4$ is known to be an equivalence relation. Determine the distinct equivalence classes.

8.48. The relation R defined on $\mathbf{Z}$ by $x \ R \ y$ if $x^3 \equiv y^3 \pmod 4$ is known to be an equivalence relation. Determine the distinct equivalence classes.

8.49. A relation R is defined on $\mathbf{Z}$ by $a \ R \ b$ if $5a \equiv 2b \pmod 3$. Prove that R is an equivalence relation. Determine the distinct equivalence classes.

8.50. A relation R is defined on $\mathbf{Z}$ by $a \ R \ b$ if $2a + 2b \equiv 0 \pmod 4$. Prove that R is an equivalence relation. Determine the distinct equivalence classes.

8.51. Let R be the relation defined on $\mathbf{Z}$ by $a \ R \ b$ if $2a + 3b \equiv 0 \pmod 5$. Prove that R is an equivalence relation and determine the distinct equivalence classes.

8.52. Let R be the relation defined on $\mathbf{Z}$ by $a \ R \ b$ if $a^2 \equiv b^2 \pmod 5$. Prove that R is an equivalence relation and determine the distinct equivalence classes.

8.53. For an integer $n \geq 2$, the relation R defined on $\mathbf{Z}$ by $a \ R \ b$ if $a \equiv b \pmod n$ is an equivalence relation. Equivalently, $a \ R \ b$ if $a - b = kn$ for some $k \in \mathbf{Z}$. Define a relation R on the set $\mathbf{R}$ of real numbers by $a \ R \ b$ if $a - b = k\pi$ for some $k \in \mathbf{Z}$. Is this relation R on $\mathbf{R}$ an equivalence relation? If not, explain why. If yes, prove this and determine $[0]$, $[\pi]$ and $[\sqrt{2}]$.

Section 8.6: The Integers Modulo n

8.54. Construct the addition and multiplication tables for $\mathbf{Z}_4$ and $\mathbf{Z}_5$.

8.55. In $\mathbf{Z}_8$, express the following sums and products as $[r]$, where $0 \leq r < 8$.
(a) $[2] + [6]$ (b) $[2] \cdot [6]$ (c) $[-13] + [138]$ (d) $[-13] \cdot [138]$

8.56. In $\mathbf{Z}_{11}$, express the following sums and products as $[r]$, where $0 \leq r < 11$.
(a) $[7] + [5]$ (b) $[7] \cdot [5]$ (c) $[-82] + [207]$ (d) $[-82] \cdot [207]$

8.57. Let $S = \mathbf{Z}$ and $T = \{4k : k \in \mathbf{Z}\}$. Thus T is a nonempty subset of S.

(a) Prove that T is closed under addition and multiplication.
(b) If $a \in S - T$ and $b \in T$, is $ab \in T$?
(c) If $a \in S - T$ and $b \in T$, is $a + b \in T$?
(d) If $a, b \in S - T$, is it possible that $ab \in T$?
(e) If $a, b \in S - T$, is it possible that $a + b \in T$?

8.58. Prove that the multiplication in $\mathbf{Z}_n$, $n \geq 2$, defined by $[a][b] = [ab]$ is well-defined. (See Result 4.11.)

8.59. (a) Let $[a], [b] \in \mathbf{Z}_8$. If $[a] \cdot [b] = [0]$, does it follow that $[a] = [0]$ or $[b] = [0]$?
(b) How is the question in (a) answered if $\mathbf{Z}_8$ is replaced by $\mathbf{Z}_9$? by $\mathbf{Z}_{10}$? by $\mathbf{Z}_{11}$?
(c) For which integers $n \geq 2$ is the following statement true? (You are only asked to make a conjecture, not to provide a proof.) Let $[a], [b] \in \mathbf{Z}_n$, $n \geq 2$. If $[a] \cdot [b] = [0]$, then $[a] = [0]$ or $[b] = [0]$.

8.60. For integers $m, n \geq 2$ consider $\mathbf{Z}_m$ and $\mathbf{Z}_n$. Let $[a] \in \mathbf{Z}_m$ where $0 \leq a \leq m - 1$. Then $a, a + m \in [a]$ in $\mathbf{Z}_m$. If $a, a + m \in [b]$ for some $[b] \in \mathbf{Z}_n$, then what can be said of m and n?

8.61. (a) For integers $m, n \geq 2$ consider $\mathbf{Z}_m$ and $\mathbf{Z}_n$. If some element of $\mathbf{Z}_m$ also belongs to $\mathbf{Z}_n$, then what can be said of $\mathbf{Z}_m$ and $\mathbf{Z}_n$?

(b) Are there examples of integers $m, n \geq 2$ for which $\mathbf{Z}_m \cap \mathbf{Z}_n = \emptyset$?

ADDITIONAL EXERCISES FOR CHAPTER 8

8.62. Prove or disprove:

(a) There exists an integer a such that $ab \equiv 0 \pmod 3$ for every integer b.

(b) If $a \in \mathbf{Z}$, then $ab \equiv 0 \pmod 3$ for every $b \in \mathbf{Z}$.

(c) For every integer a, there exists an integer b such that $ab \equiv 0 \pmod 3$.

8.63. A relation R is defined on $\mathbf{R}$ by $a \, R \, b$ if $a - b \in \mathbf{Z}$. Prove that R is an equivalence relation and determine the equivalence classes $[1/2]$ and $[\sqrt{2}]$.

8.64. A relation R is defined on $\mathbf{Z}$ by $a \, R \, b$ if $|a - 2| = |b - 2|$. Prove that R is an equivalence relation and determine the distinct equivalence classes.

8.65. Let k and ℓ be integers such that $k + \ell \equiv 0 \pmod 3$ and let $a, b \in \mathbf{Z}$. Prove that if $a \equiv b \pmod 3$, then $ka + \ell b \equiv 0 \pmod 3$.

8.66. State and prove a generalization of Exercise 8.65.

8.67. A relation R is defined on $\mathbf{Z}$ by $a \, R \, b$ if $3 \mid (a^3 - b)$. Prove or disprove the following:
(a) R is reflexive.   (b) R is transitive.

8.68. A relation R is defined on $\mathbf{Z}$ by $a \, R \, b$ if $a \equiv b \pmod 2$ and $a \equiv b \pmod 3$. Prove or disprove: R is an equivalence relation on $\mathbf{Z}$.

8.69. A relation R is defined on $\mathbf{Z}$ by $a \, R \, b$ if $a \equiv b \pmod 2$ or $a \equiv b \pmod 3$. Prove or disprove: R is an equivalence relation on $\mathbf{Z}$.

8.70. Determine each of the following.
(a) $[4]^3 = [4][4][4]$ in $\mathbf{Z}_5$ (b) $[7]^5$ in $\mathbf{Z}_{10}$

8.71. Let $S = \{(a, b) : a, b \in \mathbf{R}, a \neq 0\}$.

(a) Show that the relation R defined on S by $(a, b) \, R \, (c, d)$ if $ad = bc$ is an equivalence relation.

(b) Describe geometrically the elements of the equivalence classes $[(1, 2)]$ and $[(3, 0)]$.

8.72. In Exercise 8.19. (of this chapter), a relation R was defined on $\mathbf{Z}$ by $x \, R \, y$ if $x \cdot y \geq 0$, and we were asked to determine which of the properties reflexive, symmetric and transitive are satisfied.

(a) How would our answers have changed if $x \cdot y \geq 0$ was replaced by: (i) $x \cdot y \leq 0$, (ii) $x \cdot y > 0$, (iii) $x \cdot y \neq 0$, (iv) $x \cdot y \geq 1$, (v) $x \cdot y$ is odd, (vi) $x \cdot y$ is even, (vii) $xy \not\equiv 2 \pmod 3$?

(b) What are some additional questions you could ask?

8.73. For the following statement S and proposed proof, either (1) S is true and the proof is correct, (2) S is true and the proof is incorrect or (3) S is false (and the proof is incorrect). Explain which of these occurs.

S: Every symmetric and transitive relation on a nonempty set is an equivalence relation.

Proof   Let R be a symmetric and transitive relation defined on a nonempty set A. We need only show that R is reflexive. Let $x \in A$. We show that $x \, R \, x$. Let $y \in A$ such that $x \, R \, y$. Since R is symmetric, $y \, R \, x$. Now $x \, R \, y$ and $y \, R \, x$. Since R is transitive, $x \, R \, x$. Thus R is reflexive. $\blacksquare$

8.74. Evaluate the proposed proof of the following result.

Result   A relation R is defined on $\mathbf{Z}$ by $a \, R \, b$ if $3 \mid (a + 2b)$. Then R is an equivalence relation.

Proof   Assume that $a \mathrel{R} a$. Then $3 \mid (a + 2a)$. Since $a + 2a = 3a$ and $a \in \mathbf{Z}$, it follows that $3 \mid 3a$ or $3 \mid (a + 2a)$. Therefore, $a \mathrel{R} a$ and R is reflexive.

Next, we show that R is symmetric. Assume that $a \mathrel{R} b$. Then $3 \mid (a + 2b)$. So $a + 2b = 3x$, where $x \in \mathbf{Z}$. Hence $a = 3x - 2b$. Therefore,

$$b + 2a = b + 2(3x - 2b) = b + 6x - 4b = 6x - 3b = 3(2x - b).$$

Since $2x - b$ is an integer, $3 \mid (b + 2a)$. So $b \mathrel{R} a$ and R is symmetric.

Finally, we show that R is transitive. Assume that $a \mathrel{R} b$ and $b \mathrel{R} c$. Then $3 \mid (a + 2b)$ and $3 \mid (b + 2c)$. So $a + 2b = 3x$ and $b + 2c = 3y$, where $x, y \in \mathbf{Z}$. Adding, we have $(a + 2b) + (b + 2c) = 3x + 3y$. So

$$a + 2c = 3x + 3y - 3b = 3(x + y - b).$$

Since $x + y - b$ is an integer, $3 \mid (a + 2c)$. Hence $a \mathrel{R} c$ and R is transitive. ∎

8.75. (a) Show that the relation R defined on $\mathbf{R} \times \mathbf{R}$ by $(a, b) \mathrel{R} (c, d)$ if $|a| + |b| = |c| + |d|$ is an equivalence relation.
   (b) Describe geometrically the elements of the equivalence classes $[(1, 2)]$ and $[(3, 0)]$.

8.76. Let $x \in \mathbf{Z}_m$ and $y \in \mathbf{Z}_n$, where $m, n \geq 2$. If $x \subseteq y$, then what can be said of m and n?

8.77. Let A be a nonempty set and let B be a fixed subset of A. A relation R is defined on $\mathcal{P}(A)$ by $X \mathrel{R} Y$ if $X \cap B = Y \cap B$.

   (a) Prove that R is an equivalence relation.
   (b) Let $A = \{1, 2, 3, 4\}$ and $B = \{1, 3, 4\}$. For $X = \{2, 3, 4\}$, determine $[X]$.

8.78. Let R_1 and R_2 be equivalence relations on a nonempty set A. Prove or disprove each of the following.

   (a) If $R_1 \cap R_2$ is reflexive, then so are R_1 and R_2.
   (b) If $R_1 \cap R_2$ is symmetric, then so are R_1 and R_2.
   (c) If $R_1 \cap R_2$ is transitive, then so are R_1 and R_2.

8.79. Prove that if R is an equivalence relation on a set A, then the inverse relation R^{-1} is an equivalence relation on A.

8.80. Let R_1 and R_2 be equivalence relations on a nonempty set A. A relation $R = R_1 R_2$ is defined on A as follows: For $a, b \in A$, $a \mathrel{R} b$ if there exists $c \in A$ such that $a \mathrel{R_1} c$ and $c \mathrel{R_2} b$. Prove or disprove: R is an equivalence relation on A.

8.81. A relation R on a nonempty set S is called **sequential** if for every sequence x, y, z of elements of S (distinct or not), at least one of the ordered pairs (x, y) and (y, z) belongs to R. Prove or disprove: Every symmetric, sequential relation on a nonempty set is an equivalence relation.

8.82. Consider the subset $H = \{[3k] : k \in \mathbf{Z}\}$ of $\mathbf{Z}_{12}$.

   (a) Determine the distinct elements of H and construct an addition table for H.
   (b) A relation R on $\mathbf{Z}_{12}$ is defined by $[a] \mathrel{R} [b]$ if $[a - b] \in H$. Show that R is an equivalence relation and determine the distinct equivalence classes.

8.83. For elements $a, b \in \mathbf{Z}_n$, $n \geq 2$, $a = [c]$ and $b = [d]$ for some integers c and d. Define $a - b = [c] - [d]$ as the equivalence class $[c - d]$. Let $H = \{x_1, x_2, \ldots, x_d\}$ be a subset of $\mathbf{Z}_n$, $n \geq 2$, such that a relation R defined on $\mathbf{Z}_n$ by $a \mathrel{R} b$ if $a - b \in H$ is an equivalence relation.

   (a) For each $a \in \mathbf{Z}_n$, determine the equivalence class $[a]$ and show that $[a]$ consists of d elements.
   (b) Prove that $d \mid n$.

9

Functions

If R is a relation from a set A to a set B and x is an element of A, then either x is related to no elements of B or x is related to at least one element of B. In the latter case, it may occur that x is related to all elements of B or perhaps to exactly one element of B. If every element of A is related to no elements of B, then R is the empty set $\emptyset$. If every element of A is related to all elements of B, then R is the Cartesian product $A \times B$. However, if every element of A is related to exactly one element of B, then we have the most studied relation of all: a function. Surely, you have encountered functions before, at least in calculus and precalculus. But it is likely that you have not studied functions in the manner we are about to describe here.

9.1 The Definition of Function

Let A and B be nonempty sets. By a **function** f from A to B, written $f : A \rightarrow B$, we mean a relation from A to B with the property that every element a in A is the first coordinate of exactly one ordered pair in f. Since f is a relation, the set A in this case is the **domain** of f, denoted by $\mathrm{dom}(f)$. The set B is called the **codomain** of f.

For a function $f : A \rightarrow B$, let $(a, b) \in f$. Since f contains only one ordered pair whose first coordinate is a, it follows that b is the unique second coordinate of an ordered pair whose first coordinate is a; that is, if $(a, b) \in f$ and $(a, c) \in f$, then $b = c$. If $(a, b) \in f$, then we write $b = f(a)$ and refer to b as the **image** of a. Sometimes f is said to **map** a into b. Indeed, f itself is sometimes called a **mapping**. The set

$$\mathrm{range}(f) = \{b \in B \ : \ b \text{ is an image under } f \text{ of some element of } A\} = \{f(x) : x \in A\}$$

is the **range** of f and consists of the second coordinates of the elements of f. If A is a finite set, then the function f is a finite set and the number of elements in f is $|A|$ since there is exactly one ordered pair in f corresponding to each element of A. Throughout this chapter, as with earlier chapters, whenever we refer to cardinalities of sets, we are concerned with finite sets only.

Suppose that $f : A \rightarrow B$ and $g : A \rightarrow B$ are two functions from A to B and $a \in A$. Then f and g contain exactly one ordered pair having a as its first coordinate, say $(a, x) \in f$ and $(a, y) \in g$. If the sets f and g are equal, then (a, x) belongs to g as well. Since g contains only one ordered pair whose first coordinate is a, it follows that $(a, x) = (a, y)$. But this implies that $x = y$, that is, $f(a) = g(a)$. Hence it is natural to define two functions $f : A \to B$ and $g : A \to B$ to be **equal**, written $f = g$, if $f(a) = g(a)$ for all $a \in A$.

Let $A = \{1, 2, 3\}$ and $B = \{w, x, y, z\}$. Then $f_1 = \{(1, y), (2, w), (3, y)\}$ is a function from A to B and so we may write $f_1 : A \to B$. On the other hand, $f_2 = \{(1, x), (2, z), (3, y), (2, x)\}$ is not a function since there are two ordered pairs whose first coordinate is 2. In addition, $f_3 = \{(1, z), (3, x)\}$ is not a function from A to B either because $\text{dom}(f_3) \neq A$. On the other hand, f_3 *is* a function from $A - \{2\}$ to B.

It is often convenient to "visualize" a function $f : A \to B$ by representing the two sets A and B by diagrams and drawing an arrow (a directed line segment) from an element $x \in A$ to its image $f(x) \in B$. This is illustrated for the function f_1 described above in Figure 9.1. Therefore, in order to represent a function in this way, exactly one directed line segment must leave each element of A and proceed to an element of B.

In calculus, *functions* such as $f(x) = x^2$ are considered. This function f is from **R** to **R**, that is, $A = \mathbf{R}$ and $B = \mathbf{R}$. Although $f(x) = x^2$ is commonly referred to as a *function* in calculus and elsewhere, strictly speaking $f(x)$ is the image of a real number x under f. The function f itself is actually the set

$$f = \{(x, x^2) : x \in \mathbf{R}\}.$$

So $(2, 4)$ and $(-3, 9)$, for example, belong to f. The set $\{(x, x^2) : x \in \mathbf{R}\}$ of points in the plane is the graph of f. In this case, the graph is a parabola. Here the function $f : \mathbf{R} \to \mathbf{R}$ defined by $f(x) = x^2$ can also be thought of as defined by a rule, namely the rule that associates the number x^2 with each real number x.

Example 9.1   *Another function encountered in calculus is $g(x) = e^x$. As we mentioned above, this function is actually the set*

$$g = \{(x, e^x) : x \in \mathbf{R}\}.$$

More precisely, this is the function $g : \mathbf{R} \to \mathbf{R}$ defined by $g(x) = e^x$ for all $x \in \mathbf{R}$. In general, we will follow this latter convention for defining functions that are often described by some rule or formula. Consequently, the function $h(x) = \dfrac{1}{x - 1}$ from calculus is the

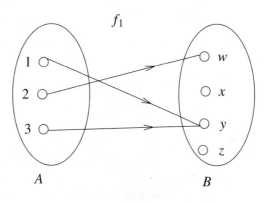

f_1

A B

Figure 9.1   A function $f_1 : A \to B$

function $h : \mathbf{R} - \{1\} \to \mathbf{R}$ defined by $h(x) = \dfrac{1}{x-1}$ for all $x \in \mathbf{R}$ with $x \neq 1$ and the function $\phi(x) = \ln x$ is the function $\phi : \mathbf{R}^+ \to \mathbf{R}$ defined by $\phi(x) = \ln x$ for all $x \in \mathbf{R}^+$, where, recall, $\mathbf{R}^+$ is the set of all positive real numbers. ◆

For a function $f : A \to B$ and a subset C of A, the **image** $f(C)$ of C is defined as

$$f(C) = \{f(x) : x \in C\}.$$

Therefore, $f(C) \subseteq B$ for each subset C of A. If $C = A$, then $f(A)$ is the range of f.

Example 9.2   *For $A = \{a, b, c, d, e\}$ and $B = \{1, 2, \ldots, 6\}$,*

$$f = \{(a, 3), (b, 5), (c, 2), (d, 3), (e, 6)\}$$

is a function from A to B. For $C_1 = \{a, b, c\}$, $C_2 = \{a, d\}$, $C_3 = \{e\}$ and $C_4 = A$,

$$f(C_1) = \{2, 3, 5\}, \ f(C_2) = \{3\}, \ f(C_3) = \{6\}, \ f(C_4) = range(f) = \{2, 3, 5, 6\}.$$ ◆

For a function $f : A \to B$ and a subset D of B, the **inverse image** $f^{-1}(D)$ of D is defined as

$$f^{-1}(D) = \{a \in A : f(a) \in D\}.$$

Therefore, $f^{-1}(D) \subseteq A$ for each subset D of B. Necessarily then, $f^{-1}(B) = A$. In particular, for an element $b \in B$,

$$f^{-1}(\{b\}) = \{a \in A : f(a) = b\}.$$

Example 9.3   *For the function $f : A = \{a, b, c, d, e\} \to B = \{1, 2, \ldots, 6\}$ defined in Example 9.2 by*

$$f = \{(a, 3), (b, 5), (c, 2), (d, 3), (e, 6)\},$$

it follows that $f^{-1}(\{3\}) = \{a, d\}$, $f^{-1}(\{1, 3\}) = \{a, d\}$, $f^{-1}(\{4\}) = \emptyset$ and $f^{-1}(B) = A$. ◆

Among the many classes of functions encountered in calculus are the polynomial functions, rational functions and exponential functions. The function $f : \mathbf{R} \to \mathbf{R}$ defined earlier by $f(x) = x^2$ for $x \in \mathbf{R}$ is a polynomial function. The function h in Example 9.1 is a rational function and g is an exponential function. Other important classes of functions encountered often in calculus are continuous functions and differentiable functions.

The definition of function that we have given is most likely not the definition you recall from calculus; in fact, you may not recall the definition of function given in calculus at all. If this is the case, then it is not surprising. The evolution of what is meant by a function has spanned hundreds of years. It was in the development of calculus that the necessity of a formal definition of function became apparent.

Early in the 18th century, the Swiss mathematician Johann Bernoulli wrote:

I call a function of a variable magnitude a quantity composed in any manner whatsoever from this variable magnitude and from constants.

Later in the 18th century, the famous Swiss mathematician Leonhard Euler studied calculus as a theory of functions and did not appeal to diagrams and geometric interpretations, as many of his predecessors had done. The definition of function that Euler gave in his work on calculus is:

A function of a variable quantity is an analytic expression composed in any way whatsoever of the variable quantity and numbers or constant quantities.

Early in the 19th century, the German mathematician Peter Dirichlet developed a more modern definition of function:

y is a function of x when to each value of x in a given interval there corresponds a unique value of y.

Dirichlet said that it didn't matter whether y depends on x according to some formula, law, or mathematical operation. He emphasized this by considering the function $f : \mathbf{R} \to \mathbf{R}$ defined by

$$f(x) = \begin{cases} 1 & \text{if } x \text{ is rational} \\ 0 & \text{if } x \text{ is irrational.} \end{cases}$$

Later in the 19th century, the German mathematician Richard Dedekind wrote:

A function ϕ on a set S is a law according to which to every determinate element s of S there belongs a determinate thing which is called the transform of s and is denoted by $\phi(s)$.

So, by this time, the modern definition of function had nearly arrived.

9.2 The Set of All Functions from A to B

For nonempty sets A and B, we denote the set of all functions from A to B by B^A. That is, $B^A = \{f : f \text{ is a function from } A \text{ to } B\}$ or, more simply,

$$B^A = \{f : f : A \to B\}.$$

Although this may seem like peculiar notation, it is actually quite logical. In particular, let's determine B^A for $A = \{a, b\}$ and $B = \{x, y, z\}$. Each function f from A to B is necessarily of the form

$$f = \{(a, \alpha), (b, \beta)\},$$

where $\alpha, \beta \in B$. Since there are three choices for α and three choices for β, the total number of such functions f is $3 \cdot 3 = 3^2 = 9$. These nine functions are listed below:

$$\begin{aligned}
f_1 &= \{(a, x), (b, x)\}, & f_2 &= \{(a, x), (b, y)\}, & f_3 &= \{(a, x), (b, z)\}, \\
f_4 &= \{(a, y), (b, x)\}, & f_5 &= \{(a, y), (b, y)\}, & f_6 &= \{(a, y), (b, z)\}, \\
f_7 &= \{(a, z), (b, x)\}, & f_8 &= \{(a, z), (b, y)\}, & f_9 &= \{(a, z), (b, z)\}.
\end{aligned}$$

Hence the number of elements in B^A is 3^2. In general, for finite sets A and B, the number of functions from A to B is

$$\left| B^A \right| = |B|^{|A|}.$$

If $B = \{0, 1\}$, then it is common to represent the set of all functions from A to B by 2^A.

9.3 One-to-One and Onto Functions

We now consider two important properties that a function may possess. A function f from a set A to a set B is called **one-to-one** or **injective** if every two distinct elements of A have distinct images in B. In symbols, a function $f : A \to B$ is one-to-one if whenever $x, y \in A$ and $x \neq y$, then $f(x) \neq f(y)$. Thus, if a function $f : A \to B$ is not one-to-one, then there exist distinct elements w and z in A such that $f(w) = f(z)$.

Let $A = \{a, b, c, d\}$, $B = \{r, s, t, u, v\}$ and $C = \{x, y, z\}$. Then

$$f_1 = \{(a, s), (b, u), (c, v), (d, r)\}$$

is a one-to-one function from A to B since distinct elements of A have distinct images in B; while the function

$$f_2 = \{(a, s), (b, t), (c, s), (d, u)\}$$

from A to B is not one-to-one since a and c have the same image, namely s. There is no one-to-one function from A to C, however.

In general, for a function $f : A \to B$ to be one-to-one, where A and B are finite sets, every two elements of A must have distinct images in B and so there must be at least as many elements in B as in A, that is, $|A| \le |B|$.

At times, the definition of a one-to-one function is difficult to work with since it deals with *unequal* elements. However, there is a useful equivalent formulation of the definition using the contrapositive:

A function $f : A \to B$ is **one-to-one** if whenever $f(x) = f(y)$,
where $x, y \in A$, then $x = y$.

We show how this formulation can be applied to functions defined by formulas.

Result 9.4   *Let the function $f : \mathbf{R} \to \mathbf{R}$ be defined by $f(x) = 3x - 5$. Then f is one-to-one.*

Proof   Assume that $f(a) = f(b)$, where $a, b \in \mathbf{R}$. Then $3a - 5 = 3b - 5$. Adding 5 to both sides, we obtain $3a = 3b$. Dividing by 3, we have $a = b$ and so f is one-to-one. ∎

Example 9.5   *Let the function $f : \mathbf{R} \to \mathbf{R}$ be defined by $f(x) = x^2 - 3x - 2$. Determine whether f is one-to-one.*

Solution   Since $f(0) = -2$ and $f(3) = -2$, it follows that f is not one-to-one. ◆

Analysis   To show that the function f defined in Example 9.5 is not one-to-one, we must show that there exist two distinct real numbers having the same image under f. This was accomplished by showing that $f(0) = f(3)$. But what if we can't find two real numbers with this property? Naturally, if we can't find two such numbers, then we might think that f is one-to-one. In that case, we should be trying to prove that f is one-to-one. We would probably begin such a proof by assuming that $f(a) = f(b)$, that is $a^2 - 3a - 2 = b^2 - 3b - 2$. We would then try to show that $a = b$. We can simplify $a^2 - 3a - 2 = b^2 - 3b - 2$ by adding 2 to both sides, producing $a^2 - 3a = b^2 - 3b$. When attempting to solve an equation, it is often convenient to collect all terms on one side of the equation with 0 on the other side. Rewriting this equation, we obtain $a^2 - 3a - b^2 + 3b = 0$. Rearranging some terms and factoring, we have

$$a^2 - 3a - b^2 + 3b = (a^2 - b^2) - 3(a - b)$$
$$= (a - b)(a + b) - 3(a - b) = (a - b)(a + b - 3) = 0.$$

Hence if $f(a) = f(b)$, then $(a - b)(a + b - 3) = 0$. Since $(a - b)(a + b - 3) = 0$, it follows that either $a - b = 0$ (and so $a = b$) or $a + b - 3 = 0$. Therefore, $f(a) = f(b)$ does *not* imply that $a = b$. It only implies that $a = b$ or $a + b = 3$. Since $0 + 3 = 3$, we now see why $f(0) = f(3)$. In fact, if a and b are *any* two real numbers where $a + b = 3$, then $f(a) = f(b)$. This tells us how to find all possible counterexamples to the statement: f is one-to-one. Looking at $f(x) = x^2 - 3x - 2$ once again, we see that $f(x) = x(x - 3) - 2$. Since $x(x - 3) = 0$ if $x = 0$ or $x = 3$, it is now more apparent why 0 and 3 are numbers for which $f(0) = f(3)$. ◆

A function $f : A \rightarrow B$ is called **onto** or **surjective** if every element of the codomain B is the image of some element of A. Equivalently, f is onto if $f(A) = B$.

A function we considered earlier was $f_1 : A \rightarrow B$, where $A = \{1, 2, 3\}, B = \{x, y, z, w\}$ and $f_1 = \{(1, y), (2, w), (3, y)\}$. This function f_1 is *not* onto since neither x nor z is an image of some element of A. You might notice that for these two sets A and B, there is *no* function from A to B that is onto since any such function has exactly three ordered pairs but B has four elements. Thus for finite sets A and B, if $f : A \rightarrow B$ is a surjective function, then $|B| \leq |A|$. The function $g : B \rightarrow A$ where $g = \{(x, 3), (y, 1), (z, 3), (w, 2)\}$ *is* a surjective function, however, since each of the elements 1, 2 and 3 is an image of some element of B. Next, we determine which of the functions defined in Result 9.4 and Example 9.5 are onto.

Result to Prove The function $f : \mathbf{R} \rightarrow \mathbf{R}$ defined by $f(x) = 3x - 5$ is onto.

PROOF STRATEGY Let's make a few observations before we begin the proof. To show that f is onto, we must show that every element in the codomain $B = \mathbf{R}$ is the image of some element in the domain $A = \mathbf{R}$. Since $f(0) = -5$ and $f(1) = -2$, certainly -5 and -2 are images of elements of $\mathbf{R}$. The real number 10 is an image as well since $f(5) = 10$. Is π an image of some real number? To answer this question, we need to determine whether there is a real number x such that $f(x) = \pi$. Since $f(x) = 3x - 5$, we need only find a solution for x to the equation $3x - 5 = \pi$. Solving this equation for x, we find that $x = (\pi + 5)/3$, which, of course, is a real number. Finally, observe that

$$f(x) = f\left(\frac{\pi + 5}{3}\right) = 3\left(\frac{\pi + 5}{3}\right) - 5 = \pi.$$

This discussion, however, gives us the information we need to prove that f is onto since for an arbitrary real number r, say, we need to find a real number x such that $f(x) = r$. However, then, $3x - 5 = r$ and $x = (r + 5)/3$. ◆

Result 9.6 *The function $f : \mathbf{R} \rightarrow \mathbf{R}$ defined by $f(x) = 3x - 5$ is onto.*

Proof Let $r \in \mathbf{R}$. We show that there exists $x \in \mathbf{R}$ such that $f(x) = r$. Choose $x = (r + 5)/3$.

Then $x \in \mathbf{R}$ and

$$f(x) = f\left(\frac{r+5}{3}\right) = 3\left(\frac{r+5}{3}\right) - 5 = r.$$ ∎

PROOF ANALYSIS   Notice that the proof itself of Result 9.6 does not include consideration of the equation $3x - 5 = r$. Our goal was to show that some real number x exists such that $f(x) = r$. How we obtain this number, though possibly interesting, is *not* part of the proof. On the other hand, it may be a good idea to accompany the proof with this information.   ◆

Let $A = \{1, 2, 3\}$, $B = \{x, y, z, w\}$ and $C = \{a, b, c\}$. Four functions $g_1 : A \to B$, $g_2 : B \to C$, $g_3 : A \to C$ and $g_4 : A \to C$ are defined as follows:

$$g_1 = \{(1, y), (2, w), (3, x)\},$$
$$g_2 = \{(x, b), (y, a), (z, c), (w, b)\},$$
$$g_3 = \{(1, a), (2, c), (3, b)\},$$
$$g_4 = \{(1, b), (2, b), (3, b)\}.$$

The functions g_1 and g_3 are one-to-one; while g_2 and g_4 are not one-to-one since $g_2(x) = g_2(w) = b$ and $g_4(1) = g_4(2) = b$. Both g_2 and g_3 are onto. The function g_1 is not onto because z is not an image of any element of A; while g_4 is not onto since neither a nor c is an image of an element of A.

9.4 Bijective Functions

We have already mentioned, for finite sets A and B, that if $f : A \to B$ is a surjective function, then $|A| \geq |B|$. Also, we mentioned that if $f : A \to B$ is one-to-one, then $|A| \leq |B|$. Hence if A and B are finite sets and there is a function $f : A \to B$ that is both one-to-one and onto, then $|A| = |B|$. What happens when A and B are infinite sets will be dealt with in detail in Chapter 10.

A function $f : A \to B$ is called **bijective** or a **one-to-one correspondence** if it is both one-to-one and onto. From what we mentioned earlier, if a function $f : A \to B$ is bijective and A and B are finite sets, then $|A| = |B|$. Perhaps it is also clear that if A and B are finite sets with $|A| = |B|$, then there exists a bijective function $f : A \to B$. A bijective function from a set A to a set B creates a pairing of the elements of A with the elements of B.

In the case where A and B are sets with $|A| = |B| = 3$, say $A = \{a, b, c\}$ and $B = \{x, y, z\}$, the bijective functions from A to B are

$$f_1 = \{(a, x), (b, y), (c, z)\}$$
$$f_2 = \{(a, y), (b, z), (c, x)\}$$
$$f_3 = \{(a, z), (b, x), (c, y)\}$$
$$f_4 = \{(a, y), (b, x), (c, z)\}$$
$$f_5 = \{(a, z), (b, y), (c, x)\}$$
$$f_6 = \{(a, x), (b, z), (c, y)\}.$$

That is, there are six bijective functions from A to B; indeed there are six bijective functions from any 3-element set to any 3-element set. More generally, we have the following. (See Exercise 9.34.)

Theorem 9.7   *If A and B are finite sets with $|A| = |B| = n$, then there are $n!$ bijective functions from A to B.*

Proof   Suppose that $A = \{a_1, a_2, \ldots, a_n\}$. Then any bijective function $f : A \to B$ can be expressed as

$$f = \{(a_1, -), (a_2, -), \ldots, (a_n, -)\},$$

where the second coordinate of each ordered pair of f belongs to B. There are n possible images for a_1 in f. Once an image for a_1 has been determined, then there are $n - 1$ possible images for a_2. Since f is one-to-one, no element of B can be the image of two elements of A. Because neither of the images of a_1 and a_2 can be an image of a_3, there are $n - 2$ possibilities for a_3. Continuing in this manner, we see that there is only one possibility for the image of a_n. It turns out that the total number of possible bijective functions f is obtained by multiplying these numbers and so there are $n(n - 1)(n - 2) \cdots 1 = n!$ bijective functions from A to B.   ∎

There is another interesting fact concerning the existence of bijective functions $f : A \to B$ for finite sets A and B with $|A| = |B|$.

Theorem 9.8   *Let A and B be finite nonempty sets such that $|A| = |B|$ and let f be a function from A to B. Then f is one-to-one if and only if f is onto.*

Proof   Let $|A| = |B| = n$. Assume first that f is one-to-one. Since the n elements of A have distinct images, there are n distinct images. Thus range$(f) = B$ and so f is onto.

For the converse, assume that f is onto. Thus each of the n elements of B is an image of some element of A. Consequently, the n elements of A have n distinct images in B, which implies that no two distinct elements of A can have the same image and so f is one-to-one.   ∎

Theorem 9.8 concerns finite sets A and B with $|A| = |B|$. Even though we have not defined cardinality for infinite sets, we would certainly expect that $|A| = |A|$ for every infinite set A. With this understanding, Theorem 9.8 is false for infinite sets A and B, even when $A = B$. For example, the function $f : \mathbf{Z} \to \mathbf{Z}$ defined by $f(n) = 2n$ is one-to-one; yet its range is the set of all even integers. That is, f is not onto, even though f is a one-to-one function from $\mathbf{Z}$ to $\mathbf{Z}$. The function $f : \mathbf{N} \to \mathbf{N}$ defined by $g(n) = n - 1$ when $n \geq 2$ and $g(1) = 1$ is onto but not one-to-one since $g(1) = g(2) = 1$. For the sets $A = \{1, 2, 3\}$, $B = \{x, y, z, w\}$ and $C = \{a, b, c\}$ described at the end of Section 9.3, no function from A to B or from B to C can be bijective. It is possible to have a bijective function from A to C, however, since $|A| = |C|$. In fact, g_3 is such a function, although other bijective functions from A to C exist. Certainly, not every function from A to C is bijective, as g_4 illustrates.

For a nonempty set A, the function $i_A : A \rightarrow A$ defined by $i_A(a) = a$ for each $a \in A$ is called the **identity function** on A. If the set A under discussion is clear, we write the identity function i_A by i. For $S = \{1, 2, 3\}$, the identity function is

$$i_S = i = \{(1, 1), (2, 2), (3, 3)\}.$$

Not only is *this* identity function bijective, the identity function i_A is bijective for *every* nonempty set A. Identity functions are important and we will see them again soon.

We give one additional example of a bijective function.

Result 9.9   *The function $f : \mathbf{R} - \{2\} \rightarrow \mathbf{R} - \{3\}$ defined by*

$$f(x) = \frac{3x}{x - 2}$$

is bijective.

Proof   Here it is necessary to show that f is both one-to-one and onto. We begin with the first of these. Assume that $f(a) = f(b)$, where $a, b \in \mathbf{R} - \{2\}$. Then $\dfrac{3a}{a - 2} = \dfrac{3b}{b - 2}$. Multiplying both sides by $(a - 2)(b - 2)$, we obtain $3a(b - 2) = 3b(a - 2)$. Simplifying, we have $3ab - 6a = 3ab - 6b$. Adding $-3ab$ to both sides and dividing by -6, we obtain $a = b$. Thus f is one-to-one.

To show that f is onto, let $r \in \mathbf{R} - \{3\}$. We show that there exists $x \in \mathbf{R} - \{2\}$ such that $f(x) = r$. Choose $x = \dfrac{2r}{r - 3}$. Then

$$f(x) = f\left(\frac{2r}{r - 3}\right) = \frac{3\left(\frac{2r}{r-3}\right)}{\frac{2r}{r-3} - 2} = \frac{6r}{2r - 2(r - 3)} = \frac{6r}{6} = r,$$

implying that f is onto. Therefore f is bijective. ∎

PROOF ANALYSIS   Some remarks concerning the proof that the function f in Result 9.9 is onto may be useful. For a given real number r in $\mathbf{R} - \{3\}$, we need to find a real number x in $\mathbf{R} - \{2\}$ such that $f(x) = r$. Since we wanted $f(x) = \dfrac{3x}{x - 2} = r$, it was required to solve this equation for x. This can be done by rewriting this equation as $3x = r(x - 2)$ and then simplifying it to obtain $rx - 3x = 2r$. Now, factoring x from $rx - 3x$ and dividing by $r - 3$, we have the desired choice of x, namely $x = 2r/(r - 3)$. Incidentally, it was perfectly permissible to divide by $r - 3$ since $r \in \mathbf{R} - \{3\}$ and so $r \neq 3$. Notice also that $x \in \mathbf{R} - \{2\}$ for if $x = 2r/(r - 3) = 2$, then $2r = 2r - 6$, which is impossible. Although solving $\dfrac{3x}{x - 2} = r$ for x is *not* part of the proof, again it may be useful to include this work in addition to the proof. ◆

Suppose that f is a function from A to B, that is, $f : A \rightarrow B$. If $f(x) = f(y)$ implies that $x = y$ for all $x, y \in A$, then f is one-to-one. It may seem obvious that if $x = y$, then $f(x) = f(y)$ for all $x, y \in A$ since this is simply a requirement of a function. In order for a relation f from a set A to a set B to be a function from A to B, the following two conditions must be satisfied:

(1) For each element $a \in A$, there is an element $b \in B$ such that $(a, b) \in f$.
(2) If $(a, b), (a, c) \in f$, then $b = c$.

Condition (1) states that the domain of f is A, that is, every element of A has an image in B; while condition (2) says that if an element of A has an image in B, then this image is unique.

Occasionally, a function f that satisfies condition (2) is called **well-defined**. Since (2) is a requirement of every function however, it follows that every function must be well-defined. There are situations though when the definition of a function f may make it unclear whether f is well-defined. This can often occur when a function is defined on the set of equivalence classes of an equivalence relation. The next result illustrates this with the equivalence classes for the relation congruence modulo 4 on the set of integers.

Result to Prove   The function $f : \mathbf{Z}_4 \to \mathbf{Z}_4$ defined by $f([x]) = [3x + 1]$ is a well-defined bijective function.

PROOF STRATEGY   To prove that this function is well-defined, we are required to prove that if $[a] = [b]$, then $f([a]) = f([b])$, that is, $[3a + 1] = [3b + 1]$. It seems reasonable to use a direct proof, so we assume that $[a] = [b]$. Since $[a]$ and $[b]$ are elements of $\mathbf{Z}_4$, to say that $[a] = [b]$ means that $a \equiv b \,(\mathrm{mod}\ 4)$. Since $a \equiv b \,(\mathrm{mod}\ 4)$, it follows that $4 \mid (a - b)$ and so $a - b = 4k$ for some integer k. To verify that $[3a + 1] = [3b + 1]$, we are required to show that $3a + 1 \equiv 3b + 1 \,(\mathrm{mod}\ 4)$ or, equivalently, that $(3a + 1) - (3b + 1) = 3a - 3b = 3(a - b)$ is a multiple of 4.

Since $\mathbf{Z}_4$ consists only of four elements, namely, $[0], [1], [2], [3]$, to prove that f is bijective, we need only observe that the elements $f([0]), f([1]), f([2]), f([3])$ are distinct.   ♦

Result 9.10   *The function $f : \mathbf{Z}_4 \to \mathbf{Z}_4$ defined by $f([x]) = [3x + 1]$ is a well-defined bijective function.*

Proof   First, we verify that this function is well-defined; that is, if $[a] = [b]$, then $f([a]) = f([b])$. Assume then that $[a] = [b]$. Thus $a \equiv b \,(\mathrm{mod}\ 4)$ and so $4 \mid (a - b)$. Hence $a - b = 4k$ for some integer k. Therefore,

$$(3a + 1) - (3b + 1) = 3(a - b) = 3(4k) = 4(3k).$$

Since $3k$ is an integer, $4 \mid [(3a + 1) - (3b + 1)]$. Thus $3a + 1 \equiv 3b + 1 \,(\mathrm{mod}\ 4)$ and $[3a + 1] = [3b + 1]$; so $f([a]) = f([b])$. Hence f is well-defined. Since $f([0]) = [1]$, $f([1]) = [0]$, $f([2]) = [3]$ and $f([3]) = [2]$, it follows that f is both one-to-one and onto; that is, f is bijective.   ∎

9.5 Composition of Functions

As it is common to define operations on certain sets of numbers (and on the set $\mathbf{Z}_n$ of equivalence classes, as we described in Chapter 8), it is possible to define operations on certain sets of functions, under suitable circumstances. For example, for functions $f : \mathbf{R} \to \mathbf{R}$ and $g : \mathbf{R} \to \mathbf{R}$, you might recall from calculus that the sum $f + g$ and product fg of f and g are defined by

$$(f + g)(x) = f(x) + g(x) \text{ and } (fg)(x) = f(x) \cdot g(x) \tag{9.1}$$

for all $x \in \mathbf{R}$. So if f is defined by $f(x) = x^2$ and g is defined by $g(x) = \sin x$, then $(f + g)(x) = x^2 + \sin x$ and $(fg)(x) = x^2 \sin x$ for all $x \in \mathbf{R}$. In calculus we are especially interested in these operations because once we have learned how to determine the derivatives of f and g, we want to know how to use this information to find the derivatives of $f + g$ and fg. The derivative of fg, for example, gives rise to the well-known product rule for derivatives:

$$(fg)'(x) = f(x) \cdot g'(x) + g(x) \cdot f'(x).$$

This later led us to study the quotient rule for derivatives.

The definitions in (9.1) of the sum $f + g$ and product fg of the functions $f : \mathbf{R} \to \mathbf{R}$ and $g : \mathbf{R} \to \mathbf{R}$ depend on the fact that the codomain of these two functions is $\mathbf{R}$, whose elements can be added and multiplied, and so $f(x) + g(x)$ and $f(x) \cdot g(x)$ make sense. On the other hand, if $f : A \to B$ and $g : A \to B$, where $B = \{a, b, c\}$, say, then $f(x) + g(x)$ and $f(x) \cdot g(x)$ have no meaning.

There is an operation that can be defined on pairs of functions satisfying appropriate conditions that has no connection with numbers. For nonempty sets A, B and C and functions $f : A \to B$ and $g : B \to C$, it is possible to create a new function from f and g, called their composition. The **composition** $g \circ f$ of f and g is the function from A to C defined by

$$(g \circ f)(a) = g(f(a)) \quad \text{for all } a \in A.$$

To illustrate this definition, let $A = \{1, 2, 3, 4\}$, $B = \{a, b, c, d\}$ and $C = \{r, s, t, u, v\}$ and define the functions $f : A \to B$ and $g : B \to C$ by

$$f = \{(1, b), (2, d), (3, a), (4, a)\},$$
$$g = \{(a, u), (b, r), (c, r), (d, s)\}.$$

We now have the correct arrangement of sets and functions to consider the composition $g \circ f$. Since $g \circ f$ is a function from A to C, it follows that $g \circ f$ has the following appearance:

$$g \circ f = \{(1, \alpha), (2, \beta), (3, \gamma), (4, \delta)\},$$

where $\alpha, \beta, \gamma, \delta \in C$. It remains only to determine the image of each element of A. First, we find the image of 1. According to the definition of $g \circ f$,

$$(g \circ f)(1) = g(f(1)) = g(b) = r,$$

so $(1, r) \in g \circ f$. Similarly, $(g \circ f)(2) = g(f(2)) = g(d) = s$ and so $(2, s) \in g \circ f$. Continuing in this manner, we obtain

$$g \circ f = \{(1, r), (2, s), (3, u), (4, u)\}.$$

A diagram that illustrates how $g \circ f$ is determined is shown in Figure 9.2. To find the image of 1 under $g \circ f$, we follow the arrow from 1 to b and then from b to r. The function $g \circ f$ is basically found by removing the set B. The fact that $g \circ f$ is defined does not necessarily imply that $f \circ g$ is also defined. Since g is a function from B to C and f is a function from A to B, the only way that $f \circ g$ would be defined is if range$(g) \subseteq A$. In the example we have just seen, $f \circ g$ is not defined since range$(g) = \{r, s, u\} \nsubseteq A$.

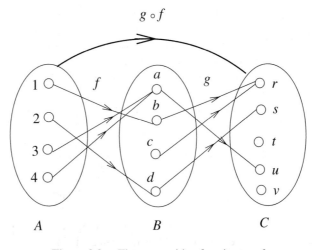

Figure 9.2   The composition function $g \circ f$

Composition of functions was also encountered in calculus. Let's consider an example of composition that you might have seen in calculus. Again, suppose that the functions $f : \mathbf{R} \to \mathbf{R}$ and $g : \mathbf{R} \to \mathbf{R}$ are defined by $f(x) = x^2$ and $g(x) = \sin x$. In this case, we can determine both $g \circ f$ and $f \circ g$; namely,

$$(g \circ f)(x) = g(f(x)) = g(x^2) = \sin\left(x^2\right)$$
$$(f \circ g)(x) = f(g(x)) = f(\sin x) = (\sin x)^2 = \sin^2 x.$$

Second, this example also serves to illustrate that even when $g \circ f$ and $f \circ g$ are both defined, they need not be equal.

The study of composition of functions in calculus led us to the well-known chain rule for differentiation:

$$(g \circ f)'(x) = g'\left(f(x)\right) \cdot f'(x).$$

There are two facts concerning properties of composition of functions that will be especially useful to us. First, if f and g are injective functions such that $g \circ f$ is defined, then $g \circ f$ is injective. The corresponding statement is also true for surjective functions.

Result to Prove   Let $f : A \to B$ and $g : B \to C$ be two functions.

(a) If f and g are injective, then so is $g \circ f$.
(b) If f and g are surjective, then so is $g \circ f$.

<u>PROOF STRATEGY</u>   To verify (a), we use a direct proof and begin by assuming that f and g are one-to-one. To show that $g \circ f$ is one-to-one, we prove that whenever $(g \circ f)(a_1) = (g \circ f)(a_2)$, then $a_1 = a_2$. However, $(g \circ f)(a_1) = (g \circ f)(a_2)$ means that $g(f(a_1)) = g(f(a_2))$. But g is one-to-one; so $g(x) = g(y)$ implies that $x = y$. The form $g(x) = g(y)$ is exactly what we have, where $x = f(a_1)$ and $y = f(a_2)$. This leads us to $f(a_1) = f(a_2)$. But we also know that f is one-to-one.

To verify (b), we need to prove that if f and g are onto, then $g \circ f$ is onto. To show that $g \circ f$ is onto, it is necessary to show that every element of C is an image of some element of A under the function $g \circ f$. So we begin with an element $c \in C$. Since g is onto, there is an element $b \in B$ such that $g(b) = c$. But f is onto; so there is an element $a \in A$ such that $f(a) = b$. This suggests considering $(g \circ f)(a)$.   ◆

Theorem 9.11   *Let $f : A \to B$ and $g : B \to C$ be two functions.*

(a) *If f and g are injective, then so is $g \circ f$.*
(b) *If f and g are surjective, then so is $g \circ f$.*

Proof   Let $f : A \to B$ and $g : B \to C$ be injective functions. Assume that $(g \circ f)(a_1) = (g \circ f)(a_2)$, where $a_1, a_2 \in A$. By definition, $g(f(a_1)) = g(f(a_2))$. Since g is injective, it follows that $f(a_1) = f(a_2)$. However, since f is injective, it follows that $a_1 = a_2$. This implies that $g \circ f$ is injective.

Next let $f : A \to B$ and $g : B \to C$ be surjective functions and let $c \in C$. Since g is surjective, there exists $b \in B$ such that $g(b) = c$. On the other hand, since f is surjective, it follows that there exists $a \in A$ such that $f(a) = b$. Hence $(g \circ f)(a) = g(f(a)) = g(b) = c$, implying that $g \circ f$ is also surjective.   ■

Combining the two parts of Theorem 9.11 produces an immediate corollary.

Corollary 9.12   *If $f : A \to B$ and $g : B \to C$ are bijective functions, then $g \circ f$ is bijective.*

For nonempty sets A, B, C and D, let $f : A \to B$, $g : B \to C$ and $h : C \to D$ be functions. Then the compositions $g \circ f : A \to C$ and $h \circ g : B \to D$ are defined, as are the compositions $h \circ (g \circ f) : A \to D$ and $(h \circ g) \circ f : A \to D$. Composition of the functions f, g and h is **associative** if the functions $h \circ (g \circ f)$ and $(h \circ g) \circ f$ are equal. This is, in fact, the case.

Theorem 9.13   *For nonempty sets A, B, C and D, let $f : A \to B$, $g : B \to C$ and $h : C \to D$ be functions. Then $(h \circ g) \circ f = h \circ (g \circ f)$.*

Proof   Let $a \in A$ and suppose that $f(a) = b$, $g(b) = c$ and $h(c) = d$. Then

$$((h \circ g) \circ f)(a) = (h \circ g)(f(a)) = (h \circ g)(b) = h(g(b)) = h(c) = d;$$

while

$$(h \circ (g \circ f))(a) = h((g \circ f)(a)) = h(g(f(a))) = h(g(b)) = h(c) = d.$$

Thus $(h \circ g) \circ f = h \circ (g \circ f)$.   ■

As we have mentioned, it is common, when considering the composition of functions, to begin with two functions f and g, where $f : A \to B$ and $g : B \to C$ and arrive at the function $g \circ f : A \to C$. Strictly speaking, however, all that is needed is for the domain of g to be a set B' where range(f) is a subset of B'. In other words, if f and g are functions with $f : A \to B$ and $g : B' \to C$, where range(f) $\subseteq B'$, then the composition $g \circ f : A \to C$ is defined.

Example 9.14   *For the sets $A = \{-3, -2, \ldots, 3\}$ and $B = \{0, 1, \ldots, 10\}$, $B' = \{0, 1, 4, 5, 8, 9\}$ and $C = \{1, 2, \ldots, 10\}$, let $f : A \to B$ and $g : B' \to C$ be functions defined by $f(n) = n^2$ for all $n \in A$ and $g(n) = n + 1$ for all $n \in B'$.*

(a) *Show that the composition $g \circ f : A \to C$ is defined.*
(b) *For $n \in A$, determine $(g \circ f)(n)$.*

Solution   (a) Since range$(f) = \{0, 1, 4, 9\}$ and range $f \subseteq B'$, it follows that the composition $g \circ f : A \to C$ is defined.

(b) For $n \in A$, $(g \circ f)(n) = g(f(n)) = g(n^2) = n^2 + 1$.   ◆

9.6 Inverse Functions

Next we describe a property possessed by all bijective functions. In preparation for doing this, we return to relations to recall a concept introduced in Chapter 8. For a relation R from a set A to a set B, the **inverse relation** R^{-1} from B to A is defined as

$$R^{-1} = \{(b, a) : (a, b) \in R\}.$$

For example, if $A = \{a, b, c, d\}$, $B = \{1, 2, 3\}$ and

$$R = \{(a, 1), (a, 3), (c, 2), (c, 3), (d, 1)\}$$

is a relation from A to B, then

$$R^{-1} = \{(1, a), (3, a), (2, c), (3, c), (1, d)\}$$

is the inverse relation of R. Of course, every function $f : A \to B$ is also a relation from A to B and so there is an inverse relation f^{-1} from B to A. This brings up a natural question: Under what conditions is the inverse relation f^{-1} from B to A also a function from B to A? If the inverse relation f^{-1} is a function from B to A, then certainly dom$(f^{-1}) = B$. This implies that f must be onto. If f is not one-to-one, then $f(a_1) = f(a_2) = b$ for some $a_1, a_2 \in A$ and $b \in B$, where $a_1 \neq a_2$. But then $(b, a_1), (b, a_2) \in f^{-1}$, which cannot occur if f^{-1} is a function. This leads us to the following theorem. In the proof, two basic facts are used repeatedly, namely

(1) $f(a) = b$ if and only if $(a, b) \in f$ and
(2) if f^{-1} is a function and $f(a) = b$, then $(b, a) \in f^{-1}$.

Theorem 9.15   *Let $f : A \to B$ be a function. Then the inverse relation f^{-1} is a function from B to A if and only if f is bijective. Furthermore, if f is bijective, then f^{-1} is also bijective.*

Proof   First, assume that f^{-1} is a function from B to A. Then we show that f is both one-to-one and onto. Assume that $f(a_1) = f(a_2) = y$, where $y \in B$. Then $(a_1, y), (a_2, y) \in f$, implying that $(y, a_1), (y, a_2) \in f^{-1}$. Since f^{-1} is a function from B to A, every element of B has a unique image under f^{-1}. Thus, in particular, y has a unique image under f^{-1}. Since $f^{-1}(y) = a_1$ and $f^{-1}(y) = a_2$, it now follows that $a_1 = a_2$ and so f is one-to-one.

To show that f is onto, let $b \in B$. Since f^{-1} is a function from B to A, there exists a unique element $a \in A$ such that $f^{-1}(b) = a$. Hence $(b, a) \in f^{-1}$, implying that $(a, b) \in f$, that is, $f(a) = b$. Therefore, f is onto.

For the converse, assume that the function $f : A \to B$ is bijective. We show that f^{-1} is a function from B to A. Let $b \in B$. Since f is onto, there exists $a \in A$ such that $(a, b) \in f$. Hence $(b, a) \in f^{-1}$. It remains to show that (b, a) is the unique element of f^{-1} whose first coordinate is b. Assume that (b, a) and (b, a') are both in f^{-1}. Then $(a, b), (a', b) \in f$, which implies that $f(a) = f(a') = b$. Since f is one-to-one, $a = a'$. Therefore, we have shown that for every $b \in B$ there exists a unique element $a \in A$ such that $(b, a) \in f^{-1}$; that is, f^{-1} is a function from B to A.

Finally, we show that if f is bijective, then f^{-1} is bijective. Assume that f is bijective. We have just seen that f^{-1} is a function from B to A. First, we show that f^{-1} is one-to-one.

Assume that $f^{-1}(b_1) = f^{-1}(b_2) = a$. Then $(b_1, a), (b_2, a) \in f^{-1}$ and so (a, b_1), $(a, b_2) \in f$. Since f is a function, $b_1 = b_2$ and f^{-1} is one-to-one. To show that f^{-1} is onto, let $a \in A$. Since f is a function, there is an element $b \in B$ such that $(a, b) \in f$. Consequently, $(b, a) \in f^{-1}$ so that $f^{-1}(b) = a$ and f^{-1} is onto. Therefore, f^{-1} is bijective. ∎

Let $f : A \to B$ be a bijective function. By Theorem 9.15 then, $f^{-1} : B \to A$ is a bijective function, which is referred to as the **inverse function** or simply the **inverse** of f. Hence both composition functions $f^{-1} \circ f$ and $f \circ f^{-1}$ are defined. In fact, $f^{-1} \circ f$ is a function from A to A and $f \circ f^{-1}$ is a function from B to B. As we are about to learn, $f^{-1} \circ f$ and $f \circ f^{-1}$ are functions we've visited earlier. Let $a \in A$ and suppose that $f(a) = b$. So $(a, b) \in f$ and therefore $(b, a) \in f^{-1}$, that is, $f^{-1}(b) = a$. Thus $\left(f^{-1} \circ f\right)(a) = f^{-1}(f(a)) = f^{-1}(b) = a$ and $\left(f \circ f^{-1}\right)(b) = f\left(f^{-1}(b)\right) = f(a) = b$. So it follows that

$$f^{-1} \circ f = i_A \quad \text{and} \quad f \circ f^{-1} = i_B$$

are the identity functions on the sets A and B. (See Figure 9.3.)

In fact, if $f : A \to B$ and $g : B \to A$ are functions for which $g \circ f = i_A$ and $f \circ g = i_B$, then f and g have some important properties.

Theorem 9.16   *If $f : A \to B$ and $g : B \to A$ are two functions such that $g \circ f = i_A$ and $f \circ g = i_B$, then f and g are bijective and $g = f^{-1}$.*

Proof   First, we show that f is one-to-one. Assume that $f(a_1) = f(a_2)$, where $a_1, a_2 \in A$. Then $g(f(a_1)) = g(f(a_2))$. Since $g \circ f = i_A$, it follows that

$$a_1 = (g \circ f)(a_1) = g(f(a_1)) = g(f(a_2)) = (g \circ f)(a_2) = a_2$$

and so f is one-to-one.

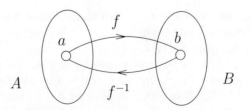

Figure 9.3   A bijective function and its inverse

Next, we show that f is onto. Let $b \in B$ and suppose that $g(b) = a$. Since $f \circ g = i_B$, it follows that $(f \circ g)(b) = b$. Therefore, $(f \circ g)(b) = f(g(b)) = f(a) = b$ and so f is onto. Hence f is bijective and so f^{-1} exists. Similarly, g is bijective.

Let $a \in A$ and suppose that $f(a) = b \in B$. Then $f^{-1}(b) = a$. Since $g \circ f = i_A$, it follows that $a = (g \circ f)(a) = g(f(a)) = g(b)$. Therefore, $g = f^{-1}$. ∎

If a bijective function f has a relatively small number of ordered pairs, then it is easy to find f^{-1}. But what if f is a bijective function that one might encounter in calculus, say? We illustrate this next with a function described in Result 9.9.

Example 9.17   *The function* $f : \mathbf{R} - \{2\} \to \mathbf{R} - \{3\}$ *defined by*

$$f(x) = \frac{3x}{x - 2}$$

is known to be bijective. Determine $f^{-1}(x)$, *where* $x \in \mathbf{R} - \{3\}$.

Solution   Since $\left(f \circ f^{-1}\right)(x) = x$ for all $x \in \mathbf{R} - \{3\}$, it follows that

$$\left(f \circ f^{-1}\right)(x) = f\left(f^{-1}(x)\right) = \frac{3f^{-1}(x)}{f^{-1}(x) - 2} = x.$$

Thus $3f^{-1}(x) = x(f^{-1}(x) - 2)$ and $3f^{-1}(x) = xf^{-1}(x) - 2x$. Collecting the terms involving $f^{-1}(x)$ on the same side of the equation and then factoring out the term $f^{-1}(x)$, we have

$$xf^{-1}(x) - 3f^{-1}(x) = 2x$$

so

$$f^{-1}(x)(x - 3) = 2x.$$

Solving for $f^{-1}(x)$, we obtain

$$f^{-1}(x) = \frac{2x}{x - 3}. \qquad \blacklozenge$$

Analysis   You might very well have dealt with the problem of finding the inverse of a function before and might recall a somewhat different approach than the one we just gave. Let's look at this example again, but from a different perspective.

When we consider functions from calculus, rather than writing $f(x) = x^2$, $g(x) = 5x + 1$ or $h(x) = x + \dfrac{1}{x}$, we sometimes write these as $y = x^2$, $y = 5x + 1$ or $y = x + \dfrac{1}{x}$. In Example 9.17, we were given $f(x) = \dfrac{3x}{x - 2}$ and found that $f^{-1}(x) = \dfrac{2x}{x - 3}$. Let's write the inverse as $y = \dfrac{2x}{x - 3}$ instead. That is, $(x, y) \in f^{-1}$, where $y = \dfrac{2x}{x - 3}$. Of course, initially, we don't know what y is. But if $(x, y) \in f^{-1}$, then $(y, x) \in f$ and we know that $x = f(y) = \dfrac{3y}{y - 2}$. Solving this equation for y, we have $x(y - 2) = 3y$, so $xy - 2x = 3y$. Collecting the terms with y on the same side of the equation and factoring out the term y, we obtain

$$xy - 3y = 2x \quad \text{and} \quad y(x - 3) = 2x.$$

Solving for y, we obtain $y = \dfrac{2x}{x - 3}$; that is,

$$f^{-1}(x) = \frac{2x}{x - 3}.$$

In short, to find f^{-1} if $f(x) = \dfrac{3x}{x - 2}$, we replace $f(x)$ by x and x by y and then solve for y. The result is $f^{-1}(x)$. Of course, the procedure we have described for finding $f^{-1}(x)$ is exactly the same as before. The only difference is the notation. You might have also noticed that the algebra performed to determine $f^{-1}(x)$ in Example 9.17 is exactly the same as the algebra performed in proving f is onto in Result 9.9.   ♦

Finding the inverse of a bijective function is not always possible by algebraic manipulation. For example, the function $f : \mathbf{R} \to (0, \infty)$ defined by $f(x) = e^x$ is bijective but $f^{-1}(x) = \ln x$. Indeed, the function $g : \mathbf{R} \to \mathbf{R}$ defined by $g(x) = 3x^7 + 5x^3 + 4x - 1$ is bijective, but there is no way to find an expression for $g^{-1}(x)$.

If $f : A \to B$ is a one-to-one function from A to B that is not onto, then, of course, f is not bijective, and, according to Theorem 9.15, f does not have an inverse (from B to A). On the other hand, if we define a new function $g : A \to \text{range}(f)$ by $g(x) = f(x)$ for all $x \in A$, then g is a bijective function and so its inverse function $g^{-1} : \text{range}(f) \to A$ exists. For example, let E denote the set of all even integers and consider the function $f : \mathbf{Z} \to \mathbf{Z}$ by $f(n) = 2n$. Then this function f is injective but not surjective and so there is no inverse function of f from $\mathbf{Z}$ to $\mathbf{Z}$. Observe that $\text{range}(f) = E$. If we define $g : \mathbf{Z} \to E$ by $g(n) = f(n)$ for all $n \in \mathbf{Z}$, then g is bijective and $g^{-1} : E \to \mathbf{Z}$ is a (bijective) function. In fact, $g^{-1}(n) = n/2$ for all $n \in E$.

9.7 Permutations

We have already mentioned that the identity function i_A defined on a nonempty set A is bijective. Normally, there are many bijective functions that can be defined on nonempty sets. Indeed, the number of bijections on an n-element set is $n!$ according to Theorem 9.7. These types of functions occur often in mathematics, especially in the area of mathematics called abstract (or modern) algebra.

A **permutation** of (or on) a nonempty set A is a bijective function on A, that is, a function from A to A that is both one-to-one and onto. By Results 9.4 and 9.6, the function $f : \mathbf{R} \to \mathbf{R}$ defined by $f(x) = 3x - 5$ is a permutation of $\mathbf{R}$. Let's consider an even simpler example. For $A = \{1, 2, 3\}$, let f be a permutation of A. Then f is completely determined once we know the images of 1, 2 and 3 under f. We saw that there are three possible choices for $f(1)$, two choices for $f(2)$ once $f(1)$ has been specified and one choice for $f(3)$ once $f(1)$ and $f(2)$ have been specified. From this, it follows that there are $3 \cdot 2 \cdot 1 = 3! = 6$ different permutations f of the set $A = \{1, 2, 3\}$. This agrees with Theorem 9.7.

One of these functions is the identity function defined on $\{1, 2, 3\}$, which we denote by α_1; that is,

$$\alpha_1 = \{(1, 1), (2, 2), (3, 3)\}.$$

Another permutation of $\{1, 2, 3\}$ is

$$\alpha_2 = \{(1, 1), (2, 3), (3, 2)\}.$$

There are other common ways to represent these permutations. A permutation of $\{1, 2, 3\}$ is also written as

$$\begin{pmatrix} 1 & 2 & 3 \\ - & - & - \end{pmatrix},$$

where the numbers immediately below 1, 2 and 3 are their images. Hence α_1, α_2 and the other four permutations of $\{1, 2, 3\}$ can be expressed as:

$$\alpha_1 = \begin{pmatrix} 1 & 2 & 3 \\ 1 & 2 & 3 \end{pmatrix} \quad \alpha_2 = \begin{pmatrix} 1 & 2 & 3 \\ 1 & 3 & 2 \end{pmatrix} \quad \alpha_3 = \begin{pmatrix} 1 & 2 & 3 \\ 3 & 2 & 1 \end{pmatrix}$$

$$\alpha_4 = \begin{pmatrix} 1 & 2 & 3 \\ 2 & 1 & 3 \end{pmatrix} \quad \alpha_5 = \begin{pmatrix} 1 & 2 & 3 \\ 2 & 3 & 1 \end{pmatrix} \quad \alpha_6 = \begin{pmatrix} 1 & 2 & 3 \\ 3 & 1 & 2 \end{pmatrix}.$$

Since each permutation α_i $(1 \leq i \leq 6)$ is a bijective function from $\{1, 2, 3\}$ to $\{1, 2, 3\}$, it follows that from Corollary 9.12 that the composition of any two permutations of $\{1, 2, 3\}$ is again a permutation of $\{1, 2, 3\}$. For example, let's consider

$$\alpha_2 \circ \alpha_5 = \begin{pmatrix} 1 & 2 & 3 \\ 1 & 3 & 2 \end{pmatrix} \circ \begin{pmatrix} 1 & 2 & 3 \\ 2 & 3 & 1 \end{pmatrix} = \begin{pmatrix} 1 & 2 & 3 \\ - & - & - \end{pmatrix}.$$

Since $(\alpha_2 \circ \alpha_5)(1) = \alpha_2(\alpha_5(1)) = \alpha_2(2) = 3$, $(\alpha_2 \circ \alpha_5)(2) = 2$ and $(\alpha_2 \circ \alpha_5)(3) = 1$, it follows that

$$\alpha_2 \circ \alpha_5 = \begin{pmatrix} 1 & 2 & 3 \\ 1 & 3 & 2 \end{pmatrix} \circ \begin{pmatrix} 1 & 2 & 3 \\ 2 & 3 & 1 \end{pmatrix} = \begin{pmatrix} 1 & 2 & 3 \\ 3 & 2 & 1 \end{pmatrix} = \alpha_3.$$

By Theorem 9.13, it follows that composition of permutations on the same nonempty set A is associative. Hence for all integers $i, j, k \in \{1, 2, \cdots, 6\}$,

$$(\alpha_i \circ \alpha_j) \circ \alpha_k = \alpha_i \circ (\alpha_j \circ \alpha_k).$$

Also, by Theorem 9.15, since a permutation is a bijective function, each permutation has an inverse, which is also a permutation. Thus for each i $(1 \leq i \leq 6)$, $\alpha_i^{-1} = \alpha_j$ for some j $(1 \leq j \leq 6)$. The inverse of a permutation can be found by interchanging the two rows and then re-ordering the columns so that the top row is in the natural order $1, 2, 3, \ldots$. Thus

$$\alpha_5^{-1} = \begin{pmatrix} 2 & 3 & 1 \\ 1 & 2 & 3 \end{pmatrix} = \begin{pmatrix} 1 & 2 & 3 \\ 3 & 1 & 2 \end{pmatrix} = \alpha_6.$$

The set of all $n!$ permutations of the set $\{1, 2, , \cdots, n\}$ is denoted by $\mathcal{S}_n$. Thus

$$\mathcal{S}_3 = \{\alpha_1, \alpha_2, \ldots, \alpha_6\}.$$

As we have seen with $\mathcal{S}_3$, the elements of $\mathcal{S}_n$ satisfy the properties of closure, associativity and the existence of inverses for every positive integer n. This will be revisited in Chapter 13.

EXERCISES FOR CHAPTER 9

Section 9.1: The Definition of Function

9.1. Let $A = \{a, b, c, d\}$ and $B = \{x, y, z\}$. Then $f = \{(a, y), (b, z), (c, y), (d, z)\}$ is a function from A to B. Determine dom(f) and range(f).

9.2. Let $A = \{1, 2, 3\}$ and $B = \{a, b, c, d\}$. Give an example of a relation R from A to B containing exactly three elements such that R is *not* a function from A to B. Explain why R is not a function.

9.3. Let A be a nonempty set. If R is a relation from A to A that is both an equivalence relation and a function, then what familiar function is R? Justify your answer.

9.4. For the given subset A_i of **R** and the relation R_i ($1 \le i \le 3$) from A_i to **R**, determine whether R_i is a function from A_i to **R**.

  (a) $A_1 = \mathbf{R}$, $R_1 = \{(x, y) : x \in A_1, y = 4x - 3\}$
  (b) $A_2 = [0, \infty)$, $R_2 = \{(x, y) : x \in A_2, (y + 2)^2 = x\}$
  (c) $A_3 = \mathbf{R}$, $R_3 = \{(x, y) : x \in A_3, (x + y)^2 = 4\}$

9.5. Let A and B be nonempty sets and let R be a nonempty relation from A to B. Show that there exists a subset A' of A and a subset f of R such that f is a function from A' to B.

9.6. In each of the following, a function $f_i : A_i \to \mathbf{R}$ ($1 \le i \le 5$) is defined, where the domain A_i consists of all real numbers x for which $f_i(x)$ is defined. In each case, determine the domain A_i and the range of f_i.

  (a) $f_1(x) = 1 + x^2$
  (b) $f_2(x) = 1 - \frac{1}{x}$
  (c) $f_3(x) = \sqrt{3x - 1}$
  (d) $f_4(x) = x^3 - 8$
  (e) $f_5(x) = \frac{x}{x-3}$.

9.7. Let $A = \{3, 17, 29, 45\}$ and $B = \{4, 6, 22, 60\}$. A relation R from A to B is defined by $a \mathrel{R} b$ if $a + b$ is a prime. Is R a function from A to B?

9.8. Let $A = \{5, 6\}$, $B = \{5, 7, 8\}$ and $S = \{n : n \ge 3 \text{ is an odd integer}\}$. A relation R from $A \times B$ to S is defined as $(a, b) \mathrel{R} s$ if $s \mid (a + b)$. Is R a function from $A \times B$ to S?

9.9. Determine which of the following five relations R_i ($i = 1, 2, \ldots, 5$) are functions.

  (a) R_1 is defined on **R** by $x \mathrel{R_1} y$ if $x^2 + y^2 = 1$.
  (b) R_2 is defined on **R** by $x \mathrel{R_2} y$ if $4x^2 + 3y^2 = 1$.
  (c) R_3 is defined from **N** to **Q** by $a \mathrel{R_3} b$ if $3a + 5b = 1$.
  (d) R_4 is defined on **R** by $x \mathrel{R_4} y$ if $y = 4 - |x - 2|$.
  (e) R_5 is defined on **R** by $x \mathrel{R_5} y$ if $|x + y| = 1$.

9.10. A function $g : \mathbf{Q} \to \mathbf{Q}$ is defined by $g(r) = 4r + 1$ for each $r \in \mathbf{Q}$.

  (a) Determine $g(\mathbf{Z})$ and $g(E)$, where E is the set of even integers.
  (b) Determine $g^{-1}(\mathbf{N})$ and $g^{-1}(D)$, where D is the set of odd integers.

9.11. Let $C = \{x \in \mathbf{R} : x \ge 1\}$ and $D = \mathbf{R}^+$. For each function f defined below, determine $f(C)$, $f^{-1}(C)$, $f^{-1}(D)$ and $f^{-1}(\{1\})$.

  (a) $f : \mathbf{R} \to \mathbf{R}$ is defined by $f(x) = x^2$.
  (b) $f : \mathbf{R}^+ \to \mathbf{R}$ is defined by $f(x) = \ln x$.
  (c) $f : \mathbf{R} \to \mathbf{R}$ is defined by $f(x) = e^x$.
  (d) $f : \mathbf{R} \to \mathbf{R}$ is defined by $f(x) = \sin x$.
  (e) $f : \mathbf{R} \to \mathbf{R}$ is defined by $f(x) = 2x - x^2$.

9.12. For a function $f : A \rightarrow B$ and subsets C and D of A and E and F of B, prove the following.

(a) $f(C \cup D) = f(C) \cup f(D)$
(b) $f(C \cap D) \subseteq f(C) \cap f(D)$
(c) $f(C) - f(D) \subseteq f(C - D)$
(d) $f^{-1}(E \cup F) = f^{-1}(E) \cup f^{-1}(F)$
(e) $f^{-1}(E \cap F) = f^{-1}(E) \cap f^{-1}(F)$
(f) $f^{-1}(E - F) = f^{-1}(E) - f^{-1}(F)$.

Section 9.2: The Set of All Functions from A to B

9.13. Let $A = \{1, 2, 3\}$ and $B = \{x, y\}$. Determine B^A.

9.14. For sets $A = \{1, 2, 3, 4\}$ and $B = \{x, y, z\}$, give an example of a function $g \in B^A$ and a function $h \in B^B$.

9.15. For $A = \{a, b, c\}$, determine 2^A.

9.16. (a) Give an example of two sets A and B such that $|B^A| = 8$.
(b) Give an example of an element in B^A for the sets A and B given in (a).

9.17. (a) For nonempty sets A, B and C, what is a possible interpretation of the notation C^{B^A}?
(b) According to the definition given in (a), determine C^{B^A} for $A = \{0, 1\}$, $B = \{a, b\}$ and $C = \{x, y\}$.

Section 9.3: One-to-One and Onto Functions

9.18. Let $A = \{w, x, y, z\}$ and $B = \{r, s, t\}$. Give an example of a function $f : A \rightarrow B$ that is neither one-to-one nor onto. Explain why f fails to have these properties.

9.19. Give an example of two finite sets A and B and two functions $f : A \rightarrow B$ and $g : B \rightarrow A$ such that f is one-to-one but not onto and g is onto but not one-to-one.

9.20. A function $f : \mathbf{Z} \rightarrow \mathbf{Z}$ is defined by $f(n) = 2n + 1$. Determine whether f is (a) injective, (b) surjective.

9.21. A function $f : \mathbf{Z} \rightarrow \mathbf{Z}$ is defined by $f(n) = n - 3$. Determine whether f is (a) injective, (b) surjective.

9.22. A function $f : \mathbf{Z} \rightarrow \mathbf{Z}$ is defined by $f(n) = 5n + 2$. Determine whether f is (a) injective, (b) surjective.

9.23. Prove or disprove: For every nonempty set A, there exists an injective function $f : A \rightarrow \mathcal{P}(A)$.

9.24. Determine whether the function $f : \mathbf{R} \rightarrow \mathbf{R}$ defined by $f(x) = x^2 + 4x + 9$ is (a) one-to-one, (b) onto.

9.25. Is there a function $f : \mathbf{R} \rightarrow \mathbf{R}$ that is onto but not one-to-one? Explain your answer.

9.26. Give an example of a function $f : \mathbf{N} \rightarrow \mathbf{N}$ that is
(a) one-to-one and onto   (b) one-to-one but not onto
(c) onto but not one-to-one   (d) neither one-to-one nor onto.

9.27. Let $A = \{2, 3, 4, 5\}$ and $B = \{6, 8, 10\}$. A relation R is defined from A to B by $a \, R \, b$ if $a \mid b$ and $b/a + 1$ is a prime.

(a) Is R a function from A to B?
(b) If R is a function from A to B, then determine whether this function is one-to-one and/or onto.

9.28. Let $A = \{2, 4, 6\}$ and $B = \{1, 3, 4, 7, 9\}$. A relation f is defined from A to B by $a \, f \, b$ if 5 divides $ab + 1$. Is f a one-to-one function?

9.29. Let f be a function with dom$(f) = A$ and let C and D be subsets of A. Prove that if f is one-to-one, then $f(C \cap D) = f(C) \cap f(D)$.

Section 9.4: Bijective Functions

9.30. Prove that the function $f : \mathbf{R} \to \mathbf{R}$ defined by $f(x) = 7x - 2$ is bijective.

9.31. Let $f : \mathbf{Z}_5 \to \mathbf{Z}_5$ be a function defined by $f([a]) = [2a + 3]$.

    (a) Show that f is well-defined.
    (b) Determine whether f bijective.

9.32. Prove that the function $f : \mathbf{R} - \{2\} \to \mathbf{R} - \{5\}$ defined by $f(x) = \frac{5x+1}{x-2}$ is bijective.

9.33. Let $A = [0, 1]$ denote the closed interval of real numbers between 0 and 1. Give an example of two different bijective functions f_1 and f_2 from A to A, neither of which is the identity function.

9.34. Give a proof of Theorem 9.7 using mathematical induction.

9.35. For two finite nonempty sets A and B, let R be a relation from A to B such that range$(R) = B$. Define the domination number $\gamma(R)$ of R as the smallest cardinality of a subset $S \subseteq A$ such that for every element y of B, there is an element $x \in S$ such that x is related to y.

    (a) Let $A = \{1, 2, 3, 4, 5, 6, 7\}$ and $B = \{a, b, c, d, e, f, g\}$ and let $R = \{(1, c), (1, e), (2, c), (2, f), (2, g), (3, b), (3, f), (4, a), (4, c), (4, g), (5, a), (5, b), (5, c), (6, d), (6, e), (7, a), (7, g)\}$. Determine $\gamma(R)$.
    (b) If R is an equivalence relation defined on a finite nonempty set A (and so $B = A$), then what is $\gamma(R)$?
    (c) If f is a bijective function from A to B, then what is $\gamma(f)$?

9.36. Let $A = \{a, b, c, d, e, f\}$ and $B = \{u, v, w, x, y, z\}$. With each element $r \in A$, there is associated a list or subset $L(r) \subseteq B$. The goal is to define a "list function" $\phi : A \to B$ with the property that $\phi(r) \in L(r)$ for each $r \in A$.

    (a) For $L(a) = \{w, x, y\}$, $L(b) = \{u, z\}$, $L(c) = \{u, v\}$, $L(d) = \{u, w\}$, $L(e) = \{u, x, y\}$, $L(f) = \{v, y\}$, does there exist a bijective list function $\phi : A \to B$ for these lists?
    (b) For $L(a) = \{u, v, x, y\}$, $L(b) = \{v, w, y\}$, $L(c) = \{v, y\}$, $L(d) = \{u, w, x, z\}$, $L(e) = \{v, w\}$, $L(f) = \{w, y\}$, does there exist a bijective list function $\phi : A \to B$ for these lists?

Section 9.5: Composition of Functions

9.37. Let $A = \{1, 2, 3, 4\}$, $B = \{a, b, c\}$ and $C = \{w, x, y, z\}$. Consider the functions $f : A \to B$ and $g : B \to C$, where $f = \{(1, b), (2, c), (3, c), (4, a)\}$ and $g = \{(a, x), (b, y), (c, x)\}$. Determine $g \circ f$.

9.38. Two functions $f : \mathbf{R} \to \mathbf{R}$ and $g : \mathbf{R} \to \mathbf{R}$ are defined by $f(x) = 3x^2 + 1$ and $g(x) = 5x - 3$ for all $x \in \mathbf{R}$. Determine $(g \circ f)(1)$ and $(f \circ g)(1)$.

9.39. Two functions $f : \mathbf{Z}_{10} \to \mathbf{Z}_{10}$ and $g : \mathbf{Z}_{10} \to \mathbf{Z}_{10}$ are defined by $f([a]) = [3a]$ and $g([a]) = [7a]$.

    (a) Determine $g \circ f$ and $f \circ g$.
    (b) What can be concluded as a result of (a)?

9.40. Let A and B be nonempty sets. Prove that if $f : A \to B$, then $f \circ i_A = f$ and $i_B \circ f = f$.

9.41. Let A be a nonempty set and let $f : A \to A$ be a function. Prove that if $f \circ f = i_A$, then f is bijective.

9.42. Prove or disprove the following:

    (a) If two functions $f : A \to B$ and $g : B \to C$ are both bijective, then $g \circ f : A \to C$ is bijective.
    (b) Let $f : A \to B$ and $g : B \to C$ be two functions. If g is onto, then $g \circ f : A \to C$ is onto.
    (c) Let $f : A \to B$ and $g : B \to C$ be two functions. If g is one-to-one, then $g \circ f : A \to C$ is one-to-one.
    (d) There exist functions $f : A \to B$ and $g : B \to C$ such that f is not onto and $g \circ f : A \to C$ is onto.
    (e) There exist functions $f : A \to B$ and $g : B \to C$ such that f is not one-to-one and $g \circ f : A \to C$ is one-to-one.

9.43. For nonempty sets A, B and C, let $f : A \to B$ and $g : B \to C$ be functions.

(a) Prove:

$$\text{If } g \circ f \text{ is one-to-one, then } f \text{ is one-to-one.}$$

using as many of the following proof techniques as possible: direct proof, proof by contrapositive, proof by contradiction.

(b) Disprove: If $g \circ f$ is one-to-one, then g is one-to-one.

9.44. Let A denote the set of integers that are multiples of 4, let B denote the set of integers that are multiples of 8 and let B' denote the set of even integers. Thus $A = \{4k : k \in \mathbf{Z}\}$, $B = \{8k : k \in \mathbf{Z}\}$ and $B' = \{2k : k \in \mathbf{Z}\}$. Let $f : A \times A \to B$ and $g : B' \to \mathbf{Z}$ be functions defined by $f((x, y)) = xy$ for $x, y \in A$ and $g(n) = n/2$ for $n \in B'$.

(a) Show that the composition function $g \circ f : A \times A \to \mathbf{Z}$ is defined.

(b) For $k, \ell \in \mathbf{Z}$, determine $(g \circ f)((4k, 4\ell))$.

9.45. Let A be the set of even integers and B the set of odd integers. A function $f : A \times B \to B \times A$ is defined by $f((a, b)) = f(a, b) = (a + b, a)$ and a function $g : B \times A \to B \times B$ is defined by $g(c, d) = (c + d, c)$.

(a) Determine $(g \circ f)(18, 11)$.

(b) Determine whether the function $g \circ f : A \times B \to B \times B$ is one-to-one.

(c) Determine whether $g \circ f$ is onto.

9.46. Let A be the set of odd integers and B the set of even integers. A function $f : A \times B \to A \times A$ is defined by $f(a, b) = (3a - b, a + b)$ and a function $g : A \times A \to B \times A$ is defined by $g(c, d) = (c - d, 2c + d)$.

(a) Determine $(g \circ f)(3, 8)$.

(b) Determine whether the function $g \circ f : A \times B \to B \times A$ is one-to-one.

(c) Determine whether $g \circ f$ is onto.

9.47. For functions f, g and h with domain and codomain $\mathbf{R}$, prove or disprove the following.

(a) $(g + h) \circ f = (g \circ f) + (h \circ f)$

(b) $f \circ (g + h) = (f \circ g) + (f \circ h)$.

9.48. The composition $g \circ f : (0, 1) \to \mathbf{R}$ of two functions f and g is given by $(g \circ f)(x) = \frac{4x - 1}{2\sqrt{x - x^2}}$, where $f : (0, 1) \to (-1, 1)$ is defined by $f(x) = 2x - 1$ for $x \in (0, 1)$. Determine the function g.

Section 9.6: Inverse Functions

9.49. Let $A = \{a, b, c\}$. Give an example of a function $f : A \to A$ such that the inverse (relation) f^{-1} is not a function.

9.50. Show that the function $f : \mathbf{R} \to \mathbf{R}$ defined by $f(x) = 4x - 3$ is bijective and determine $f^{-1}(x)$ for $x \in \mathbf{R}$.

9.51. Show that the function $f : \mathbf{R} - \{3\} \to \mathbf{R} - \{5\}$ defined by $f(x) = \frac{5x}{x - 3}$ is bijective and determine $f^{-1}(x)$ for $x \in \mathbf{R} - \{5\}$.

9.52. The functions $f : \mathbf{R} \to \mathbf{R}$ and $g : \mathbf{R} \to \mathbf{R}$ defined by $f(x) = 2x + 1$ and $g(x) = 3x - 5$ for $x \in \mathbf{R}$ are bijective. Determine the inverse function of $g \circ f^{-1}$.

9.53. Let A and B be sets with $|A| = |B| = 3$. How many functions from A to B have inverse functions?

9.54. Let the functions $f : \mathbf{R} \to \mathbf{R}$ and $g : \mathbf{R} \to \mathbf{R}$ be defined by $f(x) = 2x + 3$ and $g(x) = -3x + 5$.

(a) Show that f is one-to-one and onto.

(b) Show that g is one-to-one and onto.

(c) Determine the composition function $g \circ f$.

(d) Determine the inverse functions f^{-1} and g^{-1}.

(e) Determine the inverse function $(g \circ f)^{-1}$ of $g \circ f$ and the composition $f^{-1} \circ g^{-1}$.

9.55. Let $A = \mathbf{R} - \{1\}$ and define $f : A \to A$ by $f(x) = \dfrac{x}{x-1}$ for all $x \in A$.

(a) Prove that f is bijective.

(b) Determine f^{-1}.

(c) Determine $f \circ f \circ f$.

9.56. Let A, B and C be nonempty sets and let f, g and h be functions such that $f : A \to B$, $g : B \to C$ and $h : B \to C$. For each of the following, prove or disprove:

(a) If $g \circ f = h \circ f$, then $g = h$.

(b) If f is one-to-one and $g \circ f = h \circ f$, then $g = h$.

9.57. The function $f : \mathbf{R} \to \mathbf{R}$ is defined by

$$f(x) = \begin{cases} \frac{1}{x-1} & \text{if } x < 1 \\ \sqrt{x-1} & \text{if } x \geq 1. \end{cases}$$

(a) Show that f is a bijection.

(b) Determine the inverse f^{-1} of f.

9.58. Suppose, for a function $f : A \to B$, that there is a function $g : B \to A$ such that $f \circ g = i_B$. Prove that if g is surjective, then $g \circ f = i_A$.

9.59. Let $f : A \to B$, $g : B \to C$ and $h : B \to C$ be functions where f is a bijection. Prove that if $g \circ f = h \circ f$, then $g = h$.

Section 9.7: Permutations

9.60. Let $\alpha = \begin{pmatrix} 1\ 2\ 3\ 4\ 5 \\ 2\ 3\ 4\ 5\ 1 \end{pmatrix}$ and $\beta = \begin{pmatrix} 1\ 2\ 3\ 4\ 5 \\ 3\ 5\ 2\ 4\ 1 \end{pmatrix}$ be permutations in S_5. Determine $\alpha \circ \beta$ and β^{-1}.

9.61. Let $\alpha = \begin{pmatrix} 1\ 2\ 3\ 4\ 5\ 6 \\ 2\ 6\ 4\ 1\ 5\ 3 \end{pmatrix}$ and $\beta = \begin{pmatrix} 1\ 2\ 3\ 4\ 5\ 6 \\ 5\ 3\ 6\ 2\ 1\ 4 \end{pmatrix}$ be elements of S_6.

(a) Determine α^{-1} and β^{-1}.

(b) Determine $\alpha \circ \beta$ and $\beta \circ \alpha$.

9.62. Prove for every integer $n \geq 3$, there exist $\alpha, \beta \in S_n$ such that $\alpha \circ \beta \neq \beta \circ \alpha$.

ADDITIONAL EXERCISES FOR CHAPTER 9

9.63. Let $f : \mathbf{R} \to \mathbf{R}$ be the function defined by $f(x) = x^2 + 3x + 4$.

(a) Show that f is not injective.

(b) Find all pairs r_1, r_2 of real numbers such that $f(r_1) = f(r_2)$.

(c) Show that f is not surjective.

(d) Find the set S of all real numbers such that if $s \in S$, then there is no real number x such that $f(x) = s$.

(e) What well-known set is the set S in (d) related to?

9.64. Let $f : \mathbf{R} \to \mathbf{R}$ be the function defined by $f(x) = x^2 + ax + b$, where $a, b \in \mathbf{R}$. Show that f is not one-to-one. [Hint: It might be useful to consider the cases $a \neq 0$ and $a = 0$ separately.]

9.65. In Result 9.4, we saw that the (linear) function $f : \mathbf{R} \to \mathbf{R}$ defined by $f(x) = 3x - 5$ is one-to-one. In fact, we have seen that other linear functions are one-to-one. Prove the following generalization of this result: The function $f : \mathbf{R} \to \mathbf{R}$ defined by $f(x) = ax + b$, where $a, b \in \mathbf{R}$ and $a \neq 0$, is one-to-one.

9.66. Evaluate the proposed proof of the following result.

Result   The function $f : \mathbf{R} - \{1\} \to \mathbf{R} - \{3\}$ defined by $f(x) = \dfrac{3x}{x-1}$ is bijective.

Proof   First, we show that f is one-to-one. Assume that $f(a) = f(b)$, where $a, b \in \mathbf{R} - \{1\}$. Then $\dfrac{3a}{a-1} = \dfrac{3b}{b-1}$. Crossmultiplying, we obtain $3a(b-1) = 3b(a-1)$. Simplifying, we have $3ab - 3a = 3ab - 3b$. Subtracting $3ab$ from both sides and dividing by -3, we have $a = b$. Thus f is one-to-one.

Next, we show that f is onto. Let $f(x) = r$. Then $\dfrac{3x}{x-1} = r$; so $3x = r(x-1)$. Simplifying, we have $3x = rx - r$ and so $3x - rx = -r$. Therefore, $x(3-r) = -r$. Since $r \in \mathbf{R} - \{3\}$, we can divide by $3 - r$ and obtain $x = \dfrac{-r}{3-r} = \dfrac{r}{r-3}$. Therefore,

$$ f(x) = f\left(\frac{r}{r-3}\right) = \frac{3\left(\frac{r}{r-3}\right)}{\frac{r}{r-3}-1} = \frac{3r}{r-(r-3)} = \frac{3r}{3} = r. $$

Thus f is onto.   ∎

9.67. For each of the following functions, determine, with explanation, whether the function is one-to-one and whether it is onto.

  (a) $f : \mathbf{R} \times \mathbf{R} \to \mathbf{R} \times \mathbf{R}$, where $f(x, y) = (3x - 2, 5y + 7)$.
  (b) $g : \mathbf{Z} \times \mathbf{Z} \to \mathbf{Z} \times \mathbf{Z}$, where $g(m, n) = (n + 6, 2 - m)$.
  (c) $h : \mathbf{Z} \times \mathbf{Z} \to \mathbf{Z} \times \mathbf{Z}$, where $h(r, s) = (2r + 1, 4s + 3)$.
  (d) $\phi : \mathbf{Z} \times \mathbf{Z} \to S = \{a + b\sqrt{2} : a, b \in \mathbf{Z}\}$, where $\phi(a, b) = a + b\sqrt{2}$.
  (e) $\alpha : \mathbf{R} \to \mathbf{R} \times \mathbf{R}$, where $\alpha(x) = (x^2, 2x + 1)$.

9.68. Let S be a nonempty set. Show that there exists an injective function from $\mathcal{P}(S)$ to $\mathcal{P}(\mathcal{P}(S))$.

9.69. Let $A = \{a, b, c, d, e\}$. Then $f = \{(a, c), (b, e), (c, d), (d, b), (e, a)\}$ is a bijective function from A to A.

  (a) Show that it is possible to list the five elements of A in such a way that the image of each of the first four elements on the list is to the immediate right of the element and that the image of the last element on the list is the first element on the list.
  (b) Show that it is not possible to list elements of A as in (a) for every bijective function from A to A.

9.70. Let $A = \mathbf{R} - \{0\}$ and let $f : A \to A$ be defined by $f(x) = 1 - \frac{1}{x}$ for all $x \in \mathbf{R}$.
  (a) Show that $f \circ f \circ f = i_A$.   (b) Determine f^{-1}.

9.71. Give an example of a finite nonempty set A and a bijective function $f : A \to A$ such that (1) $f \neq i_A$, (2) $f \circ f \neq i_A$ and (3) $f \circ f \circ f = i_A$.

9.72. For nonempty sets A and B and functions $f : A \to B$ and $g : B \to A$, suppose that $g \circ f = i_A$, the identity function on A.

  (a) Prove that f is one-to-one and g is onto.
  (b) Show that f need not be onto.
  (c) Show that g need not be one-to-one.
  (d) Prove that if f is onto, then g is one-to-one.
  (e) Prove that if g is one-to-one, then f is onto.
  (f) Combine the results in (d) and (e) into a single statement.

9.73. Let $A = \{1, 2\}$, $B = \{1, -1, 2, -2\}$ and $C = \{1, 2, 3, 4\}$. Then $f = \{(1, 1), (1, -1), (2, 2), (2, -2)\}$ is a relation from A to B while $g = \{(1, 1), (-1, 1), (2, 4), (-2, 4)\}$ is a relation from B to C. Furthermore,

$$gf = \{(x, z) : (x, y) \in f \text{ and } (y, z) \in g \text{ for some } y \in B\}$$

is a relation from A to C. Observe that even though the relation f is not a function from A to B, the relation gf is a function from A to C. Explain why.

9.74. A relation f on $\mathbf{R}$ is defined by $f = \{(x, y) : x \in \mathbf{R} \text{ and } y = x \text{ or } y = -x\}$ and a function $g : \mathbf{R} \to \mathbf{R}$ is defined by $g(x) = x^2$. Then

$$gf = \{(x, z) : (x, y) \in f \text{ and } (y, z) \in g \text{ for some } y \in \mathbf{R}\}.$$

(a) Explain why f is not a function from $\mathbf{R}$ to $\mathbf{R}$.
(b) Show that gf is a function from $\mathbf{R}$ to $\mathbf{R}$ and explicitly determine it.
(c) Even though the relation f is not a function from $\mathbf{R}$ to $\mathbf{R}$, the relation gf is a function from $\mathbf{R}$ to $\mathbf{R}$. Explain why.

9.75. Let $A = \{1, 2\}$, $B = \{1, 2, 3, 4\}$ and $C = \{1, 2, 3, 4, 5, 6\}$. Give an example of a function f from A to B and a relation g from B to C that is not a function from B to C such that

$$gf = \{(x, z) : (x, y) \in f \text{ and } (y, z) \in g \text{ for some } y \in B\}$$

is a function from A to C.

9.76. Let $\mathcal{F}$ be the set of all functions with domain and codomain $\mathbf{R}$. Define a relation R on $\mathcal{F}$ by $f \, R \, g$ if there exists a constant C such that $f(x) = g(x) + C$ for all $x \in \mathbf{R}$.

(a) Show that R is an equivalence relation.
(b) Let $f \in \mathcal{F}$. If the derivative of f is defined for all $x \in \mathbf{R}$, then use this information to describe the elements in the equivalence class $[f]$.

9.77. Let S be the set of odd positive integers. A function $F : \mathbf{N} \to S$ is defined by $F(n) = k$ for each $n \in \mathbf{N}$, where k is that odd positive integer for which $3n + 1 = 2^m k$ for some nonnegative integer m. Prove or disprove the following:

(a) F is one-to-one.
(b) F is onto.

9.78. A function $F : \mathbf{N} \to \mathbf{N} \cup \{0\}$ is defined by $F(n) = m$ for each $n \in \mathbf{N}$, where m is that nonnegative integer for which $3n + 1 = 2^m k$ and k is an odd integer. Prove or disprove the following:

(a) F is one-to-one.
(b) F is onto.

9.79. Recall that the derivative of $\ln x$ is $1/x$ and that the derivative of x^n is nx^{n-1} for every integer n. In symbols, $\frac{d}{dx}(\ln x) = \frac{1}{x}$ and $\frac{d}{dx}(x^n) = nx^{n-1}$. Let $f : \mathbf{R}^+ \to \mathbf{R}$ be defined by $f(x) = \ln x$ for every $x \in \mathbf{R}^+$. Prove that the nth derivative of $f(x)$ is given by $f^{(n)}(x) = \frac{(-1)^{n+1}(n-1)!}{x^n}$ for every positive integer n.

9.80. Let $f : \mathbf{R} \to \mathbf{R}$ be defined by $f(x) = xe^{-x}$ for every $x \in \mathbf{R}$. Prove that the nth derivative of $f(x)$ is given by $f^{(n)}(x) = (-1)^n e^{-x}(x - n)$ for every positive integer n.

9.81. The function $h : \mathbf{Z}_{16} \to \mathbf{Z}_{24}$ is defined by $h([a]) = [3a]$ for $a \in \mathbf{Z}$.

(a) Prove that the function h is well-defined; that is, prove that if $[a] = [b]$ in $\mathbf{Z}_{16}$, then $h([a]) = h([b])$ in $\mathbf{Z}_{24}$.
(b) For the subsets $A = \{[0], [3], [6], [9], [12], [15]\}$ and $B = \{[0], [8]\}$ of $\mathbf{Z}_{16}$, determine the subsets $h(A)$ and $h(B)$ of $\mathbf{Z}_{24}$.
(c) For the subsets $C = \{[0], [6], [16], [18]\}$ and $D = \{[4], [8], [16]\}$ of $\mathbf{Z}_{24}$, determine $h^{-1}(C)$ and $h^{-1}(D)$.

9.82. Let $\mathcal{U}$ be some universal set and A a subset of $\mathcal{U}$. A function $g_A : \mathcal{U} \to \{0, 1\}$ is defined by

$$g_A(x) = \begin{cases} 1 & \text{if } x \in A \\ 0 & \text{if } x \notin A. \end{cases}$$

Verify each of the following.

(a) $g_{\mathcal{U}}(x) = 1$ for all $x \in \mathcal{U}$.

(b) $g_{\emptyset}(x) = 0$ for all $x \in \mathcal{U}$.

(c) For $\mathcal{U} = \mathbf{R}$ and $A = [0, \infty)$, $(g_A \circ g_A)(x) = 1$ for $x \in \mathbf{R}$.

(d) For subsets A and B of $\mathcal{U}$ and $C = A \cap B$, $g_C = (g_A) \cdot (g_B)$, where $((g_A) \cdot (g_B))(x) = g_A(x) \cdot g_B(x)$.

(e) For $A \subseteq \mathcal{U}$, $g_{\overline{A}}(x) = 1 - g_A(x)$ for each $x \in \mathcal{U}$.

9.83. (a) Let $S = \{a, b, c, d\}$ and let T be the set of all six 2-element subsets of S. Show that there exists an injective function $f : S \to \{0, 1, 2, \ldots, |T|\}$ such that the function $g : T \to \{1, 2, \ldots, |T|\}$ defined by $g(\{i, j\}) = |f(i) - f(j)|$ is bijective.

(b) Let $S = \{a, b, c, d, e\}$ and let T be the set of all ten 2-element subsets of S. Show that there exists no injective function $f : S \to \{0, 1, 2, \ldots, |T|\}$ such that the function $g : T \to \{1, 2, \ldots, |T|\}$ defined by $g(\{i, j\}) = |f(i) - f(j)|$ is bijective.

(c) For the sets S and T in (b), show that there exists an injective function $f : S \to \{0, 1, 2, \ldots, |T| + 2\}$ such that the function $g : T \to \{1, 2, \ldots, |T| + 2\}$ defined by $g(\{i, j\}) = |f(i) - f(j)|$ is injective.

(d) The results in (b) and (c) should suggest a question to you. Ask and answer such a question.

10

Cardinalities of Sets

Many consider the Italian mathematician and scientist Galileo Galilei to be the founder of modern physics. Among his major contributions was his mathematical view of the laws of motion. Early in the 17th century, Galileo applied mathematics to study the motion of the Earth. He was convinced that the Earth revolved about the sun, an opinion not shared by the Catholic Church at that time. This led him to be imprisoned for the last nine years of his life.

Galileo's two main scientific writings were *Dialogue Concerning the Two Chief World Systems* and *Discourses and Mathematical Demonstrations Concerning Two New Sciences*, the first published before he went to prison and the second published (in the Netherlands) while he was in prison. In these two works, he would often discuss scientific theories by means of a dialogue among fictional characters. It is in this manner that he could state his positions on various theories.

One topic that intrigued Galileo was infinite sets. Galileo observed that there is a one-to-one correspondence (that is, a bijective function) between the set **N** of positive integers and the subset S of **N** consisting of the squares of positive integers. This led Galileo to observe that even though there are many positive integers that are not squares, there are as many squares as there are positive integers. This led Galileo to be faced with a property of an infinite set that he found bothersome: There can be a one-to-one correspondence between a set and a proper subset of the set. While Galileo concluded correctly that the number of squares of positive integers is not less than the number of positive integers, he could not bring himself to say that these sets have the same number of elements.

Bernhard Bolzano was a Bohemian priest, philosopher and mathematician. Although best known for his work in calculus during the first half of the 19th century, he too was interested in infinite sets. His *Paradoxes of the Infinite*, published two years after his death and unnoticed for twenty years, contained many ideas of the modern theory of sets. He noted that one-to-one correspondences between an infinite set and a proper subset of itself are common and was comfortable with this fact, contrary to Galileo's feelings. The German mathematician Richard Dedekind studied under the brilliant Carl Friedrich Gauss. Dedekind had a long and productive career in mathematics and made many contributions to the study of irrational numbers. What had confused Galileo and interested Bolzano gave rise to a definition of an infinite set by Dedekind during the last part of the 19th century: A set S is *infinite* if it contains a proper subset that can be put in one-to-one correspondence with S. Certainly, then, understanding infinite sets was not an easy task, even among well-known mathematicians of the past.

We mentioned in Chapter 1 that the cardinality $|S|$ of a set S is the number of elements in S and, for the present, we would use the notation $|S|$ only when S is a finite set. A set S is **finite** if either $S = \emptyset$ or $|S| = n$ for some $n \in \mathbf{N}$; while a set is **infinite** if it is not finite. It may seem that we should write $|S| = \infty$ if S is infinite but we will soon see that this is not particularly informative. Indeed, it is considerably more difficult to give a meaning to $|S|$ if S is an infinite set; however, it is precisely this topic that we are about to explore.

10.1 Numerically Equivalent Sets

It is rather obvious that the sets $A = \{a, b, c\}$ and $B = \{x, y, z\}$ have the same cardinality since each has exactly three elements. That is, if we count the number of elements in two sets and arrive at the same value, then these two sets have the same cardinality. There is, however, another way to see that the sets A and B described above have the same cardinality without counting the elements in each set. Observe that we can pair off the elements of A and B, say as (a, x), (b, y) and (c, z). This implies that A and B have the same number of elements, that is, $|A| = |B|$. What we have actually done is describe a bijective function $f : A \to B$, namely $f = \{(a, x), (b, y), (c, z)\}$. Although it is much easier to see that $|A| = |B|$ by observing that each set has three elements than by constructing a bijective function from A to B, it is this latter method of showing that $|A| = |B|$ that can be generalized to the situation where A and B are infinite sets.

Two sets A and B (finite or infinite) are said to have the **same cardinality**, written $|A| = |B|$, if either A and B are both empty or there is a bijective function f from A to B. Two sets having the same cardinality are also referred to as **numerically equivalent sets**. Two finite sets are therefore numerically equivalent if they are both empty or if both have n elements for some positive integer n. Consequently, two nonempty sets A and B are not numerically equivalent, written $|A| \neq |B|$, if there is no bijective function f from one set to the other. The study of numerically equivalent infinite sets is more challenging but considerably more interesting than the study of numerically equivalent finite sets.

The justification for the term "numerically equivalent sets" lies in the following theorem, which combines the major concepts of Chapters 8 and 9.

Theorem 10.1   *Let S be a nonempty collection of nonempty sets. A relation R is defined on S by $A\ R\ B$ if there exists a bijective function from A to B. Then R is an equivalence relation.*

Proof   Let $A \in S$. Since the identity function $i_A : A \to A$ is bijective, it follows that $A\ R\ A$. Thus R is reflexive. Next, assume that $A\ R\ B$, where $A, B \in S$. Then there is a bijective function $f : A \to B$. By Theorem 9.15, f has an inverse function $f^{-1} : B \to A$ and, furthermore, f^{-1} is bijective. Therefore, $B\ R\ A$ and R is symmetric.

Finally, assume that $A\ R\ B$ and $B\ R\ C$, where $A, B, C \in S$. Then there are bijective functions $f : A \to B$ and $g : B \to C$. It follows by Corollary 9.12 that the composition $g \circ f : A \to C$ is bijective as well and so $A\ R\ C$. Therefore, R is transitive. Consequently, R is an equivalence relation. $\blacksquare$

According to the equivalence relation defined in Theorem 10.1, if A is a nonempty set, then the equivalence class $[A]$ consists of all those elements of S having the same cardinality as A; hence the term "numerically equivalent sets" is natural for two sets having the same cardinality.

Example 10.2   *Let $S = \{A_1, A_2, A_3, A_4, A_5, A_6\}$ where*

$$A_1 = \{1, 2, 3\}, \, A_2 = \{a, b, c, d\}, \, A_3 = \{x, y, z\},$$
$$A_4 = \{r, s, t\}, \, A_5 = \{m, n\}, \, A_6 = \{7, 8, 9, 10\}.$$

Then every two of the sets A_1, A_3 and A_4 are numerically equivalent, while A_2 and A_6 are numerically equivalent. This says that $|A_1| = |A_3| = |A_4|$ and $|A_2| = |A_6|$. The only set in S numerically equivalent to A_5 is A_5 itself. Thus,

$$[A_1] = \{A_1, A_3, A_4\}, [A_2] = \{A_2, A_6\} \text{ and } [A_5] = \{A_5\}$$

are the distinct equivalence classes of S.   ◆

While Example 10.2 deals only with finite sets, it is infinite sets in which we will be primarily interested in this chapter. In particular, we will have a special interest in sets that are numerically equivalent to **N** or to **R**.

10.2 Denumerable Sets

In order to start gaining an understanding of the cardinality of an infinite set, we begin with a particular class of infinite sets. A set A is called **denumerable** if $|A| = |\mathbf{N}|$, that is, if A has the same cardinality as the set of natural numbers. Certainly, if A is denumerable, then A is infinite. By definition, if A is a denumerable set, then there is a bijective function $f : \mathbf{N} \to A$ and so $f = \{(1, f(1)), (2, f(2)), (3, f(3)), \ldots\}$. Consequently, $A = \{f(1), f(2), f(3), \ldots\}$; that is, we can list the elements of A as $f(1), f(2), f(3), \ldots$. Equivalently, we can list the elements of A as $a_1, a_2, a_3, \ldots$, where then $a_i = f(i)$ for $i \in \mathbf{N}$. Conversely, if the elements of A can be listed as $a_1, a_2, a_3, \ldots$, where $a_i \neq a_j$ for $i \neq j$, then A is denumerable since the function $g : \mathbf{N} \to A$ defined by $g(n) = a_n$ for each $n \in \mathbf{N}$ is certainly bijective. Therefore, A is a denumerable set if and only if it is possible to list the elements of A as $a_1, a_2, a_3, \ldots$ and so $A = \{a_1, a_2, a_3, \ldots\}$.

A set is **countable** if it is either finite or denumerable. **Countably infinite** sets are then precisely the denumerable sets. Hence, if A is a nonempty countable set, then we can either write $A = \{a_1, a_2, a_3, \ldots, a_n\}$ for some $n \in \mathbf{N}$ or $A = \{a_1, a_2, a_3, \ldots\}$. A set that is not countable is called **uncountable**. An uncountable set is necessarily infinite. It may not be clear whether any set is uncountable but we will soon see that such sets do exist.

Let's look at a few examples of denumerable sets. Certainly, **N** itself is denumerable since the identity function $i_{\mathbf{N}} : \mathbf{N} \to \mathbf{N}$ is bijective. However, not only is the set of positive integers denumerable, the set of *all* integers is denumerable. The proof of this fact that we give illustrates a common technique for showing that a set is denumerable; namely, if we can list the elements of a set A as $a_1, a_2, a_3, \ldots$ such that every element of A appears exactly once in the list, then A is denumerable.

$$
\begin{array}{ccccccc}
 & 1 & 2 & 3 & 4 & 5 & \cdots \\
f: & \downarrow & \downarrow & \downarrow & \downarrow & \downarrow & \downarrow \\
 & 0 & 1 & -1 & 2 & -2 & \cdots
\end{array}
$$

Figure 10.1   A bijective function $f : \mathbf{N} \to \mathbf{Z}$

Result 10.3   *The set $\mathbf{Z}$ of integers is denumerable.*

Proof   Observe that the elements of $\mathbf{Z}$ can be listed as $0, 1, -1, 2, -2, \ldots$. Thus the function $f : \mathbf{N} \to \mathbf{Z}$ described in Figure 10.1 is bijective and so $\mathbf{Z}$ is denumerable.   ∎

The function $f : \mathbf{N} \to \mathbf{Z}$ given in Figure 10.1 can be also defined by

$$
f(n) = \frac{1 + (-1)^n (2n - 1)}{4}. \tag{10.1}
$$

Although we have already observed that this function f is bijective, Exercise 10.8. asks for a formal proof of this fact.

The fact that $\mathbf{Z}$ is denumerable illustrates what Galileo had observed centuries ago: It is possible for two sets to have the same cardinality where one is a proper subset of the other. (Such a situation could never occur with finite sets, however.) For example, $\mathbf{N} \subset \mathbf{Z}$ and $|\mathbf{N}| = |\mathbf{Z}|$. This fact serves as an illustration of a result, the proof of which is a bit intricate.

Theorem to Prove   Every infinite subset of a denumerable set is denumerable.

PROOF STRATEGY   In the proof, we begin with two sets, which we'll call A and B, where A is denumerable, $B \subseteq A$ and B is infinite. Because A is denumerable, we can write $A = \{a_1, a_2, a_3, \ldots\}$. Because our goal is to show that B is denumerable, we need to show that we can write $B = \{b_1, b_2, b_3, \ldots\}$. The question, of course, is how to do it.

Because B is an infinite subset of A, some of the elements of A belong to B (in fact, infinitely many elements of A belong to B); while, most likely, some elements of A do not belong to B. We can keep track of the elements of A that belong to B by means of a set, which we'll denote by S. If $a_1 \in B$, then $1 \in S$; if $a_1 \notin B$, then $1 \notin S$. In general, $n \in S$ if and only if $a_n \in B$. Certainly, $S \subseteq \mathbf{N}$. Since $\mathbf{N}$ is a well-ordered set (by the Well-Ordering Principle), S contains a smallest element, say s. That is, $a_s \in B$. Furthermore, if r is an integer such that $1 \leq r < s$, then $a_r \notin B$. It is the element a_s that we will call b_1. It is now logical to look at the (infinite) set $S - \{s\}$ and consider its smallest element, say t. Thus $t > s$. The element a_t will become b_2. And so on.

Since we want to give a precise and careful proof, we are already faced with two problems. First, denoting the smallest element of S by s and denoting the smallest element of $S - \{s\}$ by t will present difficulties to us. We need to use better notation. So let us denote the smallest element of S by i_1 (so $b_1 = a_{i_1}$) and the smallest element of $S - \{i_1\}$ by i_2 (so $b_2 = a_{i_2}$). This is much better notation. The other problem we have is when we wrote "And so on." Once we have the positive integers i_1 and i_2, it will follow that the positive integer i_3 is the least element of $S - \{i_1, i_2\}$. In general, once we have determined the positive integers $i_1, i_2, \ldots, i_k$, where $k \in \mathbf{N}$, the positive integer

i_{k+1} is the smallest element of $S - \{i_1, i_2, \ldots, i_k\}$. In fact, this suggests that the elements $b_1, b_2, b_3, \ldots$ can be located in A using induction.

After using induction to construct the set $\{b_1, b_2, b_3, \ldots\}$, which we will denote by B', say, then we still have one more concern. Are we certain that $B' = B$? Because each element of B' belongs to B, we know that $B' \subseteq B$. To show that $B' = B$, we must also be sure that $B \subseteq B'$. As we know, the standard way to show that $B \subseteq B'$ is to take a typical element $b \in B$ and show that $b \in B'$.

Let's now write a complete proof.   ◆

Theorem 10.4   *Every infinite subset of a denumerable set is denumerable.*

Proof   Let A be a denumerable set and let B be an infinite subset of A. Since A is denumerable, we can write $A = \{a_1, a_2, a_3, \ldots\}$. Let $S = \{i \in \mathbf{N} : a_i \in B\}$; that is, S consists of all those positive integers that are subscripts of the elements in A that also belong to B. Since B is infinite, S is infinite. First we use induction to show that B contains a denumerable subset. Since S is a nonempty subset of $\mathbf{N}$, it follows from the Well-Ordering Principle that S has a least element, say i_1. Let $b_1 = a_{i_1}$. Let $S_1 = S - \{i_1\}$. Since $S_1 \neq \emptyset$ (indeed, S_1 is infinite), S_1 has a least element, say i_2. Let $b_2 = a_{i_2}$, which, of course, is distinct from b_1. Assume that for an arbitrary integer $k \geq 2$, the (distinct) elements $b_1, b_2, \ldots, b_k$ have been defined by $b_j = a_{i_j}$ for each integer j with $1 \leq j \leq k$, where i_1 is the smallest element in S and i_j is the minimum element in $S_{j-1} = S - \{i_1, i_2, \ldots, i_{j-1}\}$ for $2 \leq j \leq k$. Now let i_{k+1} be the minimum element of $S_k = S - \{i_1, i_2, \ldots, i_k\}$ and let $b_{k+1} = a_{i_{k+1}}$. Hence it follows that for each integer $n \geq 2$, an element b_n belongs to B that is distinct from $b_1, b_2, \ldots, b_{n-1}$. Thus we have exhibited the elements $b_1, b_2, b_3, \ldots$ in B.

Let $B' = \{b_1, b_2, b_3, \ldots\}$. Certainly $B' \subseteq B$. We claim, in fact, that $B = B'$. It remains only to show that $B \subseteq B'$. Let $b \in B$. Since $B \subseteq A$, it follows that $b = a_n$ for some $n \in \mathbf{N}$ and so $n \in S$. If $n = i_1$, then $b = b_1 = a_n$ and so $b \in B'$. Thus we may assume that $n > i_1$. Let S' consist of those positive integers less than n that belong to S. Since $n > i_1$ and $i_1 \in S$, it follows that $S' \neq \emptyset$. Certainly, $1 \leq |S'| \leq n - 1$; so S' is finite. Thus $|S'| = m$ for some $m \in \mathbf{N}$. The set S' therefore consists of the m smallest integers of S, that is, $S' = \{i_1, i_2, \ldots, i_m\}$. The smallest integer that belongs to S and is greater than i_m must be i_{m+1}, of course, and $i_{m+1} \geq n$. But $n \in S$, so $n = i_{m+1}$ and $b = a_n = a_{i_{m+1}} \in B'$. Hence $B = B' = \{b_1, b_2, b_3, \ldots\}$, which is denumerable.   ∎

In order to use Theorem 10.4 to describe other denumerable sets, it is convenient to introduce some additional notation. Let $k \in \mathbf{N}$. Then the set $k\mathbf{Z}$ is defined by

$$k\mathbf{Z} = \{kn : n \in \mathbf{Z}\}.$$

Similarly,

$$k\mathbf{N} = \{kn : n \in \mathbf{N}\}.$$

Thus $1\mathbf{Z} = \mathbf{Z}$ and $1\mathbf{N} = \mathbf{N}$, while $2\mathbf{Z}$ is the set of even integers. An immediate consequence of Theorem 10.4 is stated next.

Result 10.5   *The set $2\mathbf{Z}$ of even integers is denumerable.*

Proof   Since $2\mathbf{Z}$ is infinite and $2\mathbf{Z} \subseteq \mathbf{Z}$, it follows by Theorem 10.4 that $2\mathbf{Z}$ is denumerable.   ∎

	b_1	b_2	b_3	$\cdots$
a_1	(a_1, b_1)	(a_1, b_2)	(a_1, b_3)	$\cdots$
a_2	(a_2, b_1)	(a_2, b_2)	(a_2, b_3)	$\cdots$
a_3	(a_3, b_1)	(a_3, b_2)	(a_3, b_3)	$\cdots$

(a)

	b_1	b_2	b_3	$\cdots$
a_1	(a_1, b_1)	(a_1, b_2)	(a_1, b_3)	$\cdots$
a_2	(a_2, b_1)	(a_2, b_2)	(a_2, b_3)	$\cdots$
a_3	(a_3, b_1)	(a_3, b_2)	(a_3, b_3)	$\cdots$

(b)

Figure 10.2   Constructing a bijective function $f : \mathbf{N} \to A \times B$

Of course, $k\mathbf{Z}$ is denumerable for *every* nonzero integer k. We now describe a denumerable set that can be obtained from two given sets. Recall, for sets A and B, that the Cartesian product $A \times B = \{(a, b) : a \in A, b \in B\}$.

Result 10.6   *If A and B are denumerable sets, then $A \times B$ is denumerable.*

Proof   Since A and B are denumerable sets, we can write $A = \{a_1, a_2, a_3, \ldots\}$ and $B = \{b_1, b_2, b_3, \ldots\}$. Consider the table shown in Figure 10.2(a), which has an infinite (denumerable) number of rows and columns, where the elements $a_1, a_2, a_3, \ldots$ are written along the side and $b_1, b_2, b_3, \ldots$ are written across the top. In row i, column j of the table, we place the ordered pair (a_i, b_j). Certainly, every element of $A \times B$ appears exactly once in this table. This table is reproduced in Figure 10.2(b), where the directed lines indicate the order in which we will encounter the entries is the table. That is, we encounter the elements of $A \times B$ in the order

$$(a_1, b_1), (a_1, b_2), (a_2, b_1), (a_1, b_3), (a_2, b_2), \ldots.$$

Since every element of $A \times B$ occurs in this list exactly once, this describes a bijective function $f : \mathbf{N} \to A \times B$, where

$$f(1) = (a_1, b_1), \ f(2) = (a_1, b_2), \ f(3) = (a_2, b_1), \ f(4) = (a_1, b_3), \ f(5) = (a_2, b_2), \ldots.$$

Therefore, $A \times B$ is denumerable.   ∎

We can use a technique similar to that used in proving Result 10.6 to show that another familiar set is denumerable.

Result 10.7   *The set $\mathbf{Q}^+$ of positive rational numbers is denumerable.*

Proof   Consider the table shown in Figure 10.3(a). In row i, column j, we place the rational number j/i. Certainly, then, every positive rational number appears in the table of Figure 10.3(a); indeed, it appears infinitely often. For example, the number $1/2$ appears in row 2, column 1, as well as in row 4, column 2.

The table of Figure 10.3(a) is reproduced in Figure 10.3(b), where the arrows indicate the order in which we will consider the entries in the table. That is, we now consider the positive rational numbers in the order

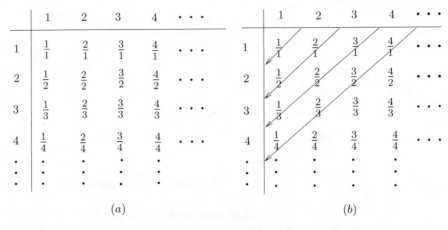

Figure 10.3   A table used to show that $\mathbf{Q}^+$ is denumerable

$$\frac{1}{1}, \ \frac{2}{1}, \ \frac{1}{2}, \ \frac{3}{1}, \ \frac{2}{2}, \ \frac{1}{3}, \ \frac{4}{1}, \ \cdots$$

With the aid of this list, we can describe a bijective function $f : \mathbf{N} \to \mathbf{Q}^+$. In particular, we define $f(1) = 1/1 = 1$, $f(2) = 2/1 = 2$, $f(3) = 1/2$ and $f(4) = 3/1 = 3$ as expected. However, since $2/2 = 1$ and we have already defined $f(1) = 1$, we do not define $f(5) = 1$ (since f must be one-to-one). We bypass $2/2 = 1$ and, following the arrows, go directly to the next number on the list, namely $1/3$. In fact, whenever we encounter a number on the list that we have previously seen, we move to the next number on the list. In this manner, the function f being described will be one-to-one. The function f is shown in Figure 10.4.

Because every element of $\mathbf{Q}^+$ is eventually encountered, f is onto as well and so f is bijective. Consequently, $\mathbf{Q}^+$ is denumerable.   ∎

The function f described in Figure 10.4 is by no means unique. There are many ways to traverse the positive rational numbers in the table described in Figure 10.3(a). The tables shown in Figure 10.5 indicate two additional methods.

Some care must be taken when proceeding about the entries in the table of Figure 10.3(a). For example, traversing the positive rational numbers by rows (see Figure 10.6) just won't do. Since the first row never ends, we will only encounter the positive integers.

The set $\mathbf{Q}^+$ can also be shown to be denumerable with the aid of the table in Figure 10.7. In the first row, all positive rational numbers j/i with $i = 1$ are shown. In the second row, all positive rational numbers j/i with $i = 2$ and such that j/i has been reduced to lowest terms are shown. This results in the rational number $(2j - 1)/2$ in row 2, column j. We continue in this manner with all other rows. In this way, every

$$
f : \quad
\begin{array}{cccccc}
1 & 2 & 3 & 4 & 5 & \cdots \\
\downarrow & \downarrow & \downarrow & \downarrow & \downarrow & \downarrow \\
1 & 2 & \frac{1}{2} & 3 & \frac{1}{3} & \cdots
\end{array}
$$

Figure 10.4   A bijective function $f : \mathbf{N} \to \mathbf{Q}^+$

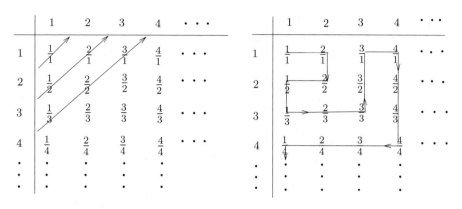

Figure 10.5   Traversing the positive rational numbers

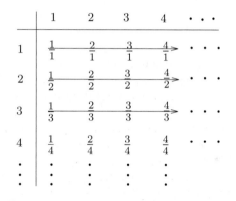

Figure 10.6   How not to traverse the positive rational numbers

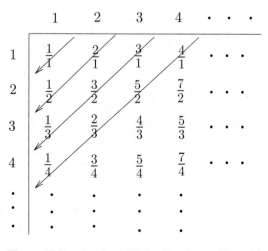

Figure 10.7   Another bijective function $g : \mathbf{N} \rightarrow \mathbf{Q}^+$

$$
\begin{array}{cccccc}
1 & 2 & 3 & 4 & 5 & \cdots
\end{array}
$$

$$
f: \quad \downarrow \; \downarrow \qquad \downarrow \qquad \downarrow \qquad \downarrow \qquad \downarrow
$$

$$
\begin{array}{cccccc}
0 & q_1 & -q_1 & q_2 & -q_2 & \cdots
\end{array}
$$

Figure 10.8   A bijective function $f : \mathbf{N} \to \mathbf{Q}$

positive rational number appears *exactly once* in the table. Thus when we proceed through the entries as the arrows indicate, we obtain the positive rational numbers in the order

$$
\tfrac{1}{1}, \tfrac{2}{1}, \tfrac{1}{2}, \tfrac{3}{1}, \tfrac{3}{2}, \tfrac{1}{3}, \tfrac{4}{1}, \cdots
$$

and the corresponding bijective function $g : \mathbf{N} \to \mathbf{Q}^+$. Therefore, $g(1) = 1$, $g(2) = 2$, $g(3) = 1/2$, $g(4) = 3$, $g(5) = 3/2$ and so on.

Now that we have shown that $\mathbf{Q}^+$ is denumerable, it is not difficult to show that the set $\mathbf{Q}$ of *all* rational numbers is denumerable.

Result 10.8   *The set $\mathbf{Q}$ of all rational numbers is denumerable.*

Proof   Since $\mathbf{Q}^+$ is denumerable, we can write $\mathbf{Q}^+ = \{q_1, q_2, q_3, \ldots\}$. Thus, $\mathbf{Q} = \{0\} \cup \{q_1, q_2, q_3, \ldots\} \cup \{-q_1, -q_2, -q_3, \ldots\}$. Therefore, $\mathbf{Q} = \{0, q_1, -q_1, q_2, -q_2, \ldots\}$, and the function $f : \mathbf{N} \to \mathbf{Q}$ shown in Figure 10.8 is bijective and so $\mathbf{Q}$ is denumerable.   ∎

10.3 Uncountable Sets

Although we have now given several examples of denumerable sets (and consequently countably infinite sets), we have yet to give an example of an uncountable set. We will do this next. First though, let's review a few facts about decimal expansions of real numbers. Every irrational number has a unique decimal expansion and this expansion is nonrepeating, while every rational number has a repeating decimal expansion. For example, $\frac{3}{11} = 0.272727 \cdots$. Some rational numbers, however, have two (repeating) decimal expansions. For example, $\frac{1}{2} = 0.5000 \cdots$ and $\frac{1}{2} = 0.4999 \cdots$. (The number $\frac{3}{11}$ has only one decimal expansion.) In particular, a rational number a/b, where $a, b \in \mathbf{N}$, that is reduced to lowest terms has two decimal expansions if and only if the only primes that divide b are 2 or 5. If a rational number has two decimal expansions, then one of the expansions repeats the digit 0 from some point on (that is, the decimal expansion terminates), while the alternate expansion repeats the digit 9 from some point on.

We are now prepared to give an example of an uncountable set. Recall that for real numbers a and b with $a < b$, the open interval (a, b) is defined by

$$
(a, b) = \{x \in \mathbf{R} : a < x < b\}.
$$

Although, as it will turn out, all open intervals (a, b) of real numbers are uncountable, we will prove now only that $(0, 1)$ is uncountable.

Theorem to Prove   The open interval $(0, 1)$ of real numbers is uncountable.

PROOF STRATEGY   Since uncountable means not countable, it is not surprising that we should try a proof by contradiction here. So the proof would begin by assuming that $(0, 1)$ is countable. Since $(0, 1)$ is an infinite set, this means that we are assuming that $(0, 1)$ is denumerable, which implies that there must exist a bijective function $f : \mathbf{N} \to (0, 1)$. Therefore, for each $n \in \mathbf{N}$, $f(n)$ is a number in the set $(0, 1)$. It might be convenient to introduce some notation for the number $f(n)$, say $f(n) = a_n$, where then $0 < a_n < 1$. Since f is assumed to be one-to-one, it follows that $a_i \neq a_j$ for distinct positive integers i and j. Each number a_n has a decimal expansion, say $a_n = 0.a_{n1}a_{n2}a_{n3} \cdots$, where a_{n1} is the first digit in the expansion, a_{n2} is the second digit in the expansion and so on. We have to be a bit careful here, however, for as we have seen, some real numbers have two decimal expansions. To avoid possible confusion, we can choose the decimal expansion that repeats the digit 0 from some point on. That is, no real number a_n has a decimal expansion that repeats 9 from some point on.

But where does this lead to a contradiction? From what we have said, $(0, 1) = \{a_1, a_2, a_3, \ldots\}$. If we can think of some real number $b \in (0, 1)$ such that $b \notin \{a_1, a_2, a_3, \ldots\}$, then this would give us a contradiction because this would say that f is not onto. So we need to find a number $b \in (0, 1)$ such that $b \neq a_n$ for each $n \in \mathbf{N}$. Since $b \in (0, 1)$, the number b has a decimal expansion, say $b = 0.b_1b_2b_3 \cdots$. How can we choose the digits $b_1, b_2, b_3, \ldots$ so that $b \neq a_n$ for every $n \in \mathbf{N}$? We could choose $b_1 \neq a_{11}, b_2 \neq a_{22}$, etc. But would this mean that $b \neq a_1, b \neq a_2$, etc? We must be careful here. For example, $0.500 \cdots$ and $0.499 \cdots$ are two equal numbers whose first digits in their expansions are not equal. Of course, the reason for this is that one is the alternate decimal expansion of the other. Thus, provided we can avoid selecting a decimal expansion for b that is the alternate decimal expansion for some number a_n, where $n \in \mathbf{N}$, we will have found a number $b \in (1, 0)$ such that $b \notin \{a_1, a_2, a_3, \ldots\}$. This will give us a contradiction. ◆

Theorem 10.9   *The open interval $(0, 1)$ of real numbers is uncountable.*

Proof   Assume, to the contrary, that $(0, 1)$ is countable. Since $(0, 1)$ is infinite, it is denumerable. Therefore, there exists a bijective function $f : \mathbf{N} \to (0, 1)$. For $n \in \mathbf{N}$, let $f(n) = a_n$. Since $a_n \in (0, 1)$, the number a_n has a decimal expansion, say $0.a_{n1}a_{n2}a_{n3} \cdots$, where $a_{ni} \in \{0, 1, 2, \ldots, 9\}$ for all $i \in \mathbf{N}$. If a_n is irrational, then its decimal expansion is unique. If $a_n \in \mathbf{Q}$, then the expansion *may* be unique. If it is not unique, then, without loss of generality, we assume that the digits of the decimal expansion $0.a_{n1}a_{n2}a_{n3} \cdots$ are 0 from some position on. For example, since f is bijective, $2/5$ is the image of exactly one positive integer and this image is written as $0.4000 \cdots$ (rather than as $0.3999 \cdots$). To summarize, we have

$$f(1) = a_1 = 0.a_{11}a_{12}a_{13} \cdots$$
$$f(2) = a_2 = 0.a_{21}a_{22}a_{23} \cdots$$
$$f(3) = a_3 = 0.a_{31}a_{32}a_{33} \cdots$$
$$\vdots \qquad \vdots \qquad \vdots$$

We show that the function f is not onto, however. Define the number $b = 0.b_1b_2b_3\cdots$, where $b_i \in \{0, 1, 2, \ldots, 9\}$ for all $i \in \mathbf{N}$, by

$$b_i = \begin{cases} 4 & \text{if } a_{ii} = 5 \\ 5 & \text{if } a_{ii} \neq 5. \end{cases}$$

(For example, let's suppose that $a_1 = 0.31717\cdots$, $a_2 = 0.151515\cdots$ and $a_3 = 0.04000\cdots$. Then the first three digits in the decimal expansion of b are 5, 4 and 5, that is, $b = 0.545\cdots$.)

For each $i \in \mathbf{N}$, the digit $b_i \neq a_{ii}$, implying that $b \neq a_n$ for all $n \in \mathbf{N}$ since b is not the alternate expansion of any rational number, as no digit in the expansion of b is 9. Thus, b is not an image of any element of $\mathbf{N}$. Therefore, f is not onto and, consequently, not bijective, producing a contradiction. ∎

In the proof of Theorem 10.9, each digit in the decimal expansion of the number b constructed is 4 or 5. We could have selected any two distinct digits that did not use 9. It is now easy to give examples of other uncountable sets with the aid of the following result.

Theorem 10.10   *Let A and B be sets such that $A \subseteq B$. If A is uncountable, then B is uncountable.*

Proof   Let A and B be two sets such that $A \subseteq B$ and A is uncountable. Necessarily then A and B are infinite. Assume, to the contrary, that B is denumerable. Since A is an infinite subset of a denumerable set, it follows by Theorem 10.4 that A is denumerable, producing a contradiction. ∎

Corollary 10.11   *The set $\mathbf{R}$ of real numbers is uncountable.*

Proof   Since $(0, 1)$ is uncountable by Theorem 10.9 and $(0, 1) \subseteq \mathbf{R}$, it follows by Theorem 10.10 that $\mathbf{R}$ is uncountable. ∎

Let's pause for a moment to review a few facts that we've discovered about infinite sets (at least about certain infinite sets). First, recall that two nonempty sets A and B are defined to have the same cardinality (same number of elements) if there exists a bijective function from A to B. We're especially interested in the situation when A and B are infinite. One family of infinite sets we've introduced is the class of denumerable sets. Recall too that a set S is denumerable if there exists a bijective function from $\mathbf{N}$ to S.

Suppose that A and B are two denumerable sets. Then there exist bijective functions $f : \mathbf{N} \to A$ and $g : \mathbf{N} \to B$. Since f is bijective, f has an inverse function $f^{-1} : A \to \mathbf{N}$, where f^{-1} is also bijective (Theorem 9.15). Since $f^{-1} : A \to \mathbf{N}$ and $g : \mathbf{N} \to B$ are bijective functions, it follows that the composition function $g \circ f^{-1} : A \to B$ is also bijective (Corollary 9.12). This tells us that $|A| = |B|$; that is, A and B have the same number of elements. We state this as a theorem for emphasis.

Theorem 10.12   *Every two denumerable sets are numerically equivalent.*

Next, let B be an uncountable set. So B is an infinite set that is not denumerable. Also, let A be a denumerable set. Therefore, there exists a bijective function $f : \mathbf{N} \to A$. We claim that $|A| \neq |B|$; that is, A and B do *not* have the same number of elements.

Let's prove this. Assume, to the contrary, that $|A| = |B|$. Hence there exists a bijective function $g : A \to B$. Since the functions $f : \mathbf{N} \to A$ and $g : A \to B$ are bijective, the composition function $g \circ f : \mathbf{N} \to B$ is bijective. But this means that B is a denumerable set, which is a contradiction. We also state this fact as a theorem.

Theorem 10.13   *If A is a denumerable set and B is an uncountable set, then A and B are not numerically equivalent.*

Theorems 10.12 and 10.13 can also be considered as consequences of Theorem 10.1. In particular, Theorem 10.13 says that $\mathbf{Z}$ and $\mathbf{R}$ are not numerically equivalent and so $|\mathbf{Z}| \ne |\mathbf{R}|$. So, here are two infinite sets that do *not* have the same number of elements. In other words, there are different sizes of infinity. This now brings up a number of questions, one of which is: Do there exist *three* infinite sets so that no two of them have the same number of elements? Also, if A is a denumerable set and B is an uncountable set, is one of these sets "bigger" than the other in some sense? In other words, we would like to be able to compare $|A|$ and $|B|$ in some precise manner. Since $|\mathbf{Z}| \ne |\mathbf{R}|$ and $\mathbf{Z} \subset \mathbf{R}$, it is tempting to conclude that $|\mathbf{Z}| < |\mathbf{R}|$ but we have yet to give a meaning to $|A| < |B|$ for sets A and B. This idea will be addressed in Section 10.4. We should remind ourselves, however, that for infinite sets C and D, it is possible that both $C \subset D$ and $|C| = |D|$. For example, $\mathbf{Z} \subset \mathbf{Q}$ and $|\mathbf{Z}| = |\mathbf{Q}|$ since $\mathbf{Z}$ and $\mathbf{Q}$ are both denumerable. Before leaving our discussion of $\mathbf{Z}$ and $\mathbf{R}$, one other observation is useful. Recall that, according to Theorem 10.4, if B is an infinite subset of a denumerable set A, then B is also denumerable. But what if A is uncountable? That is, if B is an infinite subset of an uncountable set A, can we conclude that B is uncountable? The sets $\mathbf{Z}$ and $\mathbf{R}$ answer this question since $\mathbf{Z}$ is infinite, $\mathbf{R}$ is uncountable and $\mathbf{Z} \subset \mathbf{R}$. However, $\mathbf{Z}$ is not uncountable.

We have now seen two examples of uncountable sets, namely the open interval $(0, 1)$ of real numbers and the set $\mathbf{R}$ of all real numbers. Neither of these sets has the same number of elements as any denumerable set. But how do they compare with each other? We will show, in fact, that these two sets have the same number of elements. Prior to verifying this, we show that the open interval $(-1, 1)$ and $\mathbf{R}$ have the same number of elements.

Theorem to Prove   The sets $(-1, 1)$ and $\mathbf{R}$ are numerically equivalent.

<u>**PROOF STRATEGY**</u>   The obvious approach to proving this theorem is to locate a bijective function $f : (-1, 1) \to \mathbf{R}$. Actually, there are several such functions with this property. With each such function, we are faced with the problem of determining how involved it is to show that the function is bijective. We describe one of these here. Another is given in Exercise 10.25. Consider the function $f : (-1, 1) \to \mathbf{R}$ defined by

$$f(x) = \frac{x}{1 - |x|}.$$

(See Figure 10.9.) This function is defined for all $x \in (-1, 1)$. Observe that $f(0) = 0$, $f(x) > 0$ when $0 < x < 1$ and $f(x) < 0$ when $-1 < x < 0$. This function also has the property that

$$\lim_{x \to 1^-} \frac{x}{1 - |x|} = +\infty \quad \text{and} \quad \lim_{x \to -1^+} \frac{x}{1 - |x|} = -\infty.$$

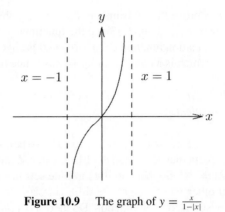

Figure 10.9   The graph of $y = \frac{x}{1-|x|}$

If you recall enough information about continuous functions from calculus, you might see that this function is continuous on the interval $(-1, 1)$. From this information, it follows that $f((-1, 1)) = \mathbf{R}$ and that f is onto. Also, the derivative of this function on the interval $(-1, 1)$ is

$$
f'(x) = \begin{cases} \frac{1}{(1-x)^2} & \text{if } x \in (0, 1) \\ 1 & \text{if } x = 0 \\ \frac{1}{(1+x)^2} & \text{if } x \in (-1, 0). \end{cases}
$$

This says that $f'(x) > 0$ for all $x \in (-1, 1)$ and so f is an increasing function on the interval $(-1, 1)$. This information tells us that f must be one-to-one and so f is bijective.

While the argument just given relies on calculus and you may not recall all of this, an argument can be given that is of the type we have been discussing.   ◆

Theorem 10.14   *The sets $(-1, 1)$ and $\mathbf{R}$ are numerically equivalent.*

Proof   Consider the function $f : (-1, 1) \to \mathbf{R}$ defined by

$$
f(x) = \frac{x}{1 - |x|}. \tag{10.2}
$$

We show that f is bijective. First, we verify that f is one-to-one. Let $f(a) = f(b)$, where $a, b \in (-1, 1)$. Then $\frac{a}{1-|a|} = \frac{b}{1-|b|}$. If $\frac{a}{1-|a|} = \frac{b}{1-|b|} = 0$, then $a = b = 0$. If $\frac{a}{1-|a|} = \frac{b}{1-|b|} > 0$, then $a > 0$ and $b > 0$. Thus $\frac{a}{1-a} = \frac{b}{1-b}$. Hence $a(1 - b) = b(1 - a)$ and so $a = b$. If $\frac{a}{1-|a|} = \frac{b}{1-|b|} < 0$, then $a < 0$ and $b < 0$. Thus $\frac{a}{1+a} = \frac{b}{1+b}$. Hence $a(1 + b) = b(1 + a)$ and so $a = b$. Therefore, f is one-to-one.

Next, we show that f is onto. Let $r \in \mathbf{R}$. Since $f(0) = 0$, we may assume that $r \neq 0$. If $r > 0$, then $\frac{r}{1+r} \in (0, 1)$ and $f(\frac{r}{1+r}) = r$. If $r < 0$, then $\frac{r}{1-r} \in (-1, 0)$ and $f(\frac{r}{1-r}) = r$. Thus f is onto.

Since f is a bijective function, the sets $(-1, 1)$ and $\mathbf{R}$ are numerically equivalent.   ∎

It is straightforward to show that the function $g : (0, 1) \to (-1, 1)$ defined by $g(x) = 2x - 1$ is bijective. For this function g and the function f in (10.2) in the proof of

Theorem 10.14, it therefore follows that $g \circ f : (0, 1) \to \mathbf{R}$ is also bijective. This gives an immediate corollary.

Corollary 10.15   *The sets $(0, 1)$ and $\mathbf{R}$ are numerically equivalent.*

Not only are $(0, 1)$ and $\mathbf{R}$ numerically equivalent, (as well as $(-1, 1)$ and $\mathbf{R}$), but every open interval (a, b) of real numbers with $a < b$ and $\mathbf{R}$ are numerically equivalent. (See Exercise 10.23.)

10.4 Comparing Cardinalities of Sets

As we know, two nonempty sets A and B have the same cardinality if there exists a bijective function $f : A \to B$. Let's illustrate this concept one more time by showing that two familiar sets associated with a given set are numerically equivalent. Recall that the power set $\mathcal{P}(A)$ of a set A is the set of all subsets of A and that 2^A is the set of all functions from A to $\{0, 1\}$. If $A = \{a, b, c\}$, then $|\mathcal{P}(A)| = 2^3 = 8$. Also, the set 2^A contains $2^{|A|} = 2^3 = 8$ functions. So in this case, $\mathcal{P}(A)$ and 2^A have the same number of elements. This is not a coincidence.

Theorem to Prove   For every nonempty set A, the sets $\mathcal{P}(A)$ and 2^A are numerically equivalent.

<u>PROOF STRATEGY</u>   If we can construct a bijective function $\phi : \mathcal{P}(A) \to 2^A$, then this will prove that $\mathcal{P}(A)$ and 2^A are numerically equivalent. We use ϕ for this function since 2^A is a set of functions and it is probably better to use more standard notation, such as f, to denote the elements of 2^A. But how can such a function ϕ be defined? Let's take a look at $\mathcal{P}(A)$ and 2^A for $A = \{a, b\}$. In this case,

$$\mathcal{P}(A) = \{\emptyset, \{a\}, \{b\}, \{a, b\}\};$$

while $2^A = \{f_1, f_2, f_2, f_4\}$, where

$$f_1 = \{(a, 0), (b, 0)\}, \quad f_2 = \{(a, 1), (b, 0)\},$$
$$f_3 = \{(a, 0), (b, 1)\}, \quad f_4 = \{(a, 1), (b, 1)\}.$$

Since each of $\mathcal{P}(A)$ and 2^A has four elements, we can easily find a bijective function from $\mathcal{P}(A)$ to 2^A. But this is not the question. What we are looking for is a bijective function $\phi : \mathcal{P}(A) \to 2^A$ for $A = \{a, b\}$ that suggests a way for us to define a bijective function from $\mathcal{P}(A)$ to 2^A for *any* set A (finite or infinite). Notice, for $A = \{a, b\}$, the connection between the following pairs of elements, the first element belonging to $\mathcal{P}(A)$ and the second belonging to 2^A:

$$\emptyset, \quad f_1 = \{(a, 0), (b, 0)\}$$
$$\{a\}, \quad f_2 = \{(a, 1), (b, 0)\}$$
$$\{b\}, \quad f_3 = \{(a, 0), (b, 1)\}$$
$$\{a, b\}, \quad f_4 = \{(a, 1), (b, 1)\}.$$

For example, the subset $\{a\}$ of $\{a, b\}$ contains a but not b, while f_2 maps a to 1 and b to 0. For an arbitrary set A, this suggests defining ϕ so that a subset S of A is mapped into the function in which 1 is the image of elements of A that belong to S and 0 is the image of elements of A that do not belong to S.   ◆

Theorem 10.16   *For every nonempty set A, the sets $\mathcal{P}(A)$ and 2^A are numerically equivalent.*

Proof   We show that there exists a bijective function ϕ from $\mathcal{P}(A)$ to 2^A. Define $\phi : \mathcal{P}(A) \to 2^A$ such that for $S \in \mathcal{P}(A)$, we have $\phi(S) = f_S$, where, for $x \in A$,

$$f_S(x) = \begin{cases} 1 & \text{if } x \in S \\ 0 & \text{if } x \notin S. \end{cases}$$

Certainly, $f_S \in 2^A$. First, we show that ϕ is one-to-one. Let $\phi(S) = \phi(T)$. Thus, $f_S = f_T$, which implies that $f_S(x) = f_T(x)$ for every $x \in A$. Therefore, $f_S(x) = 1$ if and only if $f_T(x) = 1$ for every $x \in A$; that is, $x \in S$ if and only if $x \in T$ and so $S = T$.

It remains to show that ϕ is onto. Let $f \in 2^A$. Define

$$S = \{x \in A : f(x) = 1\}.$$

Hence $f_S = f$ and so $\phi(S) = f$. Thus ϕ is onto and, consequently, ϕ is bijective.   ∎

It is clear that $A = \{x, y, z\}$ has fewer elements than $B = \{a, b, c, d, e\}$, that is, $|A| < |B|$. And it certainly seems that $|B| < |\mathbf{N}|$ and that, in general, any finite set has fewer elements than any denumerable set (or than any infinite set). Also, our discussion about countable and uncountable sets appears to suggest that uncountable sets have more elements than countable sets. But these assertions are based on intuition. We now make this more precise.

A set A is said to have **smaller cardinality** than a set B, written as $|A| < |B|$, if there exists a one-to-one function from A to B but no bijective function from A to B. That is, $|A| < |B|$ if it is possible to pair off the elements of A with some of the elements of B but not with all of the elements of B. If $|A| < |B|$, then we also write $|B| > |A|$. For example, since $\mathbf{N}$ is denumerable and $\mathbf{R}$ is uncountable, there is no bijective function from $\mathbf{N}$ to $\mathbf{R}$. Since the function $f : \mathbf{N} \to \mathbf{R}$ defined by $f(n) = n$ for all $n \in \mathbf{N}$ is injective, it follows that $|\mathbf{N}| < |\mathbf{R}|$. Moreover, $|A| \leq |B|$ means that $|A| = |B|$ or $|A| < |B|$. Hence to verify that $|A| \leq |B|$, we need only show the existence of a one-to-one function from A to B.

The cardinality of the set $\mathbf{N}$ of natural numbers is often denoted by $\aleph_o$ (often read "aleph null"); so $|\mathbf{N}| = \aleph_o$. Actually, $\aleph$ is the first letter of Hebrew alphabet. Indeed, if A is any denumerable set, then $|A| = \aleph_o$. The set $\mathbf{R}$ of real numbers is also referred to as the **continuum**, and its cardinality is denoted by c. Hence $|\mathbf{R}| = c$ and from what we have seen, $\aleph_o < c$. It was the German mathematician Georg Cantor who helped to put the theory of sets on a firm foundation. An interesting conjecture of his became known as:

The Continuum   There exists no set S such that
Hypothesis

$$\aleph_o < |S| < c.$$

Of course, if the Continuum Hypothesis were true, then this would imply that every subset of $\mathbf{R}$ is either countable or numerically equivalent to $\mathbf{R}$. However, in 1931 the

Austrian mathematician Kurt Gödel proved that it was impossible to disprove the Continuum Hypothesis from the axioms on which the theory of sets is based. In 1963 the American mathematician Paul Cohen took it one step further by showing that it was also impossible to *prove* the Continuum Hypothesis from these axioms. Thus the Continuum Hypothesis is independent of the axioms of set theory.

Another question that might occur to you is the following: Is there a set S such that $|S| > c$? This is a question we can answer, however, and the answer might be surprising.

Theorem to Prove   If A is a set, then $|A| < |\mathcal{P}(A)|$.

PROOF STRATEGY   First, it is not surprising that $|A| < |\mathcal{P}(A)|$ if A is finite, for if A has n elements, where $n \in \mathbf{N}$, then $\mathcal{P}(A)$ has 2^n elements and $2^n > n$ (which was proved by induction in Result 6.9). Of course, we must still show that $|A| < |\mathcal{P}(A)|$ when A is infinite. First we show that there exists a one-to-one function $f : A \to \mathcal{P}(A)$ for every set A. Let's give ourselves an example, say $A = \{a, b\}$. Then $\mathcal{P}(A) = \{\emptyset, \{a\}, \{b\}, \{a, b\}\}$. Although there are many injective function from A to $\mathcal{P}(A)$, there is one natural injective function:

$$f = \{(a, \{a\}), (b, \{b\})\};$$

in other words, define $f : A \to \mathcal{P}(A)$ by $f(x) = \{x\}$.

Once we have verified that this function is one-to-one, then we know that $|A| \leq |\mathcal{P}(A)|$. To show that the inequality is strict, however, we must prove that there is no bijective function from A to $\mathcal{P}(A)$. The natural technique to use for such a proof is proof by contradiction.   ◆

Theorem 10.17   *If A is a set, then $|A| < |\mathcal{P}(A)|$.*

Proof   If $A = \emptyset$, then $|A| = 0$ and $|\mathcal{P}(A)| = 1$; so $|A| < |\mathcal{P}(A)|$. Hence we may assume that $A \neq \emptyset$. First, we show that there is a one-to-one function from A to $\mathcal{P}(A)$. Define the function $f : A \to \mathcal{P}(A)$ by $f(x) = \{x\}$ for each $x \in A$. Let $f(x_1) = f(x_2)$. Then $\{x_1\} = \{x_2\}$. So $x_1 = x_2$ and f is one-to-one.

To prove that $|A| < |\mathcal{P}(A)|$, it remains to show that there is no bijective function from A to $\mathcal{P}(A)$. Assume, to the contrary, that there exists a bijective function $g : A \to \mathcal{P}(A)$. For each $x \in A$, let $g(x) = A_x$, where $A_x \subseteq A$. We show that there is a subset of A that is distinct from A_x for each $x \in A$. Define the subset B of A by

$$B = \{x \in A : x \notin A_x\}.$$

By assumption, there exists an element $y \in A$ such that $B = A_y$. If $y \in A_y$, then $y \notin B$ by the definition of B. On the other hand, if $y \notin A_y$, then, according to the definition of the set B, it follows that $y \in B$. In either case, y belongs to exactly one of A_y and B. Hence $B \neq A_y$, producing a contradiction.   ∎

According to Theorem 10.17, there is no largest set. In particular, there is a set S with $|S| > c$.

10.5 The Schröder–Bernstein Theorem

For two nonempty sets A and B, let f be a function from A to B and let D be a nonempty subset of A. By the **restriction** f_1 **of** f **to** D, we mean the function

$$f_1 = \{(x, y) \in f : x \in D\}.$$

Hence a restriction of f refers to restricting the domain of f. For example, for the sets $A = \{a, b, c, d\}$ and $B = \{1, 2, 3\}$, let $f = \{(a, 2), (b, 1), (c, 3), (d, 2)\}$ be a function from A to B. For $D = \{a, c\}$, the restriction of f to D is the function $f_1 : D \to B$ given by $\{(a, 2), (c, 3)\}$. Sometimes, we might also consider a new codomain B' for such a restriction f_1 of f. Of course, we must have range$(f_1) \subseteq B'$. Next, consider the function $g : \mathbf{R} \to [0, \infty)$ defined by $g(x) = x^2$ for $x \in \mathbf{R}$. Although g is onto, g is not one-to-one since $g(1) = g(-1) = 1$, for example. On the other hand, the restriction g_1 of g to $[0, \infty)$ is one-to-one, and so the restricted function $g_1 : [0, \infty) \to [0, \infty)$ defined by $g_1(x) = g(x) = x^2$ for all $x \in [0, \infty)$ is bijective. On the other hand, if $f : A \to B$ is a one-to-one function, then any restriction of f to a subset of A is also one-to-one.

Let $f : A \to B$ and $g : C \to D$ be functions, where A and C are disjoint sets. We define a function h from $A \cup C$ to $B \cup D$ by

$$h(x) = \begin{cases} f(x) & \text{if } x \in A \\ g(x) & \text{if } x \in C. \end{cases}$$

Recalling that a function is a *set* of ordered pairs, we see that h is the union of the two sets f and g. Of course, it is essential for A and C to be disjoint in order to be guaranteed that h is a function. If f and g are onto, then h must be onto as well; however, if f and g are one-to-one, then h need not be one-to-one. The following result does provide a sufficient condition for h to be one-to-one, however.

Lemma 10.18   *Let $f : A \to B$ and $g : C \to D$ be one-to-one functions, where $A \cap C = \emptyset$, and define $h : A \cup C \to B \cup D$ by*

$$h(x) = \begin{cases} f(x) & \text{if } x \in A \\ g(x) & \text{if } x \in C. \end{cases}$$

If $B \cap D = \emptyset$, then h is also a one-to-one function. Consequently, if f and g are bijective functions, then h is a bijective function.

Proof   Assume that $h(x_1) = h(x_2) = y$, where $x_1, x_2 \in A \cup B$. Then $y \in B \cup D$. So $y \in B$ or $y \in D$, say the former. Since $B \cap D = \emptyset$, it follows that $y \notin D$. Hence $x_1, x_2 \in A$ and so $h(x_1) = f(x_1)$ and $h(x_2) = f(x_2)$. Since $f(x_1) = f(x_2)$ and f is one-to-one, it follows that $x_1 = x_2$. ∎

Let A and B be nonempty sets such that $B \subseteq A$ and let $f : A \to B$. Thus for $x \in A$, the element $f(x) \in B$. Since $B \subseteq A$, it follows, of course, that $f(x) \in A$ and so $f(f(x)) \in B$. It is convenient to introduce some notation in this case. Let $f^1(x) = f(x)$ and let $f^2(x) = f(f(x))$. In general, for an integer $k \geq 2$, let $f^k(x) = f(f^{k-1}(x))$. Hence $f^1(x), f^2(x), f^3(x), \ldots$ is a recursively defined sequence of elements of B (and of A as well). Thus $f^n(x)$ is defined for every positive integer n.

For example, consider the function $f : \mathbf{Z} \to 2\mathbf{Z}$ defined by $f(n) = 4n$ for all $n \in \mathbf{Z}$. Then $f^1(3) = f(3) = 4 \cdot 3 = 12$ and $f^2(3) = f(f(3)) = f(12) = 4 \cdot 12 = 48$.

If A and B are nonempty sets such that $B \subseteq A$, then the function $\phi : B \to A$ defined by $\phi(x) = x$ for all $x \in B$ is injective. This gives us the expected result that $|B| \le |A|$. On the other hand, if there is an injective function from A to B, a more interesting consequence results.

Theorem 10.19   *Let A and B be nonempty sets such that $B \subseteq A$. If there exists an injective function from A to B, then there exists a bijective function from A to B.*

Proof   If $B = A$, then the identity function $i_A : A \to B = A$ is bijective. Thus we can assume that $B \subset A$ and so $A - B \ne \emptyset$. Let $f : A \to B$ be an injective function. If f is bijective, then the proof is complete. Therefore, we can assume that f is not onto. Hence range$(f) \subset B$ and so $B -$ range$(f) \ne \emptyset$.

Consider the subset B' of B defined by

$$B' = \{f^n(x) : x \in A - B, n \in \mathbf{N}\}.$$

Thus $B' \subseteq$ range(f). Hence, for each $x \in A - B$, its image $f(x)$ belongs to B'. Moreover, for $x \in A - B$, the element $f^2(x) = f(f(x)) \in B'$, $f^3(x) = f(f^2(x)) \in B'$ and so on.

Let $C = (A - B) \cup B'$ and consider the restriction $f_1 : C \to B'$ of f to C. We show that f_1 is onto. Let $y \in B'$. Then $y = f^n(x)$ for some $x \in A - B$ and some $n \in \mathbf{N}$. This implies that $y = f(x)$ for some $x \in A - B$ or $y = f(x)$ for some $x \in B'$. Therefore, $f_1(x) = y$ for some $x \in C$ and so f_1 is onto. Furthermore, since f is one-to-one, the function f_1 is also one-to-one. Hence $f_1 : C \to B'$ is bijective.

Let $D = B - B'$. Since $B -$ range$(f) \ne \emptyset$ and $B -$ range$(f) \subseteq B - B'$, it follows that $D \ne \emptyset$. Also, D and B' are disjoint, as are D and C. Certainly, the identity function $i_D : D \to D$ is bijective. Let $h : C \cup D \to B' \cup D$ be defined by

$$h(x) = \begin{cases} f_1(x) & \text{if } x \in C \\ i_D(x) & \text{if } x \in D. \end{cases}$$

By Lemma 10.18, h is bijective. However, $C \cup D = A$ and $B' \cup D = B$; so h is a bijective function from A to B.   ∎

From what we know of inequalities (of real numbers), it might seem that if A and B are sets with $|A| \le |B|$ and $|B| \le |A|$, then $|A| = |B|$. This is indeed the case. This theorem is often referred to as the Schröder–Bernstein Theorem.

Theorem 10.20   *(The Schröder–Bernstein Theorem) If A and B are sets such that $|A| \le |B|$ and $|B| \le |A|$, then $|A| = |B|$.*

Proof   Since $|A| \le |B|$ and $|B| \le |A|$, there are injective functions $f : A \to B$ and $g : B \to A$. Thus $g_1 : B \to$ range(g) defined by $g_1(x) = g(x)$ for all $x \in B$ is a bijective function. By Theorem 9.15, g_1^{-1} exists and $g_1^{-1} :$ range$(g) \to B$ is a bijective function.

Since $f : A \to B$ and $g_1 : B \to$ range(g) are injective functions, it follows by Theorem 9.11 that $g_1 \circ f : A \to$ range(g) is an injective function. Because range$(g) \subseteq A$, we have by Theorem 10.19 that there exists a bijective function $h : A \rightarrow \text{range}(g)$. Thus $h : A \rightarrow \text{range}(g)$ and $g_1^{-1} : \text{range}(g) \rightarrow B$ are bijective functions. By Corollary 9.12,

$$g_1^{-1} \circ h : A \rightarrow B$$

is a bijective function and $|A| = |B|$. ∎

The Schröder–Bernstein Theorem is referred to by some as the Cantor–Schröder–Bernstein Theorem. Although the history of this theorem has never been fully documented, there are several substantiated facts.

A mathematician who will forever be associated with the theory of sets is Georg Cantor (1845–1918). Born in Russia, Cantor studied for and obtained his Ph.D in mathematics from the University of Berlin in 1867. In 1869 he became a faculty member at the University of Halle in Germany. It was while he was there that he became interested in set theory.

In 1873 Cantor proved that the set of rational numbers is denumerable. Shortly afterward, he proved that the set of real numbers is uncountable. In this paper, he essentially introduced the idea of a one-to-one correspondence (bijective function). During the next several years, he made numerous contributions to set theory—studying sets of equal cardinality. There were, however, a number of problems that proved difficult for Cantor.

Consider the following two theorems:

Theorem A *For any two cardinal numbers a and b, exactly one of the following occurs:* (1) $a = b$, (2) $a < b$, (3) $a > b$.

Theorem B *If A and B are two sets for which there exist a one-to-one function from A to B and a one-to-one function from B to A, then $|A| = |B|$.*

Cantor observed that once Theorem A had been proved, Theorem B could be proved. On the other hand, there has never been any evidence that Cantor was able to prove Theorem A. Ernst Zermelo (1871–1953) was able to prove Theorem A in 1904. However, Zermelo's proof made use of an axiom formulated by Zermelo. This axiom, which was controversial in the mathematical world for many years, is known as the **Axiom of Choice**.

The Axiom of Choice. *For every collection of pairwise disjoint nonempty sets, there exists at least one set that contains exactly one element of each of these nonempty sets.*

As it turned out, not only can the Axiom of Choice be used to prove Theorem A, but Theorem A is true if and only if the Axiom of Choice is true.

Ernst Schröder (1841–1902), a German mathematician, was one of the important figures in mathematical logic. During 1897–1898 Schröder presented a "proof" of Theorem B, which contained a defect however. About the same time, Felix Bernstein (1878–1956) gave his own proof of Theorem B in his doctoral dissertation, which became the first complete proof of Theorem B. His proof did not require knowledge of Theorem A.

You may be surprised to learn that **R** and the power set of **N** are numerically equivalent. But how could one ever find a bijective function between these two sets? Theorem 10.20 tells us that discovering such a function is unnecessary.

Theorem 10.21   *The sets $\mathcal{P}(\mathbf{N})$ and $\mathbf{R}$ are numerically equivalent.*

Proof   First we show that there is a one-to-one function $f : (0, 1) \rightarrow \mathcal{P}(\mathbf{N})$. Recall that a real number $a \in (0, 1)$ can be expressed uniquely as $a = 0.a_1a_2a_3 \cdots$, where each $a_i \in \{0, 1, \ldots, 9\}$ and there is no positive integer N such that $a_n = 9$ for all $n \geq N$. Thus we define

$$f(a) = \{10^{n-1}a_n : n \in \mathbf{N}\} = A.$$

For example, $f(0.1234) = \{1, 20, 300, 4000\}$ and $f(1/3) = \{3, 30, 300, \ldots\}$. We now show that f is one-to-one. Assume that $f(a) = f(b)$, where $a, b \in (0, 1)$ and $a = 0.a_1a_2a_3 \cdots$ and $b = 0.b_1b_2b_3 \cdots$ with $a_i, b_i \in \{0, 1, \ldots, 9\}$ for each $i \in \mathbf{N}$ such that the decimal expansion of neither a nor b is 9 from some point on. Therefore,

$$A = \{10^{n-1}a_n : n \in \mathbf{N}\} = \{10^{n-1}b_n : n \in \mathbf{N}\} = B.$$

Consider the ith digit, namely a_i, in the decimal expansion of a. Then $10^{i-1}a_i \in A$. If $a_i \neq 0$, then $10^{i-1}a_i$ is the unique number in the interval $[10^{i-1}, 9 \cdot 10^{i-1}]$ belonging to A. Since $A = B$, it follows that $10^{i-1}a_i \in B$. However, $10^{i-1}b_i$ is the unique number in the interval $[10^{i-1}, 9 \cdot 10^{i-1}]$ belonging to B; so $10^{i-1}a_i = 10^{i-1}b_i$. Thus $a_i = b_i$. If $a_i = 0$, then $0 \in A$ and there is no number in the interval $[10^{i-1}, 9 \cdot 10^{i-1}]$ belonging to A. Since $A = B$, it follows that $0 \in B$ and there is no number in the interval $[10^{i-1}, 9 \cdot 10^{i-1}]$ belonging to B. Thus $b_i = 0$ and so $a_i = b_i$. Hence $a_i = b_i$ for all $i \in \mathbf{N}$, and so $a = b$. Therefore, f is one-to-one and $|(0, 1)| \leq |\mathcal{P}(\mathbf{N})|$.

Next we define a function $g : \mathcal{P}(\mathbf{N}) \rightarrow (0, 1)$. For $S \subseteq \mathbf{N}$, define $g(S) = 0.s_1s_2s_3 \cdots$, where

$$s_n = \begin{cases} 1 & \text{if } n \in S \\ 2 & \text{if } n \notin S. \end{cases}$$

Thus $g(S)$ is a real number in $(0, 1)$, whose decimal expansion consists only of 1s and 2s. We show that g is one-to-one. Assume that $g(S) = g(T)$, where $S, T \subseteq \mathbf{N}$. Thus

$$g(S) = s = 0.s_1s_2s_3 \cdots = 0.t_1t_2t_3 \cdots = t = g(T),$$

where

$$s_n = \begin{cases} 1 & \text{if } n \in S \\ 2 & \text{if } n \notin S \end{cases} \text{ and } t_n = \begin{cases} 1 & \text{if } n \in T \\ 2 & \text{if } n \notin T. \end{cases}$$

Since the decimal expansions of s and t contain no 0s or 9s, both s and t have unique decimal expansions. We show that $S = T$. First, we verify that $S \subseteq T$. Let $k \in S$. Then $s_k = 1$. Since $s = t$, it follows that $t_k = 1$, which implies that $k \in T$. Hence $S \subseteq T$. The proof that $T \subseteq S$ is similar and is therefore omitted. Thus $S = T$ and g is one-to-one. Therefore, $|\mathcal{P}(\mathbf{N})| \leq |(0, 1)|$. By the Schröder–Bernstein Theorem, $|\mathcal{P}(\mathbf{N})| = |(0, 1)|$. By Corollary 10.15, $|(0, 1)| = |\mathbf{R}|$. Thus, $|\mathcal{P}(\mathbf{N})| = |\mathbf{R}|$.   ∎

As a corollary to Theorems 10.16 and 10.21, we have the following result.

Corollary 10.22   *The sets $2^{\mathbf{N}}$ and $\mathbf{R}$ are numerically equivalent.*

We have already mentioned that $|A| = \aleph_o$ for every denumerable set A and that $|\mathbf{R}| = c$. If A is denumerable, then we represent the cardinality of the set 2^A by $2^{\aleph_o}$. By Corollary 10.22, $2^{\aleph_o} = c$.

EXERCISES FOR CHAPTER 10

Section 10.1: Numerically Equivalent Sets

10.1. Let $S = \{A_1, A_2, \ldots, A_5\}$ be a collection of five subsets of the set $A = \{-5, -4, \ldots, 5\}$, where
$A_1 = \{x \in A : 1 < x^2 < 10\}$
$A_2 = \{x \in A : (x + 2)(x - 4) > 0\}$
$A_3 = \{x \in A : |x + 2| + |x - 3| \le 5\}$
$A_4 = \{x \in A : \frac{1}{x^2+1} > \frac{2}{5}\}$
$A_5 = \{x \in A : \sin \frac{\pi x}{4} = 0\}$.
A relation R is defined on S by $A_i \, R \, A_j$ $(1 \le i, j \le 5)$ if A_i and A_j are numerically equivalent. According to Theorem 10.1, R is an equivalence relation. Determine the distinct equivalence classes for this equivalence relation.

10.2. (a) Let S be a collection of $n \ge 2$ numerically equivalent sets. Prove that these sets can be shown to be numerically equivalent by means of $n - 1$ bijective functions between pairs of sets in S.
(b) What other question is suggested by the problem in (a)?

Section 10.2: Denumerable Sets

10.3. Prove that if A and B are disjoint denumerable sets, then $A \cup B$ is denumerable.

10.4. Let $\mathbf{R}^+$ denote the set of positive real numbers and let A and B be denumerable subsets of $\mathbf{R}^+$. Define $C = \{x \in \mathbf{R} : -x \in B\}$. Show that $A \cup C$ is denumerable.

10.5. Prove that $|\mathbf{Z}| = |\mathbf{Z} - \{2\}|$.

10.6. (a) Prove that the function $f : \mathbf{R} - \{1\} \to \mathbf{R} - \{2\}$ defined by $f(x) = \frac{2x}{x-1}$ is bijective.
(b) Explain why $|\mathbf{R} - \{1\}| = |\mathbf{R} - \{2\}|$.

10.7. Let

$$S = \left\{ x \in \mathbf{R} : x = \frac{n^2 + \sqrt{2}}{n}, n \in \mathbf{N} \right\}.$$

Define $f : \mathbf{N} \to S$ by $f(n) = \frac{n^2 + \sqrt{2}}{n}$.

(a) List three elements that belong to S.
(b) Show that f is one-to-one.
(c) Show that f is onto.
(d) Is S denumerable? Explain.

10.8. Prove that the function $f : \mathbf{N} \to \mathbf{Z}$ defined in (10.1) by $f(n) = \frac{1+(-1)^n(2n-1)}{4}$ is bijective.

10.9. Show that every denumerable set A can be partitioned into two denumerable subsets of A.

10.10. Let A be a denumerable set and let $B = \{x, y\}$. Prove that $A \times B$ is denumerable.

10.11. Let B be a denumerable set and let A be a nonempty set of unspecified cardinality. If $f : A \to B$ is a one-to-one function, then what can be said about the cardinality of A? Explain.

10.12. Prove that the set of all 2-element subsets of $\mathbf{N}$ is denumerable.

10.13. A **Gaussian integer** is a complex number of the form $a + bi$, where $a, b \in \mathbf{Z}$ and $i = \sqrt{-1}$. Show that the set $\mathcal{G}$ of Gaussian integers is denumerable.

10.14. Prove that $S = \{(a, b) : a, b \in \mathbf{N} \text{ and } b \geq 2a\}$ is denumerable.

10.15. Let $S \subseteq \mathbf{N} \times \mathbf{N}$ be defined by $S = \{(i, j) : i \leq j\}$. Show that S is denumerable.

10.16. Let $A_1, A_2, A_3, \ldots$ be pairwise disjoint denumerable sets. Prove that $\bigcup_{i=1}^{\infty} A_i$ is denumerable.

10.17. Let $A = \{a_1, a_2, a_3, \ldots\}$. Define $B = A - \{a_{n^2} : n \in \mathbf{N}\}$. Prove that $|A| = |B|$.

10.18. A function $f : \mathbf{N} \times \mathbf{N} \to \mathbf{N}$ is defined by $f(m, n) = 2^{m-1}(2n - 1)$.

    (a) Prove that f is one-to-one and onto.
    (b) Show that $\mathbf{N} \times \mathbf{N}$ is denumerable.

10.19. Prove that every denumerable set A can be partitioned into a denumerable number of denumerable subsets of A.

Section 10.3: Uncountable Sets

10.20. Prove that the set of irrational numbers is uncountable.

10.21. Prove that the set of complex numbers is uncountable.

10.22. Prove that the open interval $(-2, 2)$ and $\mathbf{R}$ are numerically equivalent by finding a bijective function $h : (-2, 2) \to \mathbf{R}$. (Show that your function is, in fact, bijective.)

10.23. (a) Prove that the function $f : (0, 1) \to (0, 2)$, mapping the open interval $(0, 1)$ into the open interval $(0, 2)$ and defined by $f(x) = 2x$, is bijective.
    (b) Explain why $(0, 1)$ and $(0, 2)$ have the same cardinality.
    (c) Let $a, b \in \mathbf{R}$, where $a < b$. Prove that $(0, 1)$ and (a, b) have the same cardinality.

10.24. Prove that $\mathbf{R}$ and $\mathbf{R}^+$ are numerically equivalent.

10.25. Consider the function $g : (-1, 1) \to \mathbf{R}$ defined by $g(x) = \frac{x}{1-x^2}$.

    (a) Prove that g is onto.
    (b) Prove that g is one-to-one.
    (c) From the information obtained in (a) and (b), what conclusion can be made?

Section 10.4: Comparing Cardinalities of Sets

10.26. Prove or disprove the following:

    (a) If A is an uncountable set, then $|A| = |\mathbf{R}|$.
    (b) There exists a bijective function $f : \mathbf{Q} \to \mathbf{R}$.
    (c) If A, B and C are sets such that $A \subseteq B \subseteq C$ and A and C are denumerable, then B is denumerable.
    (d) The set $S = \left\{ \frac{\sqrt{2}}{n} : n \in \mathbf{N} \right\}$ is denumerable.
    (e) There exists a denumerable subset of the set of irrational numbers.
    (f) Every infinite set is a subset of some denumerable set.
    (g) If A and B are sets with the property that there exists an injective function $f : A \to B$, then $|A| = |B|$.

10.27. Let A and B be nonempty sets. Prove that $|A| \leq |A \times B|$.

10.28. Prove or disprove: If A and B are two sets such that A is countable and $|A| < |B|$, then B is uncountable.

10.29. How do the cardinalities of the sets $[0, 1]$ and $[1, 3]$ compare? Justify your answer.

10.30. Let $A = \{a, b, c\}$. Then $\mathcal{P}(A)$ consists of the following subsets of A:

$$A_a = \emptyset, \ A_b = A, \ A_c = \{a, b\}, \ A_d = \{a, c\},$$
$$A_e = \{b, c\}, \ A_f = \{a\}, \ A_g = \{b\}, \ A_h = \{c\}.$$

In one part of the proof of Theorem 10.17, it was established (using a contradiction argument) that $|A| < |\mathcal{P}(A)|$ for every nonempty set A. In this argument, the existence of a bijective function $g : A \to \mathcal{P}(A)$ is assumed, where $g(x) = A_x$ for each $x \in A$. Then a subset B of A is defined by

$$B = \{x \in A : x \notin A_x\}.$$

(a) For the sets A and $\mathcal{P}(A)$ described above, what is the set B?

(b) What does the set B in (a) illustrate?

10.31. Prove or disprove: There is no set A such that 2^A is denumerable.

Section 10.5: The Schröder–Bernstein Theorem

10.32. Prove that if A, B and C are nonempty sets such that $A \subseteq B \subseteq C$ and $|A| = |C|$, then $|A| = |B|$.

10.33. Use the Schröder–Bernstein Theorem to prove that $|(0, 1)| = |[0, 1]|$.

10.34. Prove that $|\mathbf{Q} - \{q\}| = \aleph_0$ for every rational number q and $|\mathbf{R} - \{r\}| = c$ for every real number r.

10.35. Let $\mathbf{R}^*$ be the set obtained by removing the number 0 from $\mathbf{R}$. Prove that $|\mathbf{R}^*| = |\mathbf{R}|$.

10.36. Let $f : \mathbf{Z} \to 2\mathbf{Z}$ be defined by $f(k) = 4k$ for all $k \in \mathbf{Z}$.

(a) Prove that $f^n(k) = 4^n k$ for each $k \in \mathbf{Z}$ and each $n \in \mathbf{N}$.

(b) For this function f, describe the sets B', C and D and functions f_1 and h given in Theorem 10.19.

10.37. Express each positive rational number as m/n, where $m, n \in \mathbf{N}$ and m/n is reduced to lowest terms. Let d_a denote the number of digits in $a \in \mathbf{N}$. Thus $d_2 = 1$, $d_{13} = 2$, and $d_{100} = 3$. Define the function $f : \mathbf{Q}^+ \to \mathbf{N}$ so that $f(m/n)$ is the positive integer with $2(d_m + d_n)$ digits whose first d_m digits is the integer m, whose final d_n digits is the integer n, and all of whose remaining $d_m + d_n$ digits are 0. Thus $f(2/3) = 2003$ and $f(10/271) = 1000000271$.

(a) Prove that f is one-to-one.

(b) Use the Schröder–Bernstein Theorem to prove that $\mathbf{Q}^+$ is denumerable.

ADDITIONAL EXERCISES FOR CHAPTER 10

10.38. Evaluate the proposed proof of the following result.

Result   Let A and B be two sets with $|A| = |B|$. If $a \in A$ and $b \in B$, then $|A - \{a\}| = |B - \{b\}|$.

Proof   Since A and B have the same number of elements and one element is removed from each of A and B, it follows that $|A - \{a\}| = |B - \{b\}|$. ∎

10.39. Evaluate the proposed proof of the following result.

Result   The sets $(0, \infty)$ and $[0, \infty)$ are numerically equivalent.

Proof   Define the function $f : (0, \infty) \to [0, \infty)$ by $f(x) = x$. First, we show that f is one-to-one. Assume that $f(a) = f(b)$. Then $a = b$ and so f is one-to-one.

Next, we show that f is onto. Let $r \in [0, \infty)$. Since $f(r) = r$, the function f is onto. Since f is bijective, $|(0, \infty)| = |[0, \infty)|$. ∎

10.40. For a real number x, the **floor** $\lfloor x \rfloor$ of x is the greatest integer less than or equal to x. Therefore, $\lfloor 5.5 \rfloor = 5$, $\lfloor 3 \rfloor = 3$ and $\lfloor -5.5 \rfloor = -6$. Let $f : \mathbf{N} \to \mathbf{Z}$ be defined by $f(n) = (-1)^n \lfloor n/2 \rfloor$.

(a) Prove that f is bijective.

(b) What does (a) tell us about $\mathbf{Z}$? (See Result 10.3.)

10.41. Show that the following pairs of intervals are numerically equivalent.

  (a) $(0, 1)$ and $(0, \infty)$
  (b) $(0, 1]$ and $[0, \infty)$
  (c) $[b, c)$ and $[a, \infty)$, where $a, b, c \in \mathbf{R}$ and $b < c$.

10.42. Let S and T be two sets. Prove that if $|S - T| = |T - S|$, then $|S| = |T|$.

10.43. Prove each of the following statements:

  (a) A nonempty set S is countable if and only if there exists a surjective function $f : \mathbf{N} \to S$.
  (b) A nonempty set S is countable if and only if there exists an injective function $g : S \to \mathbf{N}$.

10.44. Prove that $|A| < |\mathbf{N}|$ for every finite nonempty set A.

10.45. Let $A = (0, 1)$ be the open interval of real numbers between 0 and 1. For each number $r \in A$, let $0.r_1r_2r_3 \ldots$ denote its unique decimal expansion in which no expansion has the digit 9 from some point on. For $(a, b) \in A \times A$, let

$$f((a, b)) = f((0.a_1a_2a_3 \ldots, \ 0.b_1b_2b_3 \ldots)) = 0.a_1b_1a_2b_2 \ldots;$$

while for $a \in A$, let

$$g(a) = g(0.a_1a_2a_3 \ldots) = (0.a_1a_3a_5 \ldots, \ 0.a_2a_4a_6 \ldots).$$

With the aid of f and g, which of the following are we able to conclude? Explain your answer.

  (a) $|A \times A| \le |A|$.
  (b) $|A| \le |A \times A|$.
  (c) $|A \times A| = |A|$.
  (d) Nothing because neither $f : A \times A \to A$ nor $g : A \to A \times A$ is a function.
  (e) Nothing because both $f : A \times A \to A$ and $g : A \to A \times A$ are functions but neither is injective.
  (f) Nothing, for a reason other than those given in (d) and (e).

10.46. Prove, for every integer $n \ge 2$, that if $A_1, A_2, \ldots, A_n$ are denumerable sets, then $A_1 \times A_2 \times \cdots \times A_n$ is denumerable.

10.47. As a consequence of Exercise 10.20 in Section 10.3, the set of all irrational numbers is uncountable. Among the (many) irrational numbers are $\sqrt{2}, \sqrt{3}, \sqrt{5}, \sqrt[3]{2}, \sqrt[3]{3}, \sqrt[3]{5}$ and $\sqrt[4]{2}$. Prove that the set $S = \{\sqrt[n]{k} : \ k, n \in \mathbf{N} \text{ and } k, n \ge 2\}$ is denumerable.

10.48. Let $b, c \in \mathbf{Z}$. A number $r_{b,c}$ (real or complex) belongs to a set S if $r_{b,c}$ is a root of the polynomial $x^2 + bx + c$. Prove that S is denumerable.

10.49. We have seen that $\mathbf{R}$ is an uncountable set.

  (a) Show that $\mathbf{R}$ can be partitioned into a denumerable number of uncountable sets.
  (b) Show that $\mathbf{R}$ can be partitioned into a uncountable number of countable sets.

11

Proofs in Number Theory

Number theory is that area of mathematics dealing with integers and their properties. It is one of the oldest branches of mathematics, dating back at least to the Pythagoreans (500 B.C.). Number theory is considered to be one of the most beautiful branches of mathematics. Indeed, it has been said that mathematics is the queen of the sciences, while number theory is the queen of mathematics. In large measure, this subject is characterized by the appeal, clarity and simplicity of many of its problems and by the elegance and style exhibited in their solutions. The main goal of this chapter is to expand on some of the things we have learned in order to illustrate the kinds of proofs that occur in number theory.

11.1 Divisibility Properties of Integers

You may already know that every integer $n \geq 2$ can be expressed as a product of primes and in only one way, except for the order in which the primes are written. We will see later how to prove this fact, but first we want to return to divisibility of integers (introduced in Chapter 4) and establish several elementary divisibility properties.

Recall that a **prime** is an integer $p \geq 2$ whose only positive integer divisors are 1 and p. An integer $n \geq 2$ that is not prime is called a **composite number** (or simply **composite**). The first ten primes are 2, 3, 5, 7, 11, 13, 17, 19, 23 and 29. The first ten composite numbers are 4, 6, 8, 9, 10, 12, 14, 15, 16 and 18. If an integer $n \geq 2$ is composite, then there exist integers a and b such that $n = ab$, where $1 < a < n$ and $1 < b < n$. Certainly, if there exist integers a and b such that $n = ab$, where $1 < a < n$ and $1 < b < n$, then n is composite. We summarize these observations in the lemma below.

Lemma 11.1   *An integer $n \geq 2$ is composite if and only if there exist integers a and b such that $n = ab$, where $1 < a < n$ and $1 < b < n$.*

In Chapter 4, a number of basic divisibility properties were presented. We recall some of these in the following theorem, where the proofs are also repeated to review the proof techniques that we used.

Theorem 11.2   *Let a, b and c be integers with $a \neq 0$.*

  (i) *If $a \mid b$, then $a \mid bc$.*
 (ii) *If $a \mid b$ and $b \mid c$, where $b \neq 0$, then $a \mid c$.*
(iii) *If $a \mid b$ and $a \mid c$, then $a \mid (bx + cy)$ for all integers x and y.*

Proof   We begin with part (i). Since $a \mid b$, there exists an integer q such that $b = aq$. Therefore, $bc = a(qc)$. Since qc is an integer, $a \mid bc$.

For part (ii), let $a \mid b$ and $b \mid c$. So there exist integers q_1 and q_2 such that $b = aq_1$ and $c = bq_2$. Consequently,

$$c = bq_2 = (aq_1)q_2 = a(q_1q_2).$$

Since q_1q_2 is an integer, $a \mid c$.

For part (iii), let $a \mid b$ and $a \mid c$. Then there exist integers q_1 and q_2 such that $b = aq_1$ and $c = aq_2$. Hence, for integers x and y,

$$bx + cy = (aq_1)x + (aq_2)y = a(q_1x + q_2y).$$

Since $q_1x + q_2y$ is an integer, $a \mid (bx + cy)$.   ∎

PROOF ANALYSIS   In all three parts of the previous theorem, we were required to show that $r \mid s$ for some integers r and s, where $r \neq 0$. To do this, we showed that we could write s as rt for some integer t. Of course, this is simply the definition of what it means for r to divide s.   ◆

Proving the two parts of the next theorem relies on the definition of $r \mid s$ again, as well as making use of certain observations. For example, in the second part, we use the fact that $|xy| = |x||y|$ for every two real numbers x and y.

Theorem 11.3   *Let a and b be nonzero integers.*

  (i) *If $a \mid b$ and $b \mid a$, then $a = b$ or $a = -b$.*
 (ii) *If $a \mid b$, then $|a| \leq |b|$.*

Proof   We first prove (i). Since $a \mid b$ and $b \mid a$, it follows that $b = aq_1$ and $a = bq_2$ for some integers q_1 and q_2. Therefore, $a = bq_2 = (aq_1)q_2 = a(q_1q_2)$. Dividing by a, we obtain $1 = q_1q_2$. Hence $q_1 = q_2 = 1$ or $q_1 = q_2 = -1$. Therefore, $a = b$ or $a = -b$.

Next we prove (ii). Since $a \mid b$, it follows that $b = aq$ for some integer q. Furthermore, $q \neq 0$ since $b \neq 0$. So $|q| \geq 1$. Hence

$$|b| = |aq| = |a| \cdot |q| \geq |a| \cdot 1 = |a|.$$ ∎

11.2 The Division Algorithm

We have discussed the concept of divisibility a number of times now. Of course, when we use that term, we are referring to the statement $a \mid b$, where $a, b \in \mathbf{Z}$ and $a \neq 0$. Surely, the term *division* is more familiar to us. For positive integers a and b, it is an elementary problem to divide b by a and ask for the quotient q and remainder r. For example, for $a = 5$ and $b = 17$, we have $q = 3$ and $r = 2$; that is, if 17 is divided by 5, a quotient of

3 and a remainder of 2 result. This division can be expressed as $17 = 5 \cdot 3 + 2$. If $a = 6$ and $b = 42$, then $q = 7$ and $r = 0$; so $42 = 6 \cdot 7 + 0$ or $6 \mid 42$.

More generally, for positive integers a and b, it is always possible to write $b = aq + r$, where $0 \leq r < a$. The number q is the **quotient** and r is the **remainder** when b is divided by a. In fact, not only do the integers q and r exist, they are unique. This is the essence of a theorem called the **Division Algorithm**. Although this theorem may seem rather obvious because you've probably used it so often, it is important and its proof is not obvious.

Theorem to Prove   **(The Division Algorithm)**   For positive integers a and b, there exist unique integers q and r such that $b = aq + r$ and $0 \leq r < a$.

<u>**PROOF STRATEGY**</u>   We start the proof then with two positive integers a and b. We have two problems facing us. First, we need to show that there are integers q and r such that $b = aq + r$ and $0 \leq r < a$. Second, we must show that only one integer q and one integer r satisfy the equation $b = aq + r$ and inequality $0 \leq r < a$. How can we come up with integers q and r satisfying these conditions? Our main concern, it turns out, is showing that there exist integers q and r such that $b = aq + r$ and $0 \leq r < a$.

If $a \mid b$, then we know that $b = aq$ for some integer q. So $b = aq + 0$ and $r = 0$ satisfies $0 \leq r < a$. If $a \nmid b$, then $b \neq aq$ for *every* integer q and so $b - aq \neq 0$ for every integer q. However, when we perform the operation of dividing b by a, we obtain a quotient q and a nonzero remainder r. The integer q has the properties that $b - aq > 0$ and $b - aq$ is as small as possible. Whether $a \mid b$ or $a \nmid b$, this suggests considering the set

$$S = \{b - ax : x \in \mathbf{Z} \text{ and } b - ax \geq 0\},$$

which is a set of nonnegative integers. Once we show that $S \neq \emptyset$, then we can apply Theorem 6.7 (which states, for every integer m, that the set $\{i \in \mathbf{Z} : i \geq m\}$ is well-ordered) to conclude that S has a smallest element r, which means that there is an integer q such that $b - aq = r$. So $b = aq + r$ and we have the beginning of a proof.   ◆

Theorem 11.4   **(The Division Algorithm)** *For positive integers a and b, there exist unique integers q and r such that $b = aq + r$ and $0 \leq r < a$.*

Proof   First we show that there exist integers q and r such that $b = aq + r$ with $0 \leq r < a$. We will verify the uniqueness later. Consider the set

$$S = \{b - ax : x \in \mathbf{Z} \text{ and } b - ax \geq 0\}.$$

By letting $x = 0$, we see that $b \in S$ and S is nonempty. Therefore, by Theorem 6.7, S has a smallest element r and, necessarily, $r \geq 0$. Also, since $r \in S$, there is some integer q such that $r = b - aq$. Thus $b = aq + r$ with $r \geq 0$.

Next we show that $r < a$. Assume, to the contrary, that $r \geq a$. Let $t = r - a$. Then $t \geq 0$. Since $a > 0$, it follows that $t < r$. Moreover,

$$t = r - a = (b - aq) - a = b - (aq + a) = b - a(q + 1),$$

which implies that $t \in S$, contradicting the fact that r is the smallest element of S. Therefore, $r < a$, as desired.

It remains to show that q and r are the only integers for which $b = aq + r$ and $0 \le r < a$. Let q' and r' be integers such that $b = aq' + r'$, where $0 \le r' < a$. We show that $q = q'$ and $r = r'$. Assume, without loss of generality, that $r' \ge r$; so $r' - r \ge 0$. Since $aq + r = aq' + r'$, it follows that

$$a(q - q') = r' - r.$$

Since $q - q'$ is an integer, $a \mid (r' - r)$. Because $0 \le r' - r < a$, we must have $r' - r = 0$ and so $r' = r$. However, $a(q - q') = r' - r = 0$ and $a \ne 0$; so $q - q' = 0$ and $q = q'$. ∎

In Theorem 11.4 we restricted a and b to be positive. With minor modifications in the proof, we can remove these restrictions, although, of course, we must still require that $a \ne 0$. Exercise 11.20 asks for a proof of the following result.

Corollary 11.5   (**The Division Algorithm, General Form**) *For integers a and b with $a \ne 0$, there exist unique integers q and r such that $b = aq + r$ and $0 \le r < |a|$.*

The proof of Theorem 11.4 is an existence proof since it establishes the existence of the integers q and r but does not provide a method for finding q and r. However, there is an implicit connection between the proof of the theorem and the manner in which you were taught to divide one positive integer by another to find the quotient and remainder (as we mentioned earlier). For instance, when dividing 89 by 14 you first determine the number of times 14 goes into 89, namely 6. More formally, you found the largest nonnegative integer whose product with 14 does not exceed 89. This number is 6. You then subtracted $14 \cdot 6 = 84$ from 89 to find the remainder 5. This determines the least nonnegative value of $89 - 14q$, where q is an integer, which corresponds to the least nonnegative value of the set S for $a = 14$ and $b = 89$ in the proof of Theorem 11.4.

Example 11.6   *Consistent with the notation of Corollary 11.5, find the integers q and r for the given integers a and b.*

(i) $a = 17, \ b = 78$
(ii) $a = -17, \ b = 78$
(iii) $a = 17, \ b = -78$
(iv) $a = -17, \ b = -78$

Solutions   (i) By simple division, we see that dividing 78 by 17 results in a quotient of 4 and a remainder of 10, that is,

$$78 = 17 \cdot 4 + 10; \tag{11.1}$$

so $q = 4$ and $r = 10$.

(ii) By replacing 17 and 4 in (11.1) by -17 and -4, respectively, we obtain

$$78 = (-17)(-4) + 10;$$

so $q = -4$ and $r = 10$.

(iii) Multiplying (11.1) through by -1, we have

$$-78 = -(17 \cdot 4) + (-10) = 17(-4) + (-10). \tag{11.2}$$

Since every remainder is nonnegative, we subtract 17 from and add 17 to the right side of (11.2), producing

$$-78 = 17(-5) + 7; \tag{11.3}$$

so $q = -5$ and $r = 7$.
(iv) If, in (11.3), we replace 17 and -5 by -17 and 5, respectively, we have

$$-78 = (-17) \cdot 5 + 7;$$

so $q = 5$ and $r = 7$.   ◆

We now discuss some consequences of the Division Algorithm. If this algorithm is applied to an arbitrary integer b when $a = 2$, then we see that b must have one of the two forms $2q$ or $2q + 1$ (according to whether $r = 0$ or $r = 1$). Of course, if $b = 2q$, then b is even; while if $b = 2q + 1$, then b is odd. We have seen this earlier. This, of course, shows that every integer is even or odd.

If we apply the Division Algorithm to an arbitrary integer b when $a = 3$, then b has exactly one of the three forms $3q, 3q + 1$ or $3q + 2$ (according to whether $r = 0, 1, 2$). We saw this as well in Chapter 4. In general, for an arbitrary integer b and arbitrary positive integer a, the remainder r is one of $0, 1, 2, \ldots, a - 1$. Hence b is expressible as exactly one of $aq, aq + 1, \ldots, aq + (a - 1)$.

This last remark should also sound familiar. In Chapter 8 we considered, for an integer $n \geq 2$, a relation R defined on $\mathbf{Z}$ by $a \, R \, b$ if $a \equiv b \pmod{n}$, that is, if $n \mid (a - b)$. This relation was found to be an equivalence relation and the set

$$\mathbf{Z}_n = \{[0], [1], \ldots, [n - 1]\}$$

of distinct equivalence classes was referred to as the set of integers modulo n. With the aid of the Division Algorithm, we now take a closer look at this relation.

For each integer b, there exist, by the Division Algorithm, unique integers q and r such that $b = nq + r$, where $0 \leq r < n$. Thus $b - r = nq$ and $n \mid (b - r)$, so $b \equiv r \pmod{n}$. Since $b \, R \, r$, it follows that $b \in [r]$. However, since r is the unique integer with $0 \leq r \leq n - 1$, we see that b belongs to exactly one of the classes $[0], [1], \ldots, [n - 1]$. These observations show that

(i)   the classes $[0], [1], \ldots, [n - 1]$ are pairwise disjoint and

(ii) $\mathbf{Z} = [0] \cup [1] \cup \cdots \cup [n - 1]$,

neither of which should seem surprising once we recall that the equivalence classes always produce a partition of the set on which the equivalence relation is defined. Furthermore, for each $r = 0, 1, \ldots, n - 1$,

$$[r] = \{nq + r : q \in \mathbf{Z}\};$$

that is, $[r]$ consists of all those integers having a remainder of r when divided by n. For this reason, these equivalence classes were also referred to as residue classes modulo n.

The special case where $n = 3$ was considered in Chapter 8 and the resulting residue classes were exhibited. In the present context, these residue classes are

$$[0] = \{3q : q \in \mathbf{Z}\} = \{\ldots - 6, -3, 0, 3, 6, \ldots\}$$
$$[1] = \{3q + 1 : q \in \mathbf{Z}\} = \{\ldots - 5, -2, 1, 4, \ldots\}$$
$$[2] = \{3q + 2 : q \in \mathbf{Z}\} = \{\ldots - 4, -1, 2, 5, \ldots\}.$$

11.3 Greatest Common Divisors

We now move from the divisors of an integer to the divisors of a pair of integers. An integer $c \neq 0$ is a **common divisor** of two integers a and b if $c \mid a$ and $c \mid b$. We are primarily interested in the largest integer that is a common divisor of a and b. Formally, the **greatest common divisor** of two integers a and b, not both 0, is the greatest positive integer that is a common divisor of a and b. The requirement that a and b are not both 0 is needed since every positive integer divides 0. We denote the greatest common divisor of two integers a and b by $\gcd(a, b)$, although (a, b) is common notation as well.

If a and b are relatively small (in absolute value), it is normally easy to determine $\gcd(a, b)$. For example, it should be clear that $\gcd(8, 12) = 4$, $\gcd(4, 9) = 1$ and $\gcd(18, 54) = 18$. The definition of $\gcd(a, b)$ does not require a and b to be positive; indeed, it only requires at least one of a and b to be nonzero. For example, $\gcd(-10, -15) = 5$, $\gcd(16, -72) = 8$ and $\gcd(0, -9) = 9$.

There are two useful properties of the greatest common divisor of two integers, particularly from a theoretical point of view, that we want to mention. For integers a and b, an integer of the form $ax + by$, where $x, y \in \mathbf{Z}$, is called a **linear combination** of a and b. Using this terminology, we can now restate Theorem 11.2(iii): Every nonzero integer that divides two integers b and c divides every linear combination of b and c. Although there appears to be no apparent connection between linear combinations of a and b and $\gcd(a, b)$, we are about to see that there is, in fact, a very close connection. For example, let $a = 10$ and $b = 16$. Then 6, -4 and 0 are linear combinations of a and b since, for example,

$$6 = 10 \cdot (-1) + 16 \cdot (1), \quad -4 = 10 \cdot (-2) + 16 \cdot (1) \text{ and } 0 = 10 \cdot 0 + 16 \cdot 0.$$

The integer 4 is also a linear combination of a and b since $4 = 10(2) + 16(-1)$. Furthermore, 2 is a linear combination of a and b since $2 = 10(-3) + 16(2)$. On the other hand, no odd integer can be a linear combination of a and b since if n is a linear combination of a and b, then there exist integers x and y such that

$$n = ax + by = 10x + 16y = 2(5x + 8y).$$

Since $5x + 8y$ is an integer, n is even. Consequently, 2 is the least positive integer that is a linear combination of 10 and 16. Curiously enough, $\gcd(10, 16) = 2$. We now show that this observation is no coincidence. Again, the Well-Ordering Principle will prove to be useful.

Theorem 11.7   *Let a and b be integers that are not both 0. Then $\gcd(a, b)$ is the least positive integer that is a linear combination of a and b.*

Proof   Let S denote the set of all positive integers that are linear combinations of a and b, that is,

$$S = \{ax + by : \; x, y \in \mathbf{Z} \text{ and } ax + by > 0\}.$$

First we show that S is nonempty. By assumption, at least one of a and b is nonzero; so $a \cdot a + b \cdot b = a^2 + b^2 > 0$. Thus $a \cdot a + b \cdot b \in S$ and, as claimed, $S \neq \emptyset$.

Since S is a nonempty subset of $\mathbf{N}$, it follows by the Well-Ordering Principle that S contains a least element, which we denote by d. Thus there exist integers x_0 and y_0 such that $d = ax_0 + by_0$. We now show that $d = \gcd(a, b)$. Applying the Division Algorithm to a and d, we have $a = dq + r$, where $0 \le r < d$. Consequently,

$$r = a - dq = a - q(ax_0 + by_0) = a(1 - qx_0) + b(-qy_0);$$

that is, r is a linear combination of a and b. If $r > 0$, then necessarily $r \in S$, which would contradict the fact that d is the least element in S. Therefore, $r = 0$, which implies that $d \mid a$. By a similar argument, it follows that $d \mid b$ and so d is a common divisor of a and b.

It remains to show that d is the *greatest* common divisor of a and b. Let c be a positive integer that is also a common divisor of a and b. By Theorem 11.2 (iii), c divides every linear combination of a and b and so c divides $d = ax_0 + by_0$. Because c and d are positive and $c \mid d$, it follows that $c \le d$ and therefore $d = \gcd(a, b)$.   ∎

PROOF ANALYSIS   The proof of Theorem 11.7 illustrates a common proof technique involving divisibility of integers. At one point of the proof, we wanted to show that $d \mid a$. A common method used to show that one integer d divides another integer a is to apply the Division Algorithm and divide a by d, obtaining $a = dq + r$, where $0 \le r < |d|$ or, if $d > 0$, then $0 \le r < d$. The goal then is to show that $r = 0$.   ◆

There is another characterization of the greatest common divisor of two integers, not both 0, that is useful to know. This characterization provides an alternative definition of the greatest common divisor, which, in fact, is used as the definition on occasion.

Theorem 11.8   *Let a and b be two integers, not both 0. Then $d = \gcd(a, b)$ if and only if d is that positive integer which satisfies the following two conditions:*

(1)   *d is a common divisor of a and b;*

(2)   *if c is any common divisor of a and b, then $c \mid d$.*

Proof   First, assume that $d = \gcd(a, b)$. We show that d satisfies (1) and (2). By definition, d satisfies (1); so it only remains to show that d satisfies (2). Let c be an integer such that $c \mid a$ and $c \mid b$. Since $d = \gcd(a, b)$, there exist integers x_0 and y_0 such that $d = ax_0 + by_0$. Since $c \mid a$ and $c \mid b$, it follows by Theorem 11.2(iii) that c divides $ax_0 + by_0 = d$. Therefore, d satisfies (2).

For the converse, assume that d is a positive integer satisfying properties (1) and (2). We show that $d = \gcd(a, b)$. Since d is already a common divisor of a and b, it suffices to show that d is the greatest common divisor of a and b. Let c be any positive integer that is a common divisor of a and b. Since d satisfies (2), $c \mid d$. Since c and d are both positive, it follows by Theorem 11.3(ii) that $c \le d$, which implies that $d = \gcd(a, b)$.   ∎

11.4 The Euclidean Algorithm

Although we know the definition of the greatest common divisor d of two integers a and b, not both 0, and two characterizations, none of these is useful in computing d. For this reason, we describe an algorithm for determining $d = \gcd(a, b)$ that is attributed to the famous mathematician Euclid, best known for his work in geometry. First, we note that if $b = 0$, then $a \neq 0$ and $\gcd(a, b) = \gcd(a, 0) = |a|$. Hence we may assume that a and b are nonzero. Furthermore, since $\gcd(a, b) = \gcd(a, -b) = \gcd(-a, b) = \gcd(-a, -b)$, we can assume that a and b are positive. Therefore, $\gcd(0, -12) = \gcd(0, 12) = 12$ and $\gcd(-12, -54) = \gcd(12, -54) = \gcd(-12, 54) = \gcd(12, 54) = 6$. In general then, we may assume that $0 < a \leq b$. The procedure for computing $d = \gcd(a, b)$, which we are about to describe and which is called the **Euclidean Algorithm**, makes use of repeated applications of the Division Algorithm and the following lemma.

Lemma 11.9   *Let a and b be positive integers. If $b = aq + r$ for some integers q and r, then $\gcd(a, b) = \gcd(r, a)$.*

Proof   Let $d = \gcd(a, b)$ and $e = \gcd(r, a)$. We show that $d = e$. First, note that $b = aq + r = aq + r \cdot 1$; that is, b is a linear combination of a and r. Since $e = \gcd(r, a)$, it follows that $e \mid a$ and $e \mid r$. By Theorem 11.2 (iii), $e \mid (aq + r \cdot 1)$ and so $e \mid b$. Hence e is a common divisor of a and b. Because $d = \gcd(a, b)$, we have $e \leq d$.

Since $b = aq + r$, we can write $r = b - aq = b \cdot 1 + a(-q)$, and so r is a linear combination of a and b. From the fact that $d = \gcd(a, b)$, we obtain $d \mid (b \cdot 1 + a(-q))$; that is, $d \mid r$. So d is a common divisor of r and a. Since $e = \gcd(r, a)$, it follows that $d \leq e$. Thus $e = d$.   ∎

We are now prepared to describe the Euclidean Algorithm. We begin with two integers a and b, where $0 < a \leq b$. By the Division Algorithm,

$$b = aq_1 + r_1, \text{ where } 0 \leq r_1 < a.$$

By Lemma 11.9, $\gcd(a, b) = \gcd(r_1, a)$. So if $r_1 = 0$, then $\gcd(a, b) = \gcd(0, a) = a$. Hence we may assume that $r_1 \neq 0$ and apply the Division Algorithm to r_1 and a, obtaining

$$a = r_1 q_2 + r_2, \text{ where } 0 \leq r_2 < r_1.$$

At this point, we have $\gcd(a, b) = \gcd(r_1, a) = \gcd(r_2, r_1)$ and $0 \leq r_2 < r_1 < a$. If $r_2 = 0$, then $\gcd(a, b) = \gcd(r_1, a) = \gcd(0, r_1) = r_1$. By now we should see the usefulness of Lemma 11.9 and also what we mean by repeated applications of the Division Algorithm. We continue this process and obtain the following sequence of equalities and inequalities:

$$
\begin{array}{ll}
b = aq_1 + r_1 & 0 < r_1 < a \\
a = r_1 q_2 + r_2 & 0 \leq r_2 < r_1 \\
\;\;\vdots & \;\;\vdots \\
r_{k-1} = r_k q_{k+1} + r_{k+1} & 0 \leq r_{k+1} < r_k \\
\;\;\vdots & \;\;\vdots
\end{array}
$$

By Lemma 11.9,

$$\gcd(a, b) = \gcd(r_1, a) = \gcd(r_2, r_1) = \cdots = \gcd(r_{k+1}, r_k) = \cdots$$

and $\cdots < r_{k+1} < r_k < \cdots < r_2 < r_1 < a$. Since these remainders are nonnegative, the strictly decreasing sequence $r_1, r_2, \ldots$ of remainders contains at most a terms. Let r_{n-1}

be the last nonzero remainder. Thus $r_n = 0$. We then have:

$$b = aq_1 + r_1 \quad \text{where } 0 \le r_1 < a \tag{11.4}$$

$$a = r_1 q_2 + r_2 \quad \text{where } 0 \le r_2 < r_1$$

$$\vdots$$

$$r_{n-4} = r_{n-3} q_{n-2} + r_{n-2} \quad \text{where } 0 \le r_{n-2} < r_{n-3}$$

$$r_{n-3} = r_{n-2} q_{n-1} + r_{n-1} \quad \text{where } 0 \le r_{n-1} < r_{n-2}$$

$$r_{n-2} = r_{n-1} q_n + 0$$

and we know that

$$\gcd(a, b) = \gcd(r_1, a) = \gcd(r_2, r_1) = \cdots = \gcd(r_{n-1}, r_{n-2}) = \gcd(0, r_{n-1}) = r_{n-1}.$$

The Euclidean Algorithm can now be described. We start with two integers a and b, where $0 < a \le b$. If $a \mid b$, then $\gcd(a, b) = a$; while if $a \nmid b$, then we apply the Division Algorithm repeatedly until a remainder of 0 is obtained. In this latter case, the last nonzero remainder is then $\gcd(a, b)$.

Let's see how the Euclidean Algorithm works in practice.

Example 11.10   *Use the Euclidean Algorithm to determine $d = \gcd(374, 946)$.*

Solution   Dividing 946 by 374, we find that

$$946 = 374 \cdot 2 + 198.$$

Now, dividing 374 by 198, we have

$$374 = 198 \cdot 1 + 176.$$

Continuing in this manner, we obtain

$$198 = 176 \cdot 1 + 22$$

$$176 = 22 \cdot 8 + 0.$$

So $\gcd(374, 946) = 22.$ ◆

Given integers a and b, not both 0, we know that there exist integers s and t such that

$$\gcd(a, b) = as + bt.$$

We now describe an algorithm for finding such integers s and t, using the notation in the calculations done after the proof of Lemma 11.9. Since $\gcd(a, b) = r_{n-1}$, our goal is to find integers s and t such that $r_{n-1} = as + bt$. We begin with the equation $r_{n-3} = r_{n-2} q_{n-1} + r_{n-1}$ and rewrite it in the form

$$r_{n-1} = r_{n-3} - r_{n-2} q_{n-1}. \tag{11.5}$$

Next, using equations (11.4), we solve for r_{n-2}, obtaining

$$r_{n-2} = r_{n-4} - r_{n-3} q_{n-2}.$$

Now substituting this expression for r_{n-2} into equation (11.5), we obtain

$$r_{n-1} = r_{n-3} - q_{n-1}(r_{n-4} - r_{n-3}q_{n-2})$$
$$= (1 + q_{n-1}q_{n-2})r_{n-3} + (-q_{n-1})r_{n-4}.$$

At this point, r_{n-1} is represented as a linear combination of r_{n-3} and r_{n-4}. Notice that r_{n-2} no longer appears in the expression. As we continue with this backward substitution method, we eliminate the remainders $r_{n-3}, r_{n-4}, \ldots, r_2, r_1$ one at a time and eventually arrive at an equation of the form $r_{n-1} = as + bt$.

Example 11.11   *For $a = 374$ and $b = 946$, find integers s and t such that $as + bt = \gcd(a, b)$.*

Solution   Using the computations from Example 11.10, we have

$$22 = 198 - 176 \cdot 1 = 198 \cdot 1 + 176 \cdot (-1)$$
$$176 = 374 - 198 \cdot 1 = 374 \cdot 1 + 198 \cdot (-1)$$
$$198 = 946 - 374 \cdot 2 = 946 \cdot 1 + 374 \cdot (-2).$$

Therefore,

$$22 = 198 \cdot 1 + 176 \cdot (-1)$$
$$= 198 \cdot 1 + [374 \cdot 1 + 198 \cdot (-1)] \cdot (-1)$$
$$= 198 \cdot 1 + 374 \cdot (-1) + 198 \cdot 1$$
$$= 198 \cdot 2 + 374 \cdot (-1)$$
$$= [946 \cdot 1 + 374 \cdot (-2)] \cdot 2 + 374 \cdot (-1)$$
$$= 946 \cdot 2 + 374 \cdot (-4) + 374 \cdot (-1)$$
$$= 946 \cdot 2 + 374 \cdot (-5).$$

Hence $s = -5$ and $t = 2$. ◆

The integers s and t that we have just found are not unique. Indeed, if $\gcd(a, b) = d$ and $d = as + bt$, then $d = a(s + b) + b(t - a)$ as well.

11.5 Relatively Prime Integers

For two integers a and b, not both 0, we know that if $\gcd(a, b) = 1$, then there exist integers s and t such that $as + bt = 1$. What may be surprising is that the converse holds in this special case as well.

Theorem 11.12   *Let a and b be integers, not both 0. Then $\gcd(a, b) = 1$ if and only if there exist integers s and t such that $1 = as + bt$.*

Proof   If $\gcd(a, b) = 1$, then by Theorem 11.7 there exist integers s and t such that $as + bt = 1$. We now consider the converse. Let a and b be integers, not both 0, for which there exist integers s and t such that $as + bt = 1$. By Theorem 11.7, $\gcd(a, b)$ is the smallest positive integer that is a linear combination of a and b. Since 1 is a linear combination of a and b, it follows that $\gcd(a, b) = 1$. ∎

Two integers a and b, not both 0, are called **relatively prime** if $\gcd(a, b) = 1$. By Theorem 11.12 then, two integers a and b are relatively prime if and only if 1 is a linear combination of a and b. This fact is extremely useful, as we are about to see.

If a, b and c are integers such that $a \mid bc$, then there is no reason to believe that $a \mid b$ or $a \mid c$. For example, let $a = 4$, $b = 6$ and $c = 2$. Then $4 \mid 6 \cdot 2$ but $4 \nmid 6$ and $4 \nmid 2$. However, if a and b are relatively prime, then we can draw another conclusion. The following result is often called *Euclid's Lemma*.

Theorem to Prove   **(Euclid's Lemma)**   Let a, b and c be integers, where $a \neq 0$. If $a \mid bc$ and $\gcd(a, b) = 1$, then $a \mid c$.

<u>PROOF STRATEGY</u>   If we use a direct proof, then we assume that $a \mid bc$ and $\gcd(a, b) = 1$. In order to show that $a \mid c$, we need to show that c be can can be expressed as ar for some integer r. Because $a \mid bc$, we know that $bc = aq$ for some integer q. Also, because $\gcd(a, b) = 1$, there are integers s and t such that $as + bt = 1$. If we were to multiply $as + bt = 1$ by c, then we would have $c = acs + bct$. However, $bc = aq$ and we could factor a from $acs + bct$. This is the plan. ◆

Theorem 11.13   **(Euclid's Lemma)** *Let a, b and c be integers, where $a \neq 0$. If $a \mid bc$ and $\gcd(a, b) = 1$, then $a \mid c$.*

Proof   Since $a \mid bc$, there is some integer q such that $bc = aq$. Since a and b are relatively prime, there exist integers s and t such that $1 = as + bt$. Thus

$$c = c \cdot 1 = c(as + bt) = a(cs) + (bc)t = a(cs) + (aq)t = a(cs + qt).$$

Since $cs + qt$ is an integer, $a \mid c$. ∎

Euclid's Lemma is of special interest when the integer a is a prime.

Corollary 11.14   *Let b and c be integers and p a prime. If $p \mid bc$, then either $p \mid b$ or $p \mid c$.*

Proof   If p divides b, then the corollary is proved. Suppose then that p does not divide b. Since the only positive integer divisors of p are 1 and p, it follows that $\gcd(p, b) = 1$. Thus, by Euclid's Lemma, $p \mid c$ and the proof is complete. ∎

The preceding corollary can be extended to the case when a prime p divides any product of integers.

Corollary 11.15   *Let $a_1, a_2, \ldots, a_n$, where $n \geq 2$, be integers and let p be a prime. If*

$$p \mid a_1 a_2 \cdots a_n,$$

then $p \mid a_i$ for some integer i $(1 \leq i \leq n)$.

Proof We proceed by induction. For $n = 2$, this is simply a restatement of Corollary 11.14. Assume then that if a prime p divides the product of k integers ($k \geq 2$), then p divides at least one of the integers. Now let $a_1, a_2, \ldots, a_{k+1}$ be $k + 1$ integers, where $p \mid a_1a_2 \cdots a_{k+1}$. We show that $p \mid a_i$ for some i ($1 \leq i \leq k + 1$). Let $b = a_1a_2 \cdots a_k$. So $p \mid ba_{k+1}$. By Corollary 11.14, either $p \mid b$ or $p \mid a_{k+1}$. If $p \mid a_{k+1}$, then the proof is complete. Otherwise, $p \mid b$, that is, $p \mid a_1a_2 \cdots a_k$. However, by the induction hypothesis, $p \mid a_i$ for some i ($1 \leq i \leq k$). In any case, $p \mid a_i$ for some i ($1 \leq i \leq k + 1$).

By the Principle of Mathematical Induction, if a prime p divides the product of any $n \geq 2$ integers, then p divides at least one of the integers. ∎

There is another useful fact concerning relatively primes integers. Once again, we will have occasion to use the result that whenever two integers a and b are relatively prime, then 1 is a linear combination of a and b.

Theorem 11.16 *Let $a, b, c \in \mathbf{Z}$, where a and b are relatively prime nonzero integers. If $a \mid c$ and $b \mid c$, then $ab \mid c$.*

Proof Since $a \mid c$ and $b \mid c$, there exist integers x and y such that $c = ax$ and $c = by$. Furthermore, since a and b are relatively prime, there exist integers s and t such that $1 = as + bt$. Multiplying by c and substituting, we obtain

$$c = c \cdot 1 = c(as + bt) = c(as) + c(bt) = (by)(as) + (ax)(bt) = ab(sy + xt).$$

Since $(sy + xt)$ is an integer, $ab \mid c$. ∎

By Theorem 11.16 then, if we wish to show that 12, say, divides some integer c, we need only show that $3 \mid c$ and $4 \mid c$ since $12 = 3 \cdot 4$ and 3 and 4 are relatively prime.

11.6 The Fundamental Theorem of Arithmetic

It is a basic divisibility fact that every integer can be expressed as a product of primes. This fact is made precise in a famous theorem in number theory. Its proof serves as one of the most interesting uses of the Strong Principle of Mathematical Induction.

Theorem 11.17 **(Fundamental Theorem of Arithmetic)** *Every integer $n \geq 2$ is either prime or can be expressed as a product of primes; that is,*

$$n = p_1p_2 \cdots p_m,$$

where $p_1, p_2, \ldots, p_m$ are primes. Furthermore, this factorization is unique except possibly for the order in which the factors occur.

Proof To show the existence of such a factorization, we employ the Strong Principle of Mathematical Induction. Since 2 is a prime, the statement is certainly true for $n = 2$.

For an integer $k \geq 2$, assume that every integer i, with $2 \leq i \leq k$, is either prime or can be expressed as a product of primes. We show that $k + 1$ is either prime or can be expressed as a product of primes. Of course, if $k + 1$ is prime, then there is nothing further to prove. We may assume, then, that $k + 1$ is composite. By Lemma 11.1, there exist integers a and b such that $k + 1 = ab$, where $2 \le a \le k$ and $2 \le b \le k$. Therefore, by the induction hypothesis, each of a and b is prime or can be expressed as a product of primes. In any case, $k + 1 = ab$ is a product of primes.

By the Strong Principle of Mathematical Induction, every integer $n \ge 2$ is either prime or can be expressed as a product of primes.

To prove that such a factorization is unique, we proceed by contradiction. Assume, to the contrary, that there is an integer $n \ge 2$ that can be expressed as a product of primes in two different ways, say

$$n = p_1 p_2 \cdots p_s = q_1 q_2 \cdots q_t,$$

where in each factorization, the primes are arranged in nondecreasing order; that is, $p_1 \le p_2 \le \cdots \le p_s$ and $q_1 \le q_2 \le \cdots \le q_t$. Since the factorizations are different, there must be a smallest positive integer r such that $p_r \ne q_r$. In other words, if $r \ge 2$, then $p_i = q_i$ for every i with $1 \le i \le r - 1$. After canceling, we have

$$p_r p_{r+1} \cdots p_s = q_r q_{r+1} \cdots q_t. \tag{11.6}$$

Consider the integer p_r. Either $s = r$ and the left side of (11.6) is exactly p_r or $s > r$ and $p_{r+1} p_{r+2} \cdots p_s$ is an integer that is the product of $s - r$ primes. In either case, $p_r \mid q_r q_{r+1} \cdots q_t$. Therefore, by Corollary 11.15, $p_r \mid q_j$ for some j with $r \le j \le t$. Because q_j is prime, $p_r = q_j$. Since $q_r \le q_j$, it follows that $q_r \le p_r$. By considering the integer q_r (instead of p_r), we can show that $p_r \le q_r$. Therefore, $p_r = q_r$. But this contradicts the fact that $p_r \ne q_r$. Hence, as claimed, every integer $n \ge 2$ has a unique factorization. ∎

An immediate consequence of Theorem 11.17 is stated next.

Corollary 11.18   *Every integer exceeding 1 has a prime factor.*

In fact, we can say a bit more.

Lemma 11.19   *If n is a composite number, then n has a prime factor p such that $p \le \sqrt{n}$.*

Proof   Since n is composite, we know that $n = ab$, where $1 < a < n$ and $1 < b < n$. Assume, without loss of generality, that $a \le b$. Then $a^2 \le ab = n$ and hence $a \le \sqrt{n}$. Since $a > 1$, we know that a has a prime factor, say p. Because a is a factor of n, it follows that p is a factor of n as well and $p \le a \le \sqrt{n}$. ∎

If an integer $n \ge 2$ is expressed as a product $q_1 q_2 \cdots q_m$ of primes, then the primes $q_1, q_2, \ldots, q_m$ need not be distinct. Consequently, we can group equal prime factors and express n in the form

$$n = p_1^{a_1} p_2^{a_2} \cdots p_k^{a_k},$$

where $p_1, p_2, \ldots, p_k$ are primes such that $p_1 < p_2 < \cdots < p_k$ and each exponent a_i is a positive integer. We call this the **canonical factorization** of n. From the Fundamental Theorem of Arithmetic, every integer $n \ge 2$ has a unique canonical factorization.

For example, the canonical factorizations of 12, 210 and 1000 are $12 = 2^2 3$, $210 = 2 \cdot 3 \cdot 5 \cdot 7$ and $1000 = 2^3 5^3$. Of course, it is relatively easy to determine whether a small positive integer is prime or composite and, if it is composite, to express it as a product of primes. We mention some tests for divisibility by certain integers. These may already be known to you.

1. **Divisibility by 2, 4 and other powers of 2**:
   An integer n is divisible by 2 if and only if n is even (or the last digit of n is even). In fact, n is divisible by 4 if and only if the two-digit number consisting of the last two digits of n is divisible by 4, the integer n is divisible by 8 if and only if the three-digit number consisting of the last three digits of n is divisible by 8 and so on. Therefore, the number 14220 is divisible by 4 since 20 is divisible by 4 but it is not divisible by 8 since 220 is not divisible by 8.

2. **Divisibility by 3 and 9**:
   An integer is divisible by 3 if and only if the sum of its digits is divisible by 3. Indeed, an integer is divisible by 9 if and only if the sum of its digits is divisible by 9. This procedure stops with 9, however; that is, it doesn't extend to 27. For example, the sum of the digits of 27 itself is not divisible by 27 but certainly 27 is divisible by 27. The sum of the digits of the integer 4278 is 21, which is divisible by 3 but not by 9. Consequently, 4278 is divisible by 3 but not by 9. Clearly, 4278 is divisible by 2 but it is not divisible by 4 since 78 is not divisible by 4. Thus, 4278 is divisible by 6 by Theorem 11.16.

3. **Divisibility by 5**:
   An integer is divisible by 5 if and only if it ends in 5 or 0; that is, an integer is divisible by 5 if and only if its last digit is divisible by 5.

4. **Divisibility by 11**:
   Start with the first digit of n and sum alternate digits (every other digit). Suppose that the resulting number is a. Then sum the remaining digits, obtaining b. Then n is divisible by 11 if and only if $a - b$ is divisible by 11. For example, consider the number 71929. Observe that $a = 7 + 9 + 9 = 25$, while $b = 1 + 2 = 3$. Since $a - b = 25 - 3 = 22$ is divisible by 11, the number 71929 is divisible by 11. In fact, $71929 = 11 \cdot 6539$. However, since $(6 + 3) - (5 + 9) = -5$ is not divisible by 11, the integer 6539 is not divisible by 11; that is, 71929 is not divisible by $11^2 = 121$.

Although there are tests for divisibility by other primes such as 7 (see Exercise 11.66) and 13, none of these are sufficiently practical to merit inclusion here. If we apply the tests listed above to the number $n = 471240$, then we find that n is divisible by 5, 8 (but not 16), 9 and 11. Indeed, $n = 5 \cdot 8 \cdot 9 \cdot 11 \cdot 119 = 5 \cdot 8 \cdot 9 \cdot 11 \cdot 7 \cdot 17 = 2^3 \cdot 3^2 \cdot 5 \cdot 7 \cdot 11 \cdot 17$.

We are now in a position to describe an infinite class of irrational numbers.

Theorem 11.20   *Let n be a positive integer. Then $\sqrt{n}$ is a rational number if and only if $\sqrt{n}$ is an integer.*

Proof   Certainly, if $\sqrt{n}$ is an integer, then $\sqrt{n}$ is rational. Hence we need only verify the converse. Assume, to the contrary, that there exists some positive integer n such that $\sqrt{n}$ is a rational number but $\sqrt{n}$ is not an integer. Hence $\sqrt{n} = a/b$ for some positive integers a and b. Furthermore, we may assume that a and b have no common factors, that is, $\gcd(a, b) = 1$. Since a/b is not an integer, $b \geq 2$. Therefore, $n = a^2/b^2$ and so $a^2 = nb^2$. By Corollary 11.18, b has a prime factor p. Thus $p \mid nb^2$ and so $p \mid a^2$. By Corollary 11.14, $p \mid a$. But then $p \mid a$ and $p \mid b$, which contradicts our assumption that $\gcd(a, b) = 1$.   ∎

A consequence of this theorem is the following.

Corollary 11.21   *If p is a prime, then $\sqrt{p}$ is irrational.*

Proof   Assume, to the contrary, that there exists some prime p for which $\sqrt{p}$ is rational. By Theorem 11.20, $\sqrt{p} = n$ for some integer $n \geq 2$. Then $p = n^2$. Because n^2 is composite, this is a contradiction. ∎

Although our remarks imply that there are infinitely many primes, we have not yet proved this. We do now. Since our goal is to prove that the number of primes is *not* finite, a proof by contradiction is the expected technique.

Theorem 11.22   *The number of primes is infinite.*

Proof   Assume, to the contrary, that the number of primes is finite. Let $P = \{p_1, p_2, \ldots, p_n\}$ be the set of all primes. Consider the integer $m = p_1 p_2 \cdots p_n + 1$. Clearly, $m \geq 2$. Since m has a prime factor and every prime belongs to P, there is a prime p_i $(1 \leq i \leq n)$ such that $p_i \mid m$. Hence $m = p_i k$ for some integer k. Let $\ell = p_1 p_2 \cdots p_{i-1} p_{i+1} \cdots p_n$. Then

$$1 = m - p_1 p_2 \cdots p_n = p_i k - p_i \ell = p_i (k - \ell).$$

Since $k - \ell$ is an integer, $p_i \mid 1$, which is impossible. ∎

Two primes p and q, where $p < q$, are called **twin primes** if $q = p + 2$. Necessarily, twin primes are odd. For example, 5, 7 and 11, 13 are twin primes. Although we have just verified that there are infinitely many primes, the number of twin primes is not known.

Conjecture 11.23   *There are infinitely many twin primes.*

11.7 Concepts Involving Sums of Divisors

For an integer $n \geq 2$, a positive integer a is called a **proper divisor** of n if $a \mid n$ and $a < n$. Thus the proper divisors of 6 are 1, 2 and 3, while the proper divisors of 28 are 1, 2, 4, 7 and 14. Note also that

$$1 + 2 + 3 = 6 \quad \text{and} \quad 1 + 2 + 4 + 7 + 14 = 28.$$

A positive integer $n \geq 2$ is called **perfect** if the sum of its proper divisors is n. Hence 6 and 28 are perfect integers—indeed, they are the two smallest perfect integers. The third smallest perfect integer is 496. The largest prime divisors of 6, 28 and 496 are 3, 7 and 31, respectively. Summing the integers from 1 to each of these primes yields possibly unexpected results:

$$1 + 2 + 3 = 6$$
$$1 + 2 + 3 + 4 + 5 + 6 + 7 = 28$$
$$1 + 2 + \cdots + 31 = 496.$$

The integers 6, 28 and 496 can also be expressed as

$$6 = 2^1(2^2 - 1), \ 28 = 2^2(2^3 - 1) \text{ and } 496 = 2^4(2^5 - 1).$$

In fact, Euclid, the famous geometer who lived more than 2000 years ago, showed that whenever $2^p - 1$ is a prime, then $2^{p-1}(2^p - 1)$ is a perfect integer. During the 18th century, the brilliant Swiss mathematician Leonhard Euler proved that every even perfect integer is of the form $2^{p-1}(2^p - 1)$ where $2^p - 1$ is a prime. Prime numbers of the form $2^p - 1$ are referred to as *Mersenne primes*. As of June 2010, 47 Mersenne primes were known and therefore 47 even perfect integers were known.

Much mystery surrounds perfect numbers. Are there any odd perfect numbers? No one knows. Are there infinitely many even perfect numbers? No one knows that either.

Let's consider the first few primes—the first seven to be exact: 2, 3, 5, 7, 11, 13, 17. The 1st, 2nd, 4th and 7th primes are 2, 3, 7 and 17, and the remaining primes (the 3rd, 5th and 6th) are 5, 11 and 13. If we add the integers 1, 2, 4 and 7, we get the same result as when we add 3, 5 and 6, namely, $1 + 2 + 4 + 7 = 3 + 5 + 6 = 14$. While this fact may not appear to be anything special, it is also a fact that summing the primes that correspond to these two sets of integers also gives the same result:

$$2 + 3 + 7 + 17 = 5 + 11 + 13 = 29.$$

While the sums of these primes are equal, it is impossible for the products of these primes, namely $2 \cdot 3 \cdot 7 \cdot 17$ and $5 \cdot 11 \cdot 13$, to be equal. This is a consequence, of course, of the Fundamental Theorem of Arithmetic. On the other hand, these products are surprisingly close since

$$2 \cdot 3 \cdot 7 \cdot 17 = 714 \text{ and } 5 \cdot 11 \cdot 13 = 715.$$

The serious baseball fan will recognize these numbers. For years, the number 714 stood as the record for home runs in a career. This record *was* held by Babe Ruth. However, this record was broken in 1974 when Hank Aaron hit his 715th home run.

Two consecutive integers n, $n + 1$ are called **Ruth–Aaron pairs** of integers if the sums of their prime divisors are equal. Thus 714 and 715 are a Ruth–Aaron pair, as are 5 and 6. Although such pairs of integers may appear to be rare, the famous Hungarian mathematician Paul Erdős proved that there are, in fact, infinitely many Ruth–Aaron pairs of integers.

EXERCISES FOR CHAPTER 11

Section 11.1: Divisibility Properties of Integers

11.1. Let $a, b, c, d \in \mathbf{Z}$ with $a, c \neq 0$. Prove that if $a \mid b$ and $c \mid d$, then $ac \mid (ad + bc)$.

11.2. Let $a, b \in \mathbf{Z}$ with $a \neq 0$. Prove that if $a \mid b$, then $a \mid (-b)$ and $(-a) \mid b$.

11.3. Let $a, b, c \in \mathbf{Z}$ with $a, c \neq 0$. Prove that if $ac \mid bc$, then $a \mid b$.

11.4. Prove that $3 \mid (n^3 - n)$ for every integer n.

11.5. Prove that if $n = k^3 + 1 \geq 3$, where $k \in \mathbf{Z}$, then n is not prime. [Hint: Recall that $k^3 + 1 = (k + 1)(k^2 - k + 1)$.]

11.6. Find all primes that are 1 less than a perfect cube.

11.7. Prove that $8 \mid (5^{2n} + 7)$ for every positive integer n.

11.8. Prove that $5 \mid (3^{3n+1} + 2^{n+1})$ for every positive integer n.

11.9. Prove that for every positive integer n, there exist n consecutive positive integers, each of which is composite. [Hint: Consider the numbers

$$2 + (n+1)!, 3 + (n+1)!, \ldots, n + (n+1)!, n+1 + (n+1)!.]$$

11.10. (a) Prove that $6 \mid (5n^3 + 7n)$ for every positive integer n.
  (b) Observe that $5 + 7 = 12$ is a multiple of 6. State and prove a generalization of the problem in (a).

11.11. Prove the following: Let d be a nonzero integer. If $a_1, a_2, \ldots, a_n$ and $x_1, x_2, \ldots, x_n$ are $2n \geq 2$ integers such that $d \mid a_i$ for all i ($1 \leq i \leq n$), then $d \mid \sum_{i=1}^{n} a_i x_i$.

11.12. Let p_n denote the nth prime and c_n the nth composite number. Thus $p_1 = 2$ and $p_2 = 3$, while $c_1 = 4$ and $c_2 = 6$. Of course, $p_n \neq c_n$ for all $n \in \mathbf{N}$. Determine all positive integers n such that $|p_n - c_n| = 1$.

11.13. (a) Suppose that there are k distinct positive integers that divide an odd positive integer n. How many distinct positive integers divide $2n$? How many divide $4n$?
  (b) Suppose that there are k distinct positive integers that divide a positive integer n that is not divisible by 3. How many distinct positive integers divide $3n$? How many divide $9n$?
  (c) State and answer a question suggested by the questions in (a) and (b).

11.14. For an integer $n \geq 2$, let m be the largest positive integer less than n such that $m \mid n$. Then $n = mk$ for some positive integer k. Prove that k is a prime.

Section 11.2: The Division Algorithm

11.15. Illustrate the Division Algorithm for:

  (a) $a = 17, b = 125$ (b) $a = -17, b = 125$
  (c) $a = 8, b = 96$ (d) $a = -8, b = 96$
  (e) $a = 22, b = -17$ (f) $a = -22, b = -17$
  (g) $a = 15, b = 0$ (h) $a = -15, b = 0$.

11.16. Give an example of a prime p of each of the forms:
  (a) $4k + 1$ (b) $4k + 3$ (c) $6k + 1$ (d) $6k + 5$.

11.17. Let p be an odd prime. Prove each of the following.

  (a) p is of the form $4k + 1$ or of the form $4k + 3$ for some nonnegative integer k.
  (b) $p \geq 5$ is of the form $6k + 1$ or of the form $6k + 5$ for some nonnegative integer k.

11.18. Show that, except for 2 and 5, every prime can be expressed as $10k + 1$, $10k + 3$, $10k + 7$ or $10k + 9$, where $k \in \mathbf{Z}$.

11.19. (a) Prove that if an integer n has the form $6q + 5$ for some $q \in \mathbf{Z}$, then n has the form $3k + 2$ for some $k \in \mathbf{Z}$.
  (b) Is the converse of (a) true?

11.20. Prove the General Form of the Division Algorithm (Corollary 11.5): For integers a and b with $a \neq 0$, there exist unique integers q and r such that $b = aq + r$ and $0 \leq r < |a|$.

11.21. Prove that the square of every odd integer is of the form $4k + 1$, where $k \in \mathbf{Z}$ (that is, for each odd integer $a \in \mathbf{Z}$, there exists $k \in \mathbf{Z}$ such that $a^2 = 4k + 1$).

11.22. (a) Prove that the square of every integer that is not a multiple of 3 is of the form $3k + 1$, where $k \in \mathbf{Z}$.
  (b) Prove that the square of no integer is of the form $3m - 1$, where $m \in \mathbf{Z}$.

11.23. Complete the following statement in a best possible manner and give a proof. (See Exercise 11.22(a).) The square of an integer that is not a multiple of 5 is either of the form _____ or _____.

11.24. (a) Prove that for every integer m, one of the integers $m, m + 4, m + 8, m + 12, m + 16$ is a multiple of 5.

(b) State and prove a generalization of the result in (a).

11.25. Prove that if $a_1, a_2, \ldots, a_n$ are $n \geq 2$ integers such that $a_i \equiv 1 \pmod 3$ for every integer i $(1 \leq i \leq n)$, then $a_1 a_2 \cdots a_n \equiv 1 \pmod 3$.

11.26. Let a, b and c be integers. Prove that if $abc \equiv 1 \pmod 3$, then an odd number of a, b and c are congruent to 1 modulo 3.

11.27. Prove or disprove: If a and b are odd integers, then $4 \mid (a - b)$ or $4 \mid (a + b)$.

11.28. Prove for every positive integer n that $n^2 + 1$ is not a multiple of 6.

11.29. It is known that there are infinitely many positive integers whose square is the sum of the squares of two positive integers. For example, $5^2 = 3^2 + 4^2$ and $13^2 = 5^2 + 12^2$.

(a) Prove that there are infinitely many positive integers whose square is the sum of the squares of three positive integers. For example, $59^2 = 50^2 + 30^2 + 9^2$.

(b) Prove that there are infinitely many positive integers whose square is the sum of the squares of four positive integers.

11.30. (a) Let $n \in \mathbf{N}$. Show that for every set S of n distinct integers, there is a nonempty subset T of S such that n divides the sum of the elements of T. [Hint: Let $S = \{a_1, a_2, \ldots, a_n\}$ and consider the subsets $S_k = \{a_1, a_2, \ldots, a_k\}$ for each k $(1 \leq k \leq n)$.]

(b) Is the word "distinct" necessary in (a)?

11.31. For an integer $n \geq 2$, let S_n be the set of all positive integers m for which n is the smallest positive integer such that when m is divided by n, the remainder 1 results.

(a) What does S_2 consist of?

(b) To which set S_n does 14 belong?

(c) To which set S_n does 16 belong?

(d) Prove or disprove: For each integer $n \geq 2$, the set S_n either contains infinitely many elements or is empty.

Section 11.3: Greatest Common Divisors

11.32. Give an example of a set S of four (distinct) positive integers such that the greatest common divisor of all six pairs of elements of S is 6.

11.33. Give an example of a set S of four (distinct) positive integers such that the greatest common divisors of all six pairs of elements of S are six distinct positive integers.

11.34. Prove for $a \in \mathbf{Z}$ and $n \in \mathbf{N}$ that $\gcd(a, a + n) \mid n$.

11.35. Let a and b be two integers, not both 0, where $\gcd(a, b) = d$. Prove for a positive integer k that $\gcd(ka, kb) = kd$.

11.36. For positive integers a, b and c, the greatest common divisor $\gcd(a, b, c)$ of a, b and c is the largest positive integer that divides all of a, b and c. Let $d = \gcd(a, b, c)$, $e = \gcd(a, b)$ and $f = \gcd(e, c)$. Prove that $d = f$.

Section 11.4: The Euclidean Algorithm

11.37. Use the Euclidean Algorithm to find the greatest common divisor for each of the following pairs of integers:

(a) 51 and 288   (b) 357 and 629   (c) 180 and 252.

11.38. Determine integers x and y such that (see Exercise 11.37):

(a) $\gcd(51, 288) = 51x + 288y$

(b) $\gcd(357, 629) = 357x + 629y$

(c) $\gcd(180, 252) = 180x + 252y$.

11.39. Let a and b be integers, not both 0. Show that there are infinitely many pairs s, t of integers such that $\gcd(a, b) = as + bt$.

11.40. Let $a, b \in \mathbf{Z}$, where not both a and b are 0, and let $d = \gcd(a, b)$. Show that an integer n is a linear combination of a and b if and only if $d \mid n$.

11.41. An integer $n > 1$ has the properties that $n \mid (35m + 26)$ and $n \mid (7m + 3)$ for some integer m. What is n?

11.42. Let $a, b \in \mathbf{Z}$, where not both a and b are 0. Prove that if $d = \gcd(a, b)$, $a = a_1 d$ and $b = b_1 d$, then $\gcd(a_1, b_1) = 1$.

11.43. Prove the following: Let $a, b, c, m, n \in \mathbf{Z}$, where $m, n \geq 2$. If $a \equiv b \pmod{m}$ and $a \equiv c \pmod{n}$ where $d = \gcd(m, n)$, then $b \equiv c \pmod{d}$.

11.44. In Exercise 11.36, it was shown for positive integers a, b and c that $\gcd(a, b, c) = \gcd(\gcd(a, b), c)$. Show that there are integers x, y and z such that $\gcd(a, b, c) = ax + by + cz$.

11.45. Suppose that the Euclidean Algorithm is being applied to determine $\gcd(a, b)$ for two positive integers a and b. If, at some stage of the algorithm, we arrive at a remainder r_i that is a prime number, then what conclusion can be made about $\gcd(a, b)$?

Section 11.5: Relatively Prime Integers

11.46. (a) Let $a, b, c \in \mathbf{Z}$ such that $a \neq 0$ and $a \mid bc$. Show that if $\gcd(a, b) \neq 1$, then a need not divide c.
     (b) Let $a, b, c \in \mathbf{Z}$ such that $a, b \neq 0$, $a \mid c$ and $b \mid c$. Show that if $\gcd(a, b) \neq 1$, then ab need not divide c.

11.47. Use Corollary 11.14 to prove that $\sqrt{3}$ is irrational.

11.48. Prove that if p and q are distinct primes, then $\sqrt{pq}$ is irrational.

11.49. Let p be a prime and let $n \in \mathbf{Z}$, where $n \geq 2$. Prove that $p^{1/n}$ is irrational.

11.50. Let $n \in \mathbf{N}$. Prove or disprove each of the following:

   (a) $2n$ and $4n + 3$ are relatively prime.
   (b) $2n + 1$ and $3n + 2$ are relatively prime.

11.51. (a) Prove that every two consecutive odd positive integers are relatively prime.
     (b) State and prove a generalization of the result in (a).

11.52. Prove that if $p \geq 2$ is an integer with the property that for every pair b, c of integers $p \mid bc$ implies that $p \mid b$ or $p \mid c$, then p is prime. (This result is related to Corollary 11.14.)

11.53. Prove that if p and q are primes with $p \geq q \geq 5$, then $24 \mid (p^2 - q^2)$.

11.54. A triple (a, b, c) of positive integers such that $a^2 + b^2 = c^2$ is called a **Pythagorean triple**. A Pythagorean triple (a, b, c) is called **primitive** if $\gcd(a, b) = 1$. (In this case, it also happens that $\gcd(a, c) = \gcd(b, c) = 1$.)

   (a) Prove that if (a, b, c) is a Pythagorean triple, then (an, bn, cn) is a Pythagorean triple for every $n \in \mathbf{N}$.
   (b) In Exercise 13 of Chapter 4, it was shown that if (a, b, c) is a Pythagorean triple, then $3 \mid ab$. Use this fact and Theorem 11.16 to show that $12 \mid ab$.
   (c) Prove that if (a, b, c) is a primitive Pythagorean triple, then a and b are of opposite parity.

11.55. Prove the following: Let $a, b, m, n \in \mathbf{Z}$, where $m, n \geq 2$. If $a \equiv b \pmod{m}$ and $a \equiv b \pmod{n}$, where $\gcd(m, n) = 1$, then $a \equiv b \pmod{mn}$.

11.56. Prove the following: Let $a, b, c, n \in \mathbf{Z}$, where $n \geq 2$. If $ac \equiv bc \pmod{n}$ and $\gcd(c, n) = 1$, then $a \equiv b \pmod{n}$.

11.57. For two integers a and b, not both 0, suppose that $d = \gcd(a, b)$. Then there exist integers x and y such that $d = ax + by$; that is, d is a linear combination of a and b. This implies that d is also a linear combination of x and y. Find a necessary and sufficient condition that $d = \gcd(x, y)$.

11.58. Suppose that the Euclidean Algorithm is being applied to determine $\gcd(a, b)$ for two positive integers a and b. If, at some stage of the algorithm, we arrive at remainders r_i and r_{i+1} for which $r_{i+1} = r_i - 1$, then what conclusion can be made about $\gcd(a, b)$?

11.59. (a) Let a and b be integers different from 0 where $d = \gcd(a, b)$ and let $c \in \mathbf{Z}$. Prove that if $a \mid c$ and $b \mid c$, then $ab \mid cd$.
   (b) Show that Theorem 11.16 follows as a corollary to the result in (a).

11.60. Prove that there are infinitely many positive integers n such that each of $n, n + 1$ and $n + 2$ can be expressed as the sum of the squares of two nonnegative integers. (For example, observe that $8 = 2^2 + 2^2$, $9 = 3^2 + 0^2$, $10 = 3^2 + 1^2$ and $80 = 8^2 + 4^2$, $81 = 9^2 + 0^2$ and $82 = 9^2 + 1^2$ and that $3 = 2 + 1$ and $9 = 8 + 1$.)

11.61. (a) Give an example of integers $m, n \geq 5$ such that $x \cap y \neq \emptyset$ for each $x \in \mathbf{Z}_m$ and $y \in \mathbf{Z}_n$.
   (b) State a conjecture that provides conditions under which integers $m, n \geq 2$ have the property that $x \cap y \neq \emptyset$ for each $x \in \mathbf{Z}_m$ and $y \in \mathbf{Z}_n$.

Section 11.6: The Fundamental Theorem of Arithmetic

11.62. Find the smallest prime factor of each integer below:
   (a) 539   (b) 1575   (c) 529   (d) 1601

11.63. Find the canonical factorization of each of the following integers:
   (a) 4725   (b) 9702   (c) 180625

11.64. Prove each of the following:

   (a) Every prime of the form $3n + 1$ is also of the form $6k + 1$.
   (b) If n is a positive integer of the form $3k + 2$, then n has a prime factor of this form as well.

11.65. (a) Express each of the integers 4278 and 71929 as a product of primes.
   (b) What is $\gcd(4278, 71929)$?

11.66. Consider the periodic sequence $1, 3, 2, -1, -3, -2, 1, 3, 2, -1, -3, -2, \ldots$ which we write in reverse order:

$$\ldots, -2 - 3, -1, 2, 3, 1, -2 - 3, -1, 2, 3, 1.$$

Next, consider the 8-digit positive integer $n = a_7a_6a_5a_4a_3a_2a_1a_0$ where each a_i is a digit. It turns out that $7 \mid n$ if and only if

$$3 \cdot a_7 + 1 \cdot a_6 + (-2) \cdot a_5 + (-3) \cdot a_4 + (-1) \cdot a_3 + 2 \cdot a_2 + 3 \cdot a_1 + 1 \cdot a_0$$

is a multiple of 7. Use this to determine which of the following are multiples of 7:
   (a) 56   (b) 821,317   (c) 31,142,524.

11.67. In the proof of Theorem 11.22, it was proved that there are infinitely many primes by assuming that there were finitely many primes, say $p_1, p_2, \ldots, p_n$, where $p_1 < p_2 < \cdots < p_n$. The number $m = p_1p_2 \cdots p_n + 1$ was then considered to obtain a contradiction. Show that an alternative proof of Theorem 11.22 can be obtained by considering $p_n! + 1$ instead of m.

11.68. Determine a necessary and sufficient condition that $p_1^{a_1} p_2^{a_2} \cdots p_k^{a_k}$ is the canonical factorization of the square of some integer $n \geq 2$.

11.69. For two integers $m, n \geq 2$, let $p_1, p_2, \ldots, p_r$ be those distinct primes such that each p_i $(1 \leq i \leq r)$ divides at least one of m and n. Then m and n can be expressed as $m = p_1^{a_1} p_2^{a_2} \cdots p_r^{a_r}$ and $n = p_1^{b_1} p_2^{b_2} \cdots p_r^{b_r}$ where the integers a_i and b_i $(1 \leq i \leq r)$ are nonnegative. Let $c_i = \min(a_i, b_i)$ for $1 \leq i \leq r$. Prove that $\gcd(m, n) = p_1^{c_1} p_2^{c_2} \cdots p_r^{c_r}$.

Section 11.7: Concepts Involving Sums of Divisors

11.70. Let k be a positive integer.

    (a) Prove that if $2^k - 1$ is prime, then k is prime.
    (b) Prove that if $2^k - 1$ is prime, then $n = 2^{k-1}(2^k - 1)$ is perfect.

11.71. For a real number r, the floor $\lfloor r \rfloor$ of r is the greatest integer less than or equal to r. The greatest number of distinct positive integers whose sum is 5 is 2 ($5 = 1 + 4 = 2 + 3$), while the greatest number of distinct positive integers whose sum is 8 is 3 ($8 = 1 + 2 + 5 = 1 + 3 + 4$). Prove that the maximum number of distinct positive integers whose sum is the positive integer n is $\lfloor (\sqrt{1 + 8n} - 1)/2 \rfloor$.

ADDITIONAL EXERCISES FOR CHAPTER 11

11.72. Evaluate the proposed solution of the following problem.

Prove or disprove the following statement:
There do not exist three integers n, $n + 2$ and $n + 4$, all of which are primes.

Solution   This statement is true.

Proof   Assume, to the contrary, that there exist three integers n, $n + 2$ and $n + 4$, all of which are primes. We can write n as $3q$, $3q + 1$ or $3q + 2$, where $q \in \mathbf{Z}$. We consider these three cases.

Case 1. $n = 3q$. Then $3 \mid n$ and so n is not prime. This is a contradiction.

Case 2. $n = 3q + 1$. Then $n + 2 = 3q + 3 = 3(q + 1)$. Since $q + 1$ is an integer, $3 \mid (n + 2)$ and so $n + 2$ is not prime. Again, we have a contradiction.

Case 3. $n = 3q + 2$. Hence we have $n + 4 = 3q + 6 = 3(q + 2)$. Since $q + 2$ is an integer, $3 \mid (n + 4)$ and so $n + 4$ is not prime. This produces a contradiction. ∎

11.73. An integer $a \geq 2$ is defined to be **lucky** if $f(n) = n^2 - n + a$ is prime for every integer n with $1 \leq n \leq a - 1$. It is known that (1) 41 is lucky and (2) only nine other integers $a \geq 2$ are lucky.

    (a) Prove that if a is a lucky integer, then a is prime.
    (b) Give an example of three other lucky integers.
    (c) If a is a lucky integer, what can be said about $f(a)$?

11.74. Prove that $\log_2 3$ is irrational.

11.75. State and prove a more general result than that given in Exercise 11.74.

11.76. Given below is an incomplete result with an incomplete proof. This result is intended to determine all twin primes (primes of the form p and $q = p + 2$) such that $pq - 2$ is also prime.

Result   Let p and $q = p + 2$ be two primes. Then $pq - 2$ is prime if and only if (complete this sentence).

Proof   Let p and $q = p + 2$ be two primes such that $pq - 2$ is also prime. Since p and $p + 2$ are both primes, it follows that p is odd. By the Division Algorithm, we can write $p = 3k + r$, where $k \in \mathbf{Z}$ and $0 \leq r \leq 2$. Since p is an odd prime, $k \geq 1$. We consider three cases for p, depending on the value of r.

Case 1. $p = 3k$. Therefore, $p = $, $q = $ and $pq - 2 = $.

Case 2. $p = 3k + 1$. Hence $q = 3k + 3$. Since $k \geq 1$, it follows that $q = 3(k + 1)$. Therefore,

Case 3. $p = 3k + 2$. Then $q = 3k + 4$. Thus ∎

11.77. Exercise 11.76 should suggest another exercise to you. State a result related to Exercise 11.76 and give a proof of this result.

11.78. Assume that each positive rational number is expressed as m/n, where $m, n \in \mathbf{N}$ and m and n are relatively prime. A function $f : \mathbf{Q}^+ \to \mathbf{N}$ is defined by $f(m/n) = 2^m 3^n$.

   (a) Prove that f is one-to-one.
   (b) If you discussed the Schröder–Bernstein Theorem (Theorem 10.20) in Chapter 10, what information about $\mathbf{Q}^+$ can you obtain from part (a)?

11.79. We have seen in Section 11.7 that there is a partition of the set of the first seven primes into two subsets such that the sums of the elements in these two subsets are equal. Show that there is no such partition of the set of the first eight primes but there is such a partition of the set of the first nine primes.

11.80. (a) Show that $5039 = 5040 - 1$ is prime, while $5041 = 5040 + 1$ is not prime.
   (b) Show that, except for 5039, there is no prime between $5033 = 5040 - 7$ and $5047 = 5040 + 7$.

11.81. Let $a_0, a_1, a_2, \ldots$ be a sequence of positive integers for which

$$\text{(1) } a_0 = 1, \text{ (2) } a_{2n+1} = a_n \text{ for } n \geq 0 \text{ and (3) } a_{2n+2} = a_n + a_{n+1} \text{ for } n \geq 0.$$

Prove that a_n and a_{n+1} are relatively prime for every nonnegative integer n.

11.82. Every positive integer n can be expressed as $n = a_k a_{k-1} \cdots a_2 a_1 a_0$, that is,

$$n = a_k \cdot 10^k + a_{k-1} \cdot 10^{k-1} + \cdots + a_2 \cdot 10^2 + a_1 \cdot 10 + a_0.$$

It was mentioned in Section 11.6 that $9 \mid n$ if and only if $9 \mid (a_k + a_{k-1} + \cdots + a_2 + a_1 + a_0)$. For example, for the integer $n = 32{,}751$, $9 \mid (3 + 2 + 7 + 5 + 1)$ and so $9 \mid 32{,}751$. Verify this by using the fact that $10 = 9 + 1$ and for a positive integer r that $10^r = (9 + 1)^r = 9s + 1$ for some integer s.

11.83. Let A be the set of 2-element subsets of $\mathbf{N}$ and B the set of 3-element subsets of $\mathbf{N}$. Let $f : A \to B$ and $g : B \to A$ be functions defined by

$$f(\{i, j\}) = \{i, j, i + j\} \text{ and } g(\{i, j, k\}) = \{2^i, 3^j 5^k\},$$

where $i < j < k$. With the aid of the functions f and g and possibly the Schröder–Bernstein Theorem (Theorem 10.20 in Chapter 10), which of the following are we able to conclude?
   (a) $|A| \leq |B|$ (b) $|B| \leq |A|$ (c) $|A| = |B|$ (d) Nothing.

11.84. Let $p_1, p_2, p_3, \ldots$ be the primes, where $2 = p_1 < p_2 < p_3 < \cdots$. Let A be a denumerable set, where $A = \{a_1, a_2, a_3, \ldots\}$. For any integer $n \geq 2$, let A^n denote the Cartesian product of n copies of A; that is, A^n is the set of ordered n-tuples of elements of A. Define a function $f : A^n \to \mathbf{N}$ by
$f((a_{i_1}, a_{i_2}, \ldots, a_{i_n})) = p_1^{i_1} p_2^{i_2} \cdots p_n^{i_n}$.

   (a) Prove that f is injective.
   (b) Use (a) to show that A^n and A are numerically equivalent.
   (c) For every two denumerable sets A and B and every two integers $n, m \geq 2$, show that A^n and B^m are numerically equivalent.

11.85. Let $p_1, p_2, \ldots, p_{n+1}$ denote the first $n + 1$ primes. Suppose that $\{U, V\}$ is a partition of the set $S = \{p_1, p_2, \ldots, p_n\}$, where $U = \{q_1, q_2, \ldots, q_s\}$ and $V = \{r_1, r_2, \ldots, r_t\}$. Prove that if $M = q_1 q_2 \cdots q_s + r_1 r_2 \cdots r_t < p_{n+1}^2$, then M is a prime.

Proofs in Calculus

$\mathbf{Y}$our introduction to calculus most certainly included a study of limits—both limits of sequences (including infinite series) and limits of functions (including continuity and differentiability). While we learned methods for computing limits in these areas, the methods presented were most likely based on facts that were not carefully verified. In this chapter, some of the proofs of fundamental results from calculus will be presented. The proofs that occur in calculus are considerably different than any of those we have seen thus far. The functions encountered in calculus are real-valued functions defined on sets of real numbers. That is, each function that we study in calculus is of the type $f : X \to \mathbf{R}$, where $X \subseteq \mathbf{R}$. In the study of limits, we are often interested in such functions having the property that either (1) $X = \mathbf{N}$ and increasing values in the domain $\mathbf{N}$ result in functional values approaching some real number L or (2) the function is defined for all real numbers near some specified real number a and values approaching a result in functional values approaching some real number L. We begin with (1), where $X = \mathbf{N}$.

12.1 Limits of Sequences

A **sequence** (of real numbers) is a real-valued function defined on the set of natural numbers; that is, a **sequence** is a function $f : \mathbf{N} \to \mathbf{R}$. If $f(n) = a_n$ for each $n \in \mathbf{N}$, then $f = \{(1, a_1), (2, a_2), (3, a_3), \ldots\}$. Since only the numbers $a_1, a_2, a_3, \ldots$ are relevant in f, this sequence is often denoted by $a_1, a_2, a_3, \ldots$ or by $\{a_n\}$. The numbers a_1, a_2, a_3, etc. are called the **terms** of the sequence $\{a_n\}$, with a_1 being the first term, a_2 the second term, etc. Thus a_n is the nth term of the sequence. Hence $\left\{\dfrac{1}{n}\right\}$ is the sequence $1, 1/2, 1/3, \ldots$; while $\left\{\dfrac{n}{2n+1}\right\}$ is the sequence $1/3, 2/5, 3/7, \ldots$. In these two examples, the nth term of a sequence is given and, from this, we can easily find the first few terms and, in fact, any particular term. On the other hand, finding the nth term of a sequence whose first few terms are given can be challenging. For example, the nth term of the sequence

$$\frac{1}{2}, \frac{1}{4}, \frac{1}{6}, \ldots$$

is $1/2n$; the nth term of the sequence

$$1 + \frac{1}{2}, \; 1 + \frac{1}{4}, \; 1 + \frac{1}{8}, \; \cdots$$

is $1 + 1/2^n$; the nth term of the sequence

$$1, \; \frac{3}{5}, \; \frac{1}{2}, \; \frac{5}{11}, \; \frac{3}{7}, \; \frac{7}{17}, \; \cdots$$

is $(n + 1)/(3n - 1)$; the nth term of the sequence

$$1, \; -1, \; 1, \; -1, \; 1, \; -1, \; \ldots$$

is $(-1)^{n+1}$; while the nth term of the sequence $1, 4, 9, 16, \cdots$ is n^2.

For the sequence $\left\{ \dfrac{1}{n} \right\}$, the larger the integer n, the closer $1/n$ is to 0; and for the sequence $\left\{ \dfrac{n}{2n + 1} \right\}$, the larger the integer n, the closer $n/(2n + 1)$ is to $1/2$. On the other hand, for the sequence $\{n^2\}$, as the integer n become larger, n^2 becomes increasingly large and does not approach any real number.

When we discuss how close two numbers are to each other, we are actually considering the distance between them. We saw in Chapter 8 that the **distance** between two real numbers a and b is defined as $|a - b|$. Recall that the absolute value of a real number x is

$$|x| = \begin{cases} x & \text{if } x \geq 0 \\ -x & \text{if } x < 0. \end{cases}$$

Hence the distance between $a = 3$ and $b = 5$ is $|3 - 5| = |5 - 3| = 2$; while the distance between 0 and $1/n$, where $n \in \mathbf{N}$, is $\left| 0 - \dfrac{1}{n} \right| = \left| \dfrac{1}{n} - 0 \right| = \dfrac{1}{n}$.

For a fixed positive real number r, the inequality $|x| < r$ is equivalent to the inequalities $-r < x < r$. Hence $|x| < 3$ is equivalent to $-3 < x < 3$, while $|x - 2| < 4$ is equivalent to $-4 < x - 2 < 4$. Adding 2 throughout these inequalities, we obtain $-4 + 2 < (x - 2) + 2 < 4 + 2$ and so $-2 < x < 6$. We have seen in Exercise 4.30 and Theorem 4.17 in Chapter 4 that for real numbers x and y,

$$|xy| = |x||y| \quad \text{and} \quad |x + y| \leq |x| + |y|.$$

Both of these properties are useful throughout calculus.

We mentioned that for some sequences $\{a_n\}$, there is a real number L (or at least there appears to be a real number L) such that the larger the integer n becomes, the closer a_n is to L. We have now arrived at an important and fundamental idea in the study of sequences and are prepared to introduce a concept that describes this situation.

A sequence $\{a_n\}$ of real numbers is said to *converge* to a real number L if the larger the integer n, the closer a_n is to L. Since the words *larger* and *closer* are vague and consequently are open to interpretation, we need to make these words considerably more precise.

What we want to say then is that we can make a_n as close to L as we wish (that is, we can make $|a_n - L|$ as small as we wish) provided that n is large enough. Let ϵ (the Greek letter epsilon) denote how small we want $|a_n - L|$ to be; that is, we want $|a_n - L| < \epsilon$ by choosing n large enough. This is equivalent to $-\epsilon < a_n - L < \epsilon$, that

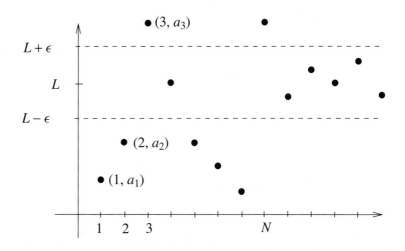

Figure 12.1   A sequence $\{a_n\}$ that converges to L

is, $L - \epsilon < a_n < L + \epsilon$. Hence we require that a_n be a number in the open interval $(L - \epsilon, L + \epsilon)$ when n is large enough. Now we need to know what we mean by "large enough." What we mean by this is that there is some positive integer N such that if n is an integer greater than N, then $a_n \in (L - \epsilon, L + \epsilon)$. If such a positive integer N can be found for every positive number ϵ, regardless of how small ϵ might be, then we say that $\{a_n\}$ converges to L. This is illustrated in Figure 12.1.

Formally then, a sequence $\{a_n\}$ of real numbers is said to **converge** to the real number L if for every real number $\epsilon > 0$, there exists a positive integer N such that if n is an integer with $n > N$, then $|a_n - L| < \epsilon$. As we indicated, the number ϵ is a measure of how close the terms a_n are required to be to the number L and N indicates a position in the sequence beyond which the required condition is satisfied. If a sequence $\{a_n\}$ converges to L, then L is referred to as the **limit** of $\{a_n\}$ and we write $\lim_{n \to \infty} a_n = L$. If a sequence does not converge, it is said to **diverge**. Consequently, if a sequence $\{a_n\}$ diverges, then there is *no* real number L such that $\lim_{n \to \infty} a_n = L$.

Before looking at a few examples, we introduce some useful notation. For a real number x, recall that $\lceil x \rceil$ denotes the smallest integer greater than or equal to x. The integer $\lceil x \rceil$ is often called the **ceiling** of x. Consequently, $\lceil 8/3 \rceil = 3$, $\left\lceil \sqrt{2} \right\rceil = 2$, $\lceil -1.6 \rceil = -1$ and $\lceil 5 \rceil = 5$. By the definition of $\lceil x \rceil$, it follows that if x is an integer, then $\lceil x \rceil = x$; while if x is not an integer, then $\lceil x \rceil > x$. In particular, if n is an integer such that $n > \lceil x \rceil$, then $n > x$.

We now show how the definition of convergent sequence is used to prove that a sequence converges to some number.

Result to Prove   The sequence $\left\{ \dfrac{1}{n} \right\}$ converges to 0.

PROOF STRATEGY   Here we are required to show, for a given real number $\epsilon > 0$, that there is a positive integer N such that if $n > N$, then $\left| \dfrac{1}{n} - 0 \right| = \left| \dfrac{1}{n} \right| = \dfrac{1}{n} < \epsilon$. The inequality $\dfrac{1}{n} < \epsilon$ is equivalent to $n > 1/\epsilon$. Hence if we let $N = \lceil 1/\epsilon \rceil$ and take n to be an integer greater than N, then $n > \dfrac{1}{\epsilon}$. We can now present a formal proof.   ◆

Result 12.1   *The sequence* $\left\{ \dfrac{1}{n} \right\}$ *converges to* 0.

Proof   Let $\epsilon > 0$. Choose $N = \lceil 1/\epsilon \rceil$ and let n be any integer such that $n > N$. Thus $n > 1/\epsilon$ and so $\left| \dfrac{1}{n} - 0 \right| = \dfrac{1}{n} < \epsilon$.   ∎

<u>PROOF ANALYSIS</u>   Although the proof of Result 12.1 is quite short, the real work in constructing the proof occurred in the proof strategy (our "scratch paper" work) that preceded the proof, but which is not part of the proof. This explains why we chose N as we did and why this choice of N was successful. In the proof of Result 12.1, we chose $N = \lceil 1/\epsilon \rceil$ and showed that with this value of N, every integer n with $n > N$ yields $\left| \dfrac{1}{n} - 0 \right| < \epsilon$, which, of course, was our goal. There is nothing unique about this choice of N, however. Indeed, we could have chosen N to be *any* integer greater than $\lceil 1/\epsilon \rceil$ or, equivalently, any integer greater than $1/\epsilon$ and reached the desired conclusion as well. We could not, however, choose N to be an integer smaller than $\lceil 1/\epsilon \rceil$. We cannot in general choose $N = 1/\epsilon$ since there is no guarantee that N is an integer.   ◆

We now consider another illustration of a convergent sequence.

Result to Prove   The sequence $\left\{ 3 + \dfrac{2}{n^2} \right\}$ converges to 3.

<u>PROOF STRATEGY</u>   Here we are required to show, for a given $\epsilon > 0$, that there exists a positive integer N such that if $n > N$, then

$$\left| \left(3 + \frac{2}{n^2}\right) - 3 \right| = \left| \frac{2}{n^2} \right| = \frac{2}{n^2} < \epsilon.$$

The inequality $\dfrac{2}{n^2} < \epsilon$ is equivalent to $\dfrac{n^2}{2} > \dfrac{1}{\epsilon}$ and $n > \sqrt{2/\epsilon}$. Therefore, if we let $N = \left\lceil \sqrt{2/\epsilon} \right\rceil$ and choose n to be an integer greater than N, then $n > \sqrt{2/\epsilon}$. We can now give a proof.   ◆

Result 12.2   *The sequence* $\left\{ 3 + \dfrac{2}{n^2} \right\}$ *converges to* 3.

Proof   Let $\epsilon > 0$. Choose $N = \left\lceil \sqrt{2/\epsilon} \right\rceil$ and let n be any integer such that $n > N$. Thus $n > \sqrt{2/\epsilon}$ and $n^2 > 2/\epsilon$. So $\dfrac{1}{n^2} < \dfrac{\epsilon}{2}$ and $\dfrac{2}{n^2} < \epsilon$. Therefore,

$$\left| \left(3 + \frac{2}{n^2}\right) - 3 \right| = \left| \frac{2}{n^2} \right| = \frac{2}{n^2} < \epsilon.$$ ∎

We now consider a somewhat more complicated example.

Result to Prove   The sequence $\left\{\dfrac{n}{2n+1}\right\}$ converges to $\dfrac{1}{2}$.

<u>**PROOF STRATEGY**</u>   Observe that

$$\left|\frac{n}{2n+1}-\frac{1}{2}\right|=\left|\frac{2n-2n-1}{2(2n+1)}\right|=\left|-\frac{1}{4n+2}\right|=\frac{1}{4n+2}.$$

The inequality $\dfrac{1}{4n+2}<\epsilon$ is equivalent to $4n+2>1/\epsilon$, which, in turn, is equivalent to $n>\dfrac{1}{4\epsilon}-\dfrac{1}{2}$. It may appear that the proper choice for N is $\left\lceil\dfrac{1}{4\epsilon}-\dfrac{1}{2}\right\rceil$; but if $\epsilon\geq 1/2$, then $N=0$, which is not acceptable since **N** is required to be a positive integer. However, notice that $\dfrac{1}{4\epsilon}>\dfrac{1}{4\epsilon}-\dfrac{1}{2}$. So if $n>\dfrac{1}{4\epsilon}$, then $n>\dfrac{1}{4\epsilon}-\dfrac{1}{2}$ as well. Hence if we choose $N=\lceil 1/4\epsilon\rceil$, then we can obtain the desired inequality.   ♦

Result 12.3   *The sequence* $\left\{\dfrac{n}{2n+1}\right\}$ *converges to* $\dfrac{1}{2}$.

Proof   Let $\epsilon>0$ be given. Choose $N=\lceil 1/4\epsilon\rceil$ and let $n>N$. Then $n>\dfrac{1}{4\epsilon}>\dfrac{1}{4\epsilon}-\dfrac{1}{2}$ and so $4n>\dfrac{1}{\epsilon}-2$ and $4n+2>1/\epsilon$. Hence $\dfrac{1}{4n+2}<\epsilon$. Thus

$$\left|\frac{n}{2n+1}-\frac{1}{2}\right|=\left|\frac{2n-2n-1}{2(2n+1)}\right|=\left|-\frac{1}{4n+2}\right|=\frac{1}{4n+2}<\epsilon.$$ ∎

Again, the choice made for N in the proof of Result 12.3 is not unique. We could choose N to be any *positive* integer greater than $\dfrac{1}{4\epsilon}$.

We mentioned that a sequence $\{a_n\}$ is said to diverge if it does not converge. To prove that a sequence $\{a_n\}$ diverges, a proof by contradiction would be anticipated. We would begin such a proof by assuming, to the contrary, that $\{a_n\}$ converges, say to some real number L. We know that for every $\epsilon>0$, there is a positive integer N such that if $n>N$, then $|a_n-L|<\epsilon$. If we could show for even one choice of $\epsilon>0$ that no such positive integer N exists, then we would have produced a contradiction and proved the desired result. Let's see how this works in two examples.

Result to Prove   The sequence $\left\{(-1)^{n+1}\right\}$ is divergent.

<u>**PROOF STRATEGY**</u>   In a proof by contradiction, we begin by assuming that $\left\{(-1)^{n+1}\right\}$ converges, to the limit L say. Our goal is to show that there is some value of $\epsilon>0$ for which there is no positive integer N that satisfies the requirement. We choose $\epsilon=1$. According to the definition of what it means for $\left\{(-1)^{n+1}\right\}$ to converge to L, there must exist a positive integer N such that if n is an integer with $n>N$, then $\left|(-1)^{n+1}-L\right|<\epsilon=1$. Let k be an odd integer such that $k>N$. Then

$$\left|(-1)^{k+1}-L\right|=|1-L|=|L-1|<1.$$

Therefore, $-1 < L - 1 < 1$ and $0 < L < 2$. Now let ℓ be an even integer such that $\ell > N$. Then

$$\left|(-1)^{\ell+1} - L\right| = |-1 - L| = |L + 1| < 1.$$

Thus, $-1 < L + 1 < 1$ and $-2 < L < 0$. So $L < 0 < L$, which, of course, is impossible. We now repeat what we have just said in a formal proof.   ♦

Result 12.4   *The sequence $\left\{(-1)^{n+1}\right\}$ is divergent.*

Proof   Assume, to the contrary, that the sequence $\left\{(-1)^{n+1}\right\}$ converges. Then $\lim\limits_{n\to\infty} (-1)^{n+1} = L$ for some real number L. Let $\epsilon = 1$. Then there exists a positive integer N such that if $n > N$, then $\left|(-1)^{n+1} - L\right| < \epsilon = 1$. Let k be an odd integer such that $k > N$. Then

$$\left|(-1)^{k+1} - L\right| = |1 - L| = |L - 1| < 1.$$

Therefore, $-1 < L - 1 < 1$ and $0 < L < 2$. Next, let ℓ be an even integer such that $\ell > N$. Then

$$\left|(-1)^{\ell+1} - L\right| = |-1 - L| = |L + 1| = |1 + L| < 1.$$

So $-1 < L + 1 < 1$ and $-2 < L < 0$. Therefore, $L < 0 < L$, which is a contradiction. ∎

PROOF ANALYSIS   One question that now occurs is how we knew to choose $\epsilon = 1$. If ϵ denotes an arbitrary positive integer, then both inequalities $|L - 1| < \epsilon$ and $|L + 1| < \epsilon$ must be satisfied, but these result in the inequalities

$$1 - \epsilon < L < 1 + \epsilon \quad \text{and} \quad -1 - \epsilon < L < -1 + \epsilon.$$

In particular, $1 - \epsilon < L < -1 + \epsilon$ and so $1 - \epsilon < -1 + \epsilon$. This is only possible if $2\epsilon > 2$ or $\epsilon > 1$. Hence if we choose ϵ to be any number such that $0 < \epsilon \leq 1$, a contradiction will be produced. We decided to choose $\epsilon = 1$.   ♦

Result to Prove   The sequence $\left\{(-1)^{n+1} \frac{n}{n+1}\right\}$ is divergent.

PROOF STRATEGY   As expected, we will attempt a proof by contradiction and assume that $\left\{(-1)^{n+1} \frac{n}{n+1}\right\}$ is a convergent sequence, with limit L say. For $\epsilon > 0$, there is a positive integer N then such that

$$\left|(-1)^{n+1} \frac{n}{n+1} - L\right| < \epsilon$$

for each integer n such that $n > N$. There are some useful observations.

First, if $n > N$ and n is odd, then

$$\left|\frac{n}{n+1} - L\right| < \epsilon \quad \text{and so} \quad -\epsilon < \frac{n}{n+1} - L < \epsilon.$$

Hence

$$L - \epsilon < \frac{n}{n+1} < L + \epsilon.$$

Second, if $n > N$ and n is even, then

$$\left| -\frac{n}{n+1} - L \right| < \epsilon \text{ and so } -\epsilon < -\frac{n}{n+1} - L < \epsilon.$$

Hence

$$L - \epsilon < -\frac{n}{n+1} < L + \epsilon.$$

Also, since $n > 1$, we have $n + n > n + 1$ and so $2n > n + 1$. Hence $\frac{n}{n+1} > \frac{1}{2}$. Depending on whether $L = 0$, $L > 0$ or $L < 0$, we are faced with the decision as to how to choose ϵ in each case to produce a contradiction. ◆

Result 12.5 *The sequence $\left\{ (-1)^{n+1} \frac{n}{n+1} \right\}$ is divergent.*

Proof Assume, to the contrary, that $\left\{ (-1)^{n+1} \frac{n}{n+1} \right\}$ converges. Then $\lim\limits_{n \to \infty} (-1)^{n+1} \frac{n}{n+1} = L$ for some real number L. We consider three cases, depending on whether $L = 0$, $L > 0$ or $L < 0$.

Case 1. $L = 0$. Let $\epsilon = \frac{1}{2}$. Then there exists a positive integer N such that if $n > N$, then $\left| (-1)^{n+1} \frac{n}{n+1} - 0 \right| < \frac{1}{2}$ or $\frac{n}{n+1} < \frac{1}{2}$. Then $2n < n + 1$ and so $n < 1$, which is a contradiction.

Case 2. $L > 0$. Let $\epsilon = \frac{L}{2}$. Then there exists a positive integer N such that if $n > N$, then $\left| (-1)^{n+1} \frac{n}{n+1} - L \right| < \frac{L}{2}$. Let n be an even integer such that $n > N$. Then

$$-\frac{L}{2} < -\frac{n}{n+1} - L < \frac{L}{2}.$$

Hence $\frac{L}{2} < -\frac{n}{n+1} < \frac{3L}{2}$, which is a contradiction.

Case 3. $L < 0$. Let $\epsilon = -\frac{L}{2}$. Then there exists a positive integer N such that if $n > N$, then $\left| (-1)^{n+1} \frac{n}{n+1} - L \right| < -\frac{L}{2}$. Let n be an odd integer such that $n > N$. Then

$$\frac{L}{2} < \frac{n}{n+1} - L < -\frac{L}{2}$$

and so $\frac{3L}{2} < \frac{n}{n+1} < \frac{L}{2}$. This is a contradiction. ■

A sequence $\{a_n\}$ may diverge because as n becomes larger, a_n becomes larger and eventually exceeds any given real number. If a sequence has this property, then $\{a_n\}$ is said to diverge to infinity. More formally, a sequence $\{a_n\}$ **diverges to infinity**, written $\lim\limits_{n \to \infty} a_n = \infty$, if for every positive number M, there exists a positive integer N such that if n is an integer such that $n > N$, then $a_n > M$. The sequence $\{(-1)^{n+1}\}$ encountered in Result 12.4, although divergent, does not diverge to infinity. However, the sequence $\{n^2 + \frac{1}{n}\}$ *does* diverge to infinity.

Result to Prove $\lim\limits_{n\to\infty} \left(n^2 + \dfrac{1}{n}\right) = \infty.$

PROOF STRATEGY   For a given positive number M, we are required to show the existence of a positive integer N such that if $n > N$, then $n^2 + \dfrac{1}{n} > M$. Notice that if $n^2 > M$, then $n^2 + \dfrac{1}{n} > n^2 > M$. Since $M > 0$, it follows that $n^2 > M$ is equivalent to $n > \sqrt{M}$. A formal proof can now be constructed.   ♦

Result 12.6 $\lim\limits_{n\to\infty} \left(n^2 + \dfrac{1}{n}\right) = \infty.$

Proof   Let M be a positive number. Choose $N = \left\lceil \sqrt{M} \right\rceil$ and let n be any integer such that $n > N$. Hence $n > \sqrt{M}$ and so $n^2 > M$. Thus $n^2 + \dfrac{1}{n} > n^2 > M$.   ∎

12.2 Infinite Series

An important concept in calculus involving sequences is infinite series. For real numbers $a_1, a_2, a_3, \ldots$, we write $\sum\limits_{k=1}^{\infty} a_k = a_1 + a_2 + a_3 + \cdots$ to denote an **infinite series** (often simply called a **series**). For example,

$$\sum_{k=1}^{\infty} \frac{1}{k^2} = 1 + \frac{1}{2^2} + \frac{1}{3^2} + \cdots \quad \text{and} \quad \sum_{k=1}^{\infty} \frac{k}{2k^2 + 1} = \frac{1}{3} + \frac{2}{9} + \frac{3}{19} + \cdots$$

are infinite series.

The numbers $a_1, a_2, a_3, \ldots$ are called the **terms** of the series $\sum\limits_{k=1}^{\infty} a_k = a_1 + a_2 + a_3 + \cdots$. The notation certainly seems to suggest that we are adding the terms $a_1, a_2, a_3, \ldots$. But what does it mean to add infinitely many numbers? A meaning must be given to this. For this reason, we construct a sequence $\{s_n\}$, called the **sequence of partial sums** of the series. Here $s_1 = a_1$, $s_2 = a_1 + a_2$, $s_3 = a_1 + a_2 + a_3$ and, in general, for $n \in \mathbf{N}$,

$$s_n = a_1 + a_2 + \cdots + a_n = \sum_{k=1}^{n} a_k.$$

Because s_n is determined by adding a finite number of terms, there is no confusion in understanding the terms of the sequence $\{s_n\}$. If the sequence $\{s_n\}$ converges, say to the number L, then the series $\sum\limits_{k=1}^{\infty} a_k$ is said to **converge** to L and we write $\sum\limits_{k=1}^{\infty} a_k = L$. This number L is called the **sum** of $\sum\limits_{k=1}^{\infty} a_k$. If $\{s_n\}$ diverges, then $\sum\limits_{k=1}^{\infty} a_k$ is said to **diverge**.

The French mathematician Augustin-Louis Cauchy was one of the most productive mathematicians of the 19th century. Among his many accomplishments was his definition of convergence of infinite series, a definition which is still used today. In his work *Cours d'Analyse*, Cauchy considered the sequence $\{s_n\}$ of partial sums of a series. He stated the following:

> *If, for increasing values of n, the sum s_n approaches indefinitely a certain limit s, the series will be called convergent and this limit in question will be called the sum of the series.*

We consider an example of a convergent series.

Result to Prove The infinite series $\displaystyle\sum_{k=1}^{\infty} \frac{1}{k(k+1)}$ converges to 1.

PROOF STRATEGY First, we consider the sequence $\{s_n\}$ of partial sums for this series. Since

$$\sum_{k=1}^{\infty} \frac{1}{k(k+1)} = \frac{1}{1 \cdot 2} + \frac{1}{2 \cdot 3} + \frac{1}{3 \cdot 4} + \cdots,$$

it follows that $s_1 = \dfrac{1}{1 \cdot 2} = \dfrac{1}{2}$, $s_2 = \dfrac{1}{1 \cdot 2} + \dfrac{1}{2 \cdot 3} = \dfrac{1}{2} + \dfrac{1}{6} = \dfrac{2}{3}$ and

$$s_3 = \frac{1}{1 \cdot 2} + \frac{1}{2 \cdot 3} + \frac{1}{3 \cdot 4} = \frac{1}{2} + \frac{1}{6} + \frac{1}{12} = \frac{3}{4}.$$

Based on these three terms, it *appears* that $s_n = \dfrac{n}{n+1}$ for every positive integer n. We prove that this is indeed the case. ◆

Lemma 12.7 *For every positive integer n,*

$$s_n = \frac{1}{1 \cdot 2} + \frac{1}{2 \cdot 3} + \frac{1}{3 \cdot 4} + \cdots + \frac{1}{n(n+1)} = \frac{n}{n+1}.$$

Proof of Lemma 12.7 We proceed by induction. For $n = 1$, we have $s_1 = \dfrac{1}{1 \cdot 2} = \dfrac{1}{1+1}$ and the result holds. Assume that $s_k = \dfrac{1}{1 \cdot 2} + \dfrac{1}{2 \cdot 3} + \dfrac{1}{3 \cdot 4} + \cdots + \dfrac{1}{k(k+1)} = \dfrac{k}{k+1}$, where k is a positive integer. We show that

$$s_{k+1} = \frac{1}{1 \cdot 2} + \frac{1}{2 \cdot 3} + \frac{1}{3 \cdot 4} + \cdots + \frac{1}{(k+1)(k+2)} = \frac{k+1}{k+2}.$$

Observe that

$$s_{k+1} = \left[\frac{1}{1 \cdot 2} + \frac{1}{2 \cdot 3} + \frac{1}{3 \cdot 4} + \cdots + \frac{1}{k(k+1)} \right] + \frac{1}{(k+1)(k+2)}$$

$$= \frac{k}{k+1} + \frac{1}{(k+1)(k+2)} = \frac{k(k+2)+1}{(k+1)(k+2)} = \frac{k^2 + 2k + 1}{(k+1)(k+2)}$$

$$= \frac{(k+1)^2}{(k+1)(k+2)} = \frac{k+1}{k+2}.$$

By the Principle of Mathematical Induction, $s_n = \dfrac{n}{n+1}$ for every positive integer n. ∎

There is another way that we might have been able to see that $s_n = \dfrac{n}{n+1}$. If we had observed that

$$a_n = \frac{1}{n(n+1)} = \frac{1}{n} - \frac{1}{n+1},$$

then $a_1 = \dfrac{1}{1 \cdot 2} = 1 - \dfrac{1}{2}$, $a_2 = \dfrac{1}{2 \cdot 3} = \dfrac{1}{2} - \dfrac{1}{3}$, etc. In particular,

$$s_n = a_1 + a_2 + a_3 + \cdots + a_n$$
$$= \left(1 - \frac{1}{2}\right) + \left(\frac{1}{2} - \frac{1}{3}\right) + \left(\frac{1}{3} - \frac{1}{4}\right) + \cdots + \left(\frac{1}{n} - \frac{1}{n+1}\right)$$
$$= 1 - \frac{1}{n+1} = \frac{n}{n+1}.$$

In any case, since we now know that $s_n = \dfrac{n}{n+1}$, it remains only to prove that

$$\lim_{n \to \infty} s_n = \lim_{n \to \infty} \frac{n}{n+1} = 1.$$

Lemma to Prove $\displaystyle\lim_{n \to \infty} \frac{n}{n+1} = 1.$

PROOF STRATEGY For a given $\epsilon > 0$, we are required to find a positive integer N such that if $n > N$, then $\left| \dfrac{n}{n+1} - 1 \right| < \epsilon$. Now

$$\left| \frac{n}{n+1} - 1 \right| = \left| \frac{n - n - 1}{n+1} \right| = \left| \frac{-1}{n+1} \right| = \frac{1}{n+1}.$$

The inequality $\dfrac{1}{n+1} < \epsilon$ is equivalent to $n + 1 > \dfrac{1}{\epsilon}$, which in turn is equivalent to $n > \dfrac{1}{\epsilon} - 1$. If $n > \dfrac{1}{\epsilon}$, then $n > \dfrac{1}{\epsilon} - 1$. We can now present a proof of this lemma. ◆

Lemma 12.8 $\displaystyle\lim_{n \to \infty} \frac{n}{n+1} = 1.$

Proof of Lemma 12.8 Let $\epsilon > 0$ be given. Choose $N = \lceil 1/\epsilon \rceil$ and let $n > N$. Then $n > \dfrac{1}{\epsilon} > \dfrac{1}{\epsilon} - 1$. So $n > \dfrac{1}{\epsilon} - 1$. Thus $n + 1 > \dfrac{1}{\epsilon}$ and $\dfrac{1}{n+1} < \epsilon$. Hence

$$\left| \frac{n}{n+1} - 1 \right| = \left| \frac{-1}{n+1} \right| = \frac{1}{n+1} < \epsilon.$$
∎

We are now prepared to give a proof of the result.

Result 12.9   *The infinite series* $\displaystyle\sum_{k=1}^{\infty} \frac{1}{k(k+1)}$ *converges to 1.*

Proof   The nth term of the sequence $\{s_n\}$ of partial sums of the series $\displaystyle\sum_{k=1}^{\infty} \frac{1}{k(k+1)}$ is

$$s_n = \frac{1}{1 \cdot 2} + \frac{1}{2 \cdot 3} + \frac{1}{3 \cdot 4} + \cdots + \frac{1}{n(n+1)}.$$

By Lemma 12.7,

$$s_n = \frac{1}{1 \cdot 2} + \frac{1}{2 \cdot 3} + \frac{1}{3 \cdot 4} + \cdots + \frac{1}{n(n+1)} = \frac{n}{n+1}$$

and so $s_n = \dfrac{n}{n+1}$. By Lemma 12.8,

$$\lim_{n \to \infty} \frac{n}{n+1} = 1.$$

Since $\lim\limits_{n \to \infty} s_n = 1$, it follows that $\displaystyle\sum_{k=1}^{\infty} \frac{1}{k(k+1)} = 1.$ ∎

We now turn to a divergent series. The series $\displaystyle\sum_{k=1}^{\infty} \frac{1}{k} = 1 + \frac{1}{2} + \frac{1}{3} + \cdots$ is famous and is called the **harmonic series**. Indeed, it is probably the best known divergent series.

Result 12.10   *The harmonic series* $\displaystyle\sum_{k=1}^{\infty} \frac{1}{k}$ *diverges.*

Proof   Assume, to the contrary, that $\displaystyle\sum_{k=1}^{\infty} \frac{1}{k}$ converges, say to the number L. For each positive integer n, let $s_n = \displaystyle\sum_{k=1}^{n} \frac{1}{k}$. Hence the sequence $\{s_n\}$ of partial sums converges to L. Therefore, for each $\epsilon > 0$, there exists a positive integer N such that if $n > N$, then $|s_n - L| < \epsilon$. Let's consider $\epsilon = 1/4$ and let n be an integer with $n > N$. Then

$$-\frac{1}{4} < s_n - L < \frac{1}{4}.$$

Since $2n > N$, it is also the case that $|s_{2n} - L| < \dfrac{1}{4}$ and so $-\dfrac{1}{4} < s_{2n} - L < \dfrac{1}{4}$. Observe that

$$s_{2n} = s_n + \frac{1}{n+1} + \frac{1}{n+2} + \cdots + \frac{1}{2n} > s_n + n\left(\frac{1}{2n}\right) = s_n + \frac{1}{2}.$$

Hence

$$\frac{1}{4} > s_{2n} - L > s_n + \frac{1}{2} - L = (s_n - L) + \frac{1}{2} > -\frac{1}{4} + \frac{1}{2} = \frac{1}{4},$$

which is impossible.   ∎

<u>PROOF ANALYSIS</u> In Result 12.10, we showed that a certain series diverges; that is, it does not converge. Consequently, it is not surprising that we proved this by contradiction. By assuming that the sequence $\{s_n\}$ converges, this meant that the sequence has a limit L. This tells us that an inequality of the type $|s_n - L| < \epsilon$ exists for every positive number ϵ and for sufficiently large integers n (which depend on ϵ). The goal, of course, was to obtain a contradiction. We did this by making a choice of ϵ ($\epsilon = 1/4$ worked!) that eventually produced a mathematical impossibility. ◆

The harmonic series $\sum_{k=1}^{\infty} \dfrac{1}{k}$ not only diverges, it diverges to infinity; that is, if $\{s_n\}$ is the sequence of partial sums for the harmonic series, then $\lim_{n \to \infty} s_n = \infty$. We also establish this fact. First, we verify a lemma, which shows once again that mathematical induction can be a useful proof technique in calculus.

Lemma 12.11 *Let* $s_n = \sum_{k=1}^{n} \dfrac{1}{k} = 1 + \dfrac{1}{2} + \cdots + \dfrac{1}{n}$, *where* $n \in \mathbf{N}$. *Then* $s_{2^n} \geq 1 + \dfrac{n}{2}$ *for every positive integer* n.

Proof We proceed by induction. For $n = 1$, $s_{2^1} = 1 + \dfrac{1}{2}$ and so the result holds for $n = 1$.

Assume that $s_{2^k} \geq 1 + \dfrac{k}{2}$, where $k \in \mathbf{N}$. We show that $s_{2^{k+1}} \geq 1 + \dfrac{k+1}{2}$. Now observe that

$$s_{2^{k+1}} = 1 + \frac{1}{2} + \cdots + \frac{1}{2^{k+1}}$$

$$= s_{2^k} + \frac{1}{2^k + 1} + \frac{1}{2^k + 2} + \cdots + \frac{1}{2^{k+1}}$$

$$\geq s_{2^k} + \frac{1}{2^{k+1}} + \frac{1}{2^{k+1}} + \cdots + \frac{1}{2^{k+1}}$$

$$= s_{2^k} + \frac{2^k}{2^{k+1}} = s_{2^k} + \frac{1}{2}$$

$$\geq 1 + \frac{k}{2} + \frac{1}{2} = 1 + \frac{k+1}{2}.$$

By the Principle of Mathematical Induction, $s_{2^n} \geq 1 + \dfrac{n}{2}$ for every positive integer n. ∎

Result 12.12 *The harmonic series* $\sum_{k=1}^{\infty} \dfrac{1}{k}$ *diverges to infinity.*

Proof For $n \in \mathbf{N}$, let $s_n = \sum_{k=1}^{n} \dfrac{1}{k}$. Thus $\{s_n\}$ is the sequence of partial sums for the harmonic series. We show that $\lim_{n \to \infty} s_n = \infty$. Let M be a positive integer and choose $N = 2^{2M}$.

Let $n > N$. Then, using Lemma 12.11, we have

$$s_n = 1 + \frac{1}{2} + \cdots + \frac{1}{N} + \frac{1}{N+1} + \cdots + \frac{1}{n}$$

$$= s_N + \frac{1}{N+1} + \frac{1}{N+2} + \cdots + \frac{1}{n}$$

$$> s_N = s_{2^{2M}} \geq 1 + \frac{2M}{2} > M. \qquad \blacksquare$$

12.3 Limits of Functions

We now turn to another common type of limit problem (perhaps the most common). Here we consider functions $f : X \to \mathbf{R}$, where $X \subseteq \mathbf{R}$, and study the behavior of such a function f near some real number (point) a. For the present, we are not concerned whether $a \in X$, but since we *are* concerned about the numbers $f(x)$ for real numbers x near a, it is necessary that f is defined in some "deleted neighborhood" of a. By a **deleted neighborhood** of a, we mean a set of the type $(a - \delta, a) \cup (a, a + \delta) = (a - \delta, a + \delta) - \{a\} \subseteq X$ for some positive real number δ (the Greek letter delta). (See Figure 12.2.) It may actually be the case that $(a - \delta, a + \delta) \subseteq X$ for some $\delta > 0$. For example, if $f : X \to \mathbf{R}$ is defined by $f(x) = \dfrac{|x|}{x}$ and we are interested in the behavior of f near 0, then $0 \notin X$. In fact, it might very well be that $X = \mathbf{R} - \{0\}$, in which case, $(-\delta, 0) \cup (0, \delta) \subseteq X$ for every positive real number δ. On the other hand, if $f : X \to \mathbf{R}$ is defined by $f(x) = \dfrac{x}{x^2 - 1}$ and, once again, we are interested in the behavior of f near 0, then $1, -1 \notin X$. A natural choice for X is $\mathbf{R} - \{1, -1\}$, in which case $(-\delta, \delta) \subseteq X$ for every real number δ such that $0 < \delta \leq 1$.

We are now prepared to present the definition of the limit of a function. Let f be a real-valued function defined on a set X of real numbers. Also, let $a \in \mathbf{R}$ such that f is defined in some deleted neighborhood of a. Then we say that the real number L is the limit of $f(x)$ as x approaches a, written $\lim_{x \to a} f(x) = L$, if the closer x is to a, the closer $f(x)$ is to L. The vagueness of the word *closer* here too requires a considerably more precise definition. Let the positive number ϵ indicate how close $f(x)$ is required to be to L; that is, we require that $|f(x) - L| < \epsilon$. Then the claim is that if x is sufficiently close to a, then $|f(x) - L| < \epsilon$. We use the positive number δ to represent how close x must be to a in order for the inequality $|f(x) - L| < \epsilon$ to be satisfied, recalling that we are not concerned about how, or even if, f is defined at a.

More precisely then, L is the **limit** of $f(x)$ as x approaches a, written $\lim_{x \to a} f(x) = L$, if for every real number $\epsilon > 0$, there exists a real number $\delta > 0$ such that for every real number x with $0 < |x - a| < \delta$, it follows that $|f(x) - L| < \epsilon$. This implies that if $0 < |x - a| < \delta$, then certainly $f(x)$ is defined. If there exists a number L such that

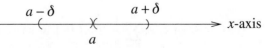

Figure 12.2   A deleted neighborhood of a

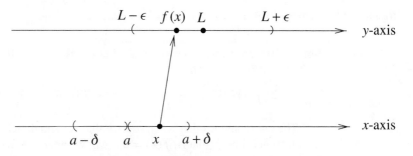

Figure 12.3 A geometric interpretation of $\lim\limits_{x \to a} f(x) = L$

$\lim\limits_{x \to a} f(x) = L$, then we say that the limit $\lim\limits_{x \to a} f(x)$ exists and is equal to L; otherwise, this limit does not exist. Thus to show that $\lim\limits_{x \to a} f(x) = L$, it is necessary to specify $\epsilon > 0$ first and then show the existence of a real number $\delta > 0$. Ordinarily, the smaller the value of ϵ, the smaller the value of δ. However, we must be certain that the number δ selected satisfies the requirement regardless of how small (or large) ϵ may be. Even though our choice of δ depends on ϵ, it should not depend on which real number x with $0 < |x - a| < \delta$ is being considered.

Accordingly, if $\lim\limits_{x \to a} f(x) = L$, then for a given $\epsilon > 0$, there exists $\delta > 0$ such that if x is any number in the open interval $(a - \delta, a + \delta)$ that is different from a, then $f(x)$ is a number in the interval $(L - \epsilon, L + \epsilon)$. This geometric interpretation of the definition of limit is illustrated in Figure 12.3.

We illustrate these ideas with an example.

Result to Prove $\lim\limits_{x \to 4} (3x - 7) = 5$.

PROOF STRATEGY Before giving a formal proof of this limit, let's discuss the procedure we will use. The proof begins by letting $\epsilon > 0$ be given. What we are required to do is to find a number $\delta > 0$ such that if $0 < |x - 4| < \delta$, then $|(3x - 7) - 5| < \epsilon$ or, equivalently, $|3(x - 4)| < \epsilon$. This is also equivalent to $|3| \cdot |x - 4| < \epsilon$ and to $|x - 4| < \epsilon/3$. This suggests our choice of δ. We can now give a proof. ◆

Result 12.13 $\lim\limits_{x \to 4} (3x - 7) = 5$.

Proof Let $\epsilon > 0$ be given. Choose $\delta = \epsilon/3$. Let $x \in \mathbf{R}$ such that $0 < |x - 4| < \delta = \epsilon/3$. Then

$$|(3x - 7) - 5| = |3x - 12| = |3(x - 4)| = 3|x - 4| < 3(\epsilon/3) = \epsilon. \qquad \blacksquare$$

Let's consider another example.

Result to Prove $\lim\limits_{x \to -3} (-2x + 1) = 7$.

PROOF STRATEGY First we do some preliminary algebra. The inequality $|(-2x + 1) - 7| < \epsilon$ is equivalent to $|-2x - 6| < \epsilon$ and to $2|x + 3| < \epsilon$. This suggests a desired value of δ. We can now give a proof. ◆

Result 12.14 $\lim\limits_{x \to -3} (-2x + 1) = 7$.

Proof   Let $\epsilon > 0$ be given and choose $\delta = \epsilon/2$. Let $x \in \mathbf{R}$ such that $0 < |x - (-3)| < \delta = \epsilon/2$, so $0 < |x + 3| < \epsilon/2$. Then

$$|(-2x + 1) - 7| = |-2(x + 3)| = |-2||x + 3| = 2|x + 3| < 2(\epsilon/2) = \epsilon. \qquad \blacksquare$$

The two examples that we have seen thus far should tell us how to proceed when the function is linear (that is, $f(x) = ax + b$, where $a, b \in \mathbf{R}$). We now present a slight variation of this.

Result to Prove $\lim\limits_{x \to \frac{3}{2}} \dfrac{4x^2 - 9}{2x - 3} = 6$.

PROOF STRATEGY   In this example, $|f(x) - L| < \epsilon$ becomes $\left|\dfrac{4x^2 - 9}{2x - 3} - 6\right| < \epsilon$ or, after simplifying, $\left|\dfrac{(2x + 3)(2x - 3)}{2x - 3} - 6\right| < \epsilon$. However, since the numbers x are in a deleted neighborhood of 3/2, it follows that $2x - 3 \neq 0$ and so $\left|\dfrac{(2x + 3)(2x - 3)}{2x - 3} - 6\right| < \epsilon$ becomes $|(2x + 3) - 6| < \epsilon$ or $|2x - 3| < \epsilon$. Therefore, $2|x - 3/2| < \epsilon$ and $|x - 3/2| < \epsilon/2$. We are now prepared to give a proof.   ◆

Result 12.15 $\lim\limits_{x \to \frac{3}{2}} \dfrac{4x^2 - 9}{2x - 3} = 6$.

Proof   Let $\epsilon > 0$ be given and choose $\delta = \epsilon/2$. Let $x \in \mathbf{R}$ such that $0 < |x - 3/2| < \delta = \epsilon/2$. So $2|x - 3/2| < \epsilon$ and $|2x - 3| < \epsilon$. Hence $|(2x + 3) - 6| < \epsilon$. Since $2x - 3 \neq 0$, it follows that $\left|\dfrac{(2x + 3)(2x - 3)}{2x - 3} - 6\right| < \epsilon$ and so $\left|\dfrac{4x^2 - 9}{2x - 3} - 6\right| < \epsilon$. $\blacksquare$

We now turn to a limit of a quadratic function.

Result to Prove $\lim\limits_{x \to 3} x^2 = 9$.

PROOF STRATEGY   Once again for a given $\epsilon > 0$, we are required to find $\delta > 0$ such that if $0 < |x - 3| < \delta$, then $|x^2 - 9| < \epsilon$. To find an appropriate choice of δ in terms of ϵ, we begin with $|x^2 - 9| < \epsilon$. We wish to work the expression $|x - 3|$ into this inequality. Actually, this is quite easy since $|x^2 - 9| < \epsilon$ is equivalent to $|x - 3||x + 3| < \epsilon$. This might make us think of writing $|x - 3| < \dfrac{\epsilon}{|x + 3|}$ and choosing $\delta = \dfrac{\epsilon}{|x + 3|}$. However, δ is required to be a positive number (a constant) that depends on ϵ but is not a function of x. The expression $|x + 3|$ can be eliminated though, as we now show. Since it is our choice how to select δ, we can certainly require $\delta \leq 1$, which we do. Thus $|x - 3| < 1$ and so $-1 < x - 3 < 1$. Hence $2 < x < 4$. Thus $5 < x + 3 < 7$ and so $|x + 3| < 7$. So, under this restriction for δ, it follows that $|x - 3||x + 3| < 7|x - 3|$. Now if $7|x - 3| < \epsilon$, that is, if $|x - 3| < \epsilon/7$, then it will certainly follow that $|x - 3||x + 3| < \epsilon$. Arriving at this inequality required that both $|x - 3| < 1$ and $|x - 3| < \epsilon/7$. This suggests an appropriate choice of δ. ◆

Result 12.16 $\lim\limits_{x \to 3} x^2 = 9$.

Proof Let $\epsilon > 0$ be given and choose $\delta = \min(1, \epsilon/7)$. Let $x \in \mathbf{R}$ such that $0 < |x - 3| < \delta = \min(1, \epsilon/7)$. Since $|x - 3| < 1$, it follows that $-1 < x - 3 < 1$ and so $5 < x + 3 < 7$. In particular, $|x + 3| < 7$. Because $|x - 3| < \epsilon/7$, it follows that

$$|x^2 - 9| = |x - 3||x + 3| < |x - 3| \cdot 7 < (\epsilon/7) \cdot 7 = \epsilon.$$ ■

We have now seen four proofs of limits of type $\lim\limits_{x \to a} f(x) = L$. In Result 12.13, we chose $\delta = \epsilon/3$ for the given $\epsilon > 0$ and in Result 12.14, we chose $\delta = \epsilon/2$. In each case, if we had considered a different value of a for the same function, then the same choice of δ would be successful. This is because the function is linear in each case. In Result 12.15, for a given $\epsilon > 0$, the selection of $\delta = \epsilon/2$ would also be successful if $a \neq 3/2$, provided $3/2 \notin (a - \delta, a + \delta)$. This is because the function f in Result 12.15 defined by $f(x) = (4x^2 - 9)/(2x - 3)$ is "nearly linear"; that is, $f(x) = 2x + 3$ if $x \neq 3/2$ and $f(3/2)$ is not defined. However, our choice of $\delta = \epsilon/7$ in the proof of Result 12.16 depended on $a = 3$; that is, if $a \neq 3$, a different choice of δ is needed. For example, if we were to prove that $\lim\limits_{x \to 4} x^2 = 16$, then for a given $\epsilon > 0$, an appropriate choice for δ is $\min(1, \epsilon/9)$.

Next we consider a limit involving a polynomial function of a higher degree.

Result to Prove $\lim\limits_{x \to 2} (x^5 - 2x^3 - 3x - 7) = 3$.

<u>**PROOF STRATEGY**</u> For a given $\epsilon > 0$, we are required to show that $|(x^5 - 2x^3 - 3x - 7) - 3| < \epsilon$ if $0 < |x - 2| < \delta$ for a suitable choice of $\delta > 0$. We then need to work $|x - 2|$ into the expression $|x^5 - 2x^3 - 3x - 10|$. Dividing $x^5 - 2x^3 - 3x - 10$ by $x - 2$, we obtain $x^5 - 2x^3 - 3x - 10 = (x - 2)(x^4 + 2x^3 + 2x^2 + 4x + 5)$. Hence we have

$$|x^5 - 2x^3 - 3x - 10| = |x - 2||x^4 + 2x^3 + 2x^2 + 4x + 5|.$$

Thus we seek an upper bound for $|x^4 + 2x^3 + 2x^2 + 4x + 5|$. To do this, we impose the restriction $\delta \leq 1$. Thus $|x - 2| < \delta \leq 1$. So $-1 < x - 2 < 1$ and $1 < x < 3$. Hence

$$|x^4 + 2x^3 + 2x^2 + 4x + 5| \leq |x^4| + |2x^3| + |2x^2| + |4x| + |5| < 170.$$

We are now prepared to prove Result 12.17. ◆

Result 12.17 $\lim\limits_{x \to 2} (x^5 - 2x^3 - 3x - 7) = 3$.

Proof Let $\epsilon > 0$ be given and choose $\delta = \min(1, \epsilon/170)$. Let $x \in \mathbf{R}$ such that $0 < |x - 2| < \delta = \min(1, \epsilon/170)$. Since $|x - 2| < 1$, it follows that $1 < x < 3$ and so

$$|x^4 + 2x^3 + 2x^2 + 4x + 5| \leq |x^4| + |2x^3| + |2x^2| + |4x| + |5| < 170.$$

Since $|x - 2| < \epsilon/170$, we have

$$\begin{aligned}|(x^5 - 2x^3 - 3x - 7) - 3| &= |x^5 - 2x^3 - 3x - 10| \\ &= |x - 2| \cdot |x^4 + 2x^3 + 2x^2 + 4x + 5| \\ &< (\epsilon/170) \cdot 170 = \epsilon. \qquad \blacksquare\end{aligned}$$

Our next example involves a rational function (the ratio of two polynomials).

Result to Prove $\displaystyle\lim_{x \to 1} \frac{x^2 + 1}{x^2 + 4} = \frac{2}{5}.$

PROOF STRATEGY   First observe that

$$\left|\frac{x^2 + 1}{x^2 + 4} - \frac{2}{5}\right| = \left|\frac{5(x^2 + 1) - 2(x^2 + 4)}{5(x^2 + 4)}\right| = \frac{|3x^2 - 3|}{5(x^2 + 4)} = \frac{3|x - 1||x + 1|}{5(x^2 + 4)}.$$

Hence it is necessary to find an upper bound for $\dfrac{3|x + 1|}{5(x^2 + 4)}$. Once again we restrict δ so that $\delta \leq 1$. Then $|x - 1| < 1$ or $0 < x < 2$. Hence $1 < x + 1 < 3$ and so $3|x + 1| < 9$. Also, since $x > 0$, it follows that $5(x^2 + 4) > 20$. Thus $\dfrac{1}{5(x^2 + 4)} < \dfrac{1}{20}$ and so

$$\frac{3|x + 1|}{5(x^2 + 4)} < 9\left(\frac{1}{20}\right) = \frac{9}{20}.$$

We now present a proof of this result. ◆

Result 12.18 $\displaystyle\lim_{x \to 1} \frac{x^2 + 1}{x^2 + 4} = \frac{2}{5}.$

Proof   Let $\epsilon > 0$ be given and choose $\delta = \min(1, 20\epsilon/9)$. Let $x \in \mathbf{R}$ such that $0 < |x - 1| < \delta$. Since $|x - 1| < 1$, we have $0 < x < 2$ and $1 < x + 1 < 3$. Hence $3|x + 1| < 3 \cdot 3 = 9$ and $5(x^2 + 4) > 20$, so $\dfrac{1}{5(x^2 + 4)} < \dfrac{1}{20}$. Therefore, $\dfrac{3|x + 1|}{5(x^2 + 4)} < 9/20$. Since $|x - 1| < 20\epsilon/9$, it follows that

$$\begin{aligned}\left|\frac{x^2 + 1}{x^2 + 4} - \frac{2}{5}\right| &= \left|\frac{5(x^2 + 1) - 2(x^2 + 4)}{5(x^2 + 4)}\right| = \frac{|3x^2 - 3|}{5(x^2 + 4)} \\ &= \frac{3|x - 1||x + 1|}{5(x^2 + 4)} < \frac{20\epsilon}{9} \cdot \frac{9}{20} = \epsilon. \qquad \blacksquare\end{aligned}$$

We now present one additional example on this topic.

Example 12.19   *Determine $\displaystyle\lim_{x \to 1} \frac{x^2 - 1}{2x - 1}$ and verify your answer.*

Solution   Since it appears that $\lim_{x \to 1}(x^2 - 1) = 0$ and $\lim_{x \to 1}(2x - 1) = 1$, we would expect that $\displaystyle\lim_{x \to 1} \frac{x^2 - 1}{2x - 1} = \frac{0}{1} = 0$. To verify this, we need to show that for a given $\epsilon > 0$, there is $\delta > 0$ such that if $0 < |x - 1| < \delta$, then

$$\left| \frac{x^2 - 1}{2x - 1} - 0 \right| = \left| \frac{x^2 - 1}{2x - 1} \right| < \epsilon.$$

Observe that

$$\left| \frac{x^2 - 1}{2x - 1} \right| = \left| \frac{(x - 1)(x + 1)}{2x - 1} \right| = \frac{|x + 1|}{|2x - 1|} |x - 1|.$$

Proceeding as before, we find an upper bound for $\dfrac{|x + 1|}{|2x - 1|}$. Ordinarily, we might restrict $\delta \leq 1$, as before, but in this situation, we have a problem. If $\delta \leq 1$, then $0 < |x - 1| < \delta$ and so $|x - 1| < 1$. Thus $0 < x < 2$ or $x \in (0, 2)$. However, this interval of real numbers includes $1/2$ and $\dfrac{|x + 1|}{|2x - 1|}$ is not defined when $x = 1/2$. Thus we place a tighter restriction on δ. The restriction $\delta \leq 1/2$ is not suitable either, for if $|x - 1| < \delta \leq 1/2$, then $1/2 < x < 3/2$. Even though $\dfrac{|x + 1|}{|2x - 1|}$ is defined for all real numbers x in this interval, this expression becomes arbitrarily large if x is arbitrarily close to $1/2$, allowing $|2x - 1|$ to be arbitrarily close to 0. That is, we cannot find an upper bound for $\dfrac{|x + 1|}{|2x - 1|}$ if $\delta = 1/2$. Hence we require that $\delta \leq 1/4$, say, and so $|x - 1| < \delta \leq 1/4$. Thus $3/4 < x < 5/4$. Hence $|x + 1| < 9/4$. Also, $|2x - 1| > 2 \left(\frac{3}{4} \right) - 1 = 1/2$ and so $\dfrac{1}{|2x - 1|} < 2$. Therefore, $\dfrac{|x + 1|}{|2x - 1|} < \dfrac{9}{4} \cdot 2 = \dfrac{9}{2}$. We now give a formal proof.   ◆

Result 12.20 $\displaystyle \lim_{x \to 1} \frac{x^2 - 1}{2x - 1} = 0.$

Proof   Let $\epsilon > 0$ be given and choose $\delta = \min(1/4, 2\epsilon/9)$. Let $x \in \mathbf{R}$ such that $0 < |x - 1| < \delta$. Since $\delta \leq 1/4$, it follows that $|x - 1| < 1/4$ and so $3/4 < x < 5/4$. Hence $|x + 1| < 5/4 + 1 = 9/4$. Also, $|2x - 1| > 2 \left(\frac{3}{4} \right) - 1 = 1/2$ and so $\dfrac{1}{|2x - 1|} < 2$. Therefore, $\dfrac{|x + 1|}{|2x - 1|} < \dfrac{9}{4} \cdot 2 = \dfrac{9}{2}$. Since $|x - 1| < \delta \leq 2\epsilon/9$, it follows that

$$\left| \frac{x^2 - 1}{2x - 1} - 0 \right| = \left| \frac{x^2 - 1}{2x - 1} \right| = \frac{|x + 1|}{|2x - 1|} |x - 1| < \frac{2\epsilon}{9} \cdot \frac{9}{2} = \epsilon.$$ ■

Next we consider a limit problem where the limit does not exist.

Result to Prove $\displaystyle \lim_{x \to 0} \frac{1}{x}$ does not exist.

PROOF STRATEGY   As expected, we will give a proof by contradiction. If $\displaystyle \lim_{x \to 0} \frac{1}{x}$ does exist, then there exists a real number L such that $\displaystyle \lim_{x \to 0} \frac{1}{x} = L$. Hence for every $\epsilon > 0$, there exists $\delta > 0$ such that if $0 < |x| < \delta$, then $\left|\dfrac{1}{x} - L\right| < \epsilon$. For numbers x "close" to 0, it certainly appears

that $\dfrac{1}{x}$ is "large" (in absolute value). Hence, regardless of the value of ϵ, it seems that

there should be a real number x with $0 < |x| < \delta$ such that $\left|\dfrac{1}{x} - L\right| \geq \epsilon$. It is our plan

to show that this is indeed the case. Thus, we choose $\epsilon = 1$, for example, and show that no desired δ can be found.   ◆

Result 12.21 $\lim\limits_{x \to 0} \dfrac{1}{x}$ *does not exist.*

Proof   Assume, to the contrary, that $\lim\limits_{x \to 0} \dfrac{1}{x}$ exists. Then there exists a real number L such that

$\lim\limits_{x \to 0} \dfrac{1}{x} = L$. Let $\epsilon = 1$. Then there exists $\delta > 0$ such that if x is a real number for which

$0 < |x| < \delta$, then $\left|\dfrac{1}{x} - L\right| < \epsilon = 1$. Choose an integer n such that $n > \lceil 1/\delta \rceil \geq 1$. Since

$n > 1/\delta$, it follows that $0 < 1/n < \delta$. We consider two cases.

Case 1. $L \leq 0$. Let $x = 1/n$. So $0 < |x| < \delta$. Since $-L \geq 0$, it follows that

$$\left|\frac{1}{x} - L\right| = |n - L| = n - L \geq n > 1 = \epsilon,$$

which is a contradiction.

Case 2. $L > 0$. Let $x = -1/n$. So $0 < |x| < \delta$. Thus

$$\left|\frac{1}{x} - L\right| = |-n - L| = |-(n + L)| = n + L > n > 1 = \epsilon,$$

producing a contradiction in this case as well.   ■

Result to Prove   Let $f(x) = |x|/x$, where $x \in \mathbf{R}$ and $x \neq 0$. Then $\lim\limits_{x \to 0} f(x)$ does not exist.

PROOF STRATEGY   The graph of this function is shown in Figure 12.4. If $x > 0$, then $f(x) = |x|/x = x/x = 1$; while if $x < 0$, then $f(x) = |x|/x = -x/x = -1$. Hence there are numbers x that are "near" 0 such that $f(x) = 1$ and numbers x that are "near" 0 such that $f(x) = -1$. This suggests a proof.   ◆

Result 12.22   *Let* $f(x) = |x|/x$, *where* $x \in \mathbf{R}$ *and* $x \neq 0$. *Then* $\lim\limits_{x \to 0} f(x)$ *does not exist.*

Proof   Assume, to the contrary, that $\lim\limits_{x \to 0} f(x)$ exists. Then there exists a real number L such that $\lim\limits_{x \to 0} f(x) = L$. Let $\epsilon = 1$. Then there exists $\delta > 0$ such that if x is a real number satisfying $0 < |x - 0| = |x| < \delta$, then $|f(x) - L| < \epsilon = 1$. We consider two cases.

Case 1. $L \geq 0$. Consider $x = -\delta/2$. Then $|x| = \delta/2 < \delta$. However, $f(x) = f(-\delta/2) = (\delta/2)/(-\delta/2) = -1$. So $|f(x) - L| = |-1 - L| = 1 + L \geq 1$, a contradiction.

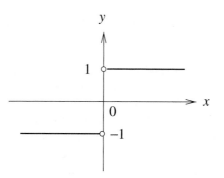

Figure 12.4   The graph of the function $f(x) = |x|/x$

Case 2. $L < 0$. Let $x = \delta/2$. Then $|x| = \delta/2 < \delta$. Also, $f(x) = f(\delta/2) = (\delta/2)/(\delta/2) = 1$. So $|f(x) - L| = |1 - L| = 1 - L > 1$, a contradiction.   ∎

12.4 Fundamental Properties of Limits of Functions

If we were to continue computing limits, then it would be essential to have some theorems at our disposal that would allow us to compute limits more rapidly. We now present some theorems that will allow us to determine limits more easily. We begin with a standard theorem on limits of sums of functions.

Theorem to Prove   If $\lim\limits_{x \to a} f(x) = L$ and $\lim\limits_{x \to a} g(x) = M$, then
$$\lim_{x \to a} (f(x) + g(x)) = L + M.$$

<u>PROOF STRATEGY</u>   In this case, we are required to show, for a given $\epsilon > 0$, that $|(f(x) + g(x)) - (L + M)| < \epsilon$ if $0 < |x - a| < \delta$ for a suitable choice of $\delta > 0$. Now
$$|(f(x) + g(x)) - (L + M)| = |(f(x) - L) + (g(x) - M)| \le |f(x) - L| + |g(x) - M|.$$

Hence if we can show that both $|f(x) - L| < \epsilon/2$ and $|g(x) - M| < \epsilon/2$, for example, then we will have obtained the desired inequality. However, because of the hypothesis, this can be accomplished. We now make all of this precise.   ♦

Theorem 12.23   *If* $\lim\limits_{x \to a} f(x) = L$ *and* $\lim\limits_{x \to a} g(x) = M$, *then*
$$\lim_{x \to a} (f(x) + g(x)) = L + M.$$

Proof   Let $\epsilon > 0$. Since $\epsilon/2 > 0$, there exists $\delta_1 > 0$ such that if $0 < |x - a| < \delta_1$, then $|f(x) - L| < \epsilon/2$. Also, there exists $\delta_2 > 0$ such that if $0 < |x - a| < \delta_2$, then $|g(x) - M| < \epsilon/2$. Choose $\delta = \min(\delta_1, \delta_2)$ and let $x \in \mathbf{R}$ such that $0 < |x - a| < \delta$. Since $0 < |x - a| < \delta$, it follows that both $0 < |x - a| < \delta_1$ and $0 < |x - a| < \delta_2$. Therefore,
$$|(f(x) + g(x)) - (L + M)| = |(f(x) - L) + (g(x) - M)|$$
$$\le |f(x) - L| + |g(x) - M| < \epsilon/2 + \epsilon/2 = \epsilon.$$ ∎

Theorem 12.23 states that the limit of the sum of two functions is the sum of their limits. Next we show that this is also true for products. Before getting to this theorem, let's see what would be involved to prove it. Let $\lim\limits_{x \to a} f(x) = L$ and $\lim\limits_{x \to a} g(x) = M$. This means that we can make the expressions $|f(x) - L|$ and $|g(x) - M|$ as small as we wish. Our goal is to show that we can make $|f(x) \cdot g(x) - LM|$ as small as we wish, say less than ϵ for every given $\epsilon > 0$. The question then becomes how to use what we know about $|f(x) - L|$ and $|g(x) - M|$ as we consider $|f(x) \cdot g(x) - LM|$. A common way to do this is to add and subtract the same quantity to and from $f(x) \cdot g(x) - LM$. For example,

$$\begin{aligned}
|f(x) \cdot g(x) - LM| &= |f(x) \cdot g(x) - f(x) \cdot M + f(x) \cdot M - LM| \\
&= |f(x)(g(x) - M) + (f(x) - L)M| \\
&\leq |f(x)||g(x) - M| + |f(x) - L||M|.
\end{aligned}$$

If we can make each of $|f(x)||g(x) - M|$ and $|f(x) - L||M|$ less than $\epsilon/2$, say, then we will have accomplished our goal. Since $|M|$ is a nonnegative constant and $|f(x) - L|$ and $|g(x) - M|$ can be made arbitrarily small, only $|f(x)|$ is in question. In fact, all that is required to show is that $f(x)$ can be bounded in a deleted neighborhood of a, that is, $|f(x)| \leq B$ for some constant $B > 0$.

Lemma 12.24   *Suppose that* $\lim\limits_{x \to a} f(x) = L$. *Then there exists* $\delta > 0$ *such that if* $0 < |x - a| < \delta$, *then* $|f(x)| < 1 + |L|$.

Proof   Let $\epsilon = 1$. Then there exists $\delta > 0$ such that if $0 < |x - a| < \delta$, then $|f(x) - L| < 1$. Thus

$$|f(x)| = |f(x) - L + L| \leq |f(x) - L| + |L| < 1 + |L|. \qquad \blacksquare$$

We are now prepared to show that the limit of the product of two functions is the product of their limits.

Theorem to Prove   If $\lim\limits_{x \to a} f(x) = L$ and $\lim\limits_{x \to a} g(x) = M$, then $\lim\limits_{x \to a} f(x) \cdot g(x) = LM$.

PROOF STRATEGY   As we discussed earlier,

$$\begin{aligned}
|f(x) \cdot g(x) - LM| &= |f(x) \cdot g(x) - f(x) \cdot M + f(x) \cdot M - LM| \\
&= |f(x)(g(x) - M) + (f(x) - L)M| \\
&\leq |f(x)||g(x) - M| + |f(x) - L||M|.
\end{aligned}$$

For a given $\epsilon > 0$, we show that each of $|f(x)||g(x) - M|$ and $|f(x) - L||M|$ can be made less than $\epsilon/2$, which will give us a proof of the result. Of course, this follows immediately for $|f(x) - L||M|$ if $M = 0$. Otherwise, we can make $|f(x) - L|$ less than $\epsilon/2|M|$. By Lemma 12.24, we can make $|f(x)|$ less than $1 + |L|$. Thus we make $|g(x) - M| < \epsilon/2(1 + |L|)$. Now, let's put all of the pieces together. $\blacklozenge$

Theorem 12.25   *If* $\lim\limits_{x \to a} f(x) = L$ *and* $\lim\limits_{x \to a} g(x) = M$, *then* $\lim\limits_{x \to a} f(x) \cdot g(x) = LM$.

Proof Let $\epsilon > 0$ be given. By Lemma 12.24, there exists $\delta_1 > 0$ such that if $0 < |x - a| < \delta_1$, then $|f(x)| < 1 + |L|$. Since $\lim_{x \to a} g(x) = M$, there exists $\delta_2 > 0$ such that if $0 < |x - a| < \delta_2$, then $|g(x) - M| < \epsilon/2(1 + |L|)$. We consider two cases.

Case 1. $M = 0$. Choose $\delta = \min(\delta_1, \delta_2)$. Let $x \in \mathbf{R}$ such that $0 < |x - a| < \delta$. Then

$$
\begin{aligned}
|f(x) \cdot g(x) - LM| &= |f(x) \cdot g(x) - f(x) \cdot M + f(x) \cdot M - LM| \\
&= |f(x)(g(x) - M) + (f(x) - L)M| \\
&\leq |f(x)||g(x) - M| + |f(x) - L||M| \\
&< (1 + |L|)\epsilon/2(1 + |L|) + 0 = \epsilon/2 < \epsilon.
\end{aligned}
$$

Case 2. $M \neq 0$. Since $\lim_{x \to a} f(x) = L$, there exists $\delta_3 > 0$ such that if $0 < |x - a| < \delta_3$, then $|f(x) - L| < \epsilon/2|M|$. In this case, we choose $\delta = \min(\delta_1, \delta_2, \delta_3)$. Now let $x \in \mathbf{R}$ such that $0 < |x - a| < \delta$. Then

$$
\begin{aligned}
|f(x) \cdot g(x) - LM| &= |f(x) \cdot g(x) - f(x) \cdot M + f(x) \cdot M - LM| \\
&= |f(x)(g(x) - M) + (f(x) - L)M| \\
&\leq |f(x)||g(x) - M| + |f(x) - L||M| \\
&< (1 + |L|)\epsilon/2(1 + |L|) + (\epsilon/2|M|)|M| \\
&= \epsilon/2 + \epsilon/2 = \epsilon.
\end{aligned}
$$ ∎

Next we consider the limit of the quotient of two functions. As before, let $\lim_{x \to a} f(x) = L$ and $\lim_{x \to a} g(x) = M$. Our goal is to show that $\lim_{x \to a} \dfrac{f(x)}{g(x)} = \dfrac{L}{M}$. Of course, this is not true if $M = 0$; so we will need to assume that $M \neq 0$. To prove that $\lim_{x \to a} \dfrac{f(x)}{g(x)} = \dfrac{L}{M}$, we are required to show that $\left| \dfrac{f(x)}{g(x)} - \dfrac{L}{M} \right|$ can be made arbitrarily small. Observe that

$$
\begin{aligned}
\left| \frac{f(x)}{g(x)} - \frac{L}{M} \right| &= \left| \frac{f(x) \cdot M - L \cdot g(x)}{g(x) \cdot M} \right| = \left| \frac{f(x) \cdot M - LM + LM - L \cdot g(x)}{g(x) \cdot M} \right| \\
&= \left| \frac{(f(x) - L)M + L(M - g(x))}{g(x) \cdot M} \right| \leq \frac{|f(x) - L||M| + |L||M - g(x)|}{|g(x)||M|} \\
&\leq \frac{|f(x) - L|}{|g(x)|} + \frac{|L||M - g(x)|}{|g(x)||M|}.
\end{aligned}
$$

Thus to show that $\left| \dfrac{f(x)}{g(x)} - \dfrac{L}{M} \right|$ can be made less than ϵ for any given positive number ϵ, it is sufficient to show that each of $\dfrac{|f(x) - L|}{|g(x)|}$ and $\dfrac{|L||M - g(x)|}{|g(x)||M|}$ can be made less than $\epsilon/2$. Only $1/|g(x)|$ requires study. In particular, we need to show that there is an upper bound for $1/|g(x)|$ in some deleted neighborhood of a.

Lemma 12.26 *If $\lim_{x \to a} g(x) = M \neq 0$, then $1/|g(x)| < 2/|M|$ for all x in some deleted neighborhood of a.*

Proof   Let $\epsilon = |M|/2$. Then there exists $\delta > 0$ such that if $0 < |x - a| < \delta$, then $|g(x) - M| < |M|/2$. Therefore,

$$|M| = |M - g(x) + g(x)| \le |M - g(x)| + |g(x)|$$

Hence $|g(x)| \ge |M| - |M - g(x)| > |M| - |M|/2 = |M|/2$. Thus $1/|g(x)| < 2/|M|$. ∎

Theorem to Prove   If $\lim\limits_{x \to a} f(x) = L$ and $\lim\limits_{x \to a} g(x) = M \ne 0$, then $\lim\limits_{x \to a} \dfrac{f(x)}{g(x)} = \dfrac{L}{M}$.

<u>PROOF STRATEGY</u>   Returning to our earlier discussion, we now have

$$\left| \frac{f(x)}{g(x)} - \frac{L}{M} \right| \le \frac{|f(x) - L|}{|g(x)|} + \frac{|L||M - g(x)|}{|g(x)||M|}$$

$$< |f(x) - L| \cdot \frac{2}{|M|} + |L||M - g(x)| \cdot \frac{2}{|M|^2}.$$

This suggests how small we must make $|f(x) - L|$ and $|g(x) - M| = |M - g(x)|$ to accomplish our goal.   ◆

Theorem 12.27   *If* $\lim\limits_{x \to a} f(x) = L$ *and* $\lim\limits_{x \to a} g(x) = M \ne 0$, *then* $\lim\limits_{x \to a} \dfrac{f(x)}{g(x)} = \dfrac{L}{M}$.

Proof   Let $\epsilon > 0$ be given. By Lemma 12.26, there exists $\delta_1 > 0$ such that if $0 < |x - a| < \delta_1$, then $1/|g(x)| < 2/|M|$. Since $\lim\limits_{x \to a} f(x) = L$, there exists $\delta_2 > 0$ such that if $0 < |x - a| < \delta_2$, then $|f(x) - L| < |M|\epsilon/4$. We consider two cases.
Case 1. $L = 0$. Define $\delta = \min(\delta_1, \delta_2)$. Let $x \in \mathbf{R}$ such that $0 < |x - a| < \delta$. Then

$$\left| \frac{f(x)}{g(x)} - \frac{L}{M} \right| \le \frac{|f(x) - L|}{|g(x)|} + \frac{|L||M - g(x)|}{|g(x)||M|}$$

$$< \frac{|M|\epsilon}{4} \cdot \frac{2}{|M|} + 0 = \frac{\epsilon}{2} < \epsilon.$$

Case 2. $L \ne 0$. Since $\lim\limits_{x \to a} g(x) = M$, there exists $\delta_3 > 0$ such that if $0 < |x - a| < \delta_3$, then $|g(x) - M| < |M|^2\epsilon/4|L|$. In this case, define $\delta = \min(\delta_1, \delta_2, \delta_3)$. Let $x \in \mathbf{R}$ such that $0 < |x - a| < \delta$. Then

$$\left| \frac{f(x)}{g(x)} - \frac{L}{M} \right| \le \frac{|f(x) - L|}{|g(x)|} + \frac{|L||M - g(x)|}{|g(x)||M|}$$

$$< \frac{|M|\epsilon}{4} \cdot \frac{2}{|M|} + \frac{|L|}{|M|} \cdot \frac{|M|^2\epsilon}{4|L|} \cdot \frac{2}{|M|} = \frac{\epsilon}{2} + \frac{\epsilon}{2} = \epsilon.$$ ∎

Now, with the aid of Theorems 12.23, 12.25, 12.27 and a few other general results, it is possible to give simpler arguments for some of the limits we have discussed. First, we present some additional results, beginning with an observation concerning constant functions and followed by limits of polynomial functions defined by $f(x) = x^n$ for some $n \in \mathbf{N}$.

Theorem 12.28   *Let* $a, c \in \mathbf{R}$. *If* $f(x) = c$ *for all* $x \in \mathbf{R}$, *then* $\lim\limits_{x \to a} f(x) = c$.

Proof Let $\epsilon > 0$ be given and choose δ to be any positive number. Let $x \in \mathbf{R}$ such that $0 < |x - a| < \delta$. Since $f(x) = c$ for all $x \in \mathbf{R}$, it follows that $|f(x) - c| = |c - c| = 0 < \epsilon$. ∎

Theorem 12.29 *Let $f(x) = x$ for all $x \in \mathbf{R}$. For each $a \in \mathbf{R}$, $\lim\limits_{x \to a} f(x) = a$.*

Proof Let $\epsilon > 0$ be given and choose $\delta = \epsilon$. Let $x \in \mathbf{R}$ such that $0 < |x - a| < \delta$. Then $|f(x) - a| = |x - a| < \delta = \epsilon$. ∎

We now extend the result in Theorem 12.29.

Theorem 12.30 *Let $n \in \mathbf{N}$ and let $f(x) = x^n$ for all $x \in \mathbf{R}$. Then for each $a \in \mathbf{R}$, $\lim\limits_{x \to a} f(x) = a^n$.*

Proof We proceed by induction. The statement is true for $n = 1$ since if $f(x) = x$, then $\lim\limits_{x \to a} f(x) = a$ by Theorem 12.29. Assume that $\lim\limits_{x \to a} x^k = a^k$, where $k \in \mathbf{N}$. We show that $\lim\limits_{x \to a} x^{k+1} = a^{k+1}$. Observe that $\lim\limits_{x \to a} x^{k+1} = \lim\limits_{x \to a} (x^k \cdot x)$. By Theorems 12.25 and 12.29 and the induction hypothesis,

$$\lim_{x \to a} x^{k+1} = \lim_{x \to a} (x^k \cdot x) = \left(\lim_{x \to a} x^k \right) \left(\lim_{x \to a} x \right) = (a^k)(a) = a^{k+1}.$$

By the Principle of Mathematical Induction, $\lim\limits_{x \to a} x^n = a^n$ for every $n \in \mathbf{N}$. ∎

It is possible to prove the following theorem by induction as well. We leave its proof as an exercise (Exercise 12.32).

Theorem 12.31 *Let $f_1, f_2, \cdots, f_n$ be functions ($n \geq 2$) such that $\lim\limits_{x \to a} f_i(x) = L_i$ for $1 \leq i \leq n$. Then*

$$\lim_{x \to a} (f_1(x) + f_2(x) + \cdots + f_n(x)) = L_1 + L_2 + \cdots + L_n.$$

With the results we have now presented, it is possible to prove that if $p(x) = c_n x^n + c_{n-1} x^{n-1} + \cdots + c_1 x + c_0$ is a polynomial, then

$$\lim_{x \to a} p(x) = c_n a^n + c_{n-1} a^{n-1} + \cdots + c_1 a + c_0 = p(a). \tag{12.1}$$

For example, applying this to Result 12.13, we have

$$\lim_{x \to 4} (3x - 7) = 3 \cdot 4 - 7.$$

Similarly, Result 12.14 can be established. Result 12.15 cannot be established directly since $\lim\limits_{x \to \frac{3}{2}} (2x - 3) = 0$. Applying what we now know to Result 12.16, we have $\lim\limits_{x \to 3} x^2 = 3^2 = 9$ and in Result 12.17,

$$\lim_{x \to 2} (x^5 - 2x^3 - 3x - 7) = 2^5 - 2 \cdot 2^3 - 3 \cdot 2 - 7 = 3.$$

Also, if r is a rational function, that is, if $r(x)$ is the ratio $p(x)/q(x)$ of two polynomials $p(x)$ and $q(x)$ such that $q(a) \neq 0$ for $a \in \mathbf{R}$, then by Theorem 12.27,

$$\lim_{x \to a} r(x) = \lim_{x \to a} \frac{p(x)}{q(x)} = \frac{\lim_{x \to a} p(x)}{\lim_{x \to a} q(x)} = \frac{p(a)}{q(a)} = r(a). \tag{12.2}$$

So, in Result 12.18, we have

$$\lim_{x \to 1} \frac{x^2 + 1}{x^2 + 4} = \frac{1^2 + 1}{1^2 + 4} = \frac{2}{5}.$$

Although it is simpler and certainly less time-consuming to verify certain limits with the aid of these theorems, we should also know how to verify limits by the $\epsilon - \delta$ definition.

12.5 Continuity

Once again, let $f : X \to \mathbf{R}$ be a function, where $X \subseteq \mathbf{R}$, and let a be a real number such that f is defined in some deleted neighborhood of a. Recall that $\lim_{x \to a} f(x) = L$ for some real number L if for every $\epsilon > 0$, there exists $\delta > 0$ such that if $x \in (a - \delta, a + \delta)$ and $x \neq a$, then $|f(x) - L| < \epsilon$. If f is defined at a and $f(a) = L$, then f is said to **continuous** at a. That is, f is continuous at a if $\lim_{x \to a} f(x) = f(a)$. Therefore, a function f is continuous at a if for every $\epsilon > 0$, there exists $\delta > 0$ such that if $|x - a| < \delta$, then $|f(x) - f(a)| < \epsilon$. (Notice that in this instance, $0 < |x - a| < \delta$ is being replaced by $|x - a| < \delta$.) Thus for f to be continuous at a, three conditions must be satisfied:

(1) f is defined at a;   (2) $\lim_{x \to a} f(x)$ exists;   (3) $\lim_{x \to a} f(x) = f(a)$.

We now illustrate this.

Problem 12.32   *A function f is defined by $f(x) = (x^2 - 3x + 2)/(x^2 - 1)$ for all $x \in \mathbf{R} - \{-1, 1\}$. Is f continuous at 1 under any of the following circumstances: (a) f is not defined at 1; (b) $f(1) = 0$; (c) $f(1) = -1/2$?*

Solution   For f to be continuous at 1, the function f must be defined at 1. So we can answer question (a) immediately. The answer is no. In order to answer questions (b) and (c), we must first determine whether $\lim_{x \to 1} f(x)$ exists. Observe that

$$f(x) = \frac{x^2 - 3x + 2}{x^2 - 1} = \frac{(x - 1)(x - 2)}{(x - 1)(x + 1)} = \frac{x - 2}{x + 1}$$

since $x \neq 1$. Because $f(x) = \dfrac{x - 2}{x + 1}$ is a rational function, we can apply (12.2) to obtain

$$\lim_{x \to 1} \frac{x - 2}{x + 1} = \frac{\lim_{x \to 1}(x - 2)}{\lim_{x \to 1}(x + 1)} = \frac{-1}{2} = -\frac{1}{2}.$$

Hence if $f(1) = -1/2$, then f is continuous at 1. Therefore, the answer to question (b) is no and the answer to (c) is yes.   ◆

For additional practice, we present an $\epsilon - \delta$ proof that $\lim_{x \to 1} \dfrac{x^2 - 3x + 2}{x^2 - 1} = -\dfrac{1}{2}$.

Result to Prove $\lim_{x \to 1} \dfrac{x^2 - 3x + 2}{x^2 - 1} = -\dfrac{1}{2}.$

PROOF STRATEGY   Observe that

$$\left| \frac{x^2 - 3x + 2}{x^2 - 1} - \left(-\frac{1}{2} \right) \right| = \left| \frac{(x-1)(x-2)}{(x-1)(x+1)} + \frac{1}{2} \right| = \left| \frac{(x-2)}{(x+1)} + \frac{1}{2} \right|$$

$$= \left| \frac{2(x-2) + (x+1)}{2(x+1)} \right| = \left| \frac{3x-3}{2(x+1)} \right| = \frac{3\,|x-1|}{2\,|x+1|}.$$

If $|x - 1| < 1$, then $0 < x < 2$ and $|x + 1| > 1$, so $1/|x + 1| < 1$. We are now prepared to prove that $\lim\limits_{x \to 1} f(x) = -1/2$. ◆

Result 12.33 $\lim\limits_{x \to 1} \dfrac{x^2 - 3x + 2}{x^2 - 1} = -\dfrac{1}{2}.$

Proof   Let $\epsilon > 0$ and choose $\delta = \min(1, 2\epsilon/3)$. Let $x \in \mathbf{R}$ such that $|x - 1| < \delta$. Since $|x - 1| < 1$, it follows that $0 < x < 2$. So $|x + 1| > 1$ and $1/|x + 1| < 1$. Hence

$$\left| \frac{x^2 - 3x + 2}{x^2 - 1} - \left(-\frac{1}{2} \right) \right| = \left| \frac{(x-1)(x-2)}{(x-1)(x+1)} + \frac{1}{2} \right| = \left| \frac{x-2}{x+1} + \frac{1}{2} \right|$$

$$= \left| \frac{3x-3}{2(x+1)} \right| = \frac{3\,|x-1|}{2\,|x+1|} < \frac{3}{2} \cdot \frac{2\epsilon}{3} = \epsilon. \qquad \blacksquare$$

Indeed, (12.2) states that if a rational function r is defined by $r(x) = p(x)/q(x)$, where $p(x)$ and $q(x)$ are polynomials such that $q(a) \neq 0$, then r is continuous at a. Also, (12.1) implies that if p is a polynomial function defined by $p(x) = c_n x^n + c_{n-1} x^{n-1} + \cdots + c_1 x + c_0$, then p is continuous at every real number a.

We now present some examples concerning continuity for functions that are neither polynomials nor rational functions.

Result to Prove   The function f defined by $f(x) = \sqrt{x}$ for $x \geq 0$ is continuous at 4.

PROOF STRATEGY   Because $f(4) = 2$, it suffices to show that $\lim\limits_{x \to 4} \sqrt{x} = 2$. Thus $|f(x) - L| = |\sqrt{x} - 2|$. To work $x - 4$ into the expression $\sqrt{x} - 2$, we multiply $\sqrt{x} - 2$ by $(\sqrt{x} + 2)/(\sqrt{x} + 2)$, obtaining

$$|\sqrt{x} - 2| = \left| \frac{(\sqrt{x} - 2)(\sqrt{x} + 2)}{\sqrt{x} + 2} \right| = \frac{|x - 4|}{\sqrt{x} + 2}.$$

First we require that $\delta \leq 1$, that is, $|x - 4| < 1$, so $3 < x < 5$. Since $\sqrt{x} + 2 > 3$, it follows that $1/(\sqrt{x} + 2) < 1/3$. Hence

$$|\sqrt{x} - 2| = \frac{|x - 4|}{\sqrt{x} + 2} < \frac{|x - 4|}{3}.$$

This suggests an appropriate choice for δ. ◆

Result 12.34   *The function f defined by $f(x) = \sqrt{x}$ for $x \geq 0$ is continuous at 4.*

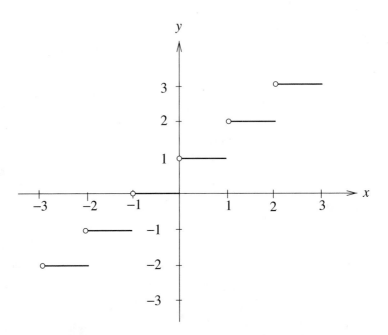

Figure 12.5   The graph of the ceiling function $f(x) = \lceil x \rceil$

Proof   Let $\epsilon > 0$ be given and choose $\delta = \min(1, 3\epsilon)$. Let $x \in \mathbf{R}$ such that $|x - 4| < \delta$. Since $|x - 4| < 1$, it follows that $3 < x < 5$ and so $\sqrt{x} + 2 > 3$. Therefore, $1/(\sqrt{x} + 2) < 1/3$. Hence

$$|\sqrt{x} - 2| = \left| \frac{(\sqrt{x} - 2)(\sqrt{x} + 2)}{\sqrt{x} + 2} \right| = \frac{|x - 4|}{\sqrt{x} + 2} < \frac{1}{3}(3\epsilon) = \epsilon. \qquad \blacksquare$$

Figure 12.5 gives the graph of the ceiling function $f : \mathbf{R} \to \mathbf{Z}$ defined by $f(x) = \lceil x \rceil$. This function is not continuous at any integer but is continuous at all other real numbers. We verify the first of these remarks and leave the proof of the second remark as an exercise (Exercise 12.37).

Result 12.35   *The ceiling function $f : \mathbf{R} \to \mathbf{Z}$ defined by $f(x) = \lceil x \rceil$ is not continuous at any integer.*

Proof   Assume, to the contrary, that there is some integer k such that f is continuous at k. Therefore, $\lim_{x \to k} f(x) = f(k) = \lceil k \rceil = k$. Hence for $\epsilon = 1$, there exists $\delta > 0$ such that if $|x - k| < \delta$, then $|f(x) - f(k)| = |f(x) - k| < \epsilon = 1$. Let $\delta_1 = \min(\delta, 1)$ and let $x_1 \in (k, k + \delta_1)$. Thus $k < x_1 < k + \delta$ and $k < x_1 < k + 1$. Hence $f(x_1) = \lceil x_1 \rceil = k + 1$ and $|f(x_1) - k| = |(k + 1) - k| = 1 < 1$, a contradiction. $\qquad \blacksquare$

12.6 Differentiability

We have discussed the existence and nonexistence of limits $\lim_{x \to a} f(x)$ for functions $f : X \to \mathbf{R}$ with $X \subseteq \mathbf{R}$, where f is defined in a deleted neighborhood of the real number a and, in the case of continuity at a, investigated whether $\lim_{x \to a} f(x) = f(a)$ if

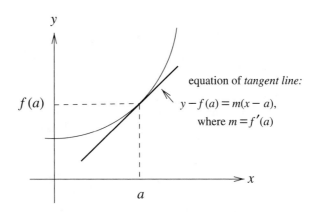

Figure 12.6   Derivatives and slopes of tangent lines

f is defined in a neighborhood of a. If f is defined in a neighborhood of a, then there is an important limit that concerns the ratio of the differences $f(x) - f(a)$ and $x - a$.

A function $f : X \to \mathbf{R}$, where $X \subseteq \mathbf{R}$, that is defined in a neighborhood of a real number a is said to be **differentiable** at a if $\displaystyle\lim_{x \to a} \frac{f(x) - f(a)}{x - a}$ exists. This limit is called the **derivative** of f at a and is denoted by $f'(a)$. Therefore,

$$f'(a) = \lim_{x \to a} \frac{f(x) - f(a)}{x - a}.$$

You probably already know that $f'(a)$ is the slope of the tangent line to the graph of $y = f(x)$ at the point $(a, f(a))$. Indeed, if $f'(a) = m$, then the equation of this line is $y - f(a) = m(x - a)$. See Figure 12.6.

We illustrate derivatives with an example.

Example 12.36   *Show that the function f defined by $f(x) = 1/x^2$ for $x \neq 0$ is differentiable at 1 and determine $f'(1)$.*

Solution   Thus we need to show that $\displaystyle\lim_{x \to 1} \frac{f(x) - f(1)}{x - 1} = \lim_{x \to 1} \frac{\frac{1}{x^2} - 1}{x - 1}$ exists. In a deleted neighborhood of 1,

$$\frac{\frac{1}{x^2} - 1}{x - 1} = \frac{\frac{1 - x^2}{x^2}}{x - 1} = \frac{1 - x^2}{x^2(x - 1)} = \frac{(1 - x)(1 + x)}{x^2(x - 1)} = -\frac{1 + x}{x^2}. \tag{12.3}$$

Since $\dfrac{1 + x}{-x^2}$ is a rational function, we can once again use (12.2) to see that

$$\lim_{x \to 1} \frac{1 + x}{-x^2} = \frac{\lim_{x \to 1}(1 + x)}{\lim_{x \to 1}(-x^2)} = \frac{2}{-1} = -2$$

and so $f'(1) = -2$.

We present an $\epsilon - \delta$ proof of this limit as well. For a given $\epsilon > 0$, we are required to find $\delta > 0$ such that if $x \in \mathbf{R}$ with $0 < |x - 1| < \delta$, then $\left| \dfrac{\frac{1}{x^2} - 1}{x - 1} - (-2) \right| < \epsilon$. Observe that

$$\left| \frac{\frac{1}{x^2} - 1}{x - 1} - (-2) \right| = \left| -\frac{1+x}{x^2} + 2 \right| = \left| \frac{2x^2 - x - 1}{x^2} \right| = \frac{|x-1||2x+1|}{x^2}.$$

If we restrict δ so that $\delta \le 1/2$, then $|x - 1| < 1/2$ and so $1/2 < x < 3/2$. Since $x > 1/2$, it follows that $x^2 > 1/4$ and $1/x^2 < 4$. Also, since $x < 3/2$, it follows that $|2x + 1| < 4$. Hence $|x - 1||2x + 1|/x^2 < 16|x - 1|$. This shows us how to select δ. We now prove that $f'(1) = -2$.   ◆

Result 12.37   *Let f be the function defined by $f(x) = 1/x^2$ for $x \ne 0$. Then $f'(1) = -2$.*

Proof   Let $\epsilon > 0$ be given and choose $\delta = \min(1/2, \epsilon/16)$. Let $x \in \mathbf{R}$ such that $0 < |x - 1| < \delta$. Since $|x - 1| < 1/2$, it follows that $1/2 < x < 3/2$. Thus $x^2 > 1/4$ and so $1/x^2 < 4$. Also, $|2x + 1| < 4$. Since $|x - 1| < \epsilon/16$, it follows that

$$\left| \frac{f(x) - f(1)}{x - 1} - (-2) \right| = \left| \frac{\frac{1}{x^2} - 1}{x - 1} - (-2) \right| = \left| -\frac{1+x}{x^2} + 2 \right|$$

$$= \left| \frac{2x^2 - x - 1}{x^2} \right| = \frac{|2x + 1|}{x^2} \cdot |x - 1| < 4 \cdot 4 \cdot \frac{\epsilon}{16} = \epsilon. \quad \blacksquare$$

From Result 12.37, it now follows that the slope of the tangent line to the graph of $y = 1/x^2$ at the point $(1, 1)$ is -2 and, consequently, that the equation of this tangent line is $y - 1 = -2(x - 1)$.

Differentiability of a function at some number a implies continuity at a, as we now show.

Theorem 12.38   *If a function f is differentiable at a, then f is continuous at a.*

Proof   Since f is differentiable at a, it follows that $\lim\limits_{x \to a} \dfrac{f(x) - f(a)}{x - a}$ exists and equals the real number $f'(a)$. To show that f is continuous at a, we need to show that $\lim\limits_{x \to a} f(x) = f(a)$. We write $f(x)$ as

$$f(x) = \frac{f(x) - f(a)}{x - a}(x - a) + f(a).$$

Now, using properties of limits, we have

$$\lim_{x \to a} f(x) = \left[\lim_{x \to a} \frac{f(x) - f(a)}{x - a} \right] \lim_{x \to a} (x - a) + \lim_{x \to a} f(a)$$

$$= f'(a) \cdot 0 + f(a) = f(a). \quad \blacksquare$$

The converse of Theorem 12.38 is not true. For example, the functions f and g defined by $f(x) = |x|$ and $g(x) = \sqrt[3]{x}$ are continuous at 0 but neither is differentiable at 0. That f is not differentiable at 0 is actually established in Example 12.22.

EXERCISES FOR CHAPTER 12

Section 12.1: Limits of Sequences

12.1. Give an example of a sequence that is not expressed in terms of trigonometry but whose terms are exactly those of the sequence of $\{\cos(n\pi)\}$.

12.2. Give an example of two sequences different from the sequence $\{n^2 - n! + |n - 2|\}$ whose first three terms are the same as those of $\{n^2 - n! + |n - 2|\}$.

12.3. Prove that the sequence $\left\{\frac{1}{2n}\right\}$ converges to 0.

12.4. Prove that the sequence $\left\{\frac{1}{n^2+1}\right\}$ converges to 0.

12.5. Prove that the sequence $\left\{1 + \frac{1}{2^n}\right\}$ converges to 1.

12.6. Prove that the sequence $\left\{\frac{n+2}{2n+3}\right\}$ converges to $\frac{1}{2}$.

12.7. By definition, $\lim_{n\to\infty} a_n = L$ if for every $\epsilon > 0$, there exists a positive integer N such that if n is an integer with $n > N$, then $|a_n - L| < \epsilon$. By taking the negation of this definition, write out the meaning of $\lim_{n\to\infty} a_n \neq L$ using quantifiers. Then write out the meaning of $\{a_n\}$ diverges using quantifiers.

12.8. Show that the sequence $\{n^4\}$ diverges to infinity.

12.9. Show that the sequence $\left\{\frac{n^5+2n}{n^2}\right\}$ diverges to infinity.

12.10. (a) Prove that $1 + \frac{1}{2} + \frac{1}{3} + \cdots + \frac{1}{n} < 2\sqrt{n}$ for every positive integer n.
    (b) Let $s_n = \frac{1}{n} + \frac{1}{2n} + \frac{1}{3n} + \cdots + \frac{1}{n^2}$ for each $n \in \mathbf{N}$. Prove that the sequence $\{s_n\}$ converges to 0.

12.11. Prove that if a sequence $\{s_n\}$ converges to L, then the sequence $\{s_{n^2}\}$ also converges to L.

Section 12.2: Infinite Series

12.12. Prove that the series $\sum_{k=1}^{\infty} \frac{1}{(3k-2)(3k+1)}$ converges and determine its sum by

(a) computing the first few terms of the sequence $\{s_n\}$ of partial sums and conjecturing a formula for s_n;
(b) using mathematical induction to verify that your conjecture in (a) is correct;
(c) completing the proof.

12.13. Prove that the series $\sum_{k=1}^{\infty} \frac{1}{2^k}$ converges and determine its sum by

(a) computing the first few terms of the sequence $\{s_n\}$ of partial sums and conjecturing a formula for s_n;
(b) using mathematical induction to verify that your conjecture in (a) is correct;
(c) completing the proof.

12.14. The terms $a_1, a_2, a_3, \cdots$ of the series $\sum_{k=1}^{\infty} a_k$ are defined recursively by $a_1 = \frac{1}{6}$ and

$$a_n = a_{n-1} - \frac{2}{n(n+1)(n+2)}$$

for $n \geq 2$. Prove that $\sum_{k=1}^{\infty} a_k$ converges and determine its value.

12.15. Prove that the series $\sum_{k=1}^{\infty} \frac{k+3}{(k+1)^2}$ diverges to infinity.

12.16. (a) Prove that if $\sum_{k=1}^{\infty} a_k$ is a convergent series, then $\lim_{n\to\infty} a_n = 0$.
    (b) Show that the converse of the result in (a) is false.

12.17. Let $\sum_{k=1}^{\infty} a_k$ be an infinite series whose sequence of partial sums is $\{s_n\}$ where $s_n = \frac{3n}{4n+2}$.

(a) What is the series $\sum_{k=1}^{\infty} a_k$?
(b) Determine the sum s of $\sum_{k=1}^{\infty} a_k$ and prove that $\sum_{k=1}^{\infty} a_k = s$.

Section 12.3: Limits of Functions

12.18. Give an $\epsilon - \delta$ proof that $\lim_{x \to 2} \left(\frac{3}{2}x + 1 \right) = 4$.

12.19. Give an $\epsilon - \delta$ proof that $\lim_{x \to -1} (3x - 5) = -8$.

12.20. Give an $\epsilon - \delta$ proof that $\lim_{x \to 2} (2x^2 - x - 5) = 1$.

12.21. Give an $\epsilon - \delta$ proof that $\lim_{x \to 2} x^3 = 8$.

12.22. Determine $\lim_{x \to 1} \frac{1}{5x-4}$ and verify that your answer is correct with an $\epsilon - \delta$ proof.

12.23. Give an $\epsilon - \delta$ proof that $\lim_{x \to 3} \frac{3x+1}{4x+3} = \frac{2}{3}$.

12.24. Determine $\lim_{x \to 3} \frac{x^2-2x-3}{x^2-8x+15}$ and verify that your answer is correct with an $\epsilon - \delta$ proof.

12.25. Show that $\lim_{x \to 0} \frac{1}{x^2}$ does not exist.

12.26. The function $f : \mathbf{R} \to \mathbf{R}$ is defined by

$$
f(x) = \begin{cases} 1 & x < 3 \\ 1.5 & x = 3 \\ 2 & x > 3. \end{cases}
$$

  (a) Determine whether $\lim_{x \to 3} f(x)$ exists and verify your answer.
  (b) Determine whether $\lim_{x \to \pi} f(x)$ exists and verify your answer.

12.27. A function $g : \mathbf{R} \to \mathbf{R}$ is **bounded** if there exists a positive real number B such that $|g(x)| < B$ for each $x \in \mathbf{R}$.

  (a) Let $g : \mathbf{R} \to \mathbf{R}$ be a bounded function and suppose that $f : \mathbf{R} \to \mathbf{R}$ and $a \in \mathbf{R}$ such that $\lim_{x \to a} f(x) = 0$. Prove that $\lim_{x \to a} f(x)g(x) = 0$.
  (b) Use the result in (a) to determine $\lim_{x \to 0} x^2 \sin \left(\frac{1}{x^2} \right)$.

12.28. Suppose that $\lim_{x \to a} f(x) = L$, where $L > 0$. Prove that $\lim_{x \to a} \sqrt{f(x)} = \sqrt{L}$.

12.29. Suppose that $f : \mathbf{R} \to \mathbf{R}$ is a function such that $\lim_{x \to 0} f(x) = L$.

  (a) Let $c \in \mathbf{R}$. Prove that $\lim_{x \to c} f(x - c) = L$.
  (b) Suppose that f also has the property that $f(a + b) = f(a) + f(b)$ for all $a, b \in \mathbf{R}$. Use the result in (a) to prove that $\lim_{x \to c} f(x)$ exists for all $c \in \mathbf{R}$.

12.30. Let $f : \mathbf{R} \to \mathbf{R}$ be a function.

  (a) Prove that if $\lim_{x \to a} f(x) = L$, then $\lim_{x \to a} |f(x)| = |L|$.
  (b) Prove or disprove: If $\lim_{x \to a} |f(x)| = |L|$, then $\lim_{x \to a} f(x)$ exists.

Section 12.4: Fundamental Properties of Limits of Functions

12.31. Use limit theorems to determine the following:

  (a) $\lim_{x \to 1} (x^3 - 2x^2 - 5x + 8)$
  (b) $\lim_{x \to 1} (4x + 7)(3x^2 - 2)$
  (c) $\lim_{x \to 2} \frac{2x^2-1}{3x^3+1}$

12.32. Use induction to prove that for every integer $n \geq 2$ and every n functions $f_1, f_2, \cdots, f_n$ such that $\lim_{x \to a} f_i(x) = L_i$ for $1 \leq i \leq n$,

$$
\lim_{x \to a} (f_1(x) + f_2(x) + \cdots + f_n(x)) = L_1 + L_2 + \cdots + L_n.
$$

12.33. Use Exercise 12.32 to prove that $\lim_{x \to a} p(x) = p(a)$ for every polynomial
$p(x) = c_n x^n + c_{n-1} x^{n-1} + \cdots + c_1 x + c_0$.

12.34. Prove that if $f_1, f_2, \ldots, f_n$ are any $n \geq 2$ functions such that $\lim_{x \to a} f_i(x) = L_i$ for $1 \leq i \leq n$, then

$$\lim_{x \to a} (f_1(x) \cdot f_2(x) \cdots f_n(x)) = L_1 \cdot L_2 \cdots L_n.$$

Section 12.5: Continuity

12.35. The function $f : \mathbf{R} - \{0, 2\} \to \mathbf{R}$ is defined by $f(x) = \frac{x^2-4}{x^3-2x^2}$. Use limit theorems to determine whether f
can be defined at 2 such that f is continuous at 2.

12.36. The function f defined by $f(x) = \frac{x^2-9}{x^2-3x}$ is not defined at 3. Is it possible to define f at 3 such that f is
continuous there? Verify your answer with an $\epsilon - \delta$ proof.

12.37. Let $f : \mathbf{R} \to \mathbf{Z}$ be the ceiling function defined by $f(x) = \lceil x \rceil$. Give an $\epsilon - \delta$ proof that if a is a real
number that is not an integer, then f is continuous at a.

12.38. Show that Exercise 12.33 implies that every polynomial is continuous at every real number.

12.39. Prove that the function $f : [1, \infty) \to [0, \infty)$ defined by $f(x) = \sqrt{x-1}$ is continuous at $x = 10$.

12.40. (a) Let $f : \mathbf{R} \to \mathbf{R}$ be defined by

$$f(x) = \begin{cases} 0 & \text{if } x \text{ is rational} \\ 1 & \text{if } x \text{ is irrational.} \end{cases}$$

In particular, $f(0) = 0$. Prove or disprove: f is continuous at $x = 0$.
(b) The problem in (a) should suggest another problem to you. State and solve such a problem.

Section 12.6: Differentiability

12.41. The function $f : \mathbf{R} \to \mathbf{R}$ is defined by $f(x) = x^2$. Determine $f'(3)$ and verify that your answer is correct
with an $\epsilon - \delta$ proof.

12.42. The function $f : \mathbf{R} - \{-2\} \to \mathbf{R}$ is defined by $f(x) = \frac{1}{x+2}$. Determine $f'(1)$ and verify that your answer
is correct with an $\epsilon - \delta$ proof.

12.43. The function $f : \mathbf{R} \to \mathbf{R}$ is defined by $f(x) = x^3$. Determine $f'(a)$ for $a \in \mathbf{R}^+$ and verify that your
answer is correct with an $\epsilon - \delta$ proof.

12.44. The function $f : \mathbf{R} \to \mathbf{R}$ is defined by

$$f(x) = \begin{cases} x^2 \sin \frac{1}{x} & \text{if } x \neq 0 \\ 0 & \text{if } x = 0. \end{cases}$$

Determine $f'(0)$ and verify that your answer is correct with an $\epsilon - \delta$ proof.

ADDITIONAL EXERCISES FOR CHAPTER 12

12.45. Prove that the sequence $\left\{ \frac{n+1}{3n-1} \right\}$ converges to $\frac{1}{3}$.

12.46. Prove that $\lim_{n \to \infty} \frac{2n^2}{4n^2+1} = \frac{1}{2}$.

12.47. Prove that the sequence $\{1 + (-2)^n\}$ diverges.

12.48. Prove that $\lim_{n \to \infty} (\sqrt{n^2 + 1} - n) = 0$.

12.49. Prove that the sequence $\left\{(-1)^{n+1}\frac{n}{2n+1}\right\}$ diverges.

12.50. Prove that $\lim\limits_{n\to\infty}\dfrac{n}{3n+1}=\dfrac{1}{3}$.

12.51. Let $a, c_0, c_1 \in \mathbf{R}$ such that $c_1 \neq 0$. Give an $\epsilon - \delta$ proof that $\lim_{x\to a}(c_1x+c_0)=c_1a+c_0$.

12.52. Evaluate the proposed solution of the following problem.
   Problem   The function $f: \mathbf{R} \to \mathbf{R}$ is defined by

$$f(x) = \begin{cases} \frac{x^2-4}{x-2} & \text{if } x \neq 2 \\ 2 & \text{if } x = 2. \end{cases}$$

   Determine whether $\lim\limits_{x\to 2} f(x)$ exists.

   Solution   Consider

$$\lim_{x\to 2} f(x) = \lim_{x\to 2}\frac{x^2-4}{x-2} = \lim_{x\to 2}\frac{(x-2)(x+2)}{x-2} = \lim_{x\to 2}(x+2) = 4.$$

   However, since $\lim\limits_{x\to 2} f(x) = 4 \neq 2 = f(2)$, the limit does not exist.   ◆

12.53. Evaluate the proposed proof of the following result.
   Result   The sequence $\left\{\dfrac{2n}{3n+5}\right\}$ converges to $\dfrac{2}{3}$.

   Proof   Let $\epsilon > 0$ be given. Choose $N = \left\lceil \dfrac{10}{9\epsilon} - \dfrac{5}{3} \right\rceil$ and let $n > N$. Then $n > \dfrac{10}{9\epsilon} - \dfrac{5}{3}$.
   Hence $9n > \dfrac{10}{\epsilon} - 15$ and $9n + 15 > \dfrac{10}{\epsilon}$. Therefore,

$$\frac{9n+15}{10} > \frac{1}{\epsilon} \quad \text{and} \quad \frac{10}{9n+15} < \epsilon.$$

   Now

$$\left|\frac{2n}{3n+5} - \frac{2}{3}\right| = \left|\frac{6n-2(3n+5)}{3(3n+5)}\right| = \frac{|-10|}{9n+15} = \frac{10}{9n+15} < \epsilon.$$ ∎

12.54. Evaluate the proposed proof of the following result.
   Result $\lim\limits_{x\to 1}\dfrac{1}{2x-3} = -1$.

   Proof   Let $\epsilon > 0$ and choose $\delta = \min(1, \frac{7\epsilon}{2})$. Let $0 < |x-1| < \delta = \min(1, \frac{7\epsilon}{2})$. Since $|x-1| < 1$, it follows that $0 < x < 2$ and so

$$|2x-3| \leq |2x| + |-3| = 2|x| + 3 < 4 + 3 = 7.$$

   Since $|x-1| < 7\epsilon/2$, we have

$$\left|\frac{1}{2x-3} + 1\right| = \left|\frac{2x-2}{2x-3}\right| = \frac{2|x-1|}{|2x-3|} < \frac{2}{7}\cdot\frac{7\epsilon}{2} = \epsilon.$$ ∎

12.55. Let $\{a_n\}$, $\{b_n\}$ and $\{c_n\}$ be sequences of real numbers such that $a_n \leq b_n \leq c_n$ for every positive integer n and $\lim\limits_{n\to\infty} a_n = \lim\limits_{n\to\infty} c_n = L$.

   (a) Prove that $\lim\limits_{n\to\infty}(c_n - a_n) = 0$.

   (b) Prove that $\lim\limits_{n\to\infty} b_n = L$.

12.56. In Chapter 10 it was shown that the set $\mathbf{Q}$ of rational numbers is denumerable and consequently can be expressed as $\mathbf{Q} = \{q_1, q_2, q_3, \ldots\}$. A function $f : \mathbf{R} \to \mathbf{R}$ is defined by

$$f(x) = \begin{cases} \frac{1}{n} & \text{if } x = q_n \ (n = 1, 2, 3, \ldots) \\ 0 & \text{if } x \text{ is irrational.} \end{cases}$$

(a) Prove that f is continuous at each irrational number.

(b) Prove that f is not continuous at any rational number.

(c) If the function f were defined as above except that $f(0) = 0$, then prove that f would be continuous at 0.

13

Proofs in Group Theory

Many of the proofs that we have seen involve familiar sets of numbers (especially integers, rational numbers and real numbers). Also, most of the theorems and examples that we have encountered concern additive properties of these numbers or multiplicative properties (or both). Many important properties of integers (or of rational or real numbers) come not from the integers themselves, but from the addition and multiplication of integers. This suggests a basic question in mathematics: For a given nonempty set S, can we describe other, less familiar methods of associating an element of S with each pair of elements of S in such a way that some interesting properties occur? The mathematical subject that deals with questions such as these is **abstract algebra** (also called **modern algebra** or simply **algebra**). In this chapter, we will look at one of the most familiar concepts in abstract algebra. First, however, we must have a clear understanding of what we mean by associating an element of a given set with each pair of elements of that set and what properties may be considered interesting.

13.1 Binary Operations

When we add two integers a and b, we perform an operation (namely addition) to produce an integer that we denote by $a + b$. Similarly, when we multiply these two integers, we perform another operation (namely multiplication) to produce an integer that we denote by $a \cdot b$ (or ab). Both of these operations do something very similar. Each takes a pair a, b of integers, actually an ordered pair (a, b) of integers, and associates with this pair a unique integer. Therefore, these operations are actually functions, namely functions from $\mathbf{Z} \times \mathbf{Z}$ to $\mathbf{Z}$. These functions are examples of a concept called a binary operation.

By a **binary operation** $*$ on a nonempty set S, we mean a function from $S \times S$ to S; that is, $*$ is a function that maps every ordered pair of elements of S to an element of S. Thus $* : S \times S \to S$. In particular, if the ordered pair (a, b) in $S \times S$ is mapped into the element c in S by a binary operation $*$ (that is, c is the image of (a, b) under $*$), then we write $c = a * b$ rather than the more awkward notation $*((a, b)) = c$.

Consequently, addition $+$ and multiplication $\cdot$ are binary operations on $\mathbf{Z}$. For example, under addition, the ordered pair $(3, 5)$ is mapped into $3 + 5 = 8$; while under multiplication, $(3, 5)$ is mapped into $3 \cdot 5 = 15$. Subtraction is also a binary operation on $\mathbf{Z}$ but it is not a binary operation on $\mathbf{N}$ since, for example, subtraction maps the ordered pair $(3, 5)$ into $3 - 5 = -2$, which does not belong to **N**. Therefore, subtraction is not a function from $\mathbf{N} \times \mathbf{N}$ to **N** since it is not defined at $(3, 5)$, as well as at many other ordered pairs of positive integers. Similarly, division is not a binary operation on **Z** or **N** because it is not defined for many ordered pairs, including $(1, 0) \in \mathbf{Z} \times \mathbf{Z}$ and $(2, 3) \in \mathbf{N} \times \mathbf{N}$ because $1/0 \notin \mathbf{Z}$ and $2/3 \notin \mathbf{N}$. However, division is a binary operation on the set $\mathbf{Q}^+$ of positive rational numbers since the quotient of two positive rational numbers is once again a positive rational number.

Not only are addition and multiplication binary operations on **Z**, they are binary operations on **Q** and **R**, as well as on $\mathbf{R}^+$ (the positive real numbers) and $\mathbf{Q}^+$. For the set $\mathbf{R}^*$ of nonzero real numbers, multiplication is a binary operation but addition is not (since $1 + (-1) = 0 \notin \mathbf{R}^*$, for example).

If $*$ is a binary operation on a set S, then, by definition, $a * b \in S$ for all $a, b \in S$. If T is a nonempty subset of S and $a, b \in T$, then certainly $a * b \in S$; however, $a * b$ need not belong to T. A nonempty subset T of S is said to be **closed** under $*$ if whenever, $a, b \in T$, then $a * b \in T$ as well. If $*$ is a binary operation on S, then certainly S is closed under $*$. Although subtraction is a binary operation on **Z**, the subset **N** of **Z** is not closed under subtraction.

Among familiar sets with familiar binary operations are:

(a) the set $\mathbf{Z}_n = \{[0], [1], \cdots, [n-1]\}$, $n \geq 2$, of residue classes modulo n, under residue class addition $[a] + [b] = [a + b]$ and under residue class multiplication $[a] \cdot [b] = [ab]$ (as defined in Chapter 8);

(b) the set $M_2(\mathbf{R})$ of all 2×2 matrices over **R** (that is, whose entries are real numbers) under matrix addition

$$\begin{bmatrix} a & b \\ c & d \end{bmatrix} + \begin{bmatrix} e & f \\ g & h \end{bmatrix} = \begin{bmatrix} a+e & b+f \\ c+g & d+h \end{bmatrix}$$

and under matrix multiplication

$$\begin{bmatrix} a & b \\ c & d \end{bmatrix} \cdot \begin{bmatrix} e & f \\ g & h \end{bmatrix} = \begin{bmatrix} ae+bg & af+bh \\ ce+dg & cf+dh \end{bmatrix};$$

(c) the set $\mathcal{F}_{\mathbf{R}} = \mathbf{R}^{\mathbf{R}}$ of functions from **R** to **R** under function addition $(f + g)(x) = f(x) + g(x)$, under function multiplication $(f \cdot g)(x) = f(x) \cdot g(x)$ and under function composition $(f \circ g)(x) = f(g(x))$;

(d) the power set $\mathcal{P}(A)$ of a set A under set union, under set intersection and under set difference.

For a more abstract example of a binary operation, let $S = \{a, b, c\}$. A binary operation $*$ on S is illustrated in the table in Figure 13.1, where, then $a * a = b$, $a * b = c$, $a * c = a$, etc. Since every element in the table belongs to S, it follows that $*$ is indeed a binary operation on S.

Although it may seem relatively clear that a binary operation is defined in each example given above, not all binary operations are so immediate.

Result 13.1   *For $a, b \in \mathbf{R} - \{-2\}$, define*

$$a * b = ab + 2a + 2b + 2,$$

*	a	b	c
a	b	c	a
b	a	c	a
c	c	a	b

Figure 13.1   A binary operation $*$ on $S = \{a, b, c\}$

where the operations indicated in $ab + 2a + 2b + 2$ are ordinary addition and multi-plication in $\mathbf{R}$. Then $*$ is a binary operation on $\mathbf{R} - \{-2\}$.

Proof   We need to show that if $a, b \in \mathbf{R} - \{-2\}$, then $a * b \in \mathbf{R} - \{-2\}$. Assume, to the contrary, that there exists some pair $x, y \in \mathbf{R} - \{-2\}$ such that $x * y \notin \mathbf{R} - \{-2\}$. Thus $x * y = xy + 2x + 2y + 2 = -2$. This equation is equivalent to $(x + 2)(y + 2) = 0$, so either $x = -2$ or $y = -2$, which is impossible. Hence $*$ is a binary operation on $\mathbf{R} - \{-2\}$. ∎

A nonempty set S with a binary operation $*$ is often denoted by $(S, *)$. We refer to $(S, *)$ as an **algebraic structure**. There are certain properties that $(S, *)$ may possess that will be of special interest to us. In particular,

G1 $(S, *)$ is **associative** if $a * (b * c) = (a * b) * c$ for all $a, b, c \in S$;
G2 $(S, *)$ has an element e, called an **identity element** (or simply an **identity**), if $a * e = e * a = a$ for each $a \in S$;
G3 $(S, *)$ has an identity e and, for each element $a \in S$, there is an element $s \in S$, called an **inverse** for a, such that $a * s = s * a = e$;
G4 $(S, *)$ is **commutative** if $a * b = b * a$ for all $a, b \in S$.

Two elements $a, b \in S$ are said to **commute** if $a * b = b * a$. If every two elements of S commute, then $(S, *)$ satisfies property G4. By property G2, an identity commutes with every element of S; and by property G3, each element of S commutes with an inverse of this element (assuming, of course, that it has an inverse).

An algebraic structure $(S, *)$ may satisfy all, some or none of the properties G1 – G4; however, $(S, *)$ cannot satisfy G3 without first satisfying G2. For elements $a, b, c \in S$, the expression $a * b * c$ is, strictly speaking, not defined. Since $*$ is a *binary* operation, it is only defined for *pairs* of elements of S. There are two standard interpretations of $a * b * c$. Namely, does $a * b * c$ mean $a * (b * c)$ or does $a * b * c$ mean $(a * b) * c$? On the other hand, if $(S, *)$ satisfies property G1 (the associative property), then $a * (b * c) = (a * b) * c$ and so either interpretation is acceptable in this case. For this reason, we often write $a * b * c$ (without any parentheses). Ordinarily, however, we will continue to write $a * (b * c)$ or $(a * b) * c$ to emphasize the importance of parentheses, even when $(S, *)$ satisfies the associative property.

Certainly $(\mathbf{Z}, +)$ satisfies properties G1 – G4, where 0 is an identity element and $-n$ is an inverse for the integer n. Moreover, $(\mathbf{R}, \cdot)$ satisfies properties G1, G2 and G4, where the integer 1 is an identity element. Turning to property G3, we see that every real number r has $1/r$ as an inverse, except 0, which has no inverse since there is no real number s such that $0 \cdot s = s \cdot 0 = 1$. Thus $(R, \cdot)$ does not satisfy G3. In the case of $(\mathbf{R}^*, \cdot)$, where, recall, $\mathbf{R}^*$ is the set of all nonzero real numbers, all four properties G1–G4 are satisfied.

The algebraic structure $(\mathbf{Z}_n, +)$, $n \geq 2$, also satisfies all of the properties G1 – G4, where $[0]$ is an identity and $[-a]$ is an inverse of $[a]$. On the other hand, $(\mathbf{Z}_n, \cdot)$ satisfies only G1, G2 and G4, where $[1]$ is an identity; however $(\mathbf{Z}_n, \cdot)$ does not satisfy G3 since, for example, there is no element $[s] \in \mathbf{Z}_n$ such that $[0][s] = [1]$ in $\mathbf{Z}_n$ and so $[0]$ does not have an inverse.

The algebraic structure $(M_2(\mathbf{R}), +)$ satisfies all of the properties G1 – G4, where $\begin{bmatrix} 0 & 0 \\ 0 & 0 \end{bmatrix}$ is an identity and $\begin{bmatrix} -a & -b \\ -c & -d \end{bmatrix}$ is an inverse of $\begin{bmatrix} a & b \\ c & d \end{bmatrix}$. The algebraic structure $(M_2(\mathbf{R}), \cdot)$ only satisfies G1 and G2, where $I = \begin{bmatrix} 1 & 0 \\ 0 & 1 \end{bmatrix}$ is an identity. For $A = \begin{bmatrix} a & b \\ c & d \end{bmatrix}$ to have an inverse, the number $ad - bc$ (the determinant of A) must be nonzero. Thus $(M_2(\mathbf{R}), \cdot)$ does not satisfy G3. Also,

$$\begin{bmatrix} 1 & 0 \\ 0 & 0 \end{bmatrix}\begin{bmatrix} 0 & 1 \\ 0 & 0 \end{bmatrix} = \begin{bmatrix} 0 & 1 \\ 0 & 0 \end{bmatrix} \neq \begin{bmatrix} 0 & 0 \\ 0 & 0 \end{bmatrix} = \begin{bmatrix} 0 & 1 \\ 0 & 0 \end{bmatrix}\begin{bmatrix} 1 & 0 \\ 0 & 0 \end{bmatrix}$$

shows that $(M_2(\mathbf{R}), \cdot)$ does not satisfy property G4.

The algebraic structure $(S, *)$ shown in Figure 13.1 is not associative since, for example, $b * (b * c) = b * a = a$, while $(b * b) * c = c * c = b$ and so $b * (b * c) \neq (b * b) * c$. Since S contains no element e such that $e * x = x * e = x$ for all $x \in S$, it follows that $(S, *)$ does not have an identity. Also, $a * c \neq c * a$ since $a * c = a$ and $c * a = c$. Consequently, $(S, *)$ has none of the properties G1–G4.

Let's look at another binary operation defined on $S = \{a, b, c\}$.

Example 13.2 *A binary operation $*$ is defined on the set $S = \{a, b, c\}$ by $x * y = x$ for all $x, y \in S$. Determine which of the properties G1 – G4 are satisfied by $(S, *)$.*

Solution Let x, y and z be any three elements of S (distinct or not). Then $x * (y * z) = x * y = x$, while $(x * y) * z = x * z = x$. Thus $(S, *)$ is associative. Now $(S, *)$ has no identity since for every element $e \in S$, it follows that $e * a = e * b = e$ and so it is impossible for $e * a = a$ and $e * b = b$. Since $(S, *)$ has no identity, the question of inverses does not apply here. Certainly, $(S, *)$ is not commutative since $a * b = a$ while $b * a = b$. ◆

The verification of the associative law in Example 13.2 probably would have looked better had we written

$$x * (y * z) = x * y = x = x * z = (x * y) * z.$$

Example 13.3 *Let $\mathbf{N}_0$ be the set of nonnegative integers and consider $(\mathbf{N}_0, *)$, where $*$ is the binary operation defined by $a * b = |a - b|$ for all $a, b \in \mathbf{N}_0$. Determine which of the four properties G1 – G4 are satisfied by $(\mathbf{N}_0, *)$.*

Solution Since $1 * (2 * 3) = 1 * |2 - 3| = 1 * 1 = |1 - 1| = 0$ and $(1 * 2) * 3 = |1 - 2| * 3 = 1 * 3 = |1 - 3| = 2$, it follows that $(1 * 2) * 3 \neq 1 * (2 * 3)$ and so $(\mathbf{N}_0, *)$ is not associative. Let $a \in \mathbf{N}_0$. Then $a * 0 = 0 * a = |a| = a$ and so 0 is an identity for $(\mathbf{N}_0, *)$. Since $a * a = |a - a| = 0$ for all $a \in \mathbf{N}_0$, it follows that a is an inverse of itself. Because $|a - b| = |b - a|$, we have $a * b = b * a$ and $(\mathbf{N}_0, *)$ is commutative. Hence $(\mathbf{N}_0, *)$ satisfies properties G2 – G4. ♦

Example 13.4   *Let $*$ be the binary operation defined on $\mathbf{Z}$ by $a * b = a + b - 1$ for $a, b \in \mathbf{Z}$, where the operations indicated in $a + b - 1$ are ordinary addition and subtraction. Determine which of the four properties G1 – G4 are satisfied by $(\mathbf{Z}, *)$.*

Solution   For integers a, b and c,

$$a * (b * c) = a * (b + c - 1) = a + (b + c - 1) - 1 = a + b + c - 2,$$

while

$$(a * b) * c = (a + b - 1) * c = (a + b - 1) + c - 1 = a + b + c - 2.$$

Thus $a * (b * c) = (a * b) * c$ and $(\mathbf{Z}, *)$ is associative. Since $a + b - 1 = b + a - 1$, it follows that $a * b = b * a$ for all $a, b \in \mathbf{Z}$ and so $(\mathbf{Z}, *)$ is commutative.

Let a be an integer. Observe that $a * 1 = a + 1 - 1 = a$. Thus 1 is an identity for $(\mathbf{Z}, *)$. For $b = -a + 2 \in \mathbf{Z}$, we have

$$a * b = a * (-a + 2) = a + (-a + 2) - 1 = 1.$$

Hence b is an inverse of a and every integer has an inverse. Therefore, $(\mathbf{Z}, *)$ satisfies all four properties G1 – G4. ♦

Analysis   Let's discuss this example a bit more. It was shown that 1 is an identity for $(\mathbf{Z}, *)$. How did we know to choose 1? Actually, this was a natural choice since we were looking for an integer e such that $e * a = a$ for every integer a. Since $e * a = a + e - 1 = a$, it follows that $e = 1$. The choice of $b = -a + 2$ for an inverse of a comes from solving $a * b = a + b - 1 = 1$ for b. ♦

13.2 Groups

One of the most elementary, yet fundamental, characteristics of the algebraic structure $(\mathbf{Z}, +)$ is the ability to solve linear equations, that is, equations of the type $a + x = b$. By this, we mean that given integers a and b, we seek an integer x for which $a + x = b$. How does one solve this equation? First, we are well aware that $(\mathbf{Z}, +)$ has an identity element 0. Also, a has $-a$ as an inverse. If we add $-a$ to $a + x$, which is the same as adding $-a$ to b since $a + x$ and b are the same integer, then we obtain

$$-a + (a + x) = -a + b. \tag{13.1}$$

Applying the associative law to (13.1), we now obtain

$$(-a + a) + x = -a + b$$

and so

$$0 + x = x = -a + b.$$

This tells us that *if a + x = b has a solution, then the only possible solution is −a + b.*
It doesn't tell us that −a + b is actually a solution, but we can easily take care of this.
Letting $x = -a + b$, we have

$$a + x = a + (-a + b) = (a + (-a)) + b = 0 + b = b.$$

Of course, since $(\mathbf{Z}, +)$ also satisfies the commutative property, the solution $-a + b$ can
also be written as $b + (-a) = b - a$.

Let's look at the related linear equation when the operation is multiplication, say
in $(\mathbf{R}^*, \cdot)$. Here, for $a, b \in \mathbf{R}^*$, we seek $x \in \mathbf{R}^*$ such that $a \cdot x = b$. Recall that 1 is an
identity element in $(\mathbf{R}^*, \cdot)$. Multiplying both sides of $a \cdot x = b$ by $\frac{1}{a}$ (which belongs to
$\mathbf{R}^*$), we obtain

$$\frac{1}{a} \cdot (a \cdot x) = \frac{1}{a} \cdot b = \frac{b}{a}. \tag{13.2}$$

Applying the associative law to (13.2), we obtain

$$\frac{1}{a} \cdot (a \cdot x) = \left(\frac{1}{a} \cdot a\right) \cdot x = 1 \cdot x = x$$

Therefore, $x = b/a$. To show that $\frac{b}{a}$ is, in fact, a solution of $a \cdot x = b$, we let $x = \frac{b}{a}$ and
obtain

$$a \cdot x = a \left(\frac{b}{a}\right) = b.$$

The solutions of the two equations that we have just discussed should look very
familiar to you. However, these have been given to illustrate a more general situation.
Suppose that we have an algebraic structure $(S, *)$ in which we would like to solve all
linear equations, that is, for $a, b \in S$, we wish to show that there exists an element $x \in S$
such that $a * x = b$. If $(S, *)$ is associative, has an identity e and a has an inverse $s \in S$,
then we have $s * (a * x) = s * b$ and so

$$s * (a * x) = (s * a) * x = e * x = x = s * b.$$

To show that $s * b$ is, in fact, a solution of the linear equation $a * x = b$, we let $x = s * b$
and obtain

$$a * x = a * (s * b) = (a * s) * b = e * b = b,$$

and so $s * b$ *is* a solution.

You might now have observed that in order to solve all linear equations in an algebraic
structure $(S, *)$, it is necessary that $(S, *)$ satisfy the three properties G1 – G3 (property
G4 is not required). Algebraic structures that satisfy properties G1 – G3 are so important
in abstract algebra that they are given a special name and will be the major emphasis of
this chapter.

A **group** is a nonempty set G together with a binary operation $*$ that satisfies the
following three properties:

G1 **Associative Law:** $a * (b * c) = (a * b) * c$ for all $a, b, c \in G$;
G2 **Existence of Identity:** There exists an element $e \in G$ such that $a * e = e * a = a$ for every $a \in G$;
G3 **Existence of Inverses:** For each element $a \in G$, there exists an element $s \in G$ such that $a * s = s * a = e$.

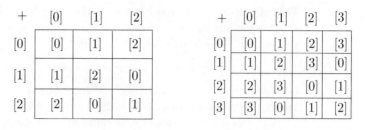

Figure 13.2   The group tables for $(\mathbf{Z}_3, +)$ and $(\mathbf{Z}_4, +)$

Hence, a group is a special kind of algebraic structure $(G, *)$, namely, one that satisfies properties G1 – G3. If the operation $*$ is clear, then we often denote this group simply by G rather than by $(G, *)$. An element $e \in G$ satisfying property G2 is called an **identity** for the group G, while an element s satisfying property G3 is called an **inverse** of a.

If a group G also satisfies the commutative property G4, then G is called an **abelian group**, a term named after the Norwegian mathematician Niels Henrik Abel. If a group G does not satisfy G4, then G is called a **nonabelian group**. We have now seen several abelian groups, namely, $(\mathbf{Z}, +)$, $(\mathbf{Q}, +)$, $(\mathbf{R}, +)$, $(\mathbf{Z}_n, +)$, $(\mathbf{Q}^+, \cdot)$, $(\mathbf{R}^+, \cdot)$ and $(\mathbf{R}^*, \cdot)$ as well as the algebraic structure $(\mathbf{Z}, *)$ described in Example 13.4.

The **order of a group** G, denoted by $|G|$, is the cardinality of G. If the order of G is finite, then G is a **finite group**; while if G has an infinite number of elements, then G is an **infinite group**. All of the groups given above are infinite groups except for $(\mathbf{Z}_n, +)$ which has order n. If a finite group G has relatively few elements, then we often describe the operation $*$ by means of a table, called a **group table** (or **operation table**). For example, the group tables for $(\mathbf{Z}_3, +)$ and $(\mathbf{Z}_4, +)$ are shown in Figure 13.2.

Although $(\mathbf{Z}_n, +)$ is a group for every integer $n \geq 2$, $(\mathbf{Z}_n, \cdot)$ is not a group for any $n \geq 2$, since, as we have mentioned, the element [0] has no inverse. This suggests considering the set $\mathbf{Z}_n^* = \mathbf{Z}_n - \{[0]\} = \{[1], [2], \cdots, [n-1]\}$, $n \geq 2$, under multiplication. For some integers $n \geq 2$, multiplication is not a binary operation on $\mathbf{Z}_n^*$. For example, $[2] \in \mathbf{Z}_4^*$ but $[2] \cdot [2] = [0] \notin \mathbf{Z}_4^*$. On the other hand, multiplication is a binary operation on $\mathbf{Z}_5^*$. In fact, $(\mathbf{Z}_5^*, \cdot)$ satisfies properties G1, G2 and G4, where [1] is an identity. Since

$$[1] \cdot [1] = [1], [2] \cdot [3] = [3] \cdot [2] = [1] \text{ and } [4] \cdot [4] = [1],$$

every element of $(\mathbf{Z}_5^*, \cdot)$ has an inverse. Therefore, $(\mathbf{Z}_5^*, \cdot)$ satisfies property G3 as well and so is an abelian group. This, of course, brings up the question of which algebraic structures $(\mathbf{Z}_n^*, \cdot)$ are groups. Perhaps the examples above suggest the answer.

Theorem to Prove   The set $\mathbf{Z}_n^*$, $n \geq 2$, is a group under multiplication if and only if n is a prime.

PROOF STRATEGY   If n is a composite number, then there exist integers a and b such that $2 \leq a, b \leq n - 1$ and $n = ab$. So $[a], [b] \in \mathbf{Z}_n^*$ and $[a][b] = [n] = [0] \notin \mathbf{Z}_n^*$ and multiplication is not a binary operation. Therefore, the only possibilities for $(\mathbf{Z}_n^*, \cdot)$ to be a group is when n is a prime.

Suppose then that p is a prime. First, we need to verify that multiplication is in fact a binary operation on $\mathbf{Z}_p^*$; that is, if $[a], [b] \in \mathbf{Z}_p^*$, then $[a][b] \in \mathbf{Z}_p^*$. If $[ab] \notin \mathbf{Z}_p^*$, then

$[ab] = [0]$; so $ab \equiv 0 \ (mod \ p)$, which implies that $p \mid ab$. By Corollary 11.14, $p \mid a$ or $p \mid b$ and so $[a] = [0]$ or $[b] = [0]$. That is, either $[a] \notin \mathbf{Z}_p^*$ or $[b] \notin \mathbf{Z}_p^*$, a contradiction.

To show that $(\mathbf{Z}_p^*, \cdot)$ is a group, it remains only to verify that property G3 is satisfied. Let r be an integer with $1 \le r \le p - 1$. We need to show that $[r]$ has an inverse; that is, there exists $[s] \in \mathbf{Z}_p^*$ such that $[r][s] = [1]$. Since p is a prime and $1 \le r \le p - 1$, the integers r and p are relatively prime. By Theorem 11.12, the integer 1 is a linear combination of r and p. So, there exist integers x and y such that $1 = rx + py$. Using the definition of addition and multiplication in $\mathbf{Z}_p$ and observing that $[p] = [0]$ in $\mathbf{Z}_p$, we have

$$[1] = [rx + py] = [rx] + [py] = [r] \cdot [x] + [p] \cdot [y]$$
$$= [r] \cdot [x] + [0] \cdot [y] = [r] \cdot [x] + [0] = [r] \cdot [x].$$

Hence $[x]$ is an inverse for $[r]$.   ◆

We now give a concise proof of the theorem.

Theorem 13.5   *The set $\mathbf{Z}_n^*$, $n \ge 2$, is a group under multiplication if and only if n is a prime.*

Proof   Assume that n is a composite number. Then there exist integers a and b such that $2 \le a, b \le n - 1$ and $n = ab$. Hence $[a], [b] \in \mathbf{Z}_n^*$ and $[a][b] = [n] = [0] \notin \mathbf{Z}_n^*$, which implies that multiplication is not a binary operation on $\mathbf{Z}_n^*$. Certainly then, $\mathbf{Z}_n^*$ is not a group under multiplication.

For the converse, assume that p is a prime. First, we show that multiplication is a binary operation on $\mathbf{Z}_p^*$. Assume, to the contrary, that it is not. Then there exist $[a], [b] \in \mathbf{Z}_p^*$ such that $[a][b] \notin \mathbf{Z}_p^*$. Since $[a][b] \notin \mathbf{Z}_p^*$, it follows that $[a][b] = [ab] = [0]$. Thus $ab \equiv 0 \ (mod \ p)$ and so $p \mid ab$. By Corollary 11.14, $p \mid a$ or $p \mid b$. Therefore, $[a] = [0]$ or $[b] = [0]$, which contradicts the fact that $[a], [b] \in \mathbf{Z}_p^*$. Here $[1]$ is the identity.

Hence $(\mathbf{Z}_p^*, \cdot)$ is an algebraic structure that satisfies properties G1 and G2. It remains to show that $(\mathbf{Z}_p^*, \cdot)$ satisfies property G3. Let $[r] \in \mathbf{Z}_p^*$, where we can assume that $1 \le r \le p - 1$. Since r and p are relatively prime, 1 is a linear combination of r and p by Theorem 11.12. Thus $1 = rx + py$ for some integers x and y. So

$$[1] = [rx + py] = [rx] + [py] = [r] \cdot [x] + [p] \cdot [y]$$
$$= [r] \cdot [x] + [0] \cdot [y] = [r] \cdot [x].$$

Thus, $[x]$ is an inverse for $[r]$ and $(\mathbf{Z}_p^*, \cdot)$ is a group.   ∎

By Theorem 13.5, for every prime p, $(\mathbf{Z}_p^*, \cdot)$ is an abelian group of order $p - 1$. Another example of an abelian group is $(G, *)$, where $G = \{a, b, c\}$ and $*$ is defined in Figure 13.3. It is not difficult to see that a is an identity for $(G, *)$ and that a, b and c are inverses for a, c and b, respectively. Since $a * b = b * a$, $a * c = c * a$ and $b * c = c * b$, it follows that G is abelian. We have one additional property to verify to show that G is a group, however, namely, the associative property. What we are required to show is that $x * (y * z) = (x * y) * z$ for all $x, y, z \in G$. Since there are three choices for each of x, y and z, we have 27 equalities to verify. Because $b * (c * b) = b * a = b$ and $(b * c) * b = a * b = b$, it follows that $b * (c * b) = (b * c) * b$. Since the remaining 26 equalities can also be verified, G is, in fact, an abelian group.

*	a	b	c
a	a	b	c
b	b	c	a
c	c	a	b

Figure 13.3 An abelian group with three elements

We mentioned earlier that the algebraic structure $(M_2(\mathbf{R}), \cdot)$ does not satisfy property G3 even though it does satisfy property G2. The matrix $I = \begin{bmatrix} 1 & 0 \\ 0 & 1 \end{bmatrix}$ is an identity for $(M_2(\mathbf{R}), \cdot)$. Furthermore, a matrix $A = \begin{bmatrix} a & b \\ c & d \end{bmatrix} \in M_2(\mathbf{R})$ has an inverse if and only if its determinant $\det A = ad - bc \neq 0$. In this case,

$$B = \frac{1}{ad - bc} \begin{bmatrix} d & -b \\ -c & a \end{bmatrix}$$

is an inverse for A and so $AB = BA = I$. Let

$$M_2^*(\mathbf{R}) = \{A \in M_2(\mathbf{R}) : \det A \neq 0\}.$$

Since $\det(AB) = \det(A) \cdot \det(B)$ for all $A, B \in M_2(\mathbf{R})$, it follows that if $A, B \in M_2^*(\mathbf{R})$, then $AB \in M_2^*(\mathbf{R})$ and so $M_2^*(\mathbf{R})$ is closed under matrix multiplication. This implies that $(M_2^*(\mathbf{R}), \cdot)$ is a group. On the other hand, since the matrices

$$A = \begin{bmatrix} 1 & 1 \\ 1 & 0 \end{bmatrix} \text{ and } B = \begin{bmatrix} 0 & 1 \\ 1 & 1 \end{bmatrix}$$

belong to $M_2^*(\mathbf{R})$ and

$$AB = \begin{bmatrix} 1 & 2 \\ 0 & 1 \end{bmatrix} \text{ and } BA = \begin{bmatrix} 1 & 0 \\ 2 & 1 \end{bmatrix},$$

it follows that $(M_2^*(\mathbf{R}), \cdot)$ is a nonabelian group.

13.3 Permutation Groups

One of the most important classes of groups concerns a concept introduced in Chapter 9. Recall that a **permutation** of a nonempty set A is a bijective function $f : A \rightarrow A$; that is, f is one-to-one and onto. In Chapter 9, it was shown that:

(1) the composition of every two permutations of A is a permutation of A;

(2) composition of permutations of A is an associative operation;

(3) the identity function $i_A : A \rightarrow A$ defined by $i_A(a) = a$ for all $a \in A$ is a permutation of A;

(4) every permutation of A has an inverse, which is also a permutation of A.

By a **permutation group**, we mean a group $(G, \circ)$, where G is a set of permutations of some set A and $\circ$ denotes composition. Let S_A denote the set of *all* permutations of A. Then by (1)–(4) above, we have the following result.

Theorem 13.6   *For every nonempty set A, the algebraic structure $(S_A, \circ)$ is a permutation group.*

The group $(S_A, \circ)$ is called the **symmetric group** on A. Therefore, every symmetric group is a permutation group.

We have already noted that for the set $\mathcal{F}_\mathbf{R}$ of all functions from $\mathbf{R}$ to $\mathbf{R}$, $(\mathcal{F}_\mathbf{R}, \circ)$ is an algebraic structure, where $\circ$ denotes function composition. By Theorem 13.6, for the set $S_\mathbf{R}$ of bijective functions from $\mathbf{R}$ to $\mathbf{R}$, the algebraic structure $(S_\mathbf{R}, \circ)$ is a group, namely, the symmetric group on $\mathbf{R}$. The identity function $i_\mathbf{R}$ in $S_\mathbf{R}$, defined by $i_\mathbf{R}(x) = x$ for all $x \in \mathbf{R}$, is an identity in the group $(S_\mathbf{R}, \circ)$.

The functions f and g defined by

$$f(x) = x + 1 \text{ and } g(x) = 2x$$

for all $x \in \mathbf{R}$ belong to $S_\mathbf{R}$; however,

$$(f \circ g)(x) = f(g(x)) = f(2x) = 2x + 1 \text{ and}$$
$$(g \circ f)(x) = g(f(x)) = g(x + 1) = 2x + 2.$$

Since $f \circ g \neq g \circ f$, it follows that $(S_\mathbf{R}, \circ)$ is a nonabelian group.

If $A = \{1, 2, \cdots, n\}$, where $n \in \mathbf{N}$, the group S_A is commonly denoted by S_n. The group $(S_n, \circ)$ has $n!$ elements (see Theorem 9.7) and is called the **symmetric group** (of degree n). The symmetric group $(S_n, \circ)$ is therefore a finite group of order $n!$. By the notation introduced in Chapter 9, $S_3 = \{\alpha_1, \alpha_2, \alpha_3, \alpha_4, \alpha_5, \alpha_6\}$, where

$$\alpha_1 = \begin{pmatrix} 1 & 2 & 3 \\ 1 & 2 & 3 \end{pmatrix} \quad \alpha_2 = \begin{pmatrix} 1 & 2 & 3 \\ 1 & 3 & 2 \end{pmatrix} \quad \alpha_3 = \begin{pmatrix} 1 & 2 & 3 \\ 3 & 2 & 1 \end{pmatrix}$$

$$\alpha_4 = \begin{pmatrix} 1 & 2 & 3 \\ 2 & 1 & 3 \end{pmatrix} \quad \alpha_5 = \begin{pmatrix} 1 & 2 & 3 \\ 2 & 3 & 1 \end{pmatrix} \quad \alpha_6 = \begin{pmatrix} 1 & 2 & 3 \\ 3 & 1 & 2 \end{pmatrix}.$$

Recall that with this notation for a permutation, an element of $\{1, 2, 3\}$ listed in the first row maps into the element in the second row directly below it. Thus α_1 is the identity of S_3. Let's consider the composition $\alpha_3 \circ \alpha_6$. For example,

$$(\alpha_3 \circ \alpha_6)(1) = \alpha_3(\alpha_6(1)) = \alpha_3(3) = 1.$$

Also, $(\alpha_3 \circ \alpha_6)(2) = 3$ and $(\alpha_3 \circ \alpha_6)(3) = 2$. Therefore,

$$\alpha_3 \circ \alpha_6 = \begin{pmatrix} 1 & 2 & 3 \\ 3 & 2 & 1 \end{pmatrix} \circ \begin{pmatrix} 1 & 2 & 3 \\ 3 & 1 & 2 \end{pmatrix} = \begin{pmatrix} 1 & 2 & 3 \\ 1 & 3 & 2 \end{pmatrix} = \alpha_2.$$

Similarly $\alpha_6 \circ \alpha_3 = \alpha_4$. Hence $\alpha_3 \circ \alpha_6 \neq \alpha_6 \circ \alpha_3$. Therefore, $(S_3, \circ)$ is a nonabelian group. This shows that there is a nonabelian group of order 6. There is no nonabelian group having order less than 6, however.

When taking the composition of two elements $\alpha, \beta \in S_n$, where $n \in \mathbf{N}$, we often write $\alpha \circ \beta$ as $\alpha\beta$ and say that we are *multiplying* α and β. With this notation, we have $\alpha_3\alpha_6 = \alpha_2$, $\alpha_6\alpha_3 = \alpha_4$, $\alpha_3^2 = \alpha_3\alpha_3 = \alpha_1$ and $\alpha_6^2 = \alpha_6\alpha_6 = \alpha_5$. The group table for $(S_3, \circ)$ is shown in Figure 13.4 (containing all 36 products!).

	α_1	α_2	α_3	α_4	α_5	α_6
α_1	α_1	α_2	α_3	α_4	α_5	α_6
α_2	α_2	α_1	α_5	α_6	α_3	α_4
α_3	α_3	α_6	α_1	α_5	α_4	α_2
α_4	α_4	α_5	α_6	α_1	α_2	α_3
α_5	α_5	α_4	α_2	α_3	α_6	α_1
α_6	α_6	α_3	α_4	α_2	α_1	α_5

Figure 13.4   The group table for $(S_3, \circ)$

A permutation group need not consist of all permutations of some set A. For example, if we consider the subsets $G_1 = \{\alpha_1, \alpha_2\}$ and $G_2 = \{\alpha_1, \alpha_5, \alpha_6\}$ of S_3, then $(G_1, \circ)$ and $(G_2, \circ)$ are both permutation groups. Their group tables are shown in Figure 13.5.

Also, let

$$\beta_1 = \begin{pmatrix} 1 & 2 & 3 & 4 \\ 1 & 2 & 3 & 4 \end{pmatrix} \quad \beta_2 = \begin{pmatrix} 1 & 2 & 3 & 4 \\ 2 & 1 & 3 & 4 \end{pmatrix}$$

$$\beta_3 = \begin{pmatrix} 1 & 2 & 3 & 4 \\ 1 & 2 & 4 & 3 \end{pmatrix} \quad \beta_4 = \begin{pmatrix} 1 & 2 & 3 & 4 \\ 2 & 1 & 4 & 3 \end{pmatrix}$$

be permutations of the set $\{1, 2, 3, 4\}$ and let $G_3 = \{\beta_1, \beta_2, \beta_3, \beta_4\}$. Then $(G_3, \circ)$ is an abelian permutation group, whose group table is shown in Figure 13.6.

During the latter part of the 18th century, a major problem in mathematics concerned whether the roots of every fifth degree polynomial with real coefficients could be

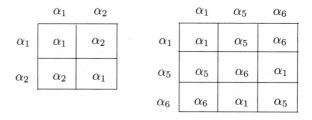

	α_1	α_2
α_1	α_1	α_2
α_2	α_2	α_1

	α_1	α_5	α_6
α_1	α_1	α_5	α_6
α_5	α_5	α_6	α_1
α_6	α_6	α_1	α_5

Figure 13.5   The group tables for $(G_1, \circ)$ and $(G_2, \circ)$

	β_1	β_2	β_3	β_4
β_1	β_1	β_2	β_3	β_4
β_2	β_2	β_1	β_4	β_3
β_3	β_3	β_4	β_1	β_2
β_4	β_4	β_3	β_2	β_1

Figure 13.6   The group table for $(G_3, \circ)$

expressed in terms of radicals and the usual operations of arithmetic. It was well known that the roots of a quadratic polynomial $ax^2 + bx + c$, where $a, b, c \in \mathbf{R}$ and $a \neq 0$, are $(-b + \sqrt{b^2 - 4ac})/2a$ and $(-b - \sqrt{b^2 - 4ac})/2a$, which is a consequence of the quadratic formula. Furthermore, it had been known since the 16th century that the roots of all third degree (cubic) and fourth degree (quartic) polynomials with real coefficients could be described in terms of radicals and the standard operations of arithmetic. But fifth degree polynomials proved to be another story. However, in 1824 Niels Henrik Abel proved that the roots of fifth degree polynomials with real coefficients could not, in general, be expressed in such a way. From this, it follows that for every integer $n \geq 5$, there exist polynomials of degree n with real coefficients whose roots cannot be expressed in terms of radicals and the standard operations of arithmetic. His work went unnoticed, however, until after his death at age 26.

Some time later the French mathematician Évariste Galois characterized those polynomials of degree 5 and greater whose roots can be expressed in terms of radicals and ordinary arithmetic. Like Abel, Galois died very early (at age 20), but in his case from a unlikely cause: a duel. Galois' work also was not recognized until 11 years after his death when Joseph Liouville addressed the Academy of Sciences in Paris: "I hope to interest the Academy in announcing that among the papers of Évariste Galois I have found a solution, as precise as it is profound, of this beautiful problem: whether or not it is solvable by radicals."

In developing his theory, Galois associated, with a given polynomial, a set G of permutations of the roots of the polynomial. This set G had the property that whenever $s, t \in G$, then the composition $s \circ t \in G$; that is, G was closed under composition. He referred to G as a *group,* a term that enjoys a permanent and prominent place in abstract algebra.

13.4 Fundamental Properties of Groups

We now consider some properties possessed by all groups. Of course, any property satisfied by all groups must be a consequence of properties G1 – G3. Unless stated otherwise, the symbol e represents an identity in the group under consideration. One simple, but important, property satisfied by every group $(G, *)$ allows us to cancel a in $a * b = a * c$ and conclude that $b = c$. Actually, since a group need not be abelian, there are two such cancellation properties.

Theorem 13.7 *Every group $(G, *)$ satisfies:*

    (*a*) **The Left Cancellation Law** *Let $a, b, c \in G$. If $a * b = a * c$, then $b = c$.*

    (*b*) **The Right Cancellation Law** *Let $a, b, c \in G$. If $b * a = c * a$, then $b = c$.*

Proof We prove (a) only. (The proof of (b) is similar. See Exercise 13.21.) Assume that $a * b = a * c$. Let s be an inverse for a. Then $s * (a * b) = s * (a * c)$. So

$$s * (a * b) = (s * a) * b = e * b = b,$$

while

$$s * (a * c) = (s * a) * c = e * c = c.$$

Therefore, $b = c$. ∎

The last two sentences of this proof could have been replaced by: Then

$$b = e * b = (s * a) * b = s * (a * b) = s * (a * c) = (s * a) * c = e * c = c.$$

The next result will come as no surprise.

Theorem 13.8 *Let $(G, *)$ be a group and let $a, b \in G$. The linear equations $a * x = b$ and $x * a = b$ have unique solutions in G.*

Proof We prove only that $a * x = b$ has a unique solution. (The remaining proof is left as Exercise 13.22.) Let e be an identity for G, let s be an inverse of a and let $x = s * b$. Then

$$a * x = a * (s * b) = (a * s) * b = e * b = b.$$

So $x = s * b$ is a solution of the equation $a * x = b$.

It remains to show that $s * b$ is the only solution of $a * x = b$. Suppose that x_1 and x_2 are solutions of $a * x = b$. Then $a * x_1 = b$ and $a * x_2 = b$. Thus $a * x_1 = a * x_2$. Applying the Left Cancellation Law (Theorem 13.7(a)), we have $x_1 = x_2$. ∎

The preceding theorem provides some interesting information for us. Suppose that we have a group table for a group G and we are looking at the row corresponding to the element a. Then this row contains the elements $a * g$ for all $g \in G$. Let $b \in G$. By Theorem 13.8, there exists $x \in G$ such that $a * x = b$. That is, the element b must appear in the row corresponding to a. This is illustrated in Figure 13.7. On the other hand, the element b cannot appear twice in this row since the equation $a * x = b$ has a unique solution. Hence we can conclude that every element of G appears exactly once in every row in the group table of G. By considering the equation $x * a = b$, we can likewise conclude that every element of G appears exactly once in every column in the group table of G.

As with composition in a symmetric group, it is customary in a group G to refer to the binary operation $*$ as *multiplication* and to indicate the *product* of elements a and b in G by ab rather than $a * b$ to simplify the notation. We thus write $a * a = aa = a^2$. The lone exception to this practice is when we have a group whose operation is addition, in which case we continue to use $+$ as the operation. It is also common practice never to use $+$ as an operation when the group is nonabelian. Let's use this newly adopted notation to present a theorem that shows that every group G has a unique identity and

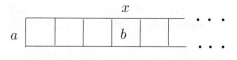

Figure 13.7 The equation $a * x = b$

every element of G has a unique inverse, two facts that you may have already suspected were true.

Theorem 13.9 *Let G be a group. Then*

(a) *G has a unique identity and*
(b) *each element in G has a unique inverse.*

Proof Assume that e and f are two identities in G. Since e is an identity, $ef = f$; and since f is an identity, $ef = e$. Thus $e = ef = f$. This verifies (a).

Next let $g \in G$ and suppose that s and t are both inverses of g. So $gs = sg = e$ and $gt = tg = e$. Thus

$$s = se = s(gt) = (sg)t = et = t,$$

which verifies (b). ∎

It is customary to denote the (unique) inverse of an element a in a group by a^{-1}. If the operation in a group under consideration is addition, then we follow the standard practice of denoting the identity by 0 and the inverse of a by $-a$. We now present two theorems involving inverses in a group.

Theorem 13.10 *Let G be a group. If $a \in G$, then*

$$\left(a^{-1}\right)^{-1} = a.$$

Proof Since $aa^{-1} = a^{-1}a = e$, the element a is the inverse of a^{-1}, that is, $\left(a^{-1}\right)^{-1} = a$. ∎

For elements a and b in a group, the next theorem establishes a connection among the inverses a^{-1}, b^{-1} and $(ab)^{-1}$.

Theorem to Prove Let G be a group. For $a, b \in G$,

$$(ab)^{-1} = b^{-1}a^{-1}.$$

PROOF STRATEGY Suppose that the group under consideration is G. For $a, b \in G$, their product $ab \in G$. Since $ab \in G$, the element ab has an inverse—indeed a unique inverse. The inverse of ab is denoted by $(ab)^{-1}$. The theorem claims that the inverse of ab is the element $b^{-1}a^{-1}$ in G. To show that an element $s \in G$ is an inverse of $x \in G$, we are required to show that $sx = xs = e$. Therefore, to show that $b^{-1}a^{-1}$ is the inverse of ab, we need to show that $(ab)\left(b^{-1}a^{-1}\right) = \left(b^{-1}a^{-1}\right)(ab) = e$. ◆

Theorem 13.11 *Let G be a group. If $a, b \in G$, then*

$$(ab)^{-1} = b^{-1}a^{-1}.$$

Proof To show that $b^{-1}a^{-1}$ is the inverse of ab, it suffices to show that $(ab)\left(b^{-1}a^{-1}\right) = e$ and $\left(b^{-1}a^{-1}\right)(ab) = e$. We verify the first of these as the proof of the second equality is similar. Observe that

$$(ab)(b^{-1}a^{-1}) = ((ab)b^{-1})a^{-1} = (a(bb^{-1}))a^{-1} = (ae)a^{-1} = aa^{-1} = e.$$

In words, what Theorem 13.11 states is that the inverse of the product of two elements in a group is the product of their inverses in reverse order. Of course, if G is an abelian group and $a, b \in G$, then $(ab)^{-1} = b^{-1}a^{-1} = a^{-1}b^{-1}$. (See Exercise 13.25.)

13.5 Subgroups

There have been occasions when we were considering a group $(G, *)$ and a subset H of G such that H is a group under the same operation $*$; that is, $(H, *)$ is also a group. If $(G, *)$ is a group and H is a subset of G such that $(H, *)$ is a group, then $(H, *)$ is called a **subgroup** of G. For example, $(\mathbf{Z}, +)$ is a subgroup of $(\mathbf{Q}, +)$, which, in turn, is a subgroup of $(\mathbf{R}, +)$. Also, the groups G_1 and G_2 in Figure 13.5 are subgroups of S_3, while the group G_3 in Figure 13.6 is a subgroup of S_4. If $(G, *)$ is a group with identity e, then $(\{e\}, *)$ and $(G, *)$ are always subgroups of $(G, *)$. Hence if G has at least two elements, then G has at least two subgroups.

The group $(2\mathbf{Z}, +)$ of even integers under addition is a subgroup of $(\mathbf{Z}, +)$. To see this, first observe that $2\mathbf{Z} \subseteq \mathbf{Z}$. Since the sum of two even integers is an even integer, $2\mathbf{Z}$ is closed under addition. Since the associative law of addition holds in $\mathbf{Z}$, it holds in $2\mathbf{Z}$ as well. The identity in $(\mathbf{Z}, +)$ is 0. Since 0 is an even integer, $0 \in 2\mathbf{Z}$. Finally, the additive inverse (the negative) of an even integer is an even integer. Thus $(2\mathbf{Z}, +)$ is a group. Therefore, showing that $(2\mathbf{Z}, +)$ is a subgroup of $(\mathbf{Z}, +)$ is simpler than showing that it is group. This observation is true for every subgroup.

Theorem 13.12   (**The Subgroup Test**) *A nonempty subset H of a group G is a subgroup of G if and only if* (1) $ab \in H$ *for all* $a, b \in H$ *and* (2) $a^{-1} \in H$ *for all* $a \in H$.

Proof   We first show that if H is a subgroup of G, then properties (1) and (2) are satisfied. Since H is closed under multiplication, property (1) is certainly satisfied. We now show that the identity e of G is also the identity of H. Let f be the identity of H. Thus $f \cdot f = f$. Since e is the identity of G, it follows that $f \cdot e = f$. Therefore, $f \cdot f = f \cdot e$. By the Left Cancellation Law (Theorem 13.7(a)), $f = e$. Hence, as claimed, the identity e of G is also the identity of H. Next, let $a \in H$. Since $a \in G$, it follows that the inverse a^{-1} of a belongs to G and so $aa^{-1} = e$. It remains to show that $a^{-1} \in H$. Since H is a subgroup of G, a has an inverse a' in H. Thus $aa' = e$ in H and $aa' = e$ in G as well. Therefore, $aa' = aa^{-1}$. By the Left Cancellation Law again, $a' = a^{-1}$ and so property (2) is satisfied.

Next, we verify the converse, namely if H is a nonempty subset of G satisfying properties (1) and (2), then H is a subgroup of G. Let $a, b, c \in H$. Since $H \subseteq G$, it follows that $a, b, c \in G$. Since G is a group, $a(bc) = (ab)c$ by property G1. Thus multiplication is associative in H as well. From property (1), it follows that H is closed under multiplication and from property (2), every element of H has an inverse. It remains only to show that H contains an identity. Since $H \neq \emptyset$, there exists an element $a \in H$. By (2), $a^{-1} \in H$; and by (1), $aa^{-1} = e \in H$. Since H contains the identity e of G, it follows that $xe = ex = x$ for all $x \in H$ and so e is the identity of H as well. ∎

We now illustrate the Subgroup Test.

Result to Prove Let $H = \left\{ \begin{bmatrix} a & b \\ c & 0 \end{bmatrix} : a, b, c \in \mathbf{R} \right\}$. Then $(H, +)$ is a group.

PROOF STRATEGY The elements of H are matrices, in fact, matrices in $M_2(\mathbf{R})$. Indeed, a matrix in $M_2(\mathbf{R})$ belongs to H if and only if the entry in row 2, column 2 is 0. We have already seen that $(M_2(\mathbf{R}), +)$ is a group. Since H uses the same operation as in $M_2(\mathbf{R})$, namely addition, it is appropriate to prove that $(H, +)$ is a group by the Subgroup Test (Theorem 13.12).

To use the Subgroup Test, we first need to know that the set H is nonempty. Since the zero matrix $\begin{bmatrix} 0 & 0 \\ 0 & 0 \end{bmatrix}$ satisfies the requirement for it to belong to H, we need only to show that conditions (1) and (2) of Theorem 13.12 are satisfied, namely, that H is closed under addition and that if A is a matrix in H, then its inverse (its negative in this case) $-A$ belongs to H as well. This will be relatively routine to show. ◆

Result 13.13 *Let $H = \left\{ \begin{bmatrix} a & b \\ c & 0 \end{bmatrix} : a, b, c \in \mathbf{R} \right\}$. Then $(H, +)$ is a group.*

Proof We show in fact that $(H, +)$ is a subgroup of $(M_2(\mathbf{R}), +)$. Certainly, H is a nonempty subset of $M_2(\mathbf{R})$ since the zero matrix $\begin{bmatrix} 0 & 0 \\ 0 & 0 \end{bmatrix}$ belongs to H. Let $A, B \in H$. Then

$$A = \begin{bmatrix} a_1 & a_2 \\ a_3 & 0 \end{bmatrix} \text{ and } B = \begin{bmatrix} b_1 & b_2 \\ b_3 & 0 \end{bmatrix},$$

where $a_i, b_i \in \mathbf{R}\,(1 \le i \le 3)$. Then $A + B = \begin{bmatrix} a_1 + b_1 & a_2 + b_2 \\ a_3 + b_3 & 0 \end{bmatrix} \in H$ and the inverse

of A is $-A = \begin{bmatrix} -a_1 & -a_2 \\ -a_3 & 0 \end{bmatrix} \in H$. Consequently, by the Subgroup Test, H is a subgroup of $M_2(\mathbf{R})$ and so $(H, +)$ is a group. ■

If G is an abelian group, then we know that every two elements of G commute. But even if G is nonabelian, we know that its identity commutes with every element of G. However, there may very well be other elements of G that commute with all elements of G. The set of all elements in a group G that commute with every element in G is called the **center** of G and, in fact, is always a subgroup of G. This subgroup is often denoted by $Z(G)$. Since $Z(G) = G$ if G is abelian, the center is most interesting when G is nonabelian.

Result to Prove For a group G, the center

$$Z(G) = \{a \in G \,:\, ga = ag \text{ for all } g \in G\}$$

is a subgroup of G.

PROOF STRATEGY To prove this result, it seems natural to use the Subgroup Test. Since $e \in Z(G)$, it follows that $Z(G) \ne \emptyset$. We are now required to show that $Z(G)$ satisfies the two properties required of the Subgroup Test.

First, we show that $Z(G)$ is closed under multiplication; that is, if $a, b \in Z(G)$, then $ab \in Z(G)$. We employ a direct proof. Let $a, b \in Z(G)$. To show that $ab \in Z(G)$, we need to show that ab commutes with every element of G. So let $g \in G$. We must show that $(ab)g = g(ab)$. This suggests starting with $(ab)g$. By the associative law, $(ab)g = a(bg)$. However, since $b \in Z(G)$, it follows that $a(bg) = a(gb)$. We can continue in this manner to complete the proof of this property.

Second, we need to show that if $a \in Z(G)$, then $a^{-1} \in Z(G)$. Again, we use a direct proof. Let $a \in Z(G)$. Then a commutes with every element of G. To show that $a^{-1} \in Z(G)$, we need to verify that a^{-1} commutes with every element of G. Let $g \in G$. We must show then that $a^{-1}g = ga^{-1}$ to complete the proof. But how do we do this? Theorem 13.11, which deals with inverses of elements, may be helpful. We know that $(xy)^{-1} = y^{-1}x^{-1}$ for all $x, y \in G$. So $(ag)^{-1} = g^{-1}a^{-1}$. This, however, involves g^{-1}. But if we start with $ag^{-1} = g^{-1}a$, then we have $\left(ag^{-1}\right)^{-1} = \left(g^{-1}a\right)^{-1}$.   ◆

Result 13.14   *For a group G, the center*

$$Z(G) = \{a \in G \; : \; ga = ag \text{ for all } g \in G\}$$

is a subgroup of G.

Proof   Since $eg = ge$ for all $g \in G$, it follows that $e \in Z(G)$ and so $Z(G)$ is nonempty. First, we show that $Z(G)$ is closed under multiplication. Let $a, b \in Z(G)$. Thus $ag = ga$ and $bg = gb$ for all $g \in G$. We show that $ab \in Z(G)$. Since

$$(ab)\, g = a\, (bg) = a\, (gb) = (ag)b = (ga)b = g\, (ab),$$

$ab \in Z(G)$. Hence $Z(G)$ is closed under multiplication. Next we show that every element of $Z(G)$ has an inverse in $Z(G)$. Let $a \in Z(G)$ and $g \in G$. We show that $a^{-1} \in Z(G)$; that is, a^{-1} and g commute. Since a commutes with all elements of G, it follows that a and g^{-1} commute and so $ag^{-1} = g^{-1}a$. Since every element of G has a unique inverse, $(ag^{-1})^{-1} = (g^{-1}a)^{-1}$. By Theorem 13.11, $\left(ag^{-1}\right)^{-1} = \left(g^{-1}\right)^{-1} a^{-1} = ga^{-1}$ and $\left(g^{-1}a\right)^{-1} = a^{-1}\left(g^{-1}\right)^{-1} = a^{-1}g$. Therefore, $a^{-1}g = ga^{-1}$.   ∎

For $A = \{1, 2, \cdots, n\}, n \geq 2$ and $k \in A$, let G_k consist of those permutations α in the symmetric group $(S_n, \circ)$ such that $\alpha(k) = k$ (that is, G_k consists of all those permutations of A that "stabilize" or fix k). The set G_k is called the **stabilizer** of k in S_n.

Result 13.15   *For integers k and n with $1 \leq k \leq n$ and $n \geq 2$, the stabilizer G_k of k in S_n is a subgroup of S_n.*

Proof   We use the Subgroup Test. Surely, the identity α_1 in S_n belongs to G_k, so $G_k \neq \emptyset$. Let $\alpha, \beta \in G_k$. Thus $\alpha(k) = \beta(k) = k$. Hence $(\alpha \circ \beta)(k) = \alpha(\beta(k)) = \alpha(k) = k$ and so $\alpha \circ \beta \in G_k$. Consider the inverse α^{-1} of α. Thus $\alpha^{-1} \circ \alpha = \alpha_1$. Therefore, $(\alpha^{-1} \circ \alpha)(k) = \alpha_1(k) = k$. Hence $(\alpha^{-1} \circ \alpha)(k) = \alpha^{-1}(\alpha(k)) = \alpha^{-1}(k) = \alpha_1(k) = k$. Thus $\alpha^{-1} \in G_k$. By the Subgroup Test, G_k is a subgroup of S_n.   ∎

The group $(G_1, \circ)$ shown in Figure 13.5 is the stabilizer of 1 in $(S_3, \circ)$. However, $(G_2, \circ)$ shown in Figure 13.5 is *not* the stabilizer of 2 in $(S_3, \circ)$.

We have already mentioned that the set $2\mathbf{Z}$ of even integers is a subgroup of $(\mathbf{Z}, +)$. In fact, for every integer $n \geq 2$, the set $n\mathbf{Z} = \{nk : k \in \mathbf{Z}\}$ of multiples of n is a subgroup of $(\mathbf{Z}, +)$ (see Exercise 13.31). In Chapter 8, we saw that the relation R defined on $\mathbf{Z}$ by $a \, R \, b$ if $a \equiv b \pmod{n}$ is an equivalence relation. This relation can also be described in another manner, namely $a \, R \, b$ if $a - b \in n\mathbf{Z}$ and so $a - b = h$ for some element $h \in n\mathbf{Z}$ or $a = b + h$. It turns out that this equivalence relation is a special case of a more general situation. Suppose that H is a subgroup of a group $(G, \cdot)$ and a relation R is defined on G by $a \, R \, b$ if $a = bh$ for some $h \in H$. (Note that $b + h$ in $\mathbf{Z}$ is replaced by bh here since the operation in G is *multiplication*.) Then this relation is also an equivalence relation.

Theorem 13.16 *Let H be a subgroup of a group $(G, \cdot)$. The relation R defined on G by $a \, R \, b$ if $a = bh$ for some $h \in H$ is an equivalence relation.*

Proof First, we show that R is reflexive. Let $a \in G$. Since $a = ae$ (where e is the identity of G and therefore of H), $a \, R \, a$ and so R is reflexive. Next, we show that R is symmetric. Assume that $a \, R \, b$, where $a, b \in G$. Then $a = bh$ for some $h \in H$. Since H is a group, $h^{-1} \in H$ and so $ah^{-1} = (bh)h^{-1} = b(hh^{-1}) = be = b$, or $b = ah^{-1}$. Therefore, $b \, R \, a$ and R is symmetric. Finally, we show that R is transitive. Assume that $a \, R \, b$ and $b \, R \, c$, where $a, b, c \in G$. Then $a = bh_1$ and $b = ch_2$ for elements h_1 and h_2 in H. Therefore, $a = bh_1 = (ch_2)h_1 = c(h_2h_1)$. Since $h_2, h_1 \in H$, it follows that $h_2h_1 \in H$ as well. Thus $a \, R \, c$ and so R is transitive. ∎

For a subgroup H of a group $(G, \cdot)$, the equivalence relation defined in Theorem 13.16 gives rise to equivalence classes. For each element $g \in G$, the equivalence class $[g]$ is defined by

$$[g] = \{x \in G : x \, R \, g\} = \{x \in G : x = gh \text{ for some } h \in H\}$$
$$= \{gh : h \in H\}.$$

The set $\{gh : h \in H\}$ is often denoted by gH and is called a **left coset** of H in G, that is, $[g] = gH$. We saw in Chapter 8 that for an equivalence relation defined on a set S, the distinct equivalence classes form a partition of S. Consequently, the equivalence relation defined in Theorem 13.16 results in a partition of G into the distinct left cosets of H in G.

An important characteristic of a left coset gH of H in G is that gH and H have the same number of elements, that is, $|gH| = |H|$. In order to see this, we show for an element $g \in G$ that there is a bijection from H to gH. Let $\phi : H \rightarrow gH$ be defined by $\phi(h) = gh$. First, we show that ϕ is one-to-one. Assume that $\phi(h_1) = \phi(h_2)$. Then $gh_1 = gh_2$. By the Left Cancellation Law, $h_1 = h_2$. Therefore, ϕ is one-to-one. Next, we show that ϕ is onto. Let $gh \in gH$. Since $\phi(h) = gh$, it follows that ϕ is onto and so ϕ is a bijection and $|gH| = |H|$. Therefore, every two left cosets of H in G have the same number of elements.

What we have just observed provides all the information that is needed to prove a fundamental theorem of group theory, one due to Joseph-Louis Lagrange, probably the greatest French mathematician of the 18th century.

Theorem 13.17   (**Lagrange's Theorem**) *If H is a subgroup of order m in a (finite) group G of order n, then $m \mid n$.*

Proof   We have already seen that the distinct left cosets of H in G form a partition of G and that every two left cosets have the same number of elements. Suppose that there are k left cosets of G. Then $n = mk$ and so $m \mid n$.   ∎

Lagrange's theorem first appeared in 1770–1771 in connection with the problem of solving the general polynomial of degree 5 or higher. While this theorem was not presented in this general form by Lagrange and in fact group theory had yet to be invented, it is universally referred to as Lagrange's theorem.

We have seen that the group consisting of the nonzero elements of $\mathbf{Z}_7$ forms a group under multiplication, that is, $G = \mathbf{Z}_7^* = \{[1], [2], \ldots, [6]\}$. Since $H = \{[1], [6]\}$ is a subgroup of order 2 in G, the distinct left cosets of H in G are

$$[1]H = H, \ [2]H = \{[2], [5]\} \text{ and } [3]H = \{[3], [4]\}.$$

13.6 Isomorphic Groups

Suppose that we were asked to give examples of two groups of order 3. One possible example is $(\mathbf{Z}_3, +)$. On the other hand, we might try to construct two groups of order 3, say $G = \{a, b, c\}$ and $H = \{x, y, z\}$. Of course, we must also describe binary operations for both G and H. Let us denote the binary operation for G by $*$ and the binary operation for H by $\circ$. So we have two groups $(G, *)$ and $(H, \circ)$, both of order 3. One of the elements of G is the identity for G and one of the elements of H is the identity for H. Suppose that we decide on a as the identity for G and x as the identity for H. Hence the operations $*$ and $\circ$ in G and H, respectively, satisfy the partial tables shown in Figure 13.8.

Because every element in each of the groups G and H must occur exactly once in every row and column in the tables shown in Figure 13.8, the complete tables for $*$ and $\circ$ must be those shown in Figure 13.9.

We can now see readily that in G we have $a^{-1} = a$, $b^{-1} = c$ and $c^{-1} = b$; while in H, $x^{-1} = x$, $y^{-1} = z$ and $z^{-1} = y$. Verifying the associative laws requires more effort, but it can be shown that the associative law holds in each case. Thus $(G, *)$ and $(H, \circ)$ are both groups, and we have just given examples of two groups of order 3. Or have we? There is something very similar about these two examples. They are not really two different groups at all. Indeed, the group $(H, \circ)$ is merely a disguised form of the group

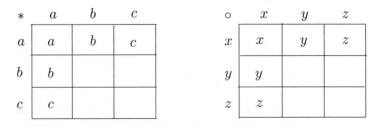

Figure 13.8   Partial tables for groups $(G, *)$ and $(H, \circ)$

$*$	a	b	c
a	a	b	c
b	b	c	a
c	c	a	b

$\circ$	x	y	z
x	x	y	z
y	y	z	x
z	z	x	y

Figure 13.9   Complete tables for groups $(G, *)$ and $(H, \circ)$

$*$	a	b	c
a	a	b	c
b	b	c	a
c	c	a	b

$\circ$	x	z	y
x	x	z	y
z	z	y	x
y	y	x	z

Figure 13.10   Groups $(G, *)$ and $(H, \circ)$

$(G, *)$. Let's describe what we mean by this. If the elements a, b, c in G are replaced by x, y, z, respectively, then we have the identical table. What's important here is not only that x, y and z in H are playing the roles of a, b and c in G but that the operations in the two groups are doing the same thing. For example, if we multiply b and c in G (obtaining the element a), then multiplying the corresponding elements y and z in H gives us the element corresponding to a, namely x.

Although it may appear that this is the natural correspondence between the elements of G and the elements of H, we should not be misled by the order in which the elements of these two groups are listed. For example, suppose that we consider the two groups $(G, *)$ and $(H, \circ)$ once again (in Figure 13.10), where the elements of H are listed in the order x, z, y.

We can see that the elements a, b, c in G also correspond to x, z, y, respectively. We consider the *two* groups $(G, *)$ and $(H, \circ)$ to be actually a single group, as these two groups have the same order (though the sets are different) and their operations perform the same functions (though different symbols are used for the operations). The technical term for this is that they are *isomorphic* groups (groups having the same structure).

In general, two groups $(G, *)$ and $(H, \circ)$ are **isomorphic** if there exists a bijective function $\phi : G \to H$ satisfying the property

$$\phi(a * b) = \phi(a) \circ \phi(b) \tag{13.3}$$

for all $a, b \in G$. Any such function ϕ that satisfies (13.3) is said to be **operation-preserving**. Thus for $(G, *)$ and $(H, \circ)$ to be isomorphic, there must be a bijective, operation-preserving function $\phi : G \to H$. If ϕ has these properties, then ϕ is called an **isomorphism**. If $\phi : G \to H$ is an isomorphism, then ϕ is also a bijective function. Thus ϕ has an inverse function $\phi^{-1} : H \to G$, which is also an isomorphism (Exercise 13.51).

For isomorphic groups $(G, *)$ and $(H, \circ)$, there are certain properties that every isomorphism from G to H must have. We look at two examples of these.

Theorem 13.18   *Let $(G, *)$ and $(H, \circ)$ be isomorphic groups, where the identity of G is e and the identity of H is f. If $\phi : G \to H$ is an isomorphism, then*

(a) $\phi(e) = f$ and
(b) $\phi\left(g^{-1}\right) = (\phi(g))^{-1}$ *for all $g \in G$.*

Proof   First we prove (a). Let $h \in H$. Since ϕ is onto, there exists $g \in G$ such that $\phi(g) = h$. Since $e * g = g * e = g$ and ϕ is operation-preserving, it follows that

$$\phi(e) \circ \phi(g) = \phi(e * g) = \phi(g) = \phi(g * e) = \phi(g) \circ \phi(e)$$

and so

$$\phi(e) \circ h = h \circ \phi(e) = h.$$

This implies that $\phi(e)$ is the identity of H and so $\phi(e) = f$.

Next we prove (b). Let $g \in G$. Since $g * g^{-1} = g^{-1} * g = e$, it follows that $\phi\left(g * g^{-1}\right) = \phi\left(g^{-1} * g\right) = \phi(e)$ and so $\phi(g) \circ \phi\left(g^{-1}\right) = \phi\left(g^{-1}\right) \circ \phi(g) = \phi(e) = f$. This says that $\phi\left(g^{-1}\right)$ is the inverse of $\phi(g)$, that is, $\phi\left(g^{-1}\right) = (\phi(g))^{-1}$.   ∎

By definition, if two groups $(G, *)$ and $(H, \circ)$ are isomorphic, then there exists an isomorphism $\phi : G \to H$. Since ϕ is a bijective function, it follows that $|G| = |H|$. Actually, it's not surprising that isomorphic groups have the same number of elements since, when we say that G and H are isomorphic, we are technically saying that these groups are the same, except for what the elements and binary operations are called. On the other hand, consider the group tables of two groups shown in Figure 13.11. You might notice that the first group is $(\mathbf{Z}_4, +)$. The second group G is abelian, has identity e and $x^{-1} = x$ for all $x \in G$. In addition, of course, G has order 4. So $\mathbf{Z}_4$ and G both have order 4. Yet $\mathbf{Z}_4$ and G are *not* isomorphic; for assume, to the contrary, that they are isomorphic. Then there exists an isomorphism $\phi : \mathbf{Z}_4 \to G$. By Theorem 13.18, we know that $\phi([0]) = e$. Let $\phi([1]) = x \in G$. So

$$\phi([2]) = \phi([1 + 1]) = \phi([1] + [1]) = \phi([1]) \cdot \phi([1]) = x \cdot x = x^2 = e.$$

Thus $\phi([2]) = \phi([0]) = e$ but this contradicts the fact that ϕ is one-to-one. Consequently, these two groups of order 4 are not isomorphic. Therefore, if two groups have the same number of elements, they need not be isomorphic. However, it turns out that any group of order 4 is isomorphic to one of the two groups of order 4 described in Figure 13.11.

+	[0]	[1]	[2]	[3]
[0]	[0]	[1]	[2]	[3]
[1]	[1]	[2]	[3]	[0]
[2]	[2]	[3]	[0]	[1]
[3]	[3]	[0]	[1]	[2]

·	e	a	b	c
e	e	a	b	c
a	a	e	c	b
b	b	c	e	a
c	c	b	a	e

Figure 13.11   Two groups of order 4

We saw in Chapter 10 that $|\mathbf{Z}| = |\mathbf{Q}|$ even though $\mathbf{Z}$ is a proper subset of $\mathbf{Q}$. However, $(\mathbf{Z}, +)$ and $(\mathbf{Q}, +)$ are not isomorphic.

Result 13.19   *The groups $(\mathbf{Z}, +)$ and $(\mathbf{Q}, +)$ are not isomorphic.*

Proof   Assume, to the contrary, that $(\mathbf{Z}, +)$ and $(\mathbf{Q}, +)$ are isomorphic. Then there exists an isomorphism $\phi : \mathbf{Z} \to \mathbf{Q}$. Let $\phi(1) = a \in \mathbf{Q}$. Since $\phi(0) = 0$, it follows that $a \neq 0$. Thus $a/2 \in \mathbf{Q}$ and $a/2 \neq 0$. Since ϕ is onto, there exists an integer $n \neq 0$ such that $\phi(n) = a/2$. Then

$$\phi(2n) = \phi(n+n) = \phi(n) + \phi(n) = \frac{a}{2} + \frac{a}{2} = a.$$

Since ϕ is one-to-one, $2n = 1$. However then, $n = 1/2 \notin \mathbf{Z}$, which is a contradiction. ∎

On the other hand, the set $2\mathbf{Z}$ of even integers is a proper subset of $\mathbf{Z}$; yet $(2\mathbf{Z}, +)$ and $(\mathbf{Z}, +)$ are isomorphic.

Result 13.20   *The groups $(2\mathbf{Z}, +)$ and $(\mathbf{Z}, +)$ are isomorphic.*

Proof   Define the function $\phi : \mathbf{Z} \to 2\mathbf{Z}$ by $\phi(n) = 2n$ for each $n \in \mathbf{Z}$. First, we show that f is one-to-one. Assume that $\phi(a) = \phi(b)$. Then $2a = 2b$. Dividing by 2, we obtain $a = b$ and so ϕ is one-to-one. Now we show that ϕ is onto. Let $n \in 2\mathbf{Z}$. Since n is even, $n = 2k$ for some integer k. Then $\phi(k) = 2k = n$. This shows that ϕ is onto. Finally, we show that ϕ is operation-preserving. Let $a, b \in \mathbf{Z}$. Then

$$\phi(a+b) = 2(a+b) = 2a + 2b = \phi(a) + \phi(b).$$ ∎

It should be obvious that every group G is isomorphic to itself. In fact, the identity function $i_G : G \to G$ defined by $i_G(g) = g$ for all $g \in G$ is an isomorphism. However, other permutations of G can be isomorphisms.

Result 13.21   *Let G a group and let $g \in G$. The function $\phi : G \to G$ defined by $\phi(a) = gag^{-1}$ for all $a \in G$ is an isomorphism.*

Proof   First, we show that f is one-to-one. Assume that $\phi(a) = \phi(b)$. Then $gag^{-1} = gbg^{-1}$. Canceling g on the left and g^{-1} on the right, we obtain $a = b$. Next, we show that ϕ is onto. Let $c \in G$. Then

$$\phi\left(g^{-1}cg\right) = g\left(g^{-1}cg\right)g^{-1} = \left(gg^{-1}\right)c\left(g^{-1}g\right) = ece = c.$$

Finally, we show that ϕ is operation-preserving. Let $a, b \in G$. Then

$$\phi(ab) = g(ab)g^{-1} = \left(gag^{-1}\right)\left(gbg^{-1}\right) = \phi(a)\phi(b).$$ ∎

EXERCISES FOR CHAPTER 13

Section 13.1: Binary Operations

13.1. Consider the algebraic structure $(S, *)$, where $S = \{x, y, z\}$ and $*$ is described in the table in Figure 13.12. Compute

(a) $x * (y * z)$ and $(x * y) * z$.
(b) $x * (x * x)$ and $(x * x) * x$.
(c) $y * (y * y)$ and $(y * y) * y$.
(d) What conclusion can you draw from (a)-(c)?

13.2. For every pair a, b of elements in the indicated sets, the element $a * b$ is defined. Which of these are binary operations? For those that are binary operations, determine which of the properties G1 – G4 are satisfied.
(a) $a * b = 1$ on the set $\mathbf{Z}$ (b) $a * b = a/b$ on the set $\mathbf{N}$
(c) $a * b = a^b$ on the set $\mathbf{N}$ (d) $a * b = \max\{a, b\}$ on the set $\mathbf{N}$
(e) $a * b = a + b + ab$ on the set $\mathbf{Z}$ (f) $a * b = a + b - 1$ on the set $\mathbf{Z}$
(g) $a * b = ab + 2a$ on the set $\mathbf{Z}$ (h) $a * b = ab - a - b + 2$ on the set $\mathbf{R} - \{1\}$
(i) $a * b = \sqrt{ab}$ on $\mathbf{Q}$ (j) $a * b = a + b$ on the set S of odd integers.

13.3. Let $T = \left\{ \begin{bmatrix} a & -b \\ b & a \end{bmatrix} : a, b \in \mathbf{R} \right\}$. Is T is closed under
(a) matrix addition?   (b) matrix multiplication?

13.4. Suppose that $*$ is an associative binary operation on a set S. Let

$$T = \{a \in S : a * x = x * a \text{ for all } x \in S\}.$$

Prove that T is closed under $*$.

13.5. Suppose that $*$ is an associative and commutative binary operation on a set S. Let

$$T = \{a \in S : a * a = a\}.$$

Prove that T is closed under $*$.

13.6. For matrices $A = \begin{bmatrix} a & b \\ c & d \end{bmatrix}$ and $B = \begin{bmatrix} e & f \\ g & h \end{bmatrix}$ in $M_2(\mathbf{R})$, the binary operation $*$ is defined on $M_2(\mathbf{R})$ by

$$A * B = \begin{bmatrix} a & b \\ c & d \end{bmatrix} * \begin{bmatrix} e & f \\ g & h \end{bmatrix} = \begin{bmatrix} a + e - 1 & b + f \\ c + g & d + h + 1 \end{bmatrix}.$$

Which of the properties G1 – G4 are satisfied?

13.7. For $n \geq 2$ and $[a], [b] \in \mathbf{Z}_n$, the binary operation $*$ is defined on $\mathbf{Z}_n$ by $[a] * [b] = [a + b + 1]$. Which of the properties G1 – G4 are satisfied?

$*$	x	y	z
x	y	z	y
y	y	x	x
z	z	z	y

Figure 13.12   A binary operation on the set $S = \{x, y, z\}$

*	a	b	c	d
a	a		c	
b				a
c		d	a	
d			b	

Figure 13.13   A binary operation on the set $S = \{a, b, c, d\}$ in Exercise 8

Section 13.2: Groups

13.8. Let $S = \{a, b, c, d\}$. Figure 13.13 shows a partially completed table for an associative binary operation $*$ defined on S.
 (a) Complete the table.
 (b) Is the algebraic structure $(S, *)$ a group?

13.9. Let $(G, *)$ be a group with $G = \{a, b, c, d\}$, where a partially completed table for $(G, *)$ is given in Figure 13.14. Complete the table.

13.10. None of the following binary operations $*$ on the given set result in a group. Which is the first property among G1, G2, G3 that fails?
 (a) Let $*$ be defined on $\mathbf{R}^+$ by $a * b = \sqrt{ab}$.
 (b) Let $*$ be defined on $\mathbf{R}^*$ by $a * b = a/b$.
 (c) Let $*$ be defined on $\mathbf{R}^+$ by $a * b = a + b + ab$.

13.11. (a) Determine whether for all $[a], [b] \in \mathbf{Z}_6^* = \{[1], [2], [3], [4], [5]\}$, there exists $[x] \in \mathbf{Z}_6^*$ such that $[a][x] = [b]$.
 (b) Why is the answer to the question posed in (a) not surprising?

13.12. Let $G = \left\{ \begin{bmatrix} a & b \\ 0 & 0 \end{bmatrix} : a, b \in \mathbf{R} \text{ and } a \neq 0 \right\}$.
 (a) Prove that G is closed under matrix multiplication.
 (b) Prove that there exists $E \in G$ such that $E \cdot A = A$ for all $A \in G$.
 (c) Prove for each $A \in G$ that there is $A' \in G$ such that $A \cdot A' = E$.
 (d) Prove or disprove: $(G, \cdot)$ is a group.

13.13. Let $*$ be an associative binary operation on the set G such that the following hold:
 (i) There exists $e \in G$ such that $g * e = g$ for all $g \in G$.
 (ii) For each $g \in G$, there exists $g' \in G$ such that $g * g' = e$.

Prove that $(G, *)$ is a group.

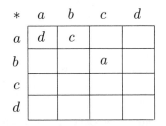

*	a	b	c	d
a	d	c		
b			a	
c				
d				

Figure 13.14   A partially completed table for $(G, *)$ in Exercise 9

Section 13.3: Permutation Groups

13.14. For $1 \leq i \leq 6$, each function f_i is a permutation on the set $\mathbf{Q} - \{0, 1\}$:
$$f_1(x) = x \qquad f_2(x) = 1 - x \qquad f_3(x) = \frac{1}{x}$$
$$f_4(x) = \frac{x-1}{x} \qquad f_5(x) = \frac{1}{1-x} \qquad f_6(x) = \frac{x}{x-1}.$$
Show that the set $F = \{f_1, f_2, \ldots, f_6\}$ is a group under composition.

13.15. Prove that if A is a set with at least three elements, then the symmetric group $(S_A, \circ)$ is nonabelian.

13.16. Give examples of the following (if they exist):

(a) a finite abelian group
(b) a finite nonabelian group
(c) an infinite abelian group
(d) an infinite nonabelian group.

13.17. Determine all elements x in the group S_3 such that $x^2 = \alpha_1$ and all those elements y in S_3 such that $y^3 = \alpha_1$.

13.18. For the permutations:
$$\gamma_1 = \begin{pmatrix} 1 & 2 & 3 & 4 \\ 1 & 2 & 3 & 4 \end{pmatrix} \qquad \gamma_2 = \begin{pmatrix} 1 & 2 & 3 & 4 \\ 2 & 3 & 4 & 1 \end{pmatrix}$$
$$\gamma_3 = \begin{pmatrix} 1 & 2 & 3 & 4 \\ 3 & 4 & 1 & 2 \end{pmatrix} \qquad \gamma_4 = \begin{pmatrix} 1 & 2 & 3 & 4 \\ 4 & 1 & 2 & 3 \end{pmatrix}$$
of the set $\{1, 2, 3, 4\}$, show that the set $G = \{\gamma_1, \gamma_2, \gamma_3, \gamma_4\}$ under composition is an abelian group.

13.19. Consider the following permutations on the set $A = \{1, 2, 3, 4, 5\}$.
$$\beta_1 = \begin{pmatrix} 1 & 2 & 3 & 4 & 5 \\ 1 & 2 & 3 & 4 & 5 \end{pmatrix} \quad \beta_2 = \begin{pmatrix} 1 & 2 & 3 & 4 & 5 \\ 2 & 3 & 1 & 4 & 5 \end{pmatrix} \quad \beta_3 = \begin{pmatrix} 1 & 2 & 3 & 4 & 5 \\ 3 & 1 & 2 & 4 & 5 \end{pmatrix}$$
$$\beta_4 = \begin{pmatrix} 1 & 2 & 3 & 4 & 5 \\ 1 & 2 & 3 & 5 & 4 \end{pmatrix} \quad \beta_5 = \begin{pmatrix} 1 & 2 & 3 & 4 & 5 \\ 2 & 3 & 1 & 5 & 4 \end{pmatrix} \quad \beta_6 = \begin{pmatrix} 1 & 2 & 3 & 4 & 5 \\ 3 & 1 & 2 & 5 & 4 \end{pmatrix}.$$
For $G = \{\beta_1, \beta_2, \ldots, \beta_6\}$, show that $(G, \circ)$ is a group of permutations on A.

13.20. For a permutation group G on a set A, a relation R is defined on A by $a\,R\,b$ if there exists $g \in G$ such that $g(a) = b$.

(a) Prove that R is an equivalence relation on A. (The equivalence classes resulting from this equivalence relation R are called the **orbits** of A under G.)
(b) For the group G in Exercise 13.19, determine the orbits of A under G.

Section 13.4: Fundamental Properties of Groups

13.21. Prove Theorem 13.7(b) (**The Right Cancellation Law**): Let $(G, *)$ be a group. If $b * a = c * a$, where $a, b, c \in G$, then $b = c$.

13.22. Prove the following (see Theorem 13.8): Let $(G, *)$ be a group and let $a, b \in G$. The linear equation $x * a = b$ has a unique solution x in G.

13.23. Let $(G, *)$ be a group and let $a, b, c \in G$. Prove that each of the following equations has a unique solution for x in G and determine the solution.
(a) $a * x * b = c$ (b) $a * b * x = c$.

13.24. Let a and b be two elements in a group G. Prove that if a and b commute, then a^{-1} and b^{-1} commute.

13.25. Let G be a group. Prove that G is abelian if and only if $(ab)^{-1} = a^{-1}b^{-1}$ for all $a, b \in G$.

+

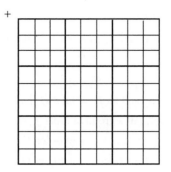

Figure 13.15   Constructing a table for the group $(\mathbf{Z}_9, +)$ in Exercise 13.26

13.26. Construct a table for the group $(\mathbf{Z}_9, +)$ by listing the elements of $\mathbf{Z}_9$ across the top of the table in some order and along the left side of the table in some order in such a way that every element of $\mathbf{Z}_9$ appears in each of the nine 3×3 regions indicated in the table in Figure 13.15.

13.27. By Theorem 13.9, every group G has a unique identity. That is, G contains only one element e such that $ae = ea = a$ for all $a \in G$. Suppose that e' is an element of G such that $e'b = b$ for some element $b \in G$. Prove or disprove: e' is the identity of G.

13.28. Let $(G, *)$ be a group. Prove that if $g * g = e$ for all $g \in G$, then G is abelian.

13.29. Let $(G, *)$ be a finite group of even order. Prove that there exists $g \in G$ such that $g \neq e$ and $g * g = e$.

13.30. Suppose that G is a finite abelian group of order n, say $G = \{g_1, g_2, \ldots, g_n\}$. Let $g = g_1 g_2 \cdots g_n g_1 g_2 \cdots g_n$. What is g?

Section 13.5: Subgroups

13.31. For an integer $n \geq 2$, prove that the set $n\mathbf{Z} = \{nk : k \in \mathbf{Z}\}$ of multiples of n is a subgroup of $(\mathbf{Z}, +)$.

13.32. Which of the following are subgroups of the given group?

(a) The subset $\mathbf{N}$ in $(\mathbf{Z}, +)$
(b) The subset $\{[0], [2], [4]\}$ in $(\mathbf{Z}_7, +)$
(c) The subset $\{[1], [2], [4]\}$ in $(\mathbf{Z}_7^*, \cdot)$
(d) The subset $\{2^n : n \in \mathbf{Z}\}$ in $(\mathbf{Q}^*, \cdot)$.

13.33. Let H and K be two subgroups of a group G. Prove or disprove:
(a) $H \cap K$ is a subgroup of G.   (b) $H \cup K$ is a subgroup of G.

13.34. For each of the following subsets H of $M_2^*(\mathbf{R})$, prove or disprove that $(H, \cdot)$ is a subgroup of $(M_2^*(\mathbf{R}), \cdot)$.

(a) $H = \left\{ \begin{bmatrix} a & b \\ c & 0 \end{bmatrix} : a, b, c \in \mathbf{R}, bc \neq 0 \right\}$

(b) $H = \left\{ \begin{bmatrix} a & b \\ 0 & c \end{bmatrix} : a, b, c \in \mathbf{R}, ac \neq 0 \right\}$.

13.35. Let $H = \{a + b\sqrt{3} : a, b \in \mathbf{Q}, a \neq 0 \text{ or } b \neq 0\}$. Prove that H is a subgroup of $(\mathbf{R}^*, \cdot)$.

13.36. For $n \in \mathbf{N}$, let T be a nonempty subset of $\{1, 2, \cdots, n\}$ and define

$$G_T = \{\alpha \in S_n : \alpha(t) = t \text{ for all } t \in T\}.$$

Prove that G_T is a subgroup of $(S_n, \circ)$.

13.37. Recall that $M_2^*(\mathbf{R}) = \{A \in M_2(\mathbf{R}) : \det(A) \neq 0\}$. Let

$$H = \{A \in M_2^*(\mathbf{R}) : \det(A) = 1 \text{ or } \det(A) = -1\}.$$

Prove that H is a subgroup of $(M_2^*(\mathbf{R}), \cdot)$.

13.38. Let G be an abelian group and let $H = \{a^2 : a \in G\}$. Prove that H is a subgroup of G.

13.39. Let G be an abelian group and let $H = \{a \in G : a^2 = e\}$. Prove that H is a subgroup of G.

13.40. What are all the subgroups of a group of order p, where p is a prime?

13.41. Prove or disprove: There exists a group of order 372 containing a subgroup of order 22.

13.42. Prove that a nonempty subset H of a group G is a subgroup of G if and only if $ab^{-1} \in H$ for all $a, b \in H$.

13.43. (a) Let $(G, *)$ be a finite group. Prove that if H is a nonempty subset of G that is closed under $*$, then H is a subgroup of G.
    (b) Show that the result in (a) is false if $(G, *)$ is an infinite group.

13.44. Let B be nonempty subset of a set A, $S = \{f \in S_A : f(B) = B\}$ and $T = \{f \in S_A : f(b) = b \text{ for each } b \in B\}$.

   (a) Prove that $(S, \circ)$ is a subgroup of $(S_A, \circ)$.
   (b) Prove that $(T, \circ)$ is a subgroup of $(S, \circ)$.

13.45. A group G has order 48. If there are six distinct left cosets of a subgroup H in G, then what is the order of H?

13.46. For the subgroup $H = \{\alpha_1, \alpha_2\}$ of $(S_3, \circ)$, where $\alpha_1 = \begin{pmatrix} 1 & 2 & 3 \\ 1 & 2 & 3 \end{pmatrix}$ and $\alpha_2 = \begin{pmatrix} 1 & 2 & 3 \\ 1 & 3 & 2 \end{pmatrix}$, determine the distinct left cosets of H in $(S_3, \circ)$.

13.47. For a subgroup H of a group $(G, \cdot)$, let gH be a left coset distinct from H. By the element $g^2 \in G$, we mean $g \cdot g$. Prove or disprove: $g^2 \in gH$.

Section 13.6: Isomorphic Groups

13.48. Let $H = \left\{ \begin{bmatrix} 1 & n \\ 0 & 1 \end{bmatrix} : n \in \mathbf{Z} \right\}$.

   (a) Prove that H is a subgroup of $(M_2^*(\mathbf{R}), \cdot)$.
   (b) Prove that the function $f : (\mathbf{Z}, +) \to (H, \cdot)$ given by $f(n) = \begin{bmatrix} 1 & n \\ 0 & 1 \end{bmatrix}$ is an isomorphism.
   (c) Parts (a) and (b) should suggest another question to you. Ask and answer a related question.

13.49. In each of the following, determine whether the function ϕ is an isomorphism from the first group to the second group.

   (a) $\phi : (\mathbf{Z}, +) \to (\mathbf{Z}, +)$ defined by $\phi(n) = 2n$.
   (b) $\phi : (\mathbf{Z}, +) \to (\mathbf{Z}, +)$ defined by $\phi(n) = n + 1$.
   (c) $\phi : (\mathbf{R}, +) \to (\mathbf{R}^+, \cdot)$ defined by $\phi(r) = 2^r$.
   (d) $\phi : (M_2^*(\mathbf{R}), \cdot) \to (\mathbf{R}^*, \cdot)$ defined by $\phi(A) = \det(A)$.

13.50. Obviously, $(\mathbf{R}^+, \cdot)$ and $(\mathbf{R}^+, \cdot)$ are isomorphic groups. Consider the function $\phi : \mathbf{R}^+ \to \mathbf{R}^+$ defined by $\phi(r) = r^2$ for all $r \in \mathbf{R}^+$. Is ϕ an isomorphism?

13.51. Let $(G, *)$ and $(H, \circ)$ be two groups. Prove that if $\phi : G \to H$ is an isomorphism, then the inverse function ϕ^{-1} of ϕ is an isomorphism from H to G.

13.52. Let G, H and K be three groups. Prove that if $\phi_1 : G \to H$ and $\phi_2 : H \to K$ are isomorphisms, then the composition $\phi_2 \circ \phi_1 : G \to K$ is an isomorphism.

13.53. Let $(G, *)$ be a group. Define a binary operation $\circ$ on G by $a \circ b = b * a$.

    (a) Prove that $(G, \circ)$ is a group.

    (b) Prove that $(G, *)$ and $(G, \circ)$ are isomorphic. [Hint: Consider the function $\phi(g) = g^{-1}$.]

13.54. Explain why the groups $(\mathbf{Q}, +)$ and $(\mathbf{R}, +)$ are not isomorphic.

13.55. We saw in Example 13.4 that with the binary operation $*$ defined on $\mathbf{Z}$ by $a * b = a + b - 1$, $(\mathbf{Z}, *)$ is an abelian group. Prove that $(\mathbf{Z}, *)$ is isomorphic to $(\mathbf{Z}, +)$.

13.56. (a) Let G and H be isomorphic groups. Prove that if G is abelian, then H is abelian.

    (b) Show that the groups $(\mathbf{Z}_6, +)$ and $(S_3, \circ)$ are not isomorphic.

13.57. Let $B = \{\frac{1}{n} : n \in \mathbf{Z} - \{0\}\}$ and let $A = \mathbf{R} - B$.

    (a) Prove for each $n \in \mathbf{Z}$ that the function $f_n : A \to A$ defined by $f_n(x) = \frac{x}{1+nx}$ is bijective.

    (b) Let $P = \{f_n : n \in \mathbf{Z}\}$. Prove that $(P, \circ)$ is a group of permutations on A.

    (c) Prove that the groups $(\mathbf{Z}, +)$ and $(P, \circ)$ are isomorphic.

13.58. Let $(G, \circ)$ and $(H, *)$ be groups with identities e and e', respectively. Suppose that $f : G \to H$ is a function with the property that $f(a \circ b) = f(a) * f(b)$ for all $a, b \in G$.

    (a) Let $M = \text{range}(f)$. Prove that $(M, *)$ is a subgroup of $(H, *)$.

    (b) Let $K = \{a \in G : f(a) = e'\}$. Prove that $(K, \circ)$ is a subgroup of $(G, \circ)$.

13.59. Let $A = \{\frac{m}{n} : m \text{ and } n \text{ are odd integers}, \gcd(m, n) = 1 \text{ and } n \geq 1\}$. Let $\mathbf{R}^* = \mathbf{R} - \{0\}$ and for $a \in A$, let $f_a : \mathbf{R}^* \to \mathbf{R}^*$ be defined by $f_a(x) = x^a$.

    (a) Prove that $(A, \cdot)$ is a subgroup of $(\mathbf{Q}^*, \cdot)$, where the product of every two elements of A is reduced to lowest terms.

    (b) Show for each $a \in A$ that f_a is a permutation on $\mathbf{R}^*$.

    (c) Let $F = \{f_a : a \in A\}$. Prove that $(F, \circ)$ is a subgroup of $(S_{\mathbf{R}^*}, \circ)$.

    (d) Prove that $(A, \cdot)$ and $(F, \circ)$ are isomorphic groups.

ADDITIONAL EXERCISES FOR CHAPTER 13

13.60. Let $(G, *)$ be a group. An element g of G is an **idempotent** for $*$ if $g * g = g$. Prove that there exists exactly one idempotent in G.

13.61. Let G be a group and let $a \in G$. The set $Z(a) = \{g \in G : ga = ag\}$ is called the **centralizer** of a. Prove that the centralizer of a is a subgroup of G.

13.62. Let $a, b \in \mathbf{Z}$, where $a, b \neq 0$ and let $H = \{am + bn : m, n \in \mathbf{Z}\}$ be the set of all linear combinations of a and b.

    (a) Prove that H is a subgroup of $(\mathbf{Z}, +)$.

    (b) Let $d = \gcd(a, b)$. Prove that $H = d\mathbf{Z}$.

13.63. Define $*$ on $\mathbf{R} - \{1\}$ by $a * b = a + b - ab$.

    (a) Prove that $(\mathbf{R} - \{1\}, *)$ is an abelian group.

    (b) Prove that $(\mathbf{R} - \{1\}, *)$ is isomorphic to $(\mathbf{R}^*, \cdot)$.

13.64. Let G be a group of order pq, where p and q are distinct primes. What are the possible orders of a subgroup of G?

13.65. Let H be a subgroup of $(\mathbf{Z}, +)$ with at least two elements and let m be the smallest positive integer in H. Prove that $H = m\mathbf{Z}$. [Hint: Use the Division Algorithm.]

13.66. Evaluate the proposed proof of the following statement.

Result   There exists no group containing exactly two distinct elements that do not commute.

Proof   Assume, to the contrary, that there exists a group G containing exactly two distinct elements, say x and y, that do not commute. Thus $xy \neq yx$. Since x and y are the only two elements of G that do not commute, x^{-1} and y do commute. Thus $x^{-1}y = yx^{-1}$. Multiplying by x on both the left and right, we obtain $x\left(x^{-1}y\right)x = x\left(yx^{-1}\right)x$. Simplifying, we have $yx = xy$. This is a contradiction. ∎

13.67. A group G of order $n \geq 2$ contains a subgroup H. In a left coset decomposition of H in G, the number of distinct left cosets is the same as the order of H. If some left coset contains p elements, where p is a prime, then what is n?

13.68. Evaluate the proposed proof of the following statement.

Result   There exists no abelian group containing exactly three distinct elements x such that $x^2 = e$.

Proof   Assume, to the contrary, that there exists an abelian group G such that $x^2 = e$ for exactly three distinct elements x of G. Certainly, $e^2 = e$, so there are two non-identity elements a and b such that $a^2 = b^2 = e$. Observe that $(ab)^2 = a^2 b^2 = ee = e$. Hence either $ab = a$, $ab = b$ or $ab = e$, which implies, respectively, that $b = e$, $a = e$ or $a = b$, producing a contradiction. ∎

13.69. Prove or disprove the following: For each odd integer $k \geq 3$, there exists no abelian group containing exactly k elements x such that $x^2 = e$.

13.70. For a function $f : \mathbf{N} \to \mathbf{N}$, we let $f^1 = f$ and $f^2 = f \circ f$. More generally, for $k \geq 2$, the function f^k is defined recursively by $f^k = f \circ f^{k-1}$. Thus $f^n : \mathbf{N} \to \mathbf{N}$ for each $n \in \mathbf{N}$. Give an example of two elements f and g in $S_{\mathbf{N}}$ such that $f^2 = g^2 = i_{\mathbf{N}}$ but $(f \circ g)^m \neq i_{\mathbf{N}}$ for all $m \in \mathbf{N}$.

Answers to Odd-Numbered Section Exercises

EXERCISES FOR CHAPTER 1

Section 1.1: Describing a Set

1.1 Only (d) and (e).

1.3 (a) $|A| = 5$, (b) $|B| = 11$, (c) $|C| = 51$, (d) $|D| = 2$, (e) $|E| = 1$, (f) $|F| = 2$

1.5 (a) $A = \{-1, -2, -3, \ldots\} = \{x \in \mathbf{Z} : x \leq -1\}$
    (b) $B = \{-3, -2, \ldots, 3\} = \{x \in \mathbf{Z} : -3 \leq x \leq 3\} = \{x \in \mathbf{Z} : |x| \leq 3\}$
    (c) $C = \{-2, -1, 1, 2\} = \{x \in \mathbf{Z} : -2 \leq x \leq 2, x \neq 0\} = \{x \in \mathbf{Z} : 0 < |x| \leq 2\}$

1.7 (a) $A = \{\cdots, -4, -1, 2, 5, 8, \cdots\} = \{3x + 2 : x \in \mathbf{Z}\}$
    (b) $B = \{\cdots, -10, -5, 0, 5, 10, \cdots\} = \{5x : x \in \mathbf{Z}\}$
    (c) $C = \{1, 8, 27, 64, 125, \cdots\} = \{x^3 : x \in \mathbf{N}\}$

1.9 $A = \{2, 3, 5, 7, 8, 10, 13\}$
    $B = \{x \in A : x = y + z, \text{ where } y, z \in A\} = \{5, 7, 8, 10, 13\}$
    $C = \{r \in B : r + s \in B \text{ for some } s \in B\} = \{5, 8\}$

Section 1.2: Subsets

1.11 Let $r = \min(c - a, b - c)$ and let $I = (c - r, c + r)$. Then I is centered at c and $I \subseteq (a, b)$.

1.13 See Figure 1.

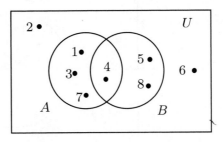

Figure 1   Answer for Exercise 1.13

1.15 $\mathcal{P}(A) = \{\emptyset, \{0\}, \{\{0\}\}, A\}$
1.17 $\mathcal{P}(A) = \{\emptyset, \{0\}, \{\emptyset\}, \{\{\emptyset\}\}, \{0, \emptyset\}, \{0, \{\emptyset\}\}, \{\emptyset, \{\emptyset\}\}, A\}; |\mathcal{P}(A)| = 8$
1.19 (a) $S = \{\emptyset, \{1\}\}$. (b) $S = \{1\}$.
 (c) $S = \{\emptyset, \{1\}, \{2\}, \{3\}, \{4, 5\}\}$. (d) $S = \{1, 2, 3, 4, 5\}$.
1.21 $B = \{1, 4, 5\}$.

Section 1.3: Set Operations

1.23 Let $A = \{1, 2, \ldots, 6\}$ and $B = \{4, 5, \ldots, 9\}$. Then
$A - B = \{1, 2, 3\}$, $B - A = \{7, 8, 9\}$ and $A \cap B = \{4, 5, 6\}$. Thus $|A - B| = |A \cap B| = |B - A| = 3$.
See Figure 2.

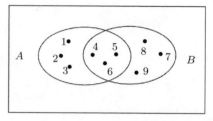

Figure 2   Answer for Exercise 1.23

1.25 (a) $A = \{1\}, B = \{\{1\}\}, C = \{1, 2\}$.
 (b) $A = \{\{1\}, 1\}, B = \{1\}, C = \{1, 2\}$.
 (c) $A = \{1\}, B = \{\{1\}\}, C = \{\{1\}, 2\}$.
1.27 Let $U = \{1, 2, \ldots, 8\}$ be a universal set, $A = \{1, 2, 3, 4\}$ and $B = \{3, 4, 5, 6\}$. Then $A - B = \{1, 2\}, B - A = \{5, 6\}, A \cap B = \{3, 4\}$ and $\overline{A \cup B} = \{7, 8\}$. See Figure 3.

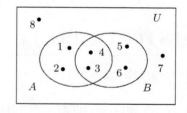

Figure 3   Answer for Exercise 1.27

1.29 (a) The sets $\emptyset$ and $\{\emptyset\}$ are elements of A. (b) $|A| = 3$.
 (c) All of $\emptyset, \{\emptyset\}$ and $\{\emptyset, \{\emptyset\}\}$ are subsets of A. (d) $\emptyset \cap A = \emptyset$.
 (e) $\{\emptyset\} \cap A = \{\emptyset\}$. (f) $\{\emptyset, \{\emptyset\}\} \cap A = \{\emptyset, \{\emptyset\}\}$.
 (g) $\emptyset \cup A = A$. (h) $\{\emptyset\} \cup A = A$. (i) $\{\emptyset, \{\emptyset\}\} \cup A = A$.
1.31 $A = \{1, 2\}, B = \{2\}, C = \{1, 2, 3\}, D = \{2, 3\}$.
1.33 $A = \{1\}, B = \{2\}$. Then $\{A \cup B, A \cap B, A - B, B - A\}$ is the power set of $\{1, 2\}$.
1.35 Let $U = \{1, 2, \ldots, 8\}, A = \{1, 2, 3, 5\}, B = \{1, 2, 4, 6\}$ and $C = \{1, 3, 4, 7\}$. See Figure 4.

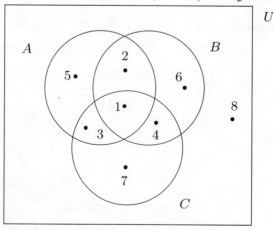

Figure 4   Answer for Exercise 1.35

Section 1.4: Indexed Collections of Sets

1.37 $\bigcup_{X \in S} X = A \cup B \cup C = \{0, 1, 2, \ldots, 5\}$ and $\bigcap_{X \in S} X = A \cap B \cap C = \{2\}$.

1.39 Since $|A| = 26$ and $|A_\alpha| = 3$ for each $\alpha \in A$, we need to have at least nine sets of cardinality 3 for their union to be A; that is, in order for $\bigcup_{\alpha \in S} A_\alpha = A$, we must have $|S| \geq 9$. However, if we let $S = \{a, d, g, j, m, p, s, v, y\}$, then $\bigcup_{\alpha \in S} A_\alpha = A$. Hence the smallest cardinality of a set S with $\bigcup_{\alpha \in S} A_\alpha = A$ is 9.

1.41 (a) $\{A_n\}_{n \in \mathbb{N}}$, where $A_n = \{x \in \mathbb{R} : 0 \leq x \leq 1/n\} = [0, 1/n]$.

   (b) $\{A_n\}_{n \in \mathbb{N}}$, where $A_n = \{a \in \mathbb{Z} : |a| \leq n\} = \{-n, -(n-1), \ldots, (n-1), n\}$.

1.43 $\bigcup_{r \in \mathbb{R}^+} A_r = \bigcup_{r \in \mathbb{R}^+}(-r, r) = \mathbb{R}$; $\bigcap_{r \in \mathbb{R}^+} A_r = \bigcap_{r \in \mathbb{R}^+}(-r, r) = \{0\}$.

1.45 $\bigcup_{n \in \mathbb{N}} A_n = \bigcup_{n \in \mathbb{N}}(-\frac{1}{n}, 2 - \frac{1}{n}) = (-1, 2)$; $\bigcap_{n \in \mathbb{N}} A_n = \bigcap_{n \in \mathbb{N}}(-\frac{1}{n}, 2 - \frac{1}{n}) = [0, 1]$.

Section 1.5: Partitions of Sets

1.47 (a) S_1 is not a partition of A since 4 belongs to no element of S_1.

   (b) S_2 is a partition of A.

   (c) S_3 is not a partition of A because 2, for example, belongs to two elements of S_3.

   (d) S_4 is not a partition of A since S_4 is not a set of subsets of A.

1.49 $A = \{1, 2, 3, 4\}$. $S_1 = \{\{1\}, \{2\}, \{3, 4\}\}$ and $S_2 = \{\{1, 2\}, \{3\}, \{4\}\}$.

1.51 Let $S = \{A_1, A_2, A_3\}$, where $A_1 = \{x \in \mathbb{Q} : x > 1\}$, $A_2 = \{x \in \mathbb{Q} : x < 1\}$ and $A_3 = \{1\}$.

1.53 Let $S = \{A_1, A_2, A_3, A_4\}$, where $A_1 = \{x \in \mathbb{Z} : x$ is odd and x is positive$\}$,
   $A_2 = \{x \in \mathbb{Z} : x$ is odd and x is negative$\}$, $A_3 = \{x \in \mathbb{Z} : x$ is even and x is nonnegative$\}$,
   $A_4 = \{x \in \mathbb{Z} : x$ is even and x is negative$\}$.

1.55 $|\mathcal{P}_1| = 2$, $|\mathcal{P}_2| = 3$, $|\mathcal{P}_3| = 5$, $|\mathcal{P}_4| = 8$, $|\mathcal{P}_5| = 13$, $|\mathcal{P}_6| = 21$.

Section 1.6: Cartesian Products of Sets

1.57 $A \times B = \{(x, x), (x, y), (y, x), (y, y), (z, x), (z, y)\}$.

1.59 $\mathcal{P}(A) = \{\emptyset, \{a\}, \{b\}, A\}$, $A \times \mathcal{P}(A) = \{(a, \emptyset), (a, \{a\}), (a, \{b\}), (a, A), (b, \emptyset), (b, \{a\}), (b, \{b\}), (b, A)\}$.

1.61 $\mathcal{P}(A) = \{\emptyset, \{1\}, \{2\}, A\}$, $\mathcal{P}(B) = \{\emptyset, B\}$, $A \times B = \{(1, \emptyset), (2, \emptyset)\}$,
   $\mathcal{P}(A) \times \mathcal{P}(B) = \{(\emptyset, \emptyset), (\emptyset, B), (\{1\}, \emptyset), (\{1\}, B), (\{2\}, \emptyset), (\{2\}, B), (A, \emptyset), (A, B)\}$.

1.63 $S = \{(3, 0), (2, 1), (1, 2), (0, 3), (-3, 0), (-2, 1), (-1, 2), (2, -1), (1, -2), (0, -3), (-2, -1), (-1, -2)\}$.
   See Figure 5.

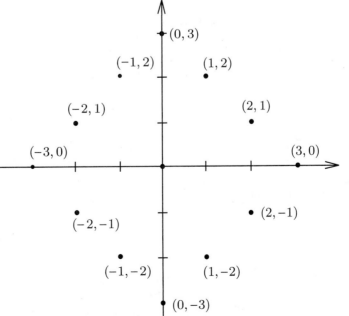

Figure 5   Answer for Exercise 1.63

1.65 $A \times B = [-1, 3] \times [2, 6]$, which is the set of all points on and within the square bounded by $x = -1$, $x = 3$, $y = 2$ and $y = 6$.

EXERCISES FOR CHAPTER 2

Section 2.1: Statements

2.1   **(a)** A false statement **(b)** A true statement **(c)** Not a statement **(d)** Not a statement (an open sentence) **(e)** Not a statement **(f)** Not a statement (an open sentence) **(g)** Not a statement

2.3   **(a)** False. $\emptyset$ has no elements. **(b)** True **(c)** True
(d) False. $\{\emptyset\}$ has $\emptyset$ as its only element. **(e)** True **(f)** False. 1 is not a set.

2.5   **(a)** $\{x \in \mathbf{Z} : x > 2\}$ **(b)** $\{x \in \mathbf{Z} : x \leq 2\}$

2.7   3, 5, 11, 17, 41, 59

2.9 $P(n) : \frac{n-1}{2}$ is even. $P(n)$ is true only for $n = 5$ and $n = 9$.

Section 2.2: The Negation of a Statement

2.11   **(a)** $\sqrt{2}$ is not a rational number.
(b)   0 is a negative integer.
(c)   111 is not a prime number.

2.13   **(a)**   The real number r is greater than $\sqrt{2}$.
(b)   The absolute value of the real number a is at least 3.
(c)   At most one angle of the triangle is 45°.
(d)   The area of the circle is less than 9π.
(e)   The sides of the triangle have different lengths.
(f)   The point P lies on or within the circle C.

Section 2.3: The Disjunction and Conjunction of Statements

2.15   See Figure 6.

P	Q	$\sim Q$	$P \wedge (\sim Q)$
T	T	F	F
T	F	T	T
F	T	F	F
F	F	T	F

Figure 6   Answer for Exercise 2.15

2.17   **(a)** $P \vee Q$: 15 is odd or 21 is prime. (True)
(b) $P \wedge Q$: 15 is odd and 21 is prime. (False)
(c) $(\sim P) \vee Q$: 15 is not odd or 21 is prime. (False)
(d) $P \wedge (\sim Q)$: 15 is odd and 21 is not prime. (True)

Section 2.4: The Implication

2.19   **(a)** $\sim P$: 17 is not even (or 17 is odd). (True)
(b) $P \vee Q$: 17 is even or 19 is prime. (True)
(c) $P \wedge Q$: 17 is even and 19 is prime. (False)
(d) $P \Rightarrow Q$: If 17 is even, then 19 is prime. (True)

2.21 (a) $P \Rightarrow Q$: If $\sqrt{2}$ is rational, then $22/7$ is rational. (True)
   (b) $Q \Rightarrow P$: If $22/7$ is rational, then $\sqrt{2}$ is rational. (False)
   (c) $(\sim P) \Rightarrow (\sim Q)$: If $\sqrt{2}$ is not rational, then $22/7$ is not rational. (False)
   (d) $(\sim Q) \Rightarrow (\sim P)$: If $22/7$ is not rational, then $\sqrt{2}$ is not rational. (True)

2.23 (a), (c), (d) are true.

2.25 (a) true. (b) false. (c) true. (d) true. (e) true.

2.27 Cindy and Don attended the talk.

2.29 Only (c) implies that $P \vee Q$ is false.

Section 2.5: More on Implications

2.31 (a) $P(x) \Rightarrow Q(x)$: If $|x| = 4$, then $x = 4$.
      $P(-4) \Rightarrow Q(-4)$ is false. $P(-3) \Rightarrow Q(-3)$ is true.
      $P(1) \Rightarrow Q(1)$ is true. $P(4) \Rightarrow Q(4)$ is true. $P(5) \Rightarrow Q(5)$ is true.
   (b) $P(x) \Rightarrow Q(x)$: If $x^2 = 16$, then $|x| = 4$. True for all $x \in S$.
   (c) $P(x) \Rightarrow Q(x)$: If $x > 3$, then $4x - 1 > 12$. True for all $x \in S$.

2.33 (a) True for $(x, y) = (3, 4)$ and $(x, y) = (5, 5)$, false for $(x, y) = (1, -1)$.
   (b) True for $(x, y) = (1, 2)$ and $(x, y) = (6, 6)$, false for $(x, y) = (2, -2)$.
   (c) True for $(x, y) \in \{(1, -1), (-3, 4), (1, 0)\}$ and false for $(x, y) = (0, -1)$.

Section 2.6: The Biconditional

2.35 $P \Leftrightarrow Q$: The integer 18 is odd if and only if 25 is even. (True)

2.37 The real number $|x - 3| < 1$ if and only if $x \in (2, 4)$.
   The condition $|x - 3| < 1$ is necessary and sufficient for $x \in (2, 4)$.

2.39 (a) True for all $x \in S - \{-4\}$. (b) True for $x \in S - \{3\}$. (c) True for $x \in S - \{-4, 0\}$.

2.41 True if $n = 3$.

2.43 $P(1) \Rightarrow Q(1)$ is false (since $P(1)$ is true and $Q(1)$ is false).
   $Q(3) \Rightarrow P(3)$ is false (since $Q(3)$ is true and $P(3)$ is false).
   $P(2) \Leftrightarrow Q(2)$ is true (since $P(2)$ and $Q(2)$ are both true).

2.45 True for all $n \in S - \{11\}$.

Section 2.7: Tautologies and Contradictions

2.47 The compound statement $(P \wedge (\sim Q)) \wedge (P \wedge Q)$ is a contradiction since it is false for all combinations of truth values for the component statements P and Q. See the truth table below.

P	Q	$\sim Q$	$P \wedge Q$	$P \wedge (\sim Q)$	$(P \wedge (\sim Q)) \wedge (P \wedge Q)$
T	T	F	T	F	F
T	F	T	F	T	F
F	T	F	F	F	F
F	F	T	F	F	F

2.49 The compound statement $((P \Rightarrow Q) \wedge (Q \Rightarrow R)) \Rightarrow (P \Rightarrow R)$ is a tautology since it is true for all combinations of truth values for the component statements P, Q, and R. See the truth table below.

P	Q	R	$P \Rightarrow Q$	$Q \Rightarrow R$	$(P \Rightarrow Q) \wedge (Q \Rightarrow R)$	$P \Rightarrow R$	$((P \Rightarrow Q) \wedge (Q \Rightarrow R)) \Rightarrow (P \Rightarrow R)$
T	T	T	T	T	T	T	T
T	F	T	F	T	F	T	T
F	T	T	T	T	T	T	T
F	F	T	T	T	T	T	T
T	T	F	T	F	F	F	T
T	F	F	F	T	F	F	T
F	T	F	T	F	F	T	T
F	F	F	T	T	T	T	T

$((P \Rightarrow Q) \wedge (Q \Rightarrow R)) \Rightarrow (P \Rightarrow R)$: If P implies Q and Q implies R, then P implies R.

Section 2.8: Logical Equivalence

2.51 **(a)** See the truth table below.

P	Q	$\sim P$	$\sim Q$	$P \Rightarrow Q$	$(\sim P) \Rightarrow (\sim Q)$
T	T	F	F	**T**	**T**
T	F	F	T	**F**	**T**
F	T	T	F	**T**	**F**
F	F	T	T	**T**	**T**

Since $P \Rightarrow Q$ and $(\sim P) \Rightarrow (\sim Q)$ do not have the same truth values for all combinations of truth values for the component statements P and Q, the compound statements $P \Rightarrow Q$ and $(\sim P) \Rightarrow (\sim Q)$ are not logically equivalent. Note that the last two columns in the truth table are not the same.

(b) The implication $Q \Rightarrow P$ is logically equivalent to $(\sim P) \Rightarrow (\sim Q)$.

2.53 **(a)** The statements $P \Rightarrow Q$ and $(P \wedge Q) \Leftrightarrow P$ are logically equivalent since they have the same truth values for all combinations of truth values for the component statements P and Q. See the truth table.

P	Q	$P \Rightarrow Q$	$P \wedge Q$	$(P \wedge Q) \Leftrightarrow P$
T	T	**T**	T	**T**
T	F	**F**	F	**F**
F	T	**T**	F	**T**
F	F	**T**	F	**T**

(b) The statements $P \Rightarrow (Q \vee R)$ and $(\sim Q) \Rightarrow ((\sim P) \vee R)$ are logically equivalent since they have the same truth values for all combinations of truth values for the component statements P, Q and R. See the truth table.

P	Q	R	$\sim P$	$\sim Q$	$Q \vee R$	$P \Rightarrow (Q \vee R)$	$(\sim P) \vee R$	$(\sim Q) \Rightarrow ((\sim P) \vee R)$
T	T	T	F	F	T	**T**	T	**T**
T	F	T	F	T	T	**T**	T	**T**
F	T	T	T	F	T	**T**	T	**T**
F	F	T	T	T	T	**T**	T	**T**
T	T	F	F	F	T	**T**	F	**T**
T	F	F	F	T	F	**F**	F	**F**
F	T	F	T	F	T	**T**	T	**T**
F	F	F	T	T	F	**T**	T	**T**

2.55 The statements $(P \vee Q) \Rightarrow R$ and $(P \Rightarrow R) \wedge (Q \Rightarrow R)$ are logically equivalent since they have the same truth values for all combinations of truth values for the component statements P, Q and R. See the truth table.

P	Q	R	$P \vee Q$	$(P \vee Q) \Rightarrow R$	$P \Rightarrow R$	$Q \Rightarrow R$	$(P \Rightarrow R) \wedge (Q \Rightarrow R)$
T	T	T	T	**T**	T	T	**T**
T	F	T	T	**T**	T	T	**T**
F	T	T	T	**T**	T	T	**T**
F	F	T	F	**T**	T	T	**T**
T	T	F	T	**F**	F	F	**F**
T	F	F	T	**F**	F	T	**F**
F	T	F	T	**F**	T	F	**F**
F	F	F	F	**T**	T	T	**T**

2.57 Since there are only four different combinations of truth values of P and Q for the second and third rows of the statements S_1, S_2, S_3, S_4, and S_5, at least two of these must have identical truth tables and so are logically equivalent.

Section 2.9: Some Fundamental Properties of Logical Equivalence

2.59 **(a)** Both $x \neq 0$ and $y \neq 0$.

(b) Either the integer a is odd or the integer b is odd.

2.61 Either $x^2 = 2$ and $x \neq \sqrt{2}$ or $x = \sqrt{2}$ and $x^2 \neq 2$.

2.63 If $3n + 4$ is odd, then $5n - 6$ is odd.

Section 2.10: Quantified Statements

2.65 $\forall x \in S$, $P(x)$: For every odd integer x, the integer $x^2 + 1$ is even.
     $\exists x \in S$, $Q(x)$: There exists an odd integer x such that x^2 is even.
2.67   **(a)** There exists a set A such that $A \cap \overline{A} \neq \emptyset$.
     (b) For every set A, we have $\overline{A} \not\subseteq A$.
2.69   **(a)** False, since $P(1)$ is false. **(b)** True, for example, $P(3)$ is true.
2.71   **(a)** $\exists a, b \in \mathbf{Z}$, $ab < 0$ and $a + b > 0$.
     (b) $\forall x, y \in \mathbf{R}$, $x \neq y$ implies that $x^2 + y^2 > 0$.
     (c) For all integers a and b either $ab \geq 0$ or $a + b \leq 0$.
        There exist real numbers x and y such that $x \neq y$ and $x^2 + y^2 \leq 0$.
     (d) $\forall a, b \in \mathbf{Z}$, $ab \geq 0$ or $a + b \leq 0$. $\exists x, y \in \mathbf{R}$, $x \neq y$ and $x^2 + y^2 \leq 0$.
2.73   **(b)** and **(c)** imply that $P(x) \Rightarrow Q(x)$ is true for all $x \in T$.
2.75   Let $S = \{3, 5, 11\}$ and $P(s, t) : st - 2$ is prime.
     (a) $\forall s, t \in S$, $P(s, t)$.
     (b) False since $P(11, 11)$ is false.
     (c) $\exists s, t \in S$, $\sim P(s, t)$.
     (d) There exist $s, t \in S$ such that $st - 2$ is not prime.
     (e) True since the statement in (a) is false.
2.77   **(a)** There exists a triangle T_1 such that for every triangle T_2, $r(T_2) \geq r(T_1)$.
     (b) $\forall T_1 \in A$, $\exists T_2 \in B$, $\sim P(T_1, T_2)$.
     (c) For every triangle T_1, there exists a triangle T_2 such that $r(T_2) < r(T_1)$.
2.79   **(a)** There exists $b \in B$ such that for every $a \in A$, $a - b < 0$.
     (b) Let $b = 10$. Then $3 - 10 = -7 < 0$, $5 - 10 = -5 < 0$ and $8 - 10 = -2 < 0$.

Section 2.11: Characterizations of Statements

2.81   An integer n is odd if and only if n^2 is odd.
2.83   **(a)** a characterization. **(b)** a characterization. **(c)** a characterization.
     (d) a characterization. (Pythagorean theorem) **(e)** not a characterization. (Every positive number is the area of some rectangle.)

EXERCISES FOR CHAPTER 3

Section 3.1: Trivial and Vacuous Proofs

3.1   *Proof*   Since $x^2 - 2x + 2 = (x - 1)^2 + 1 \geq 1$, it follows that $x^2 - 2x + 2 \neq 0$ for all $x \in \mathbf{R}$. Hence the statement is true trivially.   ∎

3.3   *Proof*   Note that $\frac{r^2 + 1}{r} = r + \frac{1}{r}$. If $r \geq 1$, then $r + \frac{1}{r} > 1$; while if $0 < r < 1$, then $\frac{1}{r} > 1$ and so $r + \frac{1}{r} > 1$. Thus $\frac{r^2 + 1}{r} \leq 1$ is false for all $r \in \mathbf{Q}^+$ and so the statement is true vacuously.   ∎

3.5   *Proof*   Since $n^2 - 2n + 1 = (n - 1)^2 \geq 0$, it follows that $n^2 + 1 \geq 2n$ and so $n + \frac{1}{n} \geq 2$. Thus the statement is true vacuously.   ∎

3.7   *Proof*   Since $(x - y)^2 + (x - z)^2 + (y - z)^2 \geq 0$, it follows that $2x^2 + 2y^2 + 2z^2 - 2xy - 2xz - 2yz \geq 0$ and so $x^2 + y^2 + z^2 \geq xy + xz + yz$. Thus, the statement is true vacuously.   ∎

Section 3.2: Direct Proofs

3.9   *Proof*   Let x be an even integer. Then $x = 2a$ for some integer a. Thus

$$5x - 3 = 5(2a) - 3 = 10a - 4 + 1 = 2(5a - 2) + 1.$$

Since $5a - 2$ is an integer, $5x - 3$ is odd.   ∎

3.11 ***Proof***   Let $1 - n^2 > 0$. Then $n = 0$. Thus $3n - 2 = 3 \cdot 0 - 2 = -2$ is an even integer.   ∎

3.13 ***Proof***   Assume that $(n + 1)^2(n + 2)^2/4$ is even, where $n \in S$. Then $n = 2$ and $(n + 2)^2(n + 3)^2/4 = 100$, which is even.   ∎

3.15 ***Proof***   Let $n \in A \cap B = \{3, 5, 7, 9\}$. Then $3^2 - 2 = 7$, $5^2 - 2 = 23$, $7^2 - 2 = 47$ and $9^2 - 2 = 79$ are all primes.   ∎

Section 3.3: Proof by Contrapositive

3.17 First, we prove a lemma. **Lemma** Let $n \in \mathbf{Z}$. If $15n$ is even, then n is even.
(Use a proof by contrapositive to verify this lemma.) Then use this lemma to prove the result.
Proof of Result   Assume that $15n$ is even. By the lemma, n is even and so $n = 2a$ for some integer a. Hence $9n = 9(2a) = 2(9a)$. Since $9a$ is an integer, $9n$ is even.   ∎
[Note: This result could also be proved by assuming that $15n$ is even (and so $15n = 2a$ for some integer a) and observing that $9n = 15n - 6n = 2a - 6n$.]

3.19 **Lemma**   Let $x \in \mathbf{Z}$. If $7x + 4$ is even, then x is even. (Use a proof by contrapositive to verify this lemma.)
Proof of Result   Assume that $7x + 4$ is even. Then by the lemma, x is even and so $x = 2a$ for some integer a. Hence
$$3x - 11 = 3(2a) - 11 = 6a - 12 + 1 = 2(3a - 6) + 1.$$
Since $3a - 6$ is an integer, $3x - 11$ is odd.   ∎

3.21 To verify the implication "If n is even, then $(n + 1)^2 - 1$ is even.", use a direct proof. For the converse, "If $(n + 1)^2 - 1$ is even, then n is even.", use a proof by contrapositive.

3.23 ***Proof***   Assume that $n \notin A \cup B$. Then $n = 3$ and $n(n - 1)(n - 2)/6 = 1$ is odd.   ∎

3.25 ***Proof***   Assume that $n \notin A$. Then $n \in B = \{2, 3, 6, 7\}$. If $n = 2$, then $(n^2 + 3n - 4)/2 = 3$ is odd. If $n = 3$, then $(n^2 + 3n - 4)/2 = 7$ is odd. If $n = 6$, then $(n^2 + 3n - 4)/2 = 25$ is odd. If $n = 7$, then $(n^2 + 3n - 4)/2 = 33$ is odd.   ∎

Section 3.4: Proof by Cases

3.27 Let $n \in \mathbf{Z}$. We consider two cases.
Case 1. n is even. Then $n = 2a$ for some integer a. Thus
$$n^3 - n = 8a^3 - 2a = 2(4a^3 - a).$$
Since $4a^3 - a$ is an integer, $n^3 - n$ is even.
Case 2. n is odd. Then $n = 2b + 1$ for some integer b. The remainder of the proof is similar to that of Case 1.

3.29 Assume that $a, b \in \mathbf{Z}$ such that ab is odd. By Exercise 3.28, a and b are both odd and so a^2 and b^2 are both odd by Theorem 3.12. Thus $a^2 + b^2$ is even.

3.31 ***Proof***   Assume that a or b is odd, say a is odd. Then $a = 2x + 1$ for some integer x. We consider two cases.
Case 1. b is even. Then $b = 2y$ for some integer y. Thus $ab = a(2y) = 2(ay)$. Since ay is an integer, ab is even. Also,
$$a + b = (2x + 1) + 2y = 2(x + y) + 1.$$
Since $x + y$ is an integer, $a + b$ is odd. Hence ab and $a + b$ are of opposite parity.
Case 2. b is odd. Then $b = 2y + 1$ for some integer y. Thus
$$a + b = (2x + 1) + (2y + 1) = 2x + 2y + 2 = 2(x + y + 1).$$
Since $x + y + 1$ is an integer, $a + b$ is even. Furthermore,
$$ab = (2x + 1)(2y + 1) = 4xy + 2x + 2y + 1 = 2(2xy + x + y) + 1.$$
Since $2xy + x + y$ is an integer, ab is odd. Hence ab and $a + b$ are of opposite parity.   ∎

3.33 ***Proof***   Assume that $n \notin A \cap B$. Then $n = 1$ or $n = 4$. If $n = 1$, then $2n^2 - 5n = -3$ is negative and odd; while if $n = 4$, then $2n^2 - 5n = 12$ is positive and even.

For the converse, assume that $n \in A \cap B$. Then $n = 2$ or $n = 3$. If $n = 2$, then $2n^2 - 5n = -2$ is negative and even; while if $n = 3$, then $2n^2 - 5n = 3$ is positive and odd. Thus if $n \in A \cap B$, then neither (a) nor (b) occurs. ∎

3.35 **Proof** Let n be a nonnegative integer. We consider two cases.
Case 1. $n = 0$. Then $2^n + 6^n = 2^0 + 6^0 = 2$, which is even.
Case 2. n *is a positive integer.* Then $n - 1$ is a nonnegative integer. Therefore,

$$2^n + 6^n = 2^n + (2 \cdot 3)^n = 2^n + 2^n \cdot 3^n = 2(2^{n-1} + 2^{n-1} \cdot 3^n).$$

Since $2^{n-1} + 2^{n-1} \cdot 3^n$ is an integer, $2^n + 6^n$ is even. ∎

Section 3.5: Proof Evaluations

3.37 (3) is proved.

3.39 The converse of the result has been proved. No proof has been given of the result itself.

3.41 From the first sentence of the proposed proof and the final sentence, it appears that the result in question is the following: Let $x, y \in \mathbf{Z}$. If x or y is even, then xy^2 is even. If this, in fact, is the result, then the proof is not correct. A proof by cases should be given, namely *Case 1. x is even.* and *Case 2. y is even.*

3.43 **Result** Let $x \in \mathbf{Z}$. If $7x - 3$ is even, then $3x + 8$ is odd.
A direct proof of the result is given with the aid of the lemma: Let $x \in \mathbf{Z}$. If $7x - 3$ is even, then x is odd.

EXERCISES FOR CHAPTER 4

Section 4.1: Proofs Involving Divisibility of Integers

4.1 **Proof** Assume that $a \mid b$. Then $b = ac$ for some integer c. Then $b^2 = (ac)^2 = a^2c^2$. Since c^2 is an integer, $a^2 \mid b^2$. ∎

4.3 **(a) Proof** Assume that $3 \mid m$. Then $m = 3q$ for some integer q. Hence $m^2 = (3q)^2 = 9q^2 = 3(3q^2)$. Since $3q^2$ is an integer, $3 \mid m^2$. ∎
   (b) Let $m \in \mathbf{Z}$. If $3 \nmid m^2$, then $3 \nmid m$.
   (c) Start with the following: Assume that $3 \nmid m$. Then $m = 3q + 1$ or $m = 3q + 2$, where $q \in \mathbf{Z}$. Consider these two cases.
   (d) Let $m \in \mathbf{Z}$. If $3 \mid m^2$, then $3 \mid m$.
   (e) Let $m \in \mathbf{Z}$. Then $3 \mid m$ if and only if $3 \mid m^2$.

4.5 **Proof** Assume that $a \mid b$ or $a \mid c$, say the latter. Then $c = ak$ for some integer k. Thus $bc = b(ak) = a(bk)$. Since bk is an integer, $a \mid bc$. ∎

4.7 For the implication "If $3 \nmid n$, then $3 \mid (2n^2 + 1)$.", use a direct proof. Assume that $3 \nmid n$. Then $n = 3q + 1$ or $n = 3q + 2$ for some integer q. Then consider these two cases.
For the converse "If $3 \mid (2n^2 + 1)$, then $3 \nmid n$." use a proof by contrapositive.

4.9 **(a) Proof** Assume that $2 \mid (x^2 - 5)$. Then $x^2 - 5 = 2y$ for some integer y and so $x^2 = 2y + 5 = 2(y + 2) + 1$. Since $y + 2 \in \mathbf{Z}$, it follows that x^2 is odd. By Theorem 3.12, x is also odd and so $x = 2a + 1$ for some integer a. Hence

$$x^2 - 5 = (2a + 1)^2 - 5 = 4a^2 + 4a - 4 = 4(a^2 + a - 1).$$

Since $a^2 + a - 1$ is an integer, $4 \mid (x^2 - 5)$. ∎
   (b) $x = 3$.

4.11 **Proof** Let $n \in \mathbf{Z}$ with $n \geq 8$. Then $n = 3q$ where $q \geq 3$, $n = 3q + 1$ where $q \geq 3$ or $n = 3q + 2$ where $q \geq 2$. We consider these three cases.
Case 1. $n = 3q$ *where* $q \geq 3$. Then $n = 3a + 5b$, where $a = q \geq 3$ and $b = 0$.
Case 2. $n = 3q + 1$ *where* $q \geq 3$. Then $n = 3(q - 3) + 10$, where $q - 3 \geq 0$. Thus $n = 3a + 5b$, where $a = q - 3 \geq 0$ and $b = 2$.
Case 3. $n = 3q + 2$ *where* $q \geq 2$. Then $n = 3(q - 1) + 5$, where $q - 1 \geq 1$. Thus $n = 3a + 5b$, where $a = q - 1 \geq 1$ and $b = 1$. ∎

4.13 First, we prove the following lemma.

Lemma   If $c \in \mathbf{Z}$, then $c^2 \equiv 0 \pmod 3$ or $c^2 \equiv 1 \pmod 3$.

Proof   If $3 \mid (c^2 - 1)$, then we have the desired conclusion. Otherwise, $3 \nmid (c^2 - 1)$ and so $3 \mid c$ by Result 4.6. Then $3 \mid c^2$ by Exercise 3 of this chapter. Thus, either $c^2 \equiv 0 \pmod 3$ or $c^2 \equiv 1 \pmod 3$.   ■

We are now prepared to prove the main result.

Proof   Assume that $3 \nmid ab$. By Result 4.5, $3 \nmid a$ and $3 \nmid b$. Thus, $3 \mid (a^2 - 1)$ and $3 \mid (b^2 - 1)$ by Result 4.6. Therefore, $3 \mid (a^2 + b^2 - 2)$. Since $c^2 \not\equiv 2 \pmod 3$ for any positive integer c by the lemma, it follows that $a^2 + b^2$ cannot equal c^2.   ■

Section 4.2: Proofs Involving Congruence of Integers

4.15 **Proof**   Assume that $a \equiv b \pmod n$ and $a \equiv c \pmod n$. Then $n \mid (a - b)$ and $n \mid (a - c)$. Hence $a - b = nx$ and $a - c = ny$, where $x, y \in \mathbf{Z}$. Thus $b = a - nx$ and $c = a - ny$. Therefore, $b - c = (a - nx) - (a - ny) = ny - nx = n(y - x)$. Since $y - x$ is an integer, $n \mid (b - c)$ and so $b \equiv c \pmod n$.   ■

4.17 **(a)** **Proof**   Assume that $a \equiv 1 \pmod 5$. Then $5 \mid (a - 1)$. So $a - 1 = 5k$ for some integer k. Thus $a = 5k + 1$ and so

$$a^2 = (5k + 1)^2 = 25k^2 + 10k + 1 = 5(5k^2 + 2k) + 1.$$

Thus

$$a^2 - 1 = 5(5k^2 + 2k).$$

Since $5k^2 + 2k$ is an integer, $5 \mid (a^2 - 1)$ and so $a^2 \equiv 1 \pmod 5$. [Note: We could also observe that $a^2 - 1 = (a - 1)(a + 1)$. This is also a consequence of Exercise 14 in this chapter.]   ■

(b) We can conclude that $b^2 \equiv 1 \pmod 5$.

4.19 **Proof**   Assume that $a \equiv 5 \pmod 6$ and $b \equiv 3 \pmod 4$. Then $6 \mid (a - 5)$ and $4 \mid (b - 3)$. Thus $a - 5 = 6x$ and $b - 3 = 4y$, where $x, y \in \mathbf{Z}$. So $a = 6x + 5$ and $b = 4y + 3$. Observe that

$$4a + 6b = 4(6x + 5) + 6(4y + 3) = 24x + 20 + 24y + 18 = 24x + 24y + 38 = 8(3x + 3y + 4) + 6.$$

Since $3x + 3y + 4$ is an integer, $8 \mid (4a + 6b - 6)$ and so $4a + 6b \equiv 6 \pmod 8$.   ■

4.21 Either $a = 3q$, $a = 3q + 1$ or $a = 3q + 2$ for some integer q. We consider these three cases.

Case 1. $a = 3q$. Then

$$a^3 - a = (3q)^3 - (3q) = 27q^3 - 3q = 3(9q^3 - q).$$

Since $9q^3 - q$ is an integer, $3 \mid (a^3 - a)$ and so $a^3 \equiv a \pmod 3$.

The other cases are handled in a similar manner.

4.23 **Proof**   Since $6 \mid a$, it follows that $a \equiv 0 \pmod 6$ and so $a = 6q$ for some $q \in \mathbf{Z}$. Therefore, $a + i \equiv i \pmod 6$ for $i = 1, 2, \ldots, 5$. First, assume that $x, y \in S = \{a, a + 1, \ldots, a + 5\}$, where one of x and y is congruent to 1 modulo 6 and the other is congruent to 5 modulo 6. We may assume that $x = a + 5$ and $y = a + 1$. Then $x = 6q + 5$ and $y = 6q + 1$. Thus

$$x^2 - y^2 = (6q + 5)^2 - (6q + 1)^2 = (36q^2 + 60q + 25) - (36q^2 + 12q + 1)$$
$$= 48q + 24 = 24(2q + 1).$$

Since $2q + 1$ is an integer, $24 \mid (x^2 - y^2)$.

For the converse, assume that x and y are distinct odd integers in S such that one of x and y is not congruent to 1 or 5 modulo 6. Since a is even, either x or y is $a + 3$. There are two cases.

Case 1. $x = a + 5$ *and* $y = a + 3$. Thus $x = 6q + 5$ and $y = 6q + 3$. Now

$$x^2 - y^2 = (6q + 5)^2 - (6q + 3)^2 = (36q^2 + 60q + 25) - (36q^2 + 36q + 9)$$
$$= 24q + 16.$$

Since q is an integer, $24 \nmid (x^2 - y^2)$.

Case 2. $x = a + 3$ *and* $y = a + 1$. (The proof here is similar to the proof of Case 1.)   ■

Section 4.3: Proofs Involving Real Numbers

4.25 **Proof** Assume that $x^2 - 4x = y^2 - 4y$ and $x \neq y$. Thus $x^2 - y^2 - 4(x - y) = 0$ and so $(x - y)[(x + y) - 4] = 0$. Since $x \neq y$, it follows that $(x + y) - 4 = 0$ and so $x + y = 4$. ∎

4.27 A proof by contrapositive can be used: Assume that $x \leq 0$. Then $3x^4 + 1 \geq 1$ and $x^7 + x^3 \leq 0$. Thus $3x^4 + 1 \geq 1 > 0 \geq x^7 + x^3$.

4.29 **Proof** Let $r \in \mathbf{R}$ such that $|r - 1| < 1$. Since $|r - 1| < 1$, it follows that $0 < r < 2$. Because $(r - 2)^2 \geq 0$, we have

$$r^2 - 4r + 4 \geq 0.$$

Thus $4 \geq 4r - r^2 = r(4 - r)$. Since $0 < r < 2$, it follows that $r(4 - r) > 0$. Dividing both sides by $r(4 - r)$, we obtain $\frac{4}{r(4-r)} \geq 1$. ∎

4.31 **Proof** Since

$$|x| = |(x + y) + (-y)| \leq |x + y| + |-y| = |x + y| + |y|,$$

it follows that $|x + y| \geq |x| - |y|$. ∎

4.33 Observe that $r^3 + s^3 + t^3 - 3rst = \frac{1}{2}(r + s + t)[(r - s)^2 + (s - t)^2 + (t - r)^2]$.

4.35 **Proof** Assume that $x(x + 1) > 2$. Then $x^2 + x > 2$ and so $x^2 + x - 2 > 0$. Thus $(x + 2)(x - 1) > 0$. Therefore, either (a) $x + 2$ and $x - 1$ are both positive or (b) $x + 2$ and $x - 1$ are both negative. If (a) occurs, then $x > 1$; while if (b) occurs, then $x < -2$. ∎

4.37 **Proof** Since $(x - y)^2 + (x - z)^2 + (y - z)^2 \geq 0$, it follows that $2x^2 + 2y^2 + 2z^2 \geq 2xy + 2xz + 2yz$. Dividing by 2 produces the desired inequality. ∎

4.39 **Proof** Observe that

$$(a^2 + c^2)(b^2 + d^2) = a^2b^2 + a^2d^2 + b^2c^2 + c^2d^2$$
$$= (ab + cd)^2 + (ad - bc)^2 \geq (ab + cd)^2. ∎$$

Section 4.4: Proofs Involving Sets

4.41 First, we show that if $A \cup B = A$, then $B \subseteq A$. Assume that $A \cup B = A$. Let $x \in B$. Then $x \in A \cup B$. Since $A \cup B = A$, it follows that $x \in A$. Thus $B \subseteq A$. Next we show that if $B \subseteq A$, then $A \cup B = A$. A proof by contrapositive begins by assuming that $A \cup B \neq A$.

4.43 **(a)** Consider $A = \{1, 2\}$, $B = \{2, 3\}$ and $C = \{2, 4\}$.
(b) Consider $A = \{1, 2\}$, $B = \{1\}$ and $C = \{2\}$.
(c) Suppose that $B \neq C$. We show that either $A \cap B \neq A \cap C$ or $A \cup B \neq A \cup C$. Since $B \neq C$, it follows that $B \not\subseteq C$ or $C \not\subseteq B$, say the former. Thus there exists $b \in B$ such that $b \notin C$. We consider two cases, according to whether $b \in A$ or $b \notin A$.
Case 1. $b \in A$. Since $b \in B$ and $b \in A$, it follows that $b \in A \cap B$. On the other hand, $b \notin C$ and so $b \notin A \cap C$. Thus $A \cap B \neq A \cap C$.
Case 2. $b \notin A$. Then show that $A \cup B \neq A \cup C$.

4.45 **Proof** Let $n \in B$. Then $n \in \mathbf{Z}$ and $n \equiv 3 \pmod 4$. So $n = 4q + 3$ for some integer q. Therefore, $n = 2(2q + 1) + 1$. Since $2q + 1 \in \mathbf{Z}$, it follows that $2 \mid (n - 1)$ and so $n \equiv 1 \pmod 2$. Thus $n \in A$. ∎

4.47 **(a)** Each element $n \in A - B$ can be written as $n = 3a + 2$ for some integer a, where n is even. This implies that a is even, say $a = 2b$ for some integer b. Thus $n = 3a + 2 = 3(2b) + 2 = 6b + 2$.
(b) **Proof** Let $n \in A \cap B$. Then $n = 3a + 2$ for some integer a and n is odd. Thus a is odd, say $a = 2b + 1$ for some integer b. Thus $n = 3a + 2 = 3(2b + 1) + 2 = 6b + 5$. Therefore,

$$n^2 - 1 = (6b + 5)^2 - 1 = 36b^2 + 60b + 24 = 12(3b^2 + 5b + 2).$$

Since $3b^2 + 5b + 2$ is an integer, $12 \mid (n^2 - 1)$ and so $n^2 \equiv 1 \pmod{12}$. ∎

4.49 First, we show that $A \subseteq (A - B) \cup (A \cap B)$. Let $x \in A$. Then $x \notin B$ or $x \in B$. If $x \notin B$, then $x \in A - B$ and $x \in (A - B) \cup (A \cap B)$. If $x \in B$, then $x \in A \cap B$ and so $x \in (A - B) \cup (A \cap B)$. Therefore, $A \subseteq (A - B) \cup (A \cap B)$. Next, show that $(A - B) \cup (A \cap B) \subseteq A$.

4.51 (e) is a necessary condition for A and B to be disjoint.

Section 4.5: Fundamental Properties of Set Operations

4.53 First, we show that $A \cap (B \cup C) \subseteq (A \cap B) \cup (A \cap C)$. Let $x \in A \cap (B \cup C)$. Then $x \in A$ and $x \in B \cup C$. Since $x \in B \cup C$, it follows that $x \in B$ or $x \in C$, say $x \in B$. Because $x \in A$ and $x \in B$, it follows that $x \in A \cap B$. Hence $x \in (A \cap B) \cup (A \cap C)$. Next, show that $(A \cap B) \cup (A \cap C) \subseteq A \cap (B \cup C)$.

4.55 We first show that $(A - B) \cap (A - C) \subseteq A - (B \cup C)$. Let $x \in (A - B) \cap (A - C)$. Then $x \in A - B$ and $x \in A - C$. Since $x \in A - B$, it follows that $x \in A$ and $x \notin B$. Because $x \in A - C$, we have $x \in A$ and $x \notin C$. Since $x \notin B$ and $x \notin C$, we have $x \notin B \cup C$. Thus $x \in A - (B \cup C)$. Next, show that $A - (B \cup C) \subseteq (A - B) \cap (A - C)$.

4.57 ***Proof***   By Theorem 4.22,

$$\overline{\overline{A} \cup (\overline{B} \cap C)} = \overline{\overline{A}} \cap \overline{(\overline{B} \cap C)} = A \cap (\overline{\overline{B} \cup \overline{C}})$$

$$= A \cap (B \cup \overline{C}) = (A \cap B) \cup (A \cap \overline{C})$$

$$= (A \cap B) \cup (A - C),$$

as desired. ∎

4.59 First, we show that $A - (B - C) \subseteq (A \cap C) \cup (A - B)$. Let $x \in A - (B - C)$. Then $x \in A$ and $x \notin B - C$. Since $x \notin B - C$, it is not the case that $x \in B$ and $x \notin C$. Thus either $x \notin B$ or $x \in C$. Since $x \in A$, either $x \in A - B$ or $x \in A \cap C$. Thus $x \in (A \cap C) \cup (A - B)$. Therefore, $A - (B - C) \subseteq (A \cap C) \cup (A - B)$. Next, show that $(A \cap C) \cup (A - B) \subseteq A - (B - C)$.

Section 4.6: Proofs Involving Cartesian Products of Sets

4.61 For $A = \{1\}$ and $B = \{2\}$, $\mathcal{P}(A) = \{\emptyset, A\}$ and $\mathcal{P}(B) = \{\emptyset, B\}$. Thus
$$\mathcal{P}(A) \times \mathcal{P}(B) = \{(\emptyset, \emptyset), (\emptyset, B), (A, \emptyset), (A, B)\}.$$
Since $A \times B = \{(1, 2)\}$, it follows that $\mathcal{P}(A \times B) = \{\emptyset, A \times B\}$.

4.63 Let A and B be sets. Then $A \times B = B \times A$ if and only if $A = B$ or one of A and B is empty.
Proof   First, we show that if $A = B$ or one of A and B is empty, then $A \times B = B \times A$. If $A = B$, then certainly $A \times B = B \times A$; while if one of A and B is empty, say $A = \emptyset$, then $A \times B = \emptyset \times B = \emptyset = B \times \emptyset = B \times A$.

For the converse, assume that A and B are nonempty sets with $A \neq B$. Since $A \neq B$, at least one of A and B is not a subset of the other, say $A \nsubseteq B$. Then there is an element $a \in A$ such that $a \notin B$. Since $B \neq \emptyset$, there exists an element $b \in B$. Then $(a, b) \in A \times B$ but $(a, b) \notin B \times A$. Hence $A \times B \neq B \times A$. ∎

4.65 First, assume that $A \times C \subseteq B \times C$. Show that $A \subseteq B$. For the converse, assume that $A \subseteq B$. Show that $A \times C \subseteq B \times C$.

4.67 We first show that $A \times (B \cap C) \subseteq (A \times B) \cap (A \times C)$. Let $(x, y) \in A \times (B \cap C)$. Then $x \in A$ and $y \in B \cap C$. Thus $y \in B$ and $y \in C$. Thus $(x, y) \in A \times B$ and $(x, y) \in A \times C$. Therefore, $(x, y) \in (A \times B) \cap (A \times C)$. It remains to show that $(A \times B) \cap (A \times C) \subseteq A \times (B \cap C)$.

4.69 ***Proof***   Let $(x, y) \in (A \times B) \cup (C \times D)$. Then $(x, y) \in A \times B$ or $(x, y) \in C \times D$. Assume, without loss of generality, that $(x, y) \in A \times B$. Thus $x \in A$ and $y \in B$. This implies that $x \in A \cup C$ and $y \in B \cup D$. Therefore, $(x, y) \in (A \cup C) \times (B \cup D)$. ∎

EXERCISES FOR CHAPTER 5

Section 5.1: Counterexamples

5.1 Let $a = b = -1$. Then $\log(ab) = \log 1 = 0$ but $\log(a)$ and $\log(b)$ are not defined. Thus $a = b = -1$ is a counterexample.

5.3 If $n = 3$, then $2n^2 + 1 = 19$. Since $3 \nmid 19$, it follows that $n = 3$ is a counterexample.

5.5 If $a = 1$ and $b = 2$, then $(a + b)^3 = 3^3 = 27$, but $a^3 + 2a^2b + 2ab + 2ab^2 + b^3 = 1 + 4 + 4 + 8 + 8 = 25$. Thus $a = 1$ and $b = 2$ form a counterexample.

5.7 There is no counterexample. It can be shown that if $a, b \in \mathbf{R}^+$ and $(a+b)\left(\frac{1}{a}+\frac{1}{b}\right) = 4$, then $a = b$.

5.9 Let $x = 3$ and $n = 2$. Then $x^n + (x+1)^n = 3^2 + 4^2 = 25 = 5^2 = (x+2)^n$. Then $x = 3, n = 2$ is a counterexample.

Section 5.2: Proof by Contradiction

5.11 Assume, to the contrary, that there exists a smallest positive irrational number r. Then $r/2$ is a positive irrational number and $r/2 < r$.

5.13 Let a and b be odd integers and assume, to the contrary, that $4 \mid (a^2 + b^2)$. Then $a^2 + b^2 = 4x$ for some integer x. Starting with this, a contradiction can be obtained.

5.15 Assume, to the contrary, that 1000 can be expressed as the sum of three integers a, b and c, an even number of which are even. There are two cases.

Case 1. None of a, b and c is even. Then $a = 2x + 1, b = 2y + 1$ and $c = 2z + 1$, where $x, y, z \in \mathbf{Z}$. Thus

$$1000 = (2x+1) + (2y+1) + (2z+1) = 2(x+y+z+1) + 1.$$

Since $x + y + z + 1$ is an integer, 1000 is odd, which is a contradiction.

Case 2. Exactly two of a, b and c are even, say a and b are even and c is odd. (The argument is similar to that in Case 1.)

5.17 *Proof*   Assume, to the contrary, that there exist an irrational number a and a nonzero rational number b such that a/b is a rational number. Then $a/b = p/q$, where $p, q \in \mathbf{Z}$ and $p, q \neq 0$. Since b is a nonzero rational number, $b = r/s$, where $r, s \in \mathbf{Z}$ and $r, s \neq 0$. Thus $a = (bp)/q = (rp)/(sq)$. Since $rp, sq \in \mathbf{Z}$ and $sq \neq 0$, it follows that a is a rational number, which is a contradiction. ∎

5.19 **Lemma**   Let a be an integer. Then $3 \mid a^2$ if and only if $3 \mid a$. (See Exercise 3 in Chapter 4.)
Proof of Result   Assume to the contrary, that $\sqrt{3}$ is rational. Then $\sqrt{3} = p/q$, where $p, q \in \mathbf{Z}$ and $q \neq 0$. We may assume that p/q has been reduced to lowest terms. Thus $3 = p^2/q^2$ or $p^2 = 3q^2$. Since $3 \mid p^2$, it follows by the lemma that $3 \mid p$. Thus $p = 3x$ for some integer x. Thus $p^2 = (3x)^2 = 9x^2 = 3q^2$. So $3x^2 = q^2$. Since x^2 is an integer, $3 \mid q^2$. By the lemma, $3 \mid q$ and so $q = 3y$, where $y \in \mathbf{Z}$. Hence $p = 3x$ and $q = 3y$, which contradicts our assumption that p/q has been reduced to lowest terms. ∎

5.21 **(a)** One possible way to prove this is to use the fact that for integers a and b, the product ab is even if and only if a is even or b is even.
   Proof   Assume, to the contrary, that $\sqrt{6}$ is rational. Then $\sqrt{6} = a/b$ for nonzero integers a and b. We can further assume that a/b has been reduced to lowest terms. Thus $6 = a^2/b^2$; so $a^2 = 6b^2 = 2(3b^2)$. Because $3b^2$ is an integer, a^2 is even. By Theorem 3.12, a is even. So $a = 2c$, where $c \in \mathbf{Z}$. Thus $(2c)^2 = 6b^2$ and so $4c^2 = 6b^2$. Therefore, $3b^2 = 2c^2$. Because c^2 is an integer, $3b^2$ is even. By Theorem 3.17, either 3 is even or b^2 is even. Since 3 is not even, b^2 is even and so b is even by Theorem 3.12. However, since a and b are both even, each has 2 as a divisor, contradicting the fact that a/b has been reduced to lowest terms. ∎
   (b) We can use an argument similar to that employed in (a) to prove that $\sqrt{2k}$ is irrational for every odd positive integer k.

5.23 *Proof*   Assume, to the contrary, that there is some integer a such that $a \equiv 5 \pmod{14}$ and $a \equiv 3 \pmod{21}$. Then $14 \mid (a-5)$ and $21 \mid (a-3)$; so $a = 5 + 14x$ and $a = 3 + 21y$ for some integers x and y. Therefore, $5 + 14x = 3 + 21y$, which implies that $2 = 21y - 14x = 7(3y - 2x)$. Since $3y - 2x$ is an integer, $7 \mid 2$, which is a contradiction. ∎

5.25 *Proof*   Suppose that there exist three distinct positive integers a, b and c such that each divides the difference of the other two. We may assume that $a < b < c$. Thus $c \mid (b - a)$. Since $0 < b - a < c$, this is a contradiction. ∎

5.27 *Proof*   Assume, to the contrary, that there exist positive real numbers x and y such that $\sqrt{x+y} = \sqrt{x} + \sqrt{y}$. Squaring both sides, we obtain $x + y = x + 2\sqrt{x}\sqrt{y} + y$ and so $2\sqrt{x}\sqrt{y} = 2\sqrt{xy} = 0$. This implies that $xy = 0$. Thus $x = 0$ or $y = 0$, which is a contradiction. ∎

5.29 Assume, to the contrary, that there exist positive integers x and y such that $x^2 - y^2 = m = 2s$. Then $(x+y)(x-y) = 2s$, where s is an odd integer. We consider two cases, according to whether x and y are of the same parity or of opposite parity. Note that if x and y are of the same parity, then both $x + y$ and $x - y$ are even, while if x and y are of opposite parity, then both $x + y$ and $x - y$ are odd. Produce a contradiction in each case.

5.31   Assume, to the contrary, that there exists an integer m such that $3 \nmid (m^2 - 1)$ and $3 \nmid m$. Thus $m = 3q + 1$ or $m = 3q + 2$ for some integer q. Produce a contradiction in each case.

5.33   **(a)** ***Proof***   Assume, to the contrary, that $x^2 - 3x + 1 = 0$ has a rational number solution p/q, where $p, q \in \mathbf{Z}$ and $q \neq 0$. We may assume that p/q is expressed in lowest terms. Thus $\frac{p^2}{q^2} - \frac{3p}{q} + 1 = 0$ and so $p^2 - 3pq + q^2 = 0$. We consider two cases.

Case 1. Exactly one of p and q is even, say p is even and q is odd. Then $p = 2r$ and $q = 2s + 1$, where $r, s \in \mathbf{Z}$. Hence

$$p^2 - 3pq + q^2 = (2r)^2 - 3(2r)(2s + 1) + (2s + 1)^2$$
$$= 4r^2 - 12rs - 6r + 4s^2 + 4s + 1$$
$$= 2(2r^2 - 6rs - 3r + 2s^2 + 2s) + 1 = 0.$$

Since $2r^2 - 6rs - 3r + 2s^2 + 2s \in \mathbf{Z}$, it follows that $p^2 - 3pq + q^2$ is odd and equals 0, which is a contradiction.

Case 2. Both p and q are odd. (The proof in this case is similar to the proof of Case 1.) ■

(b) For positive integers k and n with $k < n$ and odd integers a, b and c, the equation $ax^n + bx^k + c = 0$ has no rational number solution.

Section 5.3: A Review of Three Proof Techniques

5.35   **(a)** ***Proof***   Assume that $x - \frac{2}{x} > 1$. Since $x > 0$, it follows, by multiplying by x, that $x^2 - 2 > x$ and so $x^2 - x - 2 > 0$. Hence $(x - 2)(x + 1) > 0$. Dividing by the positive number $x + 1$, we have $x - 2 > 0$ and so $x > 2$. ■

(b) ***Proof***   Assume that $0 < x \leq 2$. Thus $x^2 - x - 2 = (x - 2)(x + 1) \leq 0$ and so $x^2 - 2 \leq x$. Dividing by the positive number x, we have $x - \frac{2}{x} \leq 1$. ■

(c) ***Proof***   Assume, to the contrary, that there exists a positive number x such that $x - \frac{2}{x} > 1$ and $x \leq 2$. Thus $x^2 - x - 2 = (x - 2)(x + 1) \leq 0$ and so $x^2 - 2 \leq x$. Dividing by the positive number x, we have $x - \frac{2}{x} \leq 1$, producing a contradiction. ■

5.37   **(a)** ***Proof***   Let $x, y \in \mathbf{R}^+$ such that $x \leq y$. Multiplying both sides by x and y, respectively, we obtain $x^2 \leq xy$ and $xy \leq y^2$. Therefore, $x^2 \leq xy \leq y^2$ and so $x^2 \leq y^2$. ■

(b) ***Proof***   Assume that $x^2 > y^2$. Thus $x^2 - y^2 > 0$ and so $(x + y)(x - y) > 0$. Dividing by the positive number $x + y$, we obtain $x - y > 0$ and $x > y$. ■

(c) ***Proof***   Assume, to the contrary, that there exist positive numbers x and y such that $x \leq y$ and $x^2 > y^2$. Since $x \leq y$, it follows that $x^2 \leq xy$ and $xy \leq y^2$. Thus $x^2 \leq y^2$, producing a contradiction. ■

5.39   ***Proof***   (Direct Proof) Let $a, b, c \in \mathbf{Z}$. We show that exactly two of ab, ac and bc cannot be odd. If all of a, b and c are odd, then ab, ac and bc are all odd; otherwise, at least one of a, b and c is even, say a is even. Then ab and ac are even. ■

Proof   (Proof by Contradiction) Let $a, b, c \in \mathbf{Z}$ and assume, to the contrary, that exactly two of ab, ac and bc are odd, say ab and ac are odd. Then a, b and c are odd, which implies that bc is odd, a contradiction. ■

Section 5.4: Existence Proofs

5.41   ***Proof***   Consider the rational number 2 and the irrational number $\frac{1}{2\sqrt{2}}$. If $2^{\frac{1}{2\sqrt{2}}}$ is irrational, then $a = 2$ and $b = \frac{1}{2\sqrt{2}}$ have the desired properties. If, on the other hand, $2^{\frac{1}{2\sqrt{2}}}$ is rational, then

$$\left(2^{\frac{1}{2\sqrt{2}}}\right)^{\sqrt{2}} = 2^{\frac{\sqrt{2}}{2\sqrt{2}}} = 2^{\frac{1}{2}} = \sqrt{2}$$

is irrational and so $a = 2^{\frac{1}{2\sqrt{2}}}$ and $b = \sqrt{2}$ have the desired properties. ■

5.43   ***Proof***   Assume, to the contrary, that there exist nonzero real numbers a and b such that $\sqrt{a^2 + b^2} = \sqrt[3]{a^3 + b^3}$.

Raising both sides to the 6th power, we obtain

$$a^6 + 3a^4b^2 + 3a^2b^4 + b^6 = a^6 + 2a^3b^3 + b^6.$$

Thus

$$3a^2 - 2ab + 3b^2 = (a - b)^2 + 2a^2 + 2b^2 = 0.$$

Since this can only occur when $a = b = 0$, we have a contradiction.   ∎

5.45   Let $W = S - T$. Since T is a proper subset of S, it follows that $\emptyset \neq W \subseteq S$. Then $R(x)$ is true for every $x \in W$, that is, $\forall x \in W$, $R(x)$ is true.

5.47   ***Proof***   Suppose that $S = \{a, b, c\}$. The nonempty subsets of S are $\{a\}, \{b\}, \{c\}, \{a, b\}, \{a, c\}, \{b, c\}$ and $\{a, b, c\}$. For each such subset A of S, σ_A is congruent to 0, 1, 2, 3, 4 or 5 modulo 6. Since there are seven nonempty subsets of S, there must be two of these seven subsets, say B and C, such that $\sigma_B \equiv \sigma_C \pmod{6}$.   ∎

Section 5.5: Disproving Existence Statements

5.49   We show that if a and b are odd integers, then $4 \nmid (3a^2 + 7b^2)$. Let a and b be odd integers. Then $a = 2x + 1$ and $b = 2y + 1$ for integers x and y. Then $3a^2 + 7b^2 = 4(3x^2 + 3x + 7y^2 + 7y + 2) + 2$. Since 2 is the remainder when $3a^2 + 7b^2$ is divided by 4, it follows that $4 \nmid (3a^2 + 7b^2)$.

5.51   Show that if n is an integer, then $n^4 + n^3 + n^2 + n$ is even. Let $n \in \mathbf{Z}$. Then n is even or n is odd. We consider these two cases.

Case 1. n is even. Then $n = 2a$ for some integer a. Then
$$n^4 + n^3 + n^2 + n = n(n + 1)(n^2 + 1) = 2a(n + 1)(n^2 + 1) = 2[a(n + 1)(n^2 + 1)].$$
Since $a(n + 1)(n^2 + 1)$ is an integer, $n^4 + n^3 + n^2 + n$ is even.

Case 2. n is odd. (The argument is similar here.)

EXERCISES FOR CHAPTER 6

Section 6.1: The Principle of Mathematical Induction

6.1   The sets in (b) and (d) are well-ordered.

6.3   ***Proof***   Let S be a nonempty set of negative integers. Let $T = \{n : -n \in S\}$. Hence T is a nonempty set of positive integers. By the Well-Ordering Principle, T has a least element m. Hence $m \leq n$ for all $n \in T$. Therefore, $-m \in S$ and $-m \geq -n$ for all $-n \in S$. Thus $-m$ is the largest element of S.   ∎

6.5   ***Proof***   We use induction. Since $1 = 2 \cdot 1^2 - 1$, the formula holds for $n = 1$. Assume that the formula holds for some integer $k \geq 1$, that is,

$$1 + 5 + 9 + \ldots + (4k - 3) = 2k^2 - k.$$

We show that

$$1 + 5 + 9 + \ldots + [4(k + 1) - 3] = 2(k + 1)^2 - (k + 1).$$

Observe that

$$1 + 5 + 9 + \ldots + [4(k + 1) - 3] = [1 + 5 + 9 + \ldots + (4k - 3)] + 4(k + 1) - 3$$
$$= (2k^2 - k) + (4k + 1) = 2k^2 + 3k + 1$$
$$= 2(k + 1)^2 - (k + 1).$$

The result then follows by the Principle of Mathematical Induction.   ∎

6.7   One possibility: $1 + 7 + 13 + \ldots + (6n - 5) = 3n^2 - 2n$ for every positive integer n.

6.9 ***Proof*** We proceed by induction. For $n = 1$, we have $1 \cdot 3 = 3 = \frac{1 \cdot (1+1)(2 \cdot +7)}{6}$, which is true. Assume that $1 \cdot 3 + 2 \cdot 4 + 3 \cdot 5 + \cdots + k(k+2) = \frac{k(k+1)(2k+7)}{6}$ where $k \in \mathbf{N}$. We then show that

$$1 \cdot 3 + 2 \cdot 4 + 3 \cdot 5 + \cdots + (k+1)(k+3) = \frac{(k+1)(k+2)[2(k+1)+7]}{6}$$

$$= \frac{(k+1)(k+2)(2k+9)}{6}.$$

Observe that

$$1 \cdot 3 + 2 \cdot 4 + 3 \cdot 5 + \cdots + (k+1)(k+3)$$
$$= [1 \cdot 3 + 2 \cdot 4 + 3 \cdot 5 + \cdots + k(k+2)] + (k+1)(k+3)$$
$$= \frac{k(k+1)(2k+7)}{6} + (k+1)(k+3)$$
$$= \frac{k(k+1)(2k+7) + 6(k+1)(k+3)}{6}$$
$$= \frac{(k+1)(2k^2 + 7k + 6k + 18)}{6} = \frac{(k+1)(2k^2 + 13k + 18)}{6}$$
$$= \frac{(k+1)(k+2)(2k+9)}{6}.$$

By the Principle of Mathematical Induction,

$$1 \cdot 3 + 2 \cdot 4 + 3 \cdot 5 + \cdots + n(n+2) = \frac{n(n+1)(2n+7)}{6}$$

for every positive integer n.

6.11 We proceed by induction. Since $\frac{1}{3 \cdot 4} = \frac{1}{3+9}$, the formula holds for $n = 1$. Assume that

$$\frac{1}{3 \cdot 4} + \frac{1}{4 \cdot 5} + \cdots + \frac{1}{(k+2)(k+3)} = \frac{k}{3k+9}$$

where k is a positive integer. Then show that

$$\frac{1}{3 \cdot 4} + \frac{1}{4 \cdot 5} + \cdots + \frac{1}{(k+3)(k+4)} = \frac{k+1}{3(k+1)+9} = \frac{k+1}{3(k+4)}.$$

6.13 ***Proof*** We proceed by induction. Since $1 \cdot 1! = 2! - 1$, the statement is true for $n = 1$. Assume that

$$1 \cdot 1! + 2 \cdot 2! + \cdots + k \cdot k! = (k+1)! - 1,$$

where $k \in \mathbf{N}$. We show that

$$1 \cdot 1! + 2 \cdot 2! + \cdots + (k+1) \cdot (k+1)! = (k+2)! - 1.$$

Now

$$1 \cdot 1! + 2 \cdot 2! + \cdots + (k+1) \cdot (k+1)! = (1 \cdot 1! + 2 \cdot 2! + \cdots + k \cdot k!) + (k+1) \cdot (k+1)!$$
$$= (k+1)! - 1 + (k+1) \cdot (k+1)!$$
$$= (k+1)!(k+2) - 1 = (k+2)! - 1.$$

By the Principle of Mathematical Induction, $1 \cdot 1! + 2 \cdot 2! + \cdots + n \cdot n! = (n+1)! - 1$ for all $n \in \mathbf{N}$.

6.15 ***Proof*** We proceed by induction. Since $\frac{1}{\sqrt{1}} \leq 2\sqrt{1} - 1$, the inequality holds for $n = 1$. Assume that

$$\frac{1}{\sqrt{1}} + \frac{1}{\sqrt{2}} + \frac{1}{\sqrt{3}} + \cdots + \frac{1}{\sqrt{k}} \leq 2\sqrt{k} - 1$$

for a positive integer k. We show that

$$\frac{1}{\sqrt{1}} + \frac{1}{\sqrt{2}} + \frac{1}{\sqrt{3}} + \cdots + \frac{1}{\sqrt{k+1}} \leq 2\sqrt{k+1} - 1.$$

Observe that

$$\frac{1}{\sqrt{1}} + \frac{1}{\sqrt{2}} + \frac{1}{\sqrt{3}} + \cdots + \frac{1}{\sqrt{k+1}} = \left(\frac{1}{\sqrt{1}} + \frac{1}{\sqrt{2}} + \frac{1}{\sqrt{3}} + \cdots + \frac{1}{\sqrt{k}} \right) + \frac{1}{\sqrt{k+1}}$$

$$\leq (2\sqrt{k} - 1) + \frac{1}{\sqrt{k+1}} = \frac{2\sqrt{k^2+k}+1}{\sqrt{k+1}} - 1.$$

Since $4(k^2 + k) \leq (2k+1)^2$, it follows that $2\sqrt{k^2+k} \leq 2k + 1$ and so $2\sqrt{k^2+k} + 1 \leq 2(k+1) = 2(\sqrt{k+1})^2$. Therefore,

$$\frac{2\sqrt{k^2+k}+1}{\sqrt{k+1}} \leq 2\sqrt{k+1}.$$

Thus

$$\frac{1}{\sqrt{1}} + \frac{1}{\sqrt{2}} + \frac{1}{\sqrt{3}} + \cdots + \frac{1}{\sqrt{k+1}} \leq 2\sqrt{k+1} - 1.$$

By the Principle of Mathematical Induction, $\frac{1}{\sqrt{1}} + \frac{1}{\sqrt{2}} + \frac{1}{\sqrt{3}} + \cdots + \frac{1}{\sqrt{n}} \leq 2\sqrt{n} - 1$ for every positive integer n. ∎

Section 6.2: A More General Principle of Mathematical Induction

6.17 *Proof* We need only show that every nonempty subset of S has a least element. So let T be a nonempty subset of S. If T is a subset of $\mathbf{N}$, then, by the Well-Ordering Principle, T has a least element. Hence we may assume that T is not a subset of $\mathbf{N}$. Thus $T - \mathbf{N}$ is a finite nonempty set and so contains a least element t. Since $t \leq 0$, it follows that $t \leq x$ for all $x \in T$; so t is a least element of T. ∎

6.19 *Proof* We use induction. We know that if a and b are two real numbers such that $ab = 0$, then $a = 0$ or $b = 0$. Thus the statement is true for $n = 2$. Assume that:

> If $a_1, a_2, \ldots, a_k$ are any $k \geq 2$ real numbers whose product is 0, then $a_i = 0$ for some integer i with $1 \leq i \leq k$.

We wish to show the statement is true in the case of $k + 1$ numbers, that is:

> If $b_1, b_2, \ldots, b_{k+1}$ are $k + 1$ real numbers such that $b_1 b_2 \cdots b_{k+1} = 0$, then $b_i = 0$ for some integer i $(1 \leq i \leq k + 1)$.

Let $b_1, b_2, \ldots, b_{k+1}$ be $k + 1$ real numbers such that $b_1 b_2 \cdots b_{k+1} = 0$. We show that $b_i = 0$ for some integer i $(1 \leq i \leq k + 1)$. Let $b = b_1 b_2 \cdots b_k$. Then

$$b_1 b_2 \cdots b_{k+1} = (b_1 b_2 \cdots b_k) b_{k+1} = b b_{k+1} = 0.$$

Therefore, either $b = 0$ or $b_{k+1} = 0$. If $b_{k+1} = 0$, then we have the desired conclusion. On the other hand, if $b = b_1 b_2 \cdots b_k = 0$, then, since b is the product of k real numbers, it follows by the inductive hypothesis that $b_i = 0$ for some integer i $(1 \leq i \leq k)$. In any case, $b_i = 0$ for some integer i $(1 \leq i \leq k + 1)$. The result then follows by the Principle of Mathematical Induction. ∎

6.21 *Proof* We proceed by induction. Since $4 \mid (5^0 - 1)$, the statement is true for $n = 0$. Assume that $4 \mid (5^k - 1)$, where k is a nonnegative integer. We show that $4 \mid (5^{k+1} - 1)$. Since $4 \mid (5^k - 1)$, it follows that $5^k = 4a + 1$ for some integer a. Observe that

$$5^{k+1} - 1 = 5 \cdot 5^k - 1 = 5(4a + 1) - 1 = 20a + 4 = 4(5a + 1).$$

Since $(5a + 1) \in \mathbf{Z}$, it follows that $4 \mid (5^{k+1} - 1)$. By the Principle of Mathematical Induction, $4 \mid (5^n - 1)$ for every nonnegative integer n. ∎

6.23 We employ mathematical induction. For $n = 0$, we have $7 \mid 0$, which is true. Assume that $7 \mid \left(3^{2k} - 2^k\right)$ for some integer $k \geq 0$. We show that $7 \mid \left(3^{2(k+1)} - 2^{(k+1)}\right)$. Since $7 \mid \left(3^{2k} - 2^k\right)$, it follows that $3^{2k} - 2^k = 7a$ for some integer a. Thus $3^{2k} = 2^k + 7a$. Then show that $3^{2(k+1)} - 2^{(k+1)} = 7(2^k + 9a)$.

6.25 ***Proof*** We use induction. Since $4! = 24 > 16 = 2^4$, the inequality holds for $n = 4$. Suppose that $k! > 2^k$ for an arbitrary integer $k \geq 4$. We show that $(k + 1)! > 2^{k+1}$. Observe that

$$(k + 1)! = (k + 1)k! > (k + 1) \cdot 2^k \geq (4 + 1)2^k = 5 \cdot 2^k > 2 \cdot 2^k = 2^{k+1}.$$

Therefore, $(k + 1)! > 2^{k+1}$. By the Principle of Mathematical Induction, $n! > 2^n$ for every integer $n \geq 4$. ∎

6.27 ***Proof*** We proceed by induction. Since $1 \leq 2 - \frac{1}{1}$, the inequality holds for $n = 1$. Assume that $1 + \frac{1}{4} + \frac{1}{9} + \cdots + \frac{1}{k^2} \leq 2 - \frac{1}{k}$ for some positive integer k. We show that $1 + \frac{1}{4} + \frac{1}{9} + \cdots + \frac{1}{(k+1)^2} \leq 2 - \frac{1}{k+1}$. Observe that

$$1 + \frac{1}{4} + \frac{1}{9} + \cdots + \frac{1}{(k + 1)^2} = \left(1 + \frac{1}{4} + \frac{1}{9} + \cdots + \frac{1}{k^2}\right) + \frac{1}{(k + 1)^2}$$

$$\leq 2 + \frac{-1}{k} + \frac{1}{(k + 1)^2} = 2 + \frac{-(k + 1)^2 + k}{k(k + 1)^2}$$

$$= 2 - \frac{k^2 + k + 1}{k(k + 1)^2} < 2 - \frac{k^2 + k}{k(k + 1)^2} = 2 - \frac{1}{k + 1}.$$

By the Principle of Mathematical Induction, $1 + \frac{1}{4} + \frac{1}{9} + \cdots + \frac{1}{n^2} \leq 2 - \frac{1}{n}$ for every positive integer n. ∎

6.29 ***Proof*** We proceed by induction. By De Morgan's law, if A and B are any two sets, then $\overline{A \cap B} = \overline{A} \cup \overline{B}$. Hence the statement is true for $n = 2$. Assume, for any k sets $A_1, A_2, \ldots, A_k$, where $k \geq 2$, that

$$\overline{A_1 \cap A_2 \cap \cdots \cap A_k} = \overline{A_1} \cup \overline{A_2} \cup \cdots \cup \overline{A_k}.$$

Now consider any $k + 1$ sets, say $B_1, B_2, \ldots, B_{k+1}$. We show that

$$\overline{B_1 \cap B_2 \cap \cdots \cap B_{k+1}} = \overline{B_1} \cup \overline{B_2} \cup \cdots \cup \overline{B_{k+1}}.$$

Let $B = B_1 \cap B_2 \cap \cdots \cap B_k$. Observe that

$$\overline{B_1 \cap B_2 \cap \cdots \cap B_{k+1}} = \overline{(B_1 \cap B_2 \cap \cdots \cap B_k) \cap B_{k+1}} = \overline{B \cap B_{k+1}}$$

$$= \overline{B} \cup \overline{B_{k+1}} = \left(\overline{B_1} \cup \overline{B_2} \cup \cdots \cup \overline{B_k}\right) \cup \overline{B_{k+1}}$$

$$= \overline{B_1} \cup \overline{B_2} \cup \cdots \cup \overline{B_{k+1}}.$$

The result then follows by the Principle of Mathematical Induction. ∎

6.31 ***Proof*** We proceed by induction. Since $a\left(\frac{1}{a}\right) = 1^2$ for every positive real number a, the inequality is true for $n = 1$. Assume for each k positive real numbers $a_1, a_2, \ldots, a_k$ that

$$\left(\sum_{i=1}^{k} a_i\right)\left(\sum_{i=1}^{k} \frac{1}{a_i}\right) \geq k^2.$$

Let $b_1, b_2, \ldots, b_{k+1}$ be $k + 1$ positive real numbers. We show that

$$\left(\sum_{i=1}^{k+1} b_i\right)\left(\sum_{i=1}^{k+1} \frac{1}{b_i}\right) \geq (k + 1)^2.$$

Observe that

$$\left(\sum_{i=1}^{k+1} b_i\right)\left(\sum_{i=1}^{k+1} \frac{1}{b_i}\right) = \left(\sum_{i=1}^{k} b_i\right)\left(\sum_{i=1}^{k} \frac{1}{b_i}\right) + b_{k+1}\left(\sum_{i=1}^{k} \frac{1}{b_i}\right) + \frac{1}{b_{k+1}}\left(\sum_{i=1}^{k} b_i\right) + b_{k+1} \cdot \frac{1}{b_{k+1}}$$

$$\geq k^2 + \sum_{i=1}^{k} \left(\frac{b_{k+1}}{b_i} + \frac{b_i}{b_{k+1}} \right) + 1.$$

Since $\frac{b_{k+1}}{b_i} + \frac{b_i}{b_{k+1}} \geq 2$ (see Exercise 4.78), it follows that

$$\left(\sum_{i=1}^{k+1} b_i \right) \left(\sum_{i=1}^{k+1} \frac{1}{b_i} \right) \geq k^2 + 2k + 1 = (k+1)^2.$$

By the Principle of Mathematical Induction, $\left(\sum_{i=1}^{n} a_i \right) \left(\sum_{i=1}^{n} \frac{1}{a_i} \right) \geq n^2$ for every n positive real numbers $a_1, a_2, \ldots, a_n$. ∎

Section 6.3: Proof by Minimum Counterexample

6.33 Assume, to the contrary, that there is a positive integer n such that $6 \nmid 7n \left(n^2 - 1 \right)$. Then there is a smallest positive integer n such that $6 \nmid 7n \left(n^2 - 1 \right)$. Let m be this integer. Since $6 \mid 0$ and $6 \mid 42$, it follows that $6 \mid 7n \left(n^2 - 1 \right)$ when $n = 1$ and $n = 2$. So $m \geq 3$ and we can write $m = k + 2$, where $1 \leq k < m$. Consequently, $6 \mid 7k \left(k^2 - 1 \right)$ and so $7k \left(k^2 - 1 \right) = 6x$ for some integer x. Then show that $7m \left(m^2 - 1 \right) = 6(x + 7k^2 + 14k + 7)$.

6.35 **Proof** Assume, to the contrary, that there is a positive integer n such that $1 + 3 + 5 + \cdots + (2n - 1) \neq n^2$. Let m be the smallest such integer. Since $2 \cdot 1 - 1 = 1^2$, it follows that $m \geq 2$. Thus m can be expressed as $m = k + 1$, where $1 \leq k < m$. Therefore, $1 + 3 + 5 + \cdots + (2k - 1) = k^2$. Then

$$
\begin{aligned}
1 + 3 + 5 + \cdots + (2m - 1) &= 1 + 3 + 5 + \cdots + (2k + 1) \\
&= [1 + 3 + 5 + \cdots + (2k - 1)] + (2k + 1) \\
&= k^2 + (2k + 1) = (k + 1)^2 = m^2,
\end{aligned}
$$

which is a contradiction. ∎

6.37 Assume, to the contrary, that there is some nonnegative integer n such that $3 \nmid \left(2^n + 2^{n+1} \right)$. Then there is a smallest nonnegative integer n such that $3 \nmid \left(2^n + 2^{n+1} \right)$. Let m be this integer. Since $2^0 + 2^1 = 3$, we have $m \geq 1$. So we can write $m = k + 1$, where $0 \leq k < m$. Thus $3 \mid \left(2^k + 2^{k+1} \right)$ and so $2^k + 2^{k+1} = 3x$ for some integer x. Then show that $2^m + 2^{m+1} = 3(2x)$.

6.39 Assume, to the contrary, that there is some positive integer n such that $12 \nmid \left(n^4 - n^2 \right)$. Then there is a smallest positive integer n such that $12 \nmid \left(n^4 - n^2 \right)$. Let m be this integer. It can be shown that if $1 \leq n \leq 6$, then $12 \mid \left(n^4 - n^2 \right)$. Therefore, $m \geq 7$. So we can write $m = k + 6$, where $1 \leq k < m$. Consider $(k + 6)^4 - (k + 6)^2$.

Section 6.4: The Strong Principle of Mathematical Induction

6.41 **Conjecture**   A sequence $\{a_n\}$ is defined recursively by $a_1 = 1$ and $a_n = 2a_{n-1}$ for $n \geq 2$. Then $a_n = 2^{n-1}$ for all $n \geq 1$.

Proof   We proceed by mathematical induction. Since $a_1 = 2^{1-1} = 2^0 = 1$, it follows that $a_n = 2^{n-1}$ when $n = 1$. Assume that $a_k = 2^{k-1}$ for some positive integer k. We show that $a_{k+1} = 2^k$. Since $k \geq 1$, it follows that $k + 1 \geq 2$. Therefore,

$$a_{k+1} = 2a_k = 2 \cdot 2^{k-1} = 2^k.$$

The result follows by the Principle of Mathematical Induction. ∎

6.43 **Conjecture**   A sequence $\{a_n\}$ is defined recursively by $a_1 = 1, a_2 = 4, a_3 = 9$ and

$$a_n = a_{n-1} - a_{n-2} + a_{n-3} + 2(2n - 3)$$

for $n \geq 4$. Then $a_n = n^2$ for all $n \geq 1$.

Proof   We proceed by the Strong Principle of Mathematical Induction. Since $a_1 = 1^2 = 1$, it follows that $a_n = n^2$ when $n = 1$. Assume that $a_i = i^2$, where $1 \leq i \leq k$ for some positive integer k. We show that

$a_{k+1} = (k + 1)^2$. Since $a_2 = a_{1+1} = (1 + 1)^2 = 4$ and $a_3 = a_{2+1} = (2 + 1)^2 = 9$, it follows that $a_{k+1} = (k + 1)^2$ for $k = 1, 2$. Hence we may assume that $k \geq 3$. Since $k + 1 \geq 4$,

$$\begin{aligned}
a_{k+1} &= a_k - a_{k-1} + a_{k-2} + 2[2(k + 1) - 3] \\
&= k^2 - (k - 1)^2 + (k - 2)^2 + (4k - 2) \\
&= k^2 - (k^2 - 2k + 1) + (k^2 - 4k + 4) + (4k - 2) \\
&= k^2 + 2k + 1 = (k + 1)^2.
\end{aligned}$$

The result then follows by the Strong Principle of Mathematical Induction. ∎

6.45 **Proof** We use the Strong Principle of Mathematical Induction. Since $12 = 3 \cdot 4 + 7 \cdot 0$, the statement is true when $n = 12$. Assume for an integer $k \geq 12$ that for every integer i with $12 \leq i \leq k$, there exist nonnegative integers a and b such that $i = 3a + 7b$. We show that there exist nonnegative integers x and y such that $k + 1 = 3x + 7y$. Since $13 = 3 \cdot 2 + 7 \cdot 1$ and $14 = 3 \cdot 0 + 7 \cdot 2$, we may assume that $k \geq 14$. Since $k - 2 \geq 12$, there exist nonnegative integers c and d such that $k - 2 = 3c + 7d$. Hence $k + 1 = 3(c + 1) + 7d$. By the Strong Principle of Mathematical Induction, for each integer $n \geq 12$, there are nonnegative integers a and b such that $n = 3a + 7b$. ∎

6.47 We show that every odd integer $n \geq 15$ can be expressed as $3a + 11b$ or as $5c + 7d$ for nonnegative integers a, b, c and d.

Proof We use the Strong Principle of Mathematical Induction. First, observe that $15 = 3 \cdot 5 + 11 \cdot 0 = 5 \cdot 3 + 7 \cdot 0$, $17 = 3 \cdot 2 + 11 \cdot 1$, $19 = 5 \cdot 1 + 7 \cdot 2$, $21 = 3 \cdot 7 + 11 \cdot 0$ and $23 = 3 \cdot 4 + 11 \cdot 1$. Thus the statement is true for 15, 17, 19, 21 and 23. Assume that the statement is true for every odd integer i with $15 \leq i \leq k$, where $k \geq 23$ is an odd integer. We show that the statement is true for the integer $k + 2$. Suppose that $k = 3a + 11b$ for some nonnegative integers a and b. Since $k \geq 23$, either $a \geq 3$ or $b \geq 2$. If $a \geq 3$, then $k + 2 = 3(a - 3) + 11(b + 1)$; while if $b \geq 2$, then $k + 2 = 3(a + 8) + 11(b - 2)$. Hence we may assume that $k = 5c + 7d$ for some nonnegative integers c and d. If $c \geq 1$, then $k + 2 = 5(c - 1) + 7(d + 1)$. The remaining situation is where $c = 0$ and so $k \geq 23$ is an odd integer multiple of 7. So $k = 7d$, where $d \geq 5$. In this case, $k + 2 = 5 \cdot 6 + 7(d - 4)$. The result then follows by the Strong Principle of Mathematical Induction. ∎

EXERCISES FOR CHAPTER 7

Section 7.1: Conjectures in Mathematics

7.1 (a) $17 + 18 + \cdots + 25 = 64 + 125$.

(b) **Conjecture** For every nonnegative integer n,
$$(n^2 + 1) + (n^2 + 2) + \cdots + (n + 1)^2 = n^3 + (n + 1)^3.$$

(c) **Proof** Observe that

$$\begin{aligned}
&(n^2 + 1) + (n^2 + 2) + \cdots + (n + 1)^2 \\
&= (n^2 + 1) + (n^2 + 2) + \cdots + (n^2 + 2n + 1) \\
&= (2n + 1)n^2 + [1 + 2 + \cdots + (2n + 1)] = (2n + 1)n^2 + (n + 1)(2n + 1) \\
&= 2n^3 + n^2 + 2n^2 + 3n + 1 = n^3 + (n^3 + 3n^2 + 3n + 1) = n^3 + (n + 1)^3.
\end{aligned}$$ ∎

7.3 (a) $a_2 = 3, a_3 = 8, a_4 = 54$.

(b) For each $n \in \mathbf{N}$, a_n is an integer.

7.5 (a) The ordered partitions of 4 are $4, 3 + 1, 1 + 3, 2 + 2, 2 + 1 + 1, 1 + 2 + 1, 1 + 1 + 2$ and $1 + 1 + 1 + 1$. So there are 8 ordered partitions of 4.

(b) **Conjecture** For each positive integer n, there are 2^{n-1} ordered partitions of n. [This conjecture is true.]

7.7 (a) $4 \cdot 2 \cdot 1 \cdot 1 = 4 + 2 + 1 + 1$.

(b) $3 \cdot 3 \cdot 1 \cdot 1 \cdot 1 = 3 + 3 + 1 + 1 + 1$.

(c) **Conjecture**   For every integer $n \geq 3$, there exist n positive integers whose sum equals their product. [This conjecture is true.]

7.9 **Conjecture**   There exists such a checkerboard if and only if mn is even. [The conjecture is true.]

Section 7.2: Revisiting Quantified Statements

7.11 (a) Let S be the set of all positive even integers and let $P(n) : 3n + 2^{n-2}$ is odd.

$\exists n \in S, P(n)$.

(b) **Proof**   For $n = 2 \in S$, $3n + 2^{n-2} = 7$ is odd.   ∎

7.13 (a) Let $P(n) : 3n^2 - 5n + 1$ is an even integer.

$\exists n \in \mathbf{Z}, P(n)$.

(b) We show the following: For all $n \in \mathbf{Z}$, $3n^2 - 5n + 1$ is odd.

This can be proved by a direct proof with two cases, namely n even and n odd.

7.15 (a) Let $P(m, n): m(n - 3) < 1$.

$\exists n \in \mathbf{Z}, \forall m \in \mathbf{Z}, P(m, n)$.

(b) **Proof**   Let $n = 3$. Then $m(n - 3) = m \cdot 0 = 0 < 1$.   ∎

7.17 (a) Let $P(m, n): -nm < 0$. $\exists n \in \mathbf{N}, \forall m \in \mathbf{Z}, P(m, n)$.

(b) $\forall n \in \mathbf{N}, \exists m \in \mathbf{Z}, \sim P(m, n)$.

(c) Let n be a positive integer. For $m = 0$, we have $-nm = -n \cdot 0 = 0$.

7.19 (a) Let $P(a, b, x): a \leq x \leq b$ and $b - a = 1$.

$\forall x \in \mathbf{R}, \exists a, b \in \mathbf{Z}, P(a, b, x)$.

(b) **Proof**   Let $x \in \mathbf{R}$. If x is an integer, then let $a = x$ and $b = x + 1$. Then $a \leq x \leq b$ and $b - a = 1$. Thus we may assume that x is not an integer. Then there exists an integer a such that $a < x < a + 1$. Let $b = a + 1$.   ∎

7.21 (a) Let S be the set of even integers, let T be the set of odd integers and let $P(a, b, c): a < c < b$ or $b < c < a$.

$\forall a \in S, \forall b \in T, \exists c \in \mathbf{Q}, P(a, b, c)$.

(b) **Proof**   For $a \in S$ and $b \in T$, let $c = (a + b)/2$. If $a < b$, then $a < c < b$; while if $b < a$, then $b < c < a$.   ∎

7.23 (a) Let S be the set of odd integers and $P(a, b, c): a + b + c = 1$.

$\exists a, b, c \in S, P(a, b, c)$.

(b) **Proof**   Let $a = 3$ and $b = c = -1$. Then $a + b + c = 1$.   ∎

7.25 (a) $\exists L \in \mathbf{R}, \forall e \in \mathbf{R}^+, \exists d \in \mathbf{R}^+, \forall x \in \mathbf{R}, P(x, d) \Rightarrow Q(x, L, e)$.

(b) **Proof**   Let $L = 0$ and let e be any positive real number. Let $d = e/3$. Let $x \in \mathbf{R}$ such that $|x| < e/3$. Then $|3x - L| = |3x| = 3|x| < 3(e/3) = e$.   ∎

7.27 **Proof**   Let $a \in \mathbf{Z}$. Then $b = a - 1$ and $c = 0$ are integers such that $|a - b| = 1 > cd = 0 \cdot d = 0$ for every integer d.   ∎

Section 7.3: Testing Statements

7.29 (a) The statement is true. **Proof.** Assume that $k^2 + 3k + 1$ is even where $k \in \mathbf{N}$. Then $k^2 + 3k + 1 = 2x$ for some integer x. Observe that

$$(k + 1)^2 + 3(k + 1) + 1 = k^2 + 2k + 1 + 3k + 3 + 1$$
$$= (k^2 + 3k + 1) + 2k + 4$$
$$= 2x + 2k + 4 = 2(x + k + 2).$$

Since $x + k + 2$ is an integer, $(k + 1)^2 + 3(k + 1) + 1$ is even.   ∎

(b) The statement is false since $P(1)$ is false.

7.31 This statement is false. Let $n = 0$ and let k be any nonnegative integer. Since $k \geq 0 = n$, the integer $n = 0$ is a counterexample.

7.33 This statement is false. Let $x = 99$ and $y = z = 1$. Then $x + y + z = 101$, while no two of x, y and z are of opposite parity. Thus, $x = 99$, $y = 1$, $z = 1$ is a counterexample.

7.35 The statement is true.
   Proof   Assume that $A \neq \emptyset$. Since $A \neq \emptyset$, there is an element $a \in A$. Let $B = \{a\}$. Then $A \cap B \neq \emptyset$. ∎

7.37 The statement is false. Let $A = \{1\}$, which is nonempty, and let B be an arbitrary set. Since $1 \in A \cup B$, it follows that $A \cup B \neq \emptyset$.

7.39 The statement is true.
   Proof   Let A be a proper subset of S and let $B = S - A$. Then $B \neq \emptyset$, $A \cup B = S$ and $A \cap B = \emptyset$. ∎

7.41 The statement is true. Observe that $0 \cdot c = 0$ for every integer c.

7.43 The statement is false. Let $x = 1$ and $y = -2$. Then $x^2 < y^2$ but $x > y$.

7.45 The statement is true.
   Proof   Let a be an odd integer. Then $a = a + 1 + (-1)$ is a sum of three odd integers. ∎

7.47 The statement is true. Let $b = c - a$.

7.49 The statement is true. Consider $r = (a + b)/2$.

7.51 The statement is false. Let $A \neq \emptyset$ and $B = \emptyset$. Then $A \cup B \neq \emptyset$.

7.53 The statement is true.
   Proof   Let a be an odd integer. Then $a + 0 = a$, where $b = 0$ is even and $c = a$ is odd. ∎

7.55 The statement is true. Let $f(x) = x^3 + x^2 - 1$. Observe that $f(0) = -1$ and $f(1) = 1$. Now apply the Intermediate Value Theorem of Calculus.

7.57 The statement is true.
   Proof   Assume that $A - B \neq \emptyset$. Then there exists $x \in A - B$. Thus $x \in A$ and $x \notin B$. Since $x \notin B$, it follows that $x \notin B - A$. Therefore, $A - B \neq B - A$. ∎

7.59 The statement is true.
   Proof   Let $b \in \mathbf{Q}^+$. Then $a = b/\sqrt{2}$ is irrational and $0 < a < b$. ∎

7.61 The statement is false. For $A = \emptyset$, $B = \{1\}$ and $C = \{1, 2\}$, we have $A \cap B = A \cap C = \emptyset$, but $B \neq C$. Thus A, B and C form a counterexample.

7.63 The statement is true. Consider $B = \emptyset$. Since $A \cup B \neq \emptyset$, this requires that $A \neq \emptyset$.

7.65 The statement is false. Note that $x^2 + x + 1 = \left(x + \frac{1}{2}\right)^2 + \frac{3}{4} \geq \frac{3}{4} > 0$ for every $x \in \mathbf{R}$.

7.67 The statement is true. For a nonzero rational number r, observe that $r = (r\sqrt{2}) \cdot \frac{1}{\sqrt{2}}$.

7.69 The statement is false. The sets $S = \{1, 2, 3\}$ and $T = \{\{1, 2\}, \{1, 3\}, \{2, 3\}\}$ form a counterexample.

7.71 The statement is false. The numbers $a = b = 0$ and $c = 1$ form a counterexample.

7.73 The statement is true. Let $a = 2$, $b = 16$ and $c = 4$.

7.75 The statement is true.
   Proof   Let $n \in \mathbf{Z}$. If $n \neq 0$, then $n = n + 0$ has the desired properties. If $n = 0$, then $n = 0 = 1 + (-1)$. ∎

7.77 The statement is false. For $n = 11$, $n^2 - n + 11 = 11^2$.

7.79 The statement is true.
   Proof   Let a and b be two consecutive integers such that $3 \nmid ab$. Since $3 \nmid ab$, it follows by Result 4.5 that $3 \nmid a$ and $3 \nmid b$. Therefore, $a \not\equiv 0 \pmod 3$ and $b \not\equiv 0 \pmod 3$. Thus $a = 3q + 1$ and $b = 3q + 2$ for some integer q and so $a + b = (3q + 1) + (3q + 2) = 6q + 3 = 3(2q + 1)$. Since $2q + 1$ is an integer, $3 \mid (a + b)$. ∎

7.81 The statement is true.
   Proof   Let $a = 6/5$, $b = 10/3$ and $c = 15/2$. Then $ab = 4$, $ac = 9$, $bc = 25$ and $abc = 30$. ∎

EXERCISES FOR CHAPTER 8

Section 8.1: Relations

8.1 $\text{dom}(R) = \{a, b\}$ and $\text{range}(R) = \{s, t\}$.

8.3 Since $A \times A = \{(0, 0), (0, 1), (1, 0), (1, 1)\}$ and $|A \times A| = 4$, the number of subsets of $A \times A$ and hence the number of relations on A is $2^4 = 16$. Four of these 16 relations are $\emptyset$, $A \times A$, $\{(0, 0)\}$, and $\{(0, 0), (0, 1), (1, 0)\}$.

8.5 $R^{-1} = \{(1, 1), (2, 1), (2, 2), (3, 1), (3, 2), (3, 3)\}$.

8.7   For $a, b \in \mathbf{N}$, $a\ R^{-1}\ b$ if and only if $b\ R\ a$, while $b\ R\ a$ if $b + 4a$ is odd. That is, $R^{-1} = \{(x, y) : \ y + 4x$ is odd$\}$.

8.9   **(a)** The statement is false. Let $A = \{1, 2, 3, 4\}$, $B = \{1, 2, 3, 5\}$, and $C = \{1, 2, 3\}$. Then $|A| = |B| = 4$. Then $R = C \times C$ is a relation from A to B with $|R| = 9$ and $R = R^{-1}$ but $A \neq B$. Thus A, B and R constitute a counterexample.

  (b) Suppose that $|R| = 9$ is replaced by $|R| = 10$. Then the statement would be true.

  Proof   Let A and B be sets with $|A| = |B| = 4$. Assume, to the contrary, that there exists a relation from A to B with $|R| = 10$ and $R = R^{-1}$ but $A \neq B$. Since A and B have the same number 4 of elements, there is an element $x \in A - B$ and an element $y \in B - A$. Since $R = R^{-1}$, it follows that x is not related to any element of B by R and no element of A is related to y. This implies that $R \subseteq (A - \{x\}) \times (B - \{y\})$. Since $|A - \{x\}| \cdot |B - \{y\}| = 3 \cdot 3 = 9$, this is a contradiction.   ∎

Section 8.2: Properties of Relations

8.11   The relation R is reflexive and transitive. Since $(a, d) \in R$ and $(d, a) \notin R$, it follows that R is not symmetric.

8.13   The relation R is transitive but neither reflexive nor symmetric.

8.15   The relation R is reflexive and symmetric. Observe that $3\ R\ 1$ and $1\ R\ 0$ but $3\ R\!\!\!/\ 0$. Thus R is not transitive.

8.17   The relation R is symmetric and transitive but not reflexive.

8.19   The relation R is reflexive and symmetric. Observe that $-1\ R\ 0$ and $0\ R\ 2$ but $-1\ R\!\!\!/\ 2$. Thus R is not transitive.

8.21   The statement is true.

  Proof   Let $A = \{a_1, a_2\}$ and suppose that R is a relation on A that has none of the properties reflexive, symmetric, and transitive. Since R is not symmetric, we may assume that $a_1\ R\ a_2$ but $a_2\ R\!\!\!/\ a_1$. Since at most one of (a_1, a_1) and (a_2, a_2) belongs to R, it follows that R is transitive, which is a contradiction.

  Since the hypothesis of the implication is false, the statement is true vacuously.   ∎

8.23   Since $a \mid a$ for every $a \in \mathbf{N}$, the relation R is reflexive. The relation R is symmetric for suppose that $a\ R\ b$. Then $a \mid b$ or $b \mid a$. This, however, says that $b\ R\ a$. The relation R is not transitive since for $a = 2$, $b = 1$, and $c = 3$, $a\ R\ b$ and $b\ R\ c$ but $a\ R\!\!\!/\ c$.

Section 8.3: Equivalence Relations

8.25   There are three distinct equivalence classes, namely $[1] = \{1, 5\}$, $[2] = \{2, 3, 6\}$, and $[4] = \{4\}$.

8.27   **Proof**   Since $a^3 = a^3$ for each $a \in \mathbf{Z}$, it follows that $a\ R\ a$ and R is reflexive. Let $a, b \in \mathbf{Z}$ such that $a\ R\ b$. Then $a^3 = b^3$ and so $b^3 = a^3$. Thus $b\ R\ a$ and R is symmetric. Let $a, b, c \in \mathbf{Z}$ such that $a\ R\ b$ and $b\ R\ c$. Thus $a^3 = b^3$ and $b^3 = c^3$. Hence $a^3 = c^3$ and so $a\ R\ c$ and R is transitive.   ∎

  Let $a, b \in \mathbf{Z}$. Note that $a^3 = b^3$ if and only if $a = b$. Thus $[a] = \{a\}$ for every $a \in \mathbf{Z}$.

8.29   **Proof**   Assume that $a\ R\ b$, $c\ R\ d$ and $a\ R\ d$. Since $a\ R\ b$ and R is symmetric, $b\ R\ a$. Similarly, $d\ R\ c$. Because $b\ R\ a$, $a\ R\ d$ and R is transitive, $b\ R\ d$. Finally, since $b\ R\ d$ and $d\ R\ c$, it follows that $b\ R\ c$, as desired.   ∎

8.31   **Proof**   First assume that R is an equivalence relation on A. Thus R is reflexive. It remains only to show that R is circular. Assume that $x\ R\ y$ and $y\ R\ z$. Since R is transitive, $x\ R\ z$. Since R is symmetric, $z\ R\ x$. Thus R is circular.

  For the converse, assume that R is a reflexive, circular relation on A. Since R is reflexive, it remains only to show that R is symmetric and transitive. Let $x, y \in A$ such that $x\ R\ y$. Since R is reflexive, $y\ R\ y$. Because (1) $x\ R\ y$ and $y\ R\ y$ and (2) R is circular, it follows that $y\ R\ x$ and so R is symmetric. Let $x, y, z \in A$ such that $x\ R\ y$ and $y\ R\ z$. Since R is circular, $z\ R\ x$. Now because R is symmetric, we have $x\ R\ z$. Thus R is transitive. Therefore, R is an equivalence relation on A.   ∎

8.33   **(a)** **Proof**   Let $a \in \mathbf{Z}$. Since $a - a = 4 \cdot 0 \in H$, it follows that $a\ R\ a$ and R is reflexive. Next, assume that $a\ R\ b$, where $a, b \in \mathbf{Z}$. Then $a - b \in H$ and so $a - b = 4k$, where $k \in \mathbf{Z}$. Then $b - a = 4(-k)$. Since $-k \in \mathbf{Z}$, it follows that $b - a \in H$ and $b\ R\ a$. Therefore, R is symmetric. Finally, assume that $a\ R\ b$ and $b\ R\ c$ where $a, b, c \in \mathbf{Z}$. Then $a - b \in H$ and $b - c \in H$. So $a - b = 4k$ and $b - c = 4\ell$ for $k, \ell \in \mathbf{Z}$. Therefore, $a - c = (a - b) + (b - c) = 4k + 4\ell = 4(k + \ell)$. Since $k + \ell \in \mathbf{Z}$, it follows that $a - c \in H$ and so $a\ R\ c$. Thus R is transitive and R is an equivalence relation.   ∎

(b) Let $a \in \mathbf{Z}$. Then

$$[a] = \{x \in \mathbf{Z} : x \, R \, a\} = \{x \in \mathbf{Z} : x - a \in H\}$$
$$= \{x \in \mathbf{Z} : x - a = 4k \text{ for some integer } k\} = \{a + 4k : k \in \mathbf{Z}\}.$$

Since every integer can be expressed as $4k + r$ where r is an integer with $0 \leq r \leq 3$, it follows that the distinct equivalence classes are $[0], [1], [2]$, and $[3]$, where $[r] = \{4k + r : k \in \mathbf{Z}\}$ for $r = 0, 1, 2, 3$.

8.35 The statement is false. Suppose that there are equivalence relations R_1 and R_2 on $S = \{a, b, c\}$ such that $R_1 \not\subseteq R_2$, $R_2 \not\subseteq R_1$ and $R_1 \cup R_2 = S \times S$. Since R_1 and R_2 are both reflexive, it follows that $(a, a), (b, b), (c, c) \in R_1 \cap R_2$. Because $R_1 \not\subseteq R_2$, there exists some element of R_1 that is not in R_2, say $(a, b) \in R_1 - R_2$. Necessarily then, $(b, a) \in R_1 - R_2$ as well. Because $R_2 \not\subseteq R_1$, there exists some element of R_2 that is not in R_1. We may assume that $(b, c) \in R_2 - R_1$. Thus $(c, b) \in R_2 - R_1$. Since $R_1 \cup R_2 = S \times S$, it follows that $(a, c) \in R_1 \cup R_2$. We may assume that $(a, c) \in R_1$. Since $(b, a) \in R_1$, it follows by the transitive property that $(b, c) \in R_1$, which is not true.

Section 8.4: Properties of Equivalence Classes

8.37 Let $a \in \mathbf{N}$. Then $a^2 + a^2 = 2(a^2)$ is an even integer and so $a \, R \, a$. Thus R is reflexive. Assume that $a \, R \, b$, where $a, b \in \mathbf{N}$. Then $a^2 + b^2$ is even. Since $b^2 + a^2 = a^2 + b^2$, it follows that $b^2 + a^2$ is even. Therefore, $b \, R \, a$ and R is symmetric. Finally, show that R is transitive.

There are two distinct equivalence classes:
$$[1] = \{x \in \mathbf{N} : x^2 + 1 \text{ is even}\} = \{x \in \mathbf{N} : x^2 \text{ is odd}\} = \{x \in \mathbf{N} : x \text{ is odd}\}$$
$$[2] = \{x \in \mathbf{N} : x^2 + 4 \text{ is even}\} = \{x \in \mathbf{N} : x^2 \text{ is even}\} = \{x \in \mathbf{N} : x \text{ is even}\}.$$

8.39 **(a)** **Proof**   First, we show that R is reflexive. Let $x \in S$. Then $x + 2x = 3x$. Since $3 \mid (x + 2x)$, it follows that $x \, R \, x$ and R is reflexive. Next, we show that R is symmetric. Let $x \, R \, y$, where $x, y \in S$. Then $3 \mid (x + 2y)$ and so $x + 2y = 3a$, where $a \in \mathbf{Z}$ and so $x = 3a - 2y$. Thus $y + 2x = y + 2(3a - 2y) = 6a - 3y = 3(2a - y)$. Since $2a - y$ is an integer, $3 \mid (y + 2x)$. Thus $y \, R \, x$ and R is symmetric.

Finally, we show that R is transitive. Let $x \, R \, y$ and $y \, R \, z$, where $x, y, z \in S$. Then $3 \mid (x + 2y)$ and $3 \mid (y + 2z)$. So $x + 2y = 3a$ and $y + 2z = 3b$, where $a, b \in \mathbf{Z}$. Thus $(x + 2y) + (y + 2z) = 3a + 3b$ and so $x + 2z = 3a + 3b - 3y = 3(a + b - y)$. Since $a + b - y$ is an integer, $3 \mid (x + 2z)$ and so $x \, R \, z$. Therefore, R is transitive.   ∎

(b) There are three distinct equivalence classes:
$$[0] = \{0, -6\}, [1] = \{1, -2, 4, 7\}, \text{ and } [-7] = \{-7, 5\}.$$

8.41 **(a)** Suppose that R_1 and R_2 are two equivalence relations defined on a set S. Let $R = R_1 \cap R_2$. First, we show that R is reflexive. Let $a \in S$. Since R_1 and R_2 are equivalence relations on S, it follows that $(a, a) \in R_1$ and $(a, a) \in R_2$. Thus $(a, a) \in R$ and so R is reflexive. Assume that $a \, R \, b$, where $a, b \in S$. Then $(a, b) \in R = R_1 \cap R_2$. Thus $(a, b) \in R_1$ and $(a, b) \in R_2$. Since R_1 and R_2 are symmetric, $(b, a) \in R_1$ and $(b, a) \in R_2$. Thus $(b, a) \in R$ and so $b \, R \, a$. Hence R is symmetric. Finally, show that R is transitive.

(b) Let $a \in \mathbf{Z}$. For $x \in \mathbf{Z}$, it follows that $x \, R_1 \, a$ if and only if $x \, R_2 \, a$ and $x \, R_3 \, a$. That is, $x \, R_1 \, a$ if and only if $x \equiv a \pmod 2$ and $x \equiv a \pmod 3$. First, suppose that $x \equiv a \pmod 2$ and $x \equiv a \pmod 3$. Hence $x = a + 2k$ and $x = a + 3\ell$ for some integers k and ℓ. Therefore, $2k = 3\ell$ and so ℓ is even. Thus $\ell = 2m$ for some integer m, implying that $x = a + 3\ell = a + 3(2m) = a + 6m$ and so $x - a = 6m$. Hence $x \equiv a \pmod 6$. If $x \equiv a \pmod 6$, then $x \equiv a \pmod 2$ and $x \equiv a \pmod 3$. Thus $[a] = \{x \in \mathbf{Z} : x \equiv a \pmod 6\}$.
$$[0] = \{\ldots, -12, -6, 0, 6, 12, \ldots\}, \quad [1] = \{\ldots, -11, -5, 1, 7, 13, \ldots\},$$
$$[2] = \{\ldots, -10, -4, 2, 8, 14, \ldots\}, \quad [3] = \{\ldots, -9, -3, 3, 9, 15, \ldots\},$$
$$[4] = \{\ldots, -8, -2, 4, 10, 16, \ldots\}, \quad [5] = \{\ldots, -7, -1, 5, 11, 17, \ldots\}.$$

8.43 **Proof**   For $a_i \in A$, $[a_i] = \{x \in A : x \, R \, a_i\} = \{x \in A : (x, a_i) \in R\}$. Therefore, $\|[a_i]\|$ counts all those ordered pairs in R for which a_i is the second coordinate and so $\sum_{i=1}^{n} \|[a_i]\|$ counts all ordered pairs in R, that is, $\sum_{i=1}^{n} \|[a_i]\| = |R|$. Since R is reflexive, $(a_i, a_i) \in R$ for $i = 1, 2, \ldots, n$ and for each pair i, j of distinct integers, either (a_i, a_j) and (a_j, a_i) both belong to R or neither belongs to R. Suppose that there are k ordered pairs (a_i, a_j) in R with $1 \leq i < j \leq n$. Then $|R| = n + 2k$ and so $\sum_{i=1}^{n} \|[a_i]\|$ is even if and only if n is even.   ∎

Section 8.5: Congruence Modulo *n*

8.45   Let $a \in \mathbf{Z}$. Since $3a + 5a = 8a$, it follows that $8 \mid (3a + 5a)$ and so $3a + 5a \equiv 0 \pmod{8}$. Hence $a \, R \, a$ and R is reflexive. Next, we show that R is symmetric. Assume that $a \, R \, b$, where $a, b \in \mathbf{Z}$. Then $3a + 5b \equiv 0 \pmod{8}$, that is, $3a + 5b = 8k$ for some integer k. Observe that $(3a + 5b) + (3b + 5a) = 8a + 8b$. Thus

$$3b + 5a = 8a + 8b - (3a + 5b) = 8a + 8b - 8k = 8(a + b - k).$$

Since $a + b - k$ is an integer,
$8 \mid (3b + 5a)$ and so $3b + 5a \equiv 0 \pmod{8}$. Hence $b \, R \, a$ and R is symmetric. Finally, show that R is transitive.

8.47   There are two distinct equivalence classes, namely, $[0] = \{0, \pm 2, \pm 4, \ldots\}$ and $[1] = \{\pm 1, \pm 3, \pm 5, \ldots\}$.

8.49   ***Proof***   Let $a \in \mathbf{Z}$.
Since $5a - 2a = 3a$, it follows that $3 \mid (5a - 2a)$ and so $5a \equiv 2a \pmod{3}$. Hence $a \, R \, a$ and R is reflexive.
    Next, we show that R is symmetric. Assume that $a \, R \, b$, where $a, b \in \mathbf{Z}$. Then $5a \equiv 2b \pmod{3}$, that is, $5a - 2b = 3k$ for some integer k. Observe that $(5a - 2b) + (5b - 2a) = 3a + 3b$. Thus

$$5b - 2a = 3a + 3b - (5a - 2b) = 3a + 3b - 3k = 3(a + b - k).$$

Since $a + b - k$ is an integer, $3 \mid (5b - 2a)$ and so $5b \equiv 2a \pmod{3}$. Hence $b \, R \, a$ and R is symmetric.
    Finally, we show that R is transitive. Assume that $a \, R \, b$ and $b \, R \, c$, where $a, b, c \in \mathbf{Z}$. Thus $5a \equiv 2b \pmod{3}$ and $5b \equiv 2c \pmod{3}$. So $5a - 2b = 3x$ and $5b - 2c = 3y$, where $x, y \in \mathbf{Z}$. Observe that

$$(5a - 2b) + (5b - 2c) = (5a - 2c) + 3b = 3x + 3y.$$

Thus $5a - 2c = 3x + 3y - 3b = 3(x + y - b)$.
Since $x + y - b$ is an integer, $3 \mid (5a - 2c)$ and $5a \equiv 2c \pmod{3}$. Therefore, $a \, R \, c$ and R is transitive. ∎
    There are three distinct equivalence classes, namely,
$[0] = \{0, \pm 3, \pm 6, \ldots\}$, $[1] = \{\ldots, -5, -2, 1, 4, \ldots\}$, and $[2] = \{\ldots, -4, -1, 2, 5, \ldots\}$.

8.51   First, we show that R is reflexive. Let $a \in \mathbf{Z}$. Since $2a + 3a = 5a$, it follows that $5 \mid (2a + 3a)$ and so $a \, R \, a$. Hence R is reflexive. Next, we show that R is symmetric. Assume that $a \, R \, b$, where $a, b \in \mathbf{Z}$. Then $2a + 3b \equiv 0 \pmod{5}$. Hence $2a + 3b = 5k$ for some integer k. Observe that $(2a + 3b) + (2b + 3a) = 5a + 5b$. Thus

$$2b + 3a = 5a + 5b - (2a + 3b) = 5a + 5b - 5k = 5(a + b - k).$$

Since $a + b - k$ is an integer, $5 \mid (2b + 3a)$ and so $2b + 3a \equiv 0 \pmod{5}$. Hence $b \, R \, a$ and R is symmetric. Finally, show that R is transitive.
    The distinct equivalence classes are $[0]$, $[1]$, $[2]$, $[3]$ and $[4]$.

8.53   The relation R is an equivalence relation.
Proof   Let $a \in \mathbf{R}$. Since $a - a = 0 = 0 \cdot \pi$, it follows that $a \, R \, a$ and R is reflexive. Next, suppose that $a \, R \, b$, where $a, b \in \mathbf{R}$. Then $a - b = k\pi$ for some $k \in \mathbf{Z}$. Since $b - a = (-k)\pi$ and $-k \in \mathbf{Z}$, it follows that $b \, R \, a$ and so R is symmetric. Finally, suppose that $a \, R \, b$ and $b \, R \, c$, where $a, b, c \in \mathbf{R}$. Then $a - b = k\pi$ and $b - c = \ell\pi$ for $k, \ell \in \mathbf{Z}$. Thus $a - c = (a - b) + (b - c) = (k + \ell)\pi$. Because $k + \ell \in \mathbf{Z}$, $a \, R \, c$ and R is transitive. Therefore, R is an equivalence relation. ∎
$[0] = \{x \in \mathbf{R} : x \, R \, 0\} = \{x \in \mathbf{R} : x = k\pi, \text{ where } k \in \mathbf{Z}\} = \{k\pi : k \in \mathbf{Z}\}$.
$[\pi] = \{x \in \mathbf{R} : x \, R \, \pi\} = \{x \in \mathbf{R} : x - \pi = k\pi, \text{ where } k \in \mathbf{Z}\} = \{(k + 1)\pi : k \in \mathbf{Z}\} = \{k\pi : k \in \mathbf{Z}\} = [0]$.
$[\sqrt{2}] = \{\sqrt{2} + k\pi : k \in \mathbf{Z}\}$.

Section 8.6: The Integers Modulo *n*

8.55   **(a)** $[2] + [6] = [8] = [0]$.   **(b)** $[2] \cdot [6] = [12] = [4]$.
    (c) $[-13] + [138] = [125] = [5]$.   **(d)** $[-13] \cdot [138] = [3][2] = [6]$.

8.57   **(a)** ***Proof***   Let $a, b \in T$. Then $a = 4k$ and $b = 4\ell$ for $k, \ell \in \mathbf{Z}$. Thus $a + b = 4(k + \ell)$
    and $ab = 4(4k\ell)$. Since $k + \ell, 4k\ell \in \mathbf{Z}$, it follows that T is closed under addition and multiplication. ∎
    (b) Yes. Let $a \in S - T$ and $b \in T$. Then $b = 4\ell$ for $\ell \in \mathbf{Z}$. Thus $ab = 4(a\ell)$. Since $a\ell \in \mathbf{Z}$, it follows that $ab \in T$.

    (c) No. For example, $a = 1 \in S - T$ and $b = 4 \in T$ but $a + b = 5 \notin T$.

    (d) Yes. For example, $a = 2$ and $b = 6$ belong to $S - T$ and $ab = 12 = 4 \cdot 3 \in T$.

    (e) Yes. For example, $a = 2$ and $b = 6$ belong to $S - T$ and $a + b = 8 = 4 \cdot 2 \in T$.

8.59 **(a)** No. Consider $[a] = [2]$ and $[b] = [4]$. Then $[a] \neq [0]$ and $[b] \neq [0]$, but $[a] \cdot [b] = [8] = [0]$.

    (b) If $\mathbf{Z}_8$ is replaced by $\mathbf{Z}_9$ or $\mathbf{Z}_{10}$, then the answer is no; while if $\mathbf{Z}_8$ is replaced by $\mathbf{Z}_{11}$, then the answer is yes.

    (c) Let $a, b \in \mathbf{Z}_n$, where $n \geq 2$ is prime. If $[a] \cdot [b] = [0]$, then $[a] = [0]$ or $[b] = [0]$.

8.61 **(a)** Suppose that an element $[a] \in \mathbf{Z}_m$ also belongs to $\mathbf{Z}_n$. That is, $[a]$ in $\mathbf{Z}_m$ is the same set as $[b]$ in $\mathbf{Z}_n$. Since $a, a + m \in [a] \in \mathbf{Z}_m$, it follows that $a, a + m \in [b] \in \mathbf{Z}_n$. Therefore, $n \mid [(a + m) - a]$ or $n \mid m$. Similarly, $m \mid n$ and so $n = m$, that is, $\mathbf{Z}_m = \mathbf{Z}_n$.

    (b) If $m, n \geq 2$ and $m \neq n$, then $\mathbf{Z}_m \cap \mathbf{Z}_n = \emptyset$. For example, $\mathbf{Z}_2 \cap \mathbf{Z}_3 = \emptyset$.

EXERCISES FOR CHAPTER 9

Section 9.1: The Definition of Function

9.1 $\mathrm{dom}(f) = \{a, b, c, d\}$ and $\mathrm{range}(f) = \{y, z\}$.

9.3 Since R is an equivalence relation, R is reflexive. So $(a, a) \in R$ for every $a \in A$. Since R is also a function from A to A, we must have $R = \{(a, a) : a \in A\}$ and so R is the identity function on A.

9.5 Let $A' = \{a \in A : (a, b) \in R$ for some $b \in B\}$. Furthermore, for each element $a' \in A'$, select exactly one element $b' \in \{b \in B : (a', b) \in R\}$. Then $f = \{(a', b') : a' \in A'\}$ is a function from A' to B.

9.7 $R = \{(3, 4), (17, 6), (29, 60), (45, 22)\}$ and so R is a function from A to B.

9.9 **(a)** Since $0 \, R_1 \, 1$ and $0 \, R_1 \, (-1)$, R_1 is not a function.

    (b) Since $0 \, R_2 \, (\frac{1}{\sqrt{3}})$ and $0 \, R_2 \, (-\frac{1}{\sqrt{3}})$, R_2 is not a function.

    (c) For each $a \in \mathbf{N}$, $b = (1 - 3a)/5 \in \mathbf{Q}$ is the unique element such that $3a + 5b = 1$. So R_3 defines a function.

    (d) For each $x \in \mathbf{R}$, $y = 4 - |x - 2|$ is a unique element of $\mathbf{R}$. So R_4 defines a function.

    (e) Since $0 \, R_5 \, 1$ and $0 \, R_5 \, (-1)$, R_5 is not a function.

9.11 **(a)** $f(C) = C$, $f^{-1}(C) = C \cup \{x \in \mathbf{R} : -x \in C\}$, $f^{-1}(D) = \mathbf{R} - \{0\}$, $f^{-1}(\{1\}) = \{1, -1\}$.

    (b) $f(C) = [0, \infty)$, $f^{-1}(C) = [e, \infty)$, $f^{-1}(D) = (1, \infty)$, $f^{-1}(\{1\}) = \{e\}$.

    (c) $f(C) = [e, \infty)$, $f^{-1}(C) = [0, \infty)$, $f^{-1}(D) = \mathbf{R}$, $f^{-1}(\{1\}) = \{0\}$.

    (d) $f(C) = [-1, 1]$, $f^{-1}(C) = \{\frac{\pi}{2} + 2n\pi : n \in \mathbf{Z}\}$, $f^{-1}(D) = \cup_{n \in \mathbf{Z}}(2n\pi, (2n + 1)\pi)$, $f^{-1}(\{1\}) = \{\frac{\pi}{2} + 2n\pi : n \in \mathbf{Z}\}$.

    (e) $f(C) = (-\infty, 1]$, $f^{-1}(C) = \{1\}$, $f^{-1}(D) = (0, 2)$, $f^{-1}(\{1\}) = \{1\}$.

Section 9.2: The Set of All Functions from A to B

9.13 $B^A = \{f_1, f_2, \ldots, f_8\}$, where $f_1 = \{(1, x), (2, x), (3, x)\}$, $f_2 = \{(1, x), (2, x), (3, y)\}$, $f_3 = \{(1, x), (2, y), (3, x)\}$, $f_4 = \{(1, x), (2, y), (3, y)\}$. By interchanging x and y in f_1, f_2, f_3, f_4, we obtain f_5, f_6, f_7, f_8.

9.15 For $A = \{a, b, c\}$ and $B = \{0, 1\}$, there are eight different functions from A to B, namely
$$f_1 = \{(a, 0), (b, 0), (c, 0)\}, \quad f_2 = \{(a, 0), (b, 0), (c, 1)\},$$
$$f_3 = \{(a, 0), (b, 1), (c, 0)\}, \quad f_4 = \{(a, 0), (b, 1), (c, 1)\},$$
$$f_5 = \{(a, 1), (b, 0), (c, 0)\}, \quad f_6 = \{(a, 1), (b, 0), (c, 1)\},$$
$$f_7 = \{(a, 1), (b, 1), (c, 0)\}, \quad f_8 = \{(a, 1), (b, 1), (c, 1)\},$$

9.17 **(a)** A reasonable interpretation of C^{B^A} is $\{f : f : B^A \to C\}$.

    (b) For $A = \{0, 1\}$ and $B = \{a, b\}$, $B^A = \{f_1, f_2, f_3, f_4\}$, where
$$f_1 = \{(0, a), (1, a)\}, f_2 = \{(0, a), (1, b)\},$$
$$f_3 = \{(0, b), (1, a)\} \text{ and } f_4 = \{(0, b), (1, b)\}.$$
Then for $C = \{x, y\}$, $C^{B^A} = \{g_1, g_2, \ldots, g_{16}\}$, where
$$g_1 = \{(f_1, x), (f_2, x), (f_3, x), (f_4, x)\}, g_2 = \{(f_1, x), (f_2, x), (f_3, x), (f_4, y)\}, \ldots,$$
$$g_{16} = \{(f_1, y), (f_2, y), (f_3, y), (f_4, y)\}.$$

Section 9.3: One-to-One and Onto Functions

9.19 Let $A = \{1, 2\}$ and $B = \{3, 4, 5\}$. Then $f = \{(1, 3), (2, 4)\}$ and $g = \{(3, 1), (4, 2), (5, 2)\}$ have the desired properties.

9.21 **(a)** The function f is injective.

   Proof Assume that $f(a) = f(b)$, where $a, b \in \mathbf{Z}$. Then $a - 3 = b - 3$. Adding 3 to both sides, we obtain $a = b$. ∎

   (b) The function f is surjective.

   Proof Let $n \in \mathbf{Z}$. Then $n + 3 \in \mathbf{Z}$ and $f(n + 3) = (n + 3) - 3 = n$. ∎

9.23 The statement is true. The function $f : A \to \mathcal{P}(A)$ defined by $f(a) = \{a\}$ has the desired property.

9.25 Consider the function $f : \mathbf{R} \to \mathbf{R}$ defined by $f(x) = x^3 - x = (x + 1)x(x - 1)$. Since $f(0) = f(1)$, it follows that f is not one-to-one. One way to show that f is onto is to use the Intermediate Value Theorem.

   Method #1. Let $r \in \mathbf{R}$. Since

   $$\lim_{x \to \infty}(x^3 - x) = \infty \text{ and } \lim_{x \to -\infty}(x^3 - x) = -\infty,$$

   there exist real numbers a and b such that $f(a) < r < f(b)$. Since f is continuous on the closed interval $[a, b]$, there exists c such that $a < c < b$ and $f(c) = r$.

   Method #2. Let $r \in \mathbf{R}$. If $r = 0$, then $f(0) = 0 = r$. Suppose that $r > 0$. Then $r + 1 > 1$ and $r + 2 > 1$; so $f(r + 1) = r(r + 1)(r + 2) > r$. Since $f(0) < r < f(r + 1)$, it follows by the Intermediate Value Theorem that there exists $c \in (0, r + 1)$ such that $f(c) = r$. If $r < 0$, then $s = -r > 0$ and, as we just saw, there exists $c \in (0, s + 1)$ such that $f(c) = s$. Then $f(-c) = -s = r$.

9.27 **(a)** $R = \{(2, 8), (3, 6), (4, 8), (5, 10)\}$. The relation R is a function from A to B.

   (b) Since range$(R) = B$, the function R is onto. However, since $2 \, R \, 8$ and $4 \, R \, 8$, R is not one-to-one.

9.29 **Proof** By Exercise 9.12(b), $f(C \cap D) \subseteq f(C) \cap f(D)$. So it remains to show that $f(C) \cap f(D) \subseteq f(C \cap D)$ under the added hypothesis that f is one-to-one. Let $y \in f(C) \cap f(D)$. Then $y \in f(C)$ and $y \in f(D)$. Since $y \in f(C)$, there exists $x \in C$ such that $y = f(x)$. Furthermore, since $y \in f(D)$, there exists $z \in D$ such that $y = f(z)$. Since f is one-to-one, $z = x$. Thus $x \in C \cap D$ and so $y \in f(C \cap D)$. Therefore, $f(C) \cap f(D) \subseteq f(C \cap D)$ and so $f(C \cap D) = f(C) \cap f(D)$. ∎

Section 9.4: Bijective Functions

9.31 **(a)** **Proof** Let $[a], [b] \in \mathbf{Z}_5$ such that $[a] = [b]$. We show that $f([a]) = f([b])$, that is, $[2a + 3] = [2b + 3]$. Since $[a] = [b]$, it follows that $a \equiv b \pmod{5}$ and so $a - b = 5x$ for some integer x. Observe that

   $$(2a + 3) - (2b + 3) = 2(a - b) = 2(5x) = 5(2x).$$

   Since $2x$ is an integer, $5 \mid [(2a + 3) - (2b + 3)]$. Therefore, $2a + 3 \equiv 2b + 3 \pmod{5}$ and so $[2a + 3] = [2b + 3]$. ∎

   (b) Since $f([0]) = [3]$, $f([1]) = [0]$, $f([2]) = [2]$, $f([3]) = [4]$, and $f([4]) = [1]$, it follows that f is one-to-one and onto and so f is bijective.

9.33 Define $f_1(x) = x^2$ for $x \in A$ and $f_2(x) = \sqrt{x}$ for $x \in A$. ($f_3(x) = 1 - x$ is another example.)

9.35 **(a)** Consider $S = \{2, 5, 6\}$. Observe that for each $y \in B$, there exists $x \in S$ such that x is related to y. This says that $\gamma(R) \le 3$. On the other hand, let $S' \subseteq A$ such that for every element y of B, there is an element $x \in S'$ such that x is related to y. Observe that S' must contain 6, at least one of 2 and 3 and at least one of 4, 5, and 7. Thus $|S'| \ge 3$. Therefore, $\gamma(R) = 3$.

   (b) If R is an equivalence relation defined on a finite nonempty set A, then $\gamma(R)$ is the number of distinct equivalence classes of R.

   (c) If f is a bijective function from A to B, then $\gamma(f) = |A|$.

Section 9.5: Composition of Functions

9.37 $g \circ f = \{(1, y), (2, x), (3, x), (4, x)\}$.

9.39 **(a)** $(g \circ f)([a]) = g(f([a])) = g([3a]) = [21a] = [a]$. $(f \circ g)([a]) = f(g([a])) = f([7a]) = [21a] = [a]$.

   (b) Each of $g \circ f$ and $f \circ g$ is the identity function on $\mathbf{Z}_{10}$.

9.41   *Proof*   We first show that f is one-to-one. Let $a, b \in A$ such that $f(a) = f(b)$. Now

$$a = i_A(a) = (f \circ f)(a) = f(f(a)) = f(f(b))$$
$$= (f \circ f)(b) = i_A(b) = b.$$

Thus f is one-to-one.

   Next, we show that f is onto. Let $c \in A$. Suppose that $f(c) = d \in A$. Observe that
$$f(d) = f(f(c)) = (f \circ f)(c) = i_A(c) = c.$$
Thus f is onto.   ∎

9.43   **(a)** (i)   **Direct Proof.**   Assume that $g \circ f$ is one-to-one. We show that f is one-to-one. Let $f(x) = f(y)$, where $x, y \in A$. Since $g(f(x)) = g(f(y))$, it follows that $(g \circ f)(x) = (g \circ f)(y)$. Since $g \circ f$ is one-to-one, $x = y$.   ∎

   (ii)   **Proof by Contrapositive.**   Assume that f is not one-to-one. Hence there exist distinct elements $a, b \in A$ such that $f(a) = f(b)$. Since

$$(g \circ f)(a) = g(f(a)) = g(f(b)) = (g \circ f)(b),$$

it follows that $g \circ f$ is not one-to-one.   ∎

   (iii)   **Proof by Contradiction.**   Assume, to the contrary, that there exist functions $f : A \to B$ and $g : B \to C$ such that $g \circ f$ is one-to-one and f is not one-to-one. Since f is not one-to-one, there exist distinct elements $a, b \in A$ such that $f(a) = f(b)$. However then,

$$(g \circ f)(a) = g(f(a)) = g(f(b)) = (g \circ f)(b),$$

contradicting our assumption that $g \circ f$ is one-to-one.   ∎

(b) Let $A = \{1, 2, 3\}$, $B = \{w, x, y, z\}$, and $C = \{a, b, c\}$. Define $f : A \to B$ by
$$f = \{(1, w), (2, x), (3, y)\}$$

and $g : B \to C$ by
$$g = \{(w, a), (x, b), (y, c), (z, c)\}.$$
Then $g \circ f = \{(1, a), (2, b), (3, c)\}$ is one-to-one, but g is not one-to-one.

9.45   **(a)** $(g \circ f)(18, 11) = g(f(18, 11)) = g(29, 18) = (47, 29)$.

(b) The function $g \circ f : A \times B \to B \times B$ is one-to-one.

   Proof   Assume that $(g \circ f)(a, b) = (g \circ f)(c, d)$, where $(a, b), (c, d) \in A \times B$. Then $g(f(a, b)) = g(f(c, d))$ and so $g(a + b, a) = g(c + d, c)$. Therefore, $(2a + b, a + b) = (2c + d, c + d)$, which implies that $2a + b = 2c + d$ and $a + b = c + d$. Solving these equations, we find that $a = c$ and $b = d$; so $(a, b) = (c, d)$. Thus $g \circ f$ is one-to-one.   ∎

(c) The function $g \circ f : A \times B \to B \times B$ is onto.

   Proof   Let $(m, n) \in B \times B$. Then m and n are odd integers. Therefore, $a = m - n \in A$ and $b = 2n - m \in B$. Hence $(g \circ f)(a, b) = g(f(a, b)) = g(f(m - n, 2n - m)) = g(n, m - n) = (m, n)$. Thus $g \circ f$ is onto.   ∎

9.47   **(a)** The statement is true.

   Proof   Let $a \in \mathbf{R}$. Suppose that $f(a) = b$, $g(b) = c$, and $h(b) = d$. Then $((g + h) \circ f)(a) = (g + h)(f(a)) = (g + h)(b) = g(b) + h(b) = c + d$; while $[(g \circ f) + (h \circ f)](a) = (g \circ f)(a) + (h \circ f)(a) = g(f(a)) + h(f(a)) = g(b) + h(b) = c + d$. Therefore, $(g + h) \circ f = (g \circ f) + (h \circ f)$.   ∎

(b) The statement is false. For example, suppose that $f(x) = x^2$, $g(x) = x$, and $h(x) = x$ for $x \in \mathbf{R}$. So $(g + h)(x) = 2x$. Then $[f \circ (g + h)](1) = f((g + h)(1)) = f(2) = 4$; while $[(f \circ g) + (f \circ h)](1) = (f \circ g)(1) + (f \circ h)(1) = f(g(1)) + f(h(1)) = f(1) + f(1) = 1 + 1 = 2$. Thus $f \circ (g + h) \neq (f \circ g) + (f \circ h)$ in general.

Section 9.6: Inverse Functions

9.49 Let $f = \{(a, a), (b, a), (c, b)\}$. Then f is a function from A to A. But the inverse relation $f^{-1} = \{(a, a), (a, b), (b, c)\}$ is not a function.

9.51 ***Proof*** First, we show that f is one-to-one. Assume that $f(a) = f(b)$, where $a, b \in \mathbf{R} - \{3\}$. Then $\dfrac{5a}{a-3} = \dfrac{5b}{b-3}$. Multiplying both sides by $(a-3)(b-3)$, we obtain $5a(b-3) = 5b(a-3)$. Simplifying, we have $5ab - 15a = 5ab - 15b$. Adding $-5ab$ to both sides and dividing by -15, we obtain $a = b$. Thus f is one-to-one.

To show that f is onto, let $r \in \mathbf{R} - \{5\}$. We show that there exists $x \in \mathbf{R} - \{3\}$ such that $f(x) = r$. Consider $x = \dfrac{3r}{r-5}$. (Since $\dfrac{3r}{r-5} \neq 3$, it follows that $x \in \mathbf{R} - \{3\}$.) Then

$$f(x) = f\left(\frac{3r}{r-5}\right) = \frac{5\left(\frac{3r}{r-5}\right)}{\frac{3r}{r-5} - 3} = \frac{15r}{3r - 3(r-5)} = \frac{15r}{15} = r,$$

implying that f is onto. Therefore, f is bijective. ∎

Since $\left(f \circ f^{-1}\right)(x) = x$ for all $x \in \mathbf{R} - \{5\}$, it follows that

$$\left(f \circ f^{-1}\right)(x) = f\left(f^{-1}(x)\right) = \frac{5f^{-1}(x)}{f^{-1}(x) - 3} = x.$$

Thus $5f^{-1}(x) = x(f^{-1}(x) - 3)$ and $5f^{-1}(x) = xf^{-1}(x) - 3x$. Collecting the terms involving $f^{-1}(x)$ on the same side of the equation and then factoring $f^{-1}(x)$ from this expression, we have $xf^{-1}(x) - 5f^{-1}(x) = 3x$; so $f^{-1}(x)(x-5) = 3x$. Solving for $f^{-1}(x)$, we obtain

$$f^{-1}(x) = \frac{3x}{x-5}.$$

9.53 Since there are $6 = 3!$ bijective functions from A to B, there are six functions A to B that have inverses.

9.55 **(a)** The proof is similar to that in Exercise 9.51.
(b) $f = f^{-1}$.
(c) $f \circ f \circ f = f$.

9.57 **(a)** ***Proof*** Observe that $f(x) \geq 0$ if and only if $x \geq 1$ and that $f(x) < 0$ if and only if $x < 1$. First, we show that f is one-to-one. Assume that $f(a) = f(b)$. We consider two cases.
Case 1. $f(a) = f(b) \geq 0$. Then $\sqrt{a-1} = \sqrt{b-1}$. Squaring both sides, we get $a - 1 = b - 1$ and so $a = b$.
Case 2. $f(a) = f(b) < 0$. Then $\frac{1}{a-1} = \frac{1}{b-1}$. Therefore, $a - 1 = b - 1$ and so $a = b$.
Hence f is one-to-one. Next, we show that f is onto. Let $r \in \mathbf{R}$. We consider two cases.
Case 1. $r \geq 0$. Then $f(r^2 + 1) = \sqrt{(r^2 + 1) - 1} = r$.
Case 2. $r < 0$. Then $f(\frac{r+1}{r}) = \frac{1}{\frac{r+1}{r} - 1} = r$. Therefore, f is onto and thus a bijection. ∎

(b) $f^{-1}(x) = \begin{cases} \frac{x+1}{x} & \text{if } x < 0 \\ x^2 + 1 & \text{if } x \geq 0 \end{cases}$

9.59 ***Proof*** First, observe that $g \circ f : A \to C$ and $h \circ f : A \to C$. Let $b \in B$. Since f is bijective, there is a unique element $a \in A$ such that $f(a) = b$. Since $g \circ f = h \circ f$, it follows that $(g \circ f)(a) = (h \circ f)(a)$ and so $g(f(a)) = h(f(a))$. Therefore, $g(b) = h(b)$ and so $g = h$. ∎

Section 9.7: Permutations

9.61 **(a)** $\alpha^{-1} = \begin{pmatrix} 1 & 2 & 3 & 4 & 5 & 6 \\ 4 & 1 & 6 & 3 & 5 & 2 \end{pmatrix}$ and $\beta^{-1} = \begin{pmatrix} 1 & 2 & 3 & 4 & 5 & 6 \\ 5 & 4 & 2 & 6 & 1 & 3 \end{pmatrix}$.

(b) $\alpha \circ \beta = \begin{pmatrix} 1 & 2 & 3 & 4 & 5 & 6 \\ 5 & 4 & 3 & 6 & 2 & 1 \end{pmatrix}$ and $\beta \circ \alpha = \begin{pmatrix} 1 & 2 & 3 & 4 & 5 & 6 \\ 3 & 4 & 2 & 5 & 1 & 6 \end{pmatrix}$.

EXERCISES FOR CHAPTER 10

Section 10.1: Numerically Equivalent Sets

10.1 Since $A_1 = \{-3, -2, 2, 3\}$, $A_2 = \{-5, -4, -3, 5\}$, $A_3 = \{-2, -1, 0, 1, 2, 3\}$, $A_4 = \{-1, 0, 1\}$, and $A_5 = \{-4, 0, 4\}$, it follows that $|A_1| = |A_2| = 4$, $|A_3| = 6$, and $|A_4| = |A_5| = 3$. So the distinct equivalence classes for R are $[A_1] = \{A_1, A_2\}$, $[A_3] = \{A_3\}$, and $[A_4] = \{A_4, A_5\}$.

Section 10.2: Denumerable Sets

10.3 *Proof* Since A and B are denumerable, the sets A and B can be expressed as
$$A = \{a_1, a_2, a_3, \ldots\} \text{ and } B = \{b_1, b_2, b_3, \ldots\}.$$
The function $f : \mathbf{N} \to A \cup B$ defined by

$$\begin{array}{cccccccc} 1 & 2 & 3 & 4 & 5 & 6 & \cdots \\ \downarrow & \downarrow & \downarrow & \downarrow & \downarrow & \downarrow & \cdots \\ a_1 & b_1 & a_2 & b_2 & a_3 & b_3 & \cdots \end{array}$$

is bijective. Therefore, $A \cup B$ is denumerable. ∎

10.5 *Proof* Since $\mathbf{Z} - \{2\}$ is an infinite subset of the denumerable set $\mathbf{Z}$, it follows by Theorem 10.4 that $\mathbf{Z} - \{2\}$ is denumerable and so $|\mathbf{Z}| = |\mathbf{Z} - \{2\}|$. ∎

10.7 **(a)** $1 + \sqrt{2}$, $(4 + \sqrt{2})/2$, $(9 + \sqrt{2})/3$.

(b) *Proof* Assume that $f(a) = f(b)$, where $a, b \in \mathbf{N}$. Then $\frac{a^2 + \sqrt{2}}{a} = \frac{b^2 + \sqrt{2}}{b}$. Multiplying by ab, we obtain $a^2 b + \sqrt{2} b = ab^2 + \sqrt{2} a$. Thus $a^2 b - ab^2 + \sqrt{2} b - \sqrt{2} a = ab(a - b) - \sqrt{2}(a - b) = (a - b)(ab - \sqrt{2}) = 0$. Thus $a = b$ or $ab = \sqrt{2}$. Since $ab \in \mathbf{N}$ and $\sqrt{2}$ is irrational, $ab \neq \sqrt{2}$. Therefore, $a = b$ and f is one-to-one. ∎

(c) *Proof* Let $x \in S$. Then $x = (n^2 + \sqrt{2})/n$ for some $n \in \mathbf{N}$. Then $f(n) = x$. ∎

(d) Yes, since $\mathbf{N}$ is denumerable and $f : \mathbf{N} \to S$ is a bijection by (b) and (c).

10.9 Let A be a denumerable set. Then we can write $A = \{a_1, a_2, a_3, \ldots\}$. Since $A_1 = \{a_1, a_3, a_5, \ldots\}$ and $A_2 = \{a_2, a_4, a_6, \ldots\}$ are denumerable sets, $\{A_1, A_2\}$ is a partition of A.

10.11 Either $|A| = |B|$ and A is denumerable or $|A|$ is finite. Therefore, the set A is countable.

10.13 Define $f : \mathcal{G} \to \mathbf{Z} \times \mathbf{Z}$ by $f(a + bi) = (a, b)$. Then f is bijective and so $|\mathcal{G}| = |\mathbf{Z} \times \mathbf{Z}|$. Since $\mathbf{Z} \times \mathbf{Z}$ is denumerable, $\mathcal{G}$ is denumerable.

10.15 Note that S is an infinite subset of the set $\mathbf{N} \times \mathbf{N}$. The result follows by Theorem 10.4 and Result 10.6.

10.17 Since A is denumerable and B is an infinite subset of A, it follows that B is denumerable by Theorem 10.4.

10.19 Let $A = \{a_1, a_2, a_3, \ldots\}$ be a denumerable set and place the elements of A in a table, as shown below. For $i \in \mathbf{N}$, let A_i be the set of elements in the ith row of the table. In particular,
$A_1 = \{a_1, a_3, a_6, a_{10}, \ldots\}$,
$A_2 = \{a_2, a_5, a_9, a_{14}, \ldots\}$,
$A_3 = \{a_4, a_8, a_{13}, a_{19}, \ldots\}$.
Then each set A_i is a denumerable set and $\{A_1, A_2, A_3, \ldots\}$ is a partition of A into a denumerable number of denumerable sets.

A_1	a_1	a_3	a_6	a_{10}	a_{15}	a_{21}	$\cdots$
A_2	a_2	a_5	a_9	a_{14}	a_{20}	$\cdots$	
A_3	a_4	a_8	a_{13}	a_{19}	$\cdots$		
A_4	a_7	a_{12}	a_{18}	$\cdots$			
A_5	a_{11}	a_{17}	$\cdots$				
A_6	a_{16}	$\cdots$					
$\vdots$	$\vdots$	$\vdots$					

Section 10.3: Uncountable Sets

10.21 Since the set $\mathbf{C}$ of complex numbers contains $\mathbf{R}$ as a subset and $\mathbf{R}$ is uncountable, it follows by Theorem 10.10 that $\mathbf{C}$ is uncountable.

10.23 (a) **Proof** Assume that $f(a) = f(b)$, where $a, b \in (0, 1)$. Then $2a = 2b$ and so $a = b$. Hence f is one-to-one. For each $r \in (0, 2)$, $x = r/2 \in (0, 1)$ and $f(x) = r$. Therefore, f is onto. Thus f is a bijective function from $(0, 1)$ to $(0, 2)$. $\blacksquare$

(b) This follows from (a).

(c) Define the function $g : (0, 1) \to (a, b)$ by $g(x) = (b - a)x + a$. Then g is bijective and so $(0, 1)$ and (a, b) have the same cardinality.

10.25 (a) **Proof** Let $r \in \mathbf{R}$. We show that there is $x \in (-1, 1)$ such that $g(x) = r$. If $r = 0$, then $g(0) = 0$. Hence we may assume that $r \neq 0$. [Solving $g(x) = \frac{x}{1-x^2} = r$ for x, we find that $x = (-1 \pm \sqrt{1 + 4r^2})/2r$.] If $r > 0$, then $0 < -1 + \sqrt{1 + 4r^2} < 2r$ and so $(-1 + \sqrt{1 + 4r^2})/2r \in (0, 1)$. If $r < 0$, then $0 < -1 + \sqrt{1 + 4r^2} < -2r$ and so $-1 < (-1 + \sqrt{1 + 4r^2})/2r < 0$. Thus $(-1 + \sqrt{1 + 4r^2})/2r \in (-1, 0)$. Since $g((-1 + \sqrt{1 + 4r^2})/2r) = r$, the function g is onto. $\blacksquare$

(b) **Proof** Assume that $g(a) = g(b)$. Then $\frac{a}{1-a^2} = \frac{b}{1-b^2}$ and so $a(1 - b^2) = b(1 - a^2)$. Simplifying this equation and then factoring, we have $(a - b)(ab + 1) = 0$. In order for $ab = -1$, one of a and b is at least 1 or at most -1. In either case, this is impossible. Therefore, $ab \neq -1$ and so $a = b$. Hence f is one-to-one. $\blacksquare$

(c) Since f is one-to-one and onto, f is a bijective function, which implies that $|(-1, 1)| = |\mathbf{R}|$. Since $\mathbf{R}$ is uncountable, so is $(-1, 1)$.

Section 10.4: Comparing Cardinalities of Sets

10.27 Let $b \in B$. Then the function $f : A \to A \times B$ defined by $f(a) = (a, b)$ for each $a \in A$ is one-to-one. Thus $|A| \leq |A \times B|$.

10.29 The cardinalities of these sets are the same. Consider $f : [0, 1] \to [1, 3]$ defined by $f(x) = 2x + 1$ for all $x \in [0, 1]$.

10.31 The statement is true.

Proof Let A be a set. Then A is finite, denumerable, or uncountable. If A is finite, say $|A| = n \in \mathbf{Z}, n \geq 0$, then $|2^A| = 2^n$ and so 2^A is a finite set. If A is denumerable, then since $|2^A| > |A|$, 2^A is not denumerable. If A is uncountable, then since $|2^A| > |A|$, 2^A is also an uncountable set. $\blacksquare$

Section 10.5: The Schröder–Bernstein Theorem

10.33 **Proof** Since $(0, 1) \subseteq [0, 1]$, the function $i : (0, 1) \to [0, 1]$ defined by $i(x) = x$ is an injective function. The function $f : [0, 1] \to (0, 1)$ defined by $f(x) = \frac{1}{2}x + \frac{1}{4}$ is also injective. It then follows by the Schröder–Bernstein Theorem that $|(0, 1)| = |[0, 1]|$. $\blacksquare$

10.35 **Proof** Since the function $f : \mathbf{R}^* \to \mathbf{R}$ defined by $f(x) = x$ is one-to-one, it follows that $|\mathbf{R}^*| \leq |\mathbf{R}|$. By Corollary 10.15, the sets $(0, 1)$ and $\mathbf{R}$ are numerically equivalent and so there exists a bijective function $g : \mathbf{R} \to (0, 1)$. This function can be used to define a one-to-one function $h : \mathbf{R} \to \mathbf{R}^*$ where $h(x) = g(x)$ for each $x \in \mathbf{R}$. Thus $|\mathbf{R}| \leq |\mathbf{R}^*|$. By Theorem 10.20, $|\mathbf{R}^*| = |\mathbf{R}|$. $\blacksquare$

10.37 (a) **Proof** Assume that $f(m/n) = f(r/s)$. Since $f(m/n)$ has $2k$ digits for some integer $k \geq 2$, the integer $f(m/n)$ contains at least k consecutive 0's. Then the digits to the rightmost block of k consecutive 0's make up n while the digits to the left of this block make up m. Since $f(r/s) = f(m/n)$, it follows by the same argument that $r = m$ and $s = n$. So $m/n = r/s$. $\blacksquare$

(b) **Proof** The function $g : \mathbf{N} \to \mathbf{Q}^+$ defined by $g(n) = n$ is injective. Combining this with the function f in (a) gives us, by the Schröder–Bernstein Theorem, $|\mathbf{Q}^+| = |\mathbf{N}|$ and so $\mathbf{Q}^+$ is denumerable. $\blacksquare$

EXERCISES FOR CHAPTER 11

Section 11.1: Divisibility Properties of Integers

11.1 **Proof** Assume that $a \mid b$ and $c \mid d$. Then $b = ax$ and $d = cy$ for integers x and y. Then $ad + bc = a(cy) + (ax)c = ac(y + x)$. Since $y + x$ is an integer, $ac \mid (ad + bc)$. ∎

11.3 **Proof** Assume that $ac \mid bc$. Then $bc = (ac)x = c(ax)$ for some integer x. Since $c \neq 0$, we can divide by c, obtaining $b = ax$. So $a \mid b$. ∎

11.5 **Proof** Assume, to the contrary, that there exists a prime $n \geq 3$ that can be expressed as $k^3 + 1 \geq 3$ for some integer k. Since $n = k^3 + 1 = (k + 1)(k^2 - k + 1)$, it follows that $k + 1 = 1$ or $k^2 - k + 1 = 1$, which implies that $k = 0$ or $k = 1$. Thus $n = 1$ or $n = 2$, which is a contradiction. ∎

11.7 **Proof** We employ induction. For $n = 1$, we have $5^{2 \cdot 1} + 7 = 32$ and $8 \mid 32$. Thus the result is true for $n = 1$. Assume that

$$8 \mid \left(5^{2k} + 7\right)$$

for some positive integer k. We show that

$$8 \mid \left(5^{2(k+1)} + 7\right).$$

Since $8 \mid \left(5^{2k} + 7\right)$, it follows that $5^{2k} + 7 = 8a$ for some integer a and so $5^{2k} = 8a - 7$. Thus

$$5^{2(k+1)} + 7 = 5^2 \cdot 5^{2k} + 7 = 25(8a - 7) + 7$$
$$= 200a - 175 + 7 = 200a - 168 = 8(25a - 21).$$

Since $25a - 21$ is an integer, $8 \mid \left(5^{2(k+1)} + 7\right)$. The result then follows by the Principle of Mathematical Induction. ∎

11.9 Consider the n numbers $2 + (n + 1)!, 3 + (n + 1)!, \ldots, n + (n + 1)!, (n + 1) + (n + 1)!$. Observe for $2 \leq k \leq n + 1$ that k divides $k + (n + 1)!$. Thus these n numbers are composite.

11.11 **Proof** Since $d \mid a_i$ for $i = 1, 2, \ldots, n$, there exist integers d_i $(1 \leq i \leq n)$ such that $a_i = dd_i$. Thus

$$\sum_{i=1}^{n} a_i x_i = \sum_{i=1}^{n} (dd_i) x_i = d \sum_{i=1}^{n} d_i x_i.$$

Since $\sum_{i=1}^{n} d_i x_i$ is an integer, $d \mid \sum_{i=1}^{n} a_i x_i$. ∎

11.13 (a) Let $a_1, a_2, \ldots, a_k$ be the distinct positive integers that divide n. Then $a_1, a_2, \ldots, a_k, 2a_1, 2a_2, \ldots, 2a_k$ divide $2n$. So $2k$ integers divide $2n$. In addition to these $2k$ integers, $4a_1, 4a_2, \ldots, 4a_k$ also divide $4n$. So $3k$ integers divide $4n$.

(b) Let $a_1, a_2, \ldots, a_k$ be the distinct positive integers that divide n. Then $a_1, a_2, \ldots, a_k, 3a_1, 3a_2, \ldots, 3a_k$ divide $3n$. In addition to these $2k$ integers, $9a_1, 9a_2, \ldots, 9a_k$ also divide $9n$. So $3k$ integers divide $9n$.

(c) Let n be a positive integer and let p be a prime such that $p \nmid n$. If k integers divide n, how many integers divide pn? How many integers divide $p^a n$, where $a \in \mathbf{N}$? Answer: $(a + 1)k$.

Section 11.2: The Division Algorithm

11.15 (a) $125 = 17 \cdot 7 + 6$ $(q = 7, r = 6)$
(b) $125 = (-17) \cdot (-7) + 6$ $(q = -7, r = 6)$
(c) $96 = 8 \cdot 12 + 0$ $(q = 12, r = 0)$
(d) $96 = (-8) \cdot (-12) + 0$ $(q = -12, r = 0)$
(e) $-17 = 22 \cdot (-1) + 5$ $(q = -1, r = 5)$
(f) $-17 = (-22) \cdot 1 + 5$ $(q = 1, r = 5)$

(g) $0 = 15 \cdot 0 + 0 \;\; (q = 0, r = 0)$

(h) $0 = (-15) \cdot 0 + 0 \;\; (q = 0, r = 0)$

11.17 **(a)** ***Proof*** Let p be an odd prime. Then $p = 2a + 1$ for some integer a. We consider two cases, depending on whether a is even or a is odd.

Case 1. a is even. Then $a = 2k$, where $k \in \mathbf{Z}$. Thus $p = 2a + 1 = 2(2k) + 1 = 4k + 1$.

Case 2. a is odd. Then $a = 2k + 1$, where $k \in \mathbf{Z}$. Thus $p = 2a + 1 = 2(2k + 1) + 1 = 4k + 3$. $\blacksquare$

(b) ***Proof*** Let $p \geq 5$ be an odd prime. Then $p = 2a + 1$ for some integer a. We consider three cases, depending on whether $a = 3k$, $a = 3k + 1$, or $a = 3k + 2$ for some integer k.

Case 1. a = 3k. Then $p = 2a + 1 = 2(3k) + 1 = 6k + 1$.

Case 2. a = 3k + 1. Then $p = 2a + 1 = 2(3k + 1) + 1 = 6k + 3 = 3(2k + 1)$. Since $2k + 1$ is an integer, $3 \mid p$, which is impossible as $p \geq 5$ is a prime. Thus this case cannot occur.

Case 3. a = 3k + 2. Then $p = 2a + 1 = 2(3k + 2) + 1 = 6k + 5$. $\blacksquare$

11.19 **(a)** Observe that $n = 6q + 5 = 3(2q) + 3 + 2 = 3(2q + 1) + 2$. Letting $k = 2q + 1$, we see that $n = 3k + 2$.

(b) The converse is false. The integer $2 = 3 \cdot 0 + 2$ is of the form $3k + 2$, but 2 is not of the form $6q + 5$ since $6q + 5 = 2(3q + 2) + 1$ is always odd.

11.21 ***Proof*** Let a be an odd integer. Then $a = 2b + 1$ for some integer b. Thus

$$a^2 = (2b + 1)^2 = 4b^2 + 4b + 1 = 4(b^2 + b) + 1.$$

Since $k = b^2 + b$ is an integer, $a = 4k + 1$. $\blacksquare$

11.23 **Result** The square of an integer that is not a multiple of 5 is either of the form $5k + 1$ or $5k + 4$ for some integer k.

Proof Let n be an integer that is not a multiple of 5. Then $n = 5q + r$ for some integers q and r with $1 \leq r \leq 4$. We consider these four cases.

Case 1. $n = 5q + 1$. Then

$$n^2 = (5q + 1)^2 = 25q^2 + 10q + 1 = 5(5q^2 + 2q) + 1,$$

where $k = 5q^2 + 2q \in \mathbf{Z}$.

(The other three cases are handled similarly.) $\blacksquare$

11.25 ***Proof*** We proceed by induction. By Result 4.11, the statement is true for $n = 2$. Assume that if $a_1, a_2, \ldots, a_k$ are $k \geq 2$ integers such that $a_i \equiv 1 \pmod{3}$ for each i $(1 \leq i \leq k)$, then $a_1 a_2 \cdots a_k \equiv 1 \pmod{3}$. Now let $b_1, b_2, \ldots, b_{k+1}$ be $k + 1$ integers such that $b_i \equiv 1 \pmod{3}$ for all i $(1 \leq i \leq k + 1)$. We show that $b_1 b_2 \cdots b_{k+1} \equiv 1 \pmod{3}$. Let $b = b_1 b_2 \cdots b_k$. By the induction hypothesis, $b \equiv 1 \pmod{3}$. Since $b \equiv 1 \pmod{3}$ and $b_{k+1} \equiv 1 \pmod{3}$, it follows by Result 4.11 that $b_1 b_2 \cdots b_{k+1} = bb_{k+1} \equiv 1 \pmod{3}$. The result then follows by the Principle of Mathematical Induction. $\blacksquare$

11.27 The statement is true.

Proof Since a and b are odd integers, $a = 2x + 1$ and $b = 2y + 1$, where $x, y \in \mathbf{Z}$. If $4 \mid (a - b)$, then we have the desired result. Thus we may assume that $4 \nmid (a - b)$. Then $a - b = 2(x - y)$, where $x - y$ is an odd integer. Let $x - y = 2z + 1$, where $z \in \mathbf{Z}$. Thus $a = b + 2(x - y) = b + 4z + 2$ and

$$a + b = 2b + 4z + 2 = 2(2y + 1) + 4z + 2$$

$$= 4(y + z + 1).$$

Since $y + z + 1 \in \mathbf{Z}$, it follows that $4 \mid (a + b)$. $\blacksquare$

11.29 **(a)** ***Proof*** Let $x, y \in \mathbf{N}$ and let $a = 2x^2 + y^2$, $b = 2x^2$, $c = 2xy$, and $d = y^2$. Then
$a^2 = (2x^2 + y^2)^2 = 4x^4 + 4x^2 y^2 + y^4 = b^2 + c^2 + d^2$. $\blacksquare$

(b) ***Proof*** Let $x \in \mathbf{N}$ and let $a = 2x$ and $b = c = d = e = x$. Then
$a^2 = 4x^2 = x^2 + x^2 + x^2 + x^2 = b^2 + c^2 + d^2 + e^2$. $\blacksquare$

[Note: Observe that for $x \in \mathbf{N}$, $(4x)^2 = (x)^2 + (x)^2 + (x)^2 + (2x)^2 + (3x)^2$,
$(5x)^2 = (x)^2 + (x)^2 + (x)^2 + (2x)^2 + (3x)^2 + (3x)^2$ and
$(6x)^2 = (x)^2 + (x)^2 + (x)^2 + (2x)^2 + (2x)^2 + (3x)^2 + (4x)^2$.]

11.31   **(a)** S_2 is a set of positive odd integers.

   (b) $14 \in S_{13}$.

   (c) $16 \in S_3$.

   (d) The statement is true.

   Proof   Let $n \geq 2$ be an integer. Since $n \geq 2$, the integer n is either prime or composite. We consider these two cases.

   Case 1. n is prime. Consider the set $S = \{n^k + 1 : k \in \mathbf{N}\}$. Since n is a prime, n is the smallest positive integer such that when any element of S is divided by n, a remainder of 1 results. Since S is an infinite set and $S \subseteq S_n$, it follows that S_n is infinite.

   Case 2. n is composite. Let m be any integer that results in a remainder of 1 when divided by n. Then $m = nq + 1$ for some integer q. Since n is composite, it follows by Lemma 11.1 that there are integers a and b with $1 < a < n$ and $1 < b < n$ such that $n = ab$. Then $m = a(bq) + 1$. Hence when m is divided by a, a remainder 1 results and so $m \notin S_n$. Consequently, $S_n = \emptyset$.   ∎

Section 11.3: Greatest Common Divisors

11.33 $S = \{2 \cdot 3 \cdot 5,\ 2 \cdot 3 \cdot 7,\ 2 \cdot 5 \cdot 7,\ 3 \cdot 5 \cdot 7\}$.

11.35   ***Proof***   Let $\gcd(ka, kb) = e$. We show that $e = kd$. Since $d \mid a$ and $d \mid b$, it follows that $a = dr$ and $b = ds$ for integers r and s. Then $ka = (kd)r$ and $kb = (kd)s$. Since r and s are integers, $kd \mid ka$ and $kd \mid kb$. Because e is the greatest positive integer that divides both ka and kb, we have $kd \leq e$. Also, there exist integers x and y such that $d = ax + by$ and so $kd = (ka)x + (kb)y$. Since $e \mid ka$ and $e \mid kb$, it follows that $e \mid kd$ and so $e \leq kd$. Therefore, $e = kd$.   ∎

Section 11.4: The Euclidean Algorithm

11.37   **(a)** $\gcd(51, 288) = 3$.

   (b) $\gcd(357, 629) = 17$.

   (c) $\gcd(180, 252) = 36$.

11.39   Observe that if $d = as + bt$ and $k \in \mathbf{Z}$, then $d = a(s + kb) + b(t - ka)$.

11.41   Since $n \mid (7m + 3)$, it follows that $n \mid 5(7m + 3)$. Hence $n \mid [(35m + 26) - (35m + 15)]$. Thus $n = 11$.

11.43   ***Proof***   Since $a \equiv b \pmod{m}$ and $a \equiv c \pmod{n}$, it follows that $a = b + mx$ and $a = c + ny$ for some integers x and y. Hence $b + mx = c + ny$ and so $b - c = ny - mx$. Since $d = \gcd(m, n)$, it follows that $d \mid m$ and $d \mid n$. Thus $m = dr$ and $n = ds$, where $r, s \in \mathbf{Z}$. Therefore,

$$b - c = ny - mx = (ds)y - (dr)x = d(sy - rx).$$

   Since $sy - rx$ is an integer, $d \mid (b - c)$ and so $b \equiv c \pmod{d}$.   ∎

11.45   Since $\gcd(a, b) = \gcd(r_{i-1}, r_i)$ and r_i is a prime number, $\gcd(a, b) \mid r_i$ and so $\gcd(a, b)$ is either r_i or 1.   ∎

Section 11.5: Relatively Prime Integers

11.47   ***Proof***   Assume, to the contrary, that $\sqrt{3}$ is rational. Then $\sqrt{3} = a/b$, where a and b are nonzero integers. We may assume that a/b has been reduced to lowest terms. Thus $a^2 = 3b^2$. Since b^2 is an integer, $3 \mid a^2$. It then follows by Corollary 11.14 that $3 \mid a$. Thus $a = 3x$ for some integer x. So $a^2 = (3x)^2 = 3(3x^2) = 3b^2$ and so $3x^2 = b^2$. Since x^2 is an integer, $3 \mid b^2$ and so $3 \mid b$ by Corollary 11.14. However, 3 is a common factor of a and b, contradicting the fact that a/b has been reduced to lowest terms.   ∎

11.49   ***Proof***   Assume, to the contrary, that $p^{1/n}$ is rational. Then $p^{1/n} = a/b$, where a and b are nonzero integers. We may assume that a/b has been reduced to lowest terms. Thus $a^n/b^n = p$ and so $a^n = pb^n$. Since b^n is an integer, $p \mid a^n$. Since p is a prime, it follows by Corollary 11.15 that $p \mid a$. Since $p \mid a$, it follows that $a = pc$ for some integer c. Thus $a^n = (pc)^n = p^n c^n = pb^n$. Hence $b^n = p^{n-1}c^n = p(p^{n-2}c^n)$. Since $n \geq 2$, we have that $p^{n-2}c^n$ is an integer and so $p \mid b^n$. By Corollary 11.15, $p \mid b$. This contradicts our assumption that a/b has been reduced to lowest terms.   ∎

11.51 **(a)** ***Proof*** Let a and b be two consecutive odd positive integers. Then $a = 2k + 1$ and $b = 2k + 3$ for some integer k. Since

$$1 = (2k + 1) \cdot (k + 1) + (2k + 3) \cdot (-k)$$

is a linear combination of $2k + 1$ and $2k + 3$, the integers $2k + 1$ and $2k + 3$ are relatively prime. ∎

(b) One possibility: Every two consecutive integers k and $k + 1$ are relatively prime since 1 can be expressed as a linear combination of k and $k + 1$, namely, $1 = (k + 1) \cdot 1 + k \cdot (-1)$. In part (a), we saw that every two consecutive odd positive integers $a = 2k + 1$ and $b = 2k + 3$ are relatively prime by writing $1 = ax + by$, where $x = k + 1$ and $y = -k$. (Note the values of x and y.) The integers $a = 3k + 2$ and $b = 3k + 5$ are relatively prime as well since we can write $1 = ax + by$, where $x = 2k + 3$ and $y = -(2k + 1)$. (Again, note the values of x and y.) More generally, we have:

Result For every positive integer n and every integer k, the integers $a = nk + (n - 1)$ and $b = nk + (2n - 1)$ are relatively prime.

Proof Observe that $1 = ax + by$, where $x = (n - 1)k + (2n - 3)$ and $y = -[(n - 1)k + (n - 2)]$. ∎

11.53 Let p and q be primes with $p \geq q \geq 5$. By Exercise 11.17(b), $p = 6a \pm 1$ and $q = 6b \pm 1$ for some integers a and b. Hence

$$p^2 - q^2 = (36a^2 \pm 12a + 1) - (36b^2 \pm 12b + 1) = 12(3a^2 \pm a) - 12(3b^2 \pm b).$$

By Theorem 3.12, a^2 and a (and b^2 and b) are of the same parity. Thus $3a^2 \pm a$ and $3b^2 \pm b$ are both even and we can write $p^2 - q^2 = 24k$ for some integer k.

11.55 ***Proof*** Assume that $a \equiv b \pmod{m}$ and $a \equiv b \pmod{n}$, where $\gcd(m, n) = 1$. Thus $m \mid (a - b)$ and $n \mid (a - b)$. By Theorem 11.16, $mn \mid (a - b)$. Hence $a \equiv b \pmod{mn}$. ∎

11.57 Claim: $\gcd(x, y) = d$ if and only if $d = 1$.

Proof Let $d = \gcd(a, b)$. Then there exist integers x and y such that $d = ax + by$. We show that $\gcd(x, y) = 1$. Suppose that $\gcd(x, y) = e$. Since $d \mid a$ and $d \mid b$, there exist integers r and s such that $a = dr$ and $b = ds$. Because $e \mid x$ and $e \mid y$, there exist integers w and z such that $x = ew$ and $y = ez$. Therefore, $d = ax + by = (dr)(ew) + (ds)(ez) = de(rw) + de(sz)$. So $1 = e(rw + sz)$. Since $rw + sz$ is an integer, $e = 1$. ∎

11.59 **(a)** ***Proof*** Assume that $a \mid c$ and $b \mid c$, where $\gcd(a, b) = d$. Then there exist integers x and y such that $d = ax + by$. Hence $cd = acx + bcy$. Because $a \mid c$ and $b \mid c$, there exist integers r and s such that $c = ar$ and $c = bs$. Hence $cd = a(bs)x + b(ar)y = ab(sx) + ab(ry) = ab(sx + ry)$. Since $sx + ry$ is an integer, $ab \mid cd$. ∎

(b) Let $a, b, c \in \mathbf{Z}$ such that $a \mid c$ and $b \mid c$ and a and b are relatively prime. Then $\gcd(a, b) = 1$. Letting $d = 1$ in (a), we obtain Theorem 11.16.

11.61 **(a)** $m = 5, n = 6$.

(b) Consider the pairs $\{m, n\} = \{4, 9\}, \{4, 6\}, \{9, 15\}$.

Section 11.6: The Fundamental Theorem of Arithmetic

11.63 **(a)** $4725 = 3^2 \cdot 5^2 \cdot 7$

(b) $9702 = 2 \cdot 3^2 \cdot 7^2 \cdot 11$

(c) $180625 = 5^4 \cdot 17^2$.

11.65 **(a)** $4278 = 2 \cdot 3 \cdot 23 \cdot 31$ and $71929 = 11 \cdot 13 \cdot 503$.

(b) $\gcd(4278, 71929) = 1$

11.67 ***Proof*** Assume, to the contrary, that the number of primes is finite. Let $P = \{p_1, p_2, \ldots, p_n\}$ be the set of all primes, where $p_1 < p_2 < \cdots < p_n$. Let $m = p_n! + 1$. Then $m \geq p_n + 1$ and so m is not a prime. Since m has a prime factor and every prime belongs to P, there is a prime p_i ($1 \leq i \leq n$) such that $p_i \mid m$. Hence $m = p_i k$ for some integer k. Since p_i is a factor of $p_n!$ and $1 = m - p_n!$, it follows that $p_i \mid 1$, which is a contradiction. ∎

11.69 ***Proof*** Let $d = p_1^{c_1} p_2^{c_2} \cdots p_r^{c_r}$. We show that $\gcd(m, n) = d$. Since $c_i \leq a_i$ and $c_i \leq b_i$ for each i ($1 \leq i \leq r$), it follows that $d \mid m$ and $d \mid n$. We claim that $\gcd(m, n)$ can be expressed as $p_1^{k_1} p_2^{k_2} \cdots p_r^{k_r}$ for nonnegative integers k_i ($1 \leq i \leq r$). Suppose that some prime p distinct from $p_1, p_2, \ldots, p_r$ divides d. Then $p \mid m$ and

$p \mid n$, which is impossible. Thus, as claimed, $\gcd(m, n)$ can be expressed as $p_1^{k_1} p_2^{k_2} \cdots p_r^{k_r}$ for nonnegative integers k_i $(1 \leq i \leq r)$. If $d \neq p_1^{k_1} p_2^{k_2} \cdots p_r^{k_r}$, then $p_1^{k_1} p_2^{k_2} \cdots p_r^{k_r} > p_1^{c_1} p_2^{c_2} \cdots p_r^{c_r}$, which implies that $p_s^{k_s} > p_s^{c_s}$ for some s $(1 \leq s \leq r)$. Then $k_s > c_s$ and so $p_s^{k_s} \nmid m$ or $p_s^{k_s} \nmid n$, a contradiction. ∎

Section 11.7: Concepts Involving Sums of Divisors

11.71 **Proof** Let k be the maximum number of distinct positive integers whose sum is n. Since the k smallest distinct positive integers are $1, 2, \ldots, k$, it follows that $1 + 2 + \cdots + k \leq n$. Therefore, k is the largest integer such that $k(k + 1)/2 \leq n$ and so $k^2 + k - 2n \leq 0$. Hence k is the largest integer that is at most $(\sqrt{1 + 8n} - 1)/2$. That is, $k = \lfloor (\sqrt{1 + 8n} - 1)/2 \rfloor$. ∎

EXERCISES FOR CHAPTER 12

Section 12.1: Limits of Sequences

12.1 Since $\cos n\pi = -1$ when n is odd and $\cos n\pi = 1$ when n is even, the terms of the sequence $\{(-1)^n\}$ are exactly the same as $\{\cos n\pi\}$.

12.3 **Proof** Let $\epsilon > 0$ be given. Choose $N = \lceil 1/2\epsilon \rceil$ and let $n > N$. Thus $n > 1/2\epsilon$ and so $\left| \frac{1}{2n} - 0 \right| = \frac{1}{2n} < \epsilon$. ∎

12.5 **Proof** Let $\epsilon > 0$ be given. Choose $N = \max \left(1, \lceil \log_2 \left(\frac{1}{\epsilon} \right) \rceil \right)$ and let $n > N$. Thus $n > \log_2 \left(\frac{1}{\epsilon} \right)$ and so $2^n > 1/\epsilon$ and $1/2^n < \epsilon$. Therefore, $\left| \left(1 + \frac{1}{2^n}\right) - 1 \right| = \frac{1}{2^n} < \epsilon$. ∎

12.7 There exists a real number $\epsilon > 0$ such that for each positive integer N, there exists an integer $n > N$ such that $|a_n - L| \geq \epsilon$.

For each real number L, there exists $\epsilon > 0$ such that for each positive integer N, there exists $n > N$ such that $|a_n - L| \geq \epsilon$.

Let $P(L, \epsilon, n) : |a_n - L| \geq \epsilon$. $\exists \epsilon \in \mathbf{R}^+, \forall N \in \mathbf{N}, \exists n \in \mathbf{N}, n > N, P(L, \epsilon, n)$.

12.9 **Proof** Let M be a positive number. Choose $N = \left\lceil \sqrt[3]{M} \right\rceil$ and let n be any integer such that $n > N$. Hence $n > \sqrt[3]{M}$ and so $n^3 > M$. Thus $\dfrac{n^5 + 2n}{n^2} = n^3 + \dfrac{2}{n} > n^3 > M$. ∎

12.11 **Proof** Let $\epsilon > 0$ be given. Since $\lim_{n \to \infty} s_n = L$, there is a positive integer N such that $|s_n - L| < \epsilon$ for each integer $n > N$. Since $n^2 \geq n$ for each $n \in \mathbf{N}$, it follows that $|s_{n^2} - L| < \epsilon$ for all $n^2 > N$. ∎

Section 12.2: Infinite Series

12.13 Let $s_n = \sum_{i=1}^{n} \frac{1}{2^i}$ for each integer $n \geq 1$.

(a) $s_1 = \frac{1}{2}$, $s_2 = \frac{1}{2} + \frac{1}{2^2} = \frac{1}{2} + \frac{1}{4} = \frac{3}{4}$, $s_3 = \frac{1}{2} + \frac{1}{2^2} + \frac{1}{2^3} = \frac{1}{2} + \frac{1}{4} + \frac{1}{8} = \frac{7}{8}$.

Conjecture $s_n = 1 - \frac{1}{2^n}$ for all $n \in \mathbf{N}$.

(b) **Proof** We proceed by induction. Since $s_1 = \frac{1}{2} = 1 - \frac{1}{2^1}$, the formula s_n holds for $n = 1$. Thus the statement is true for $n = 1$. Assume that $s_k = 1 - \frac{1}{2^k}$ for a positive integer k. We show that $s_{k+1} = 1 - \frac{1}{2^{k+1}}$. Observe that

$$\sum_{i=1}^{k+1} \frac{1}{2^i} = \left(\sum_{i=1}^{k} \frac{1}{2^i} \right) + \frac{1}{2^{k+1}} = 1 - \frac{1}{2^k} + \frac{1}{2^{k+1}}$$

$$= 1 - \left(\frac{1}{2^k} - \frac{1}{2^{k+1}} \right) = 1 - \frac{2 - 1}{2^{k+1}} = 1 - \frac{1}{2^{k+1}}.$$

By the Principle of Mathematical Induction, $s_n = 1 - \frac{1}{2^n}$ for all $n \in \mathbf{N}$. ∎

(c) The proof that $\lim_{n \to \infty} (1 - \frac{1}{2^n}) = 1$ is similar to the one in Exercise 12.5.

12.15 **Proof** Let M be a positive integer. By Result 12.12, there exists a positive integer N such that if $n > N$, then $1 + \frac{1}{2} + \frac{1}{3} + \cdots + \frac{1}{n} > M$. Since $n^2 + 3n \geq n^2 + 2n + 1$ for every positive integer n, it follows that

$\frac{n+3}{(n+1)^2} \geq \frac{1}{n}$. Hence

$$\frac{4}{2^2} + \frac{5}{3^2} + \cdots + \frac{n+3}{(n+1)^2} \geq 1 + \frac{1}{2} + \frac{1}{3} + \cdots + \frac{1}{n} > M$$

and so $\sum_{i=1}^{\infty} \frac{k+3}{(k+1)^2}$ diverges to infinity.

12.17 (a) Since $s_n = \frac{3n}{4n+2}$, $s_{n-1} = \frac{3n-3}{4n-2}$ and so

$$a_n = s_n - s_{n-1} = \frac{3n}{4n+2} - \frac{3n-3}{4n-2} = \frac{6}{16n^2-4} = \frac{3}{8n^2-2}.$$

Therefore, the series is $\sum_{k=1}^{\infty} \frac{3}{8k^2-2}$.

(b) The sum s of the series is $s = \lim_{n \to \infty} \frac{3n}{4n+2}$. We claim that $\lim_{n \to \infty} \frac{3n}{4n+2} = \frac{3}{4}$.

Proof Let $\epsilon > 0$ be given. Let $N = \max\{1, \lceil \frac{3}{8\epsilon} - \frac{1}{2} \rceil\}$ and let $n > N$. Then $n > \frac{3}{8\epsilon} - \frac{1}{2}$ and so $\frac{3}{8n+4} < \epsilon$. Thus

$$\left| \frac{3n}{4n+2} - \frac{3}{4} \right| = \left| \frac{-6}{16n+8} \right| = \frac{3}{8n+4} < \epsilon,$$

completing the proof.

Section 12.3: Limits of Functions

12.19 Proof Let $\epsilon > 0$ be given. Choose $\delta = \epsilon/3$. Let $x \in \mathbf{R}$ such that $0 < |x+1| < \delta = \epsilon/3$. Then

$$|(3x-5)-(-8)| = |3x+3| = 3|x+1| < 3\delta = 3(\epsilon/3) = \epsilon,$$

as desired.

12.21 Proof Let $\epsilon > 0$ be given and choose $\delta = \min(1, \epsilon/19)$. Let $x \in \mathbf{R}$ such that $0 < |x-2| < \delta = \min(1, \epsilon/19)$. Since $|x-2| < 1$, it follows that $-1 < x-2 < 1$ and so $1 < x < 3$. Thus $|x^2+2x+4| < 19$. Because $|x-2| < \epsilon/19$, it follows that $|x^3 - 8| = |x-2||x^2+2x+4| < |x-2| \cdot 19 < (\epsilon/19) \cdot 19 = \epsilon$.

12.23 Proof Let $\epsilon > 0$ be given. Choose $\delta = \min(1, 33\epsilon)$. Let $x \in \mathbf{R}$ such that $0 < |x-3| < \delta$. Since $|x-3| < \delta \leq 1$, it follow that $2 < x < 4$. Thus $11 < 4x+3 < 19$ and so $|4x+3| > 11$. Hence $\frac{1}{|4x+3|} < \frac{1}{11}$. Therefore,

$$\left| \frac{3x+1}{4x+3} - \frac{2}{3} \right| = \left| \frac{x-3}{12x+9} \right| = \frac{|x-3|}{3|4x+3|} < \frac{|x-3|}{3\cdot 11} < \frac{\delta}{33} \leq \frac{1}{33}(33\epsilon) = \epsilon,$$

as desired.

12.25 Proof Assume, to the contrary, that $\lim_{x \to 0} \frac{1}{x^2}$ exists. Then there exists a real number L such that $\lim_{x \to 0} \frac{1}{x^2} = L$.

Let $\epsilon = 1$. There exists $\delta > 0$ such that if $0 < |x| < \delta$, then $\left| \frac{1}{x^2} - L \right| < \epsilon = 1$. Let n be an integer such that $n > \lceil 1/\delta^2 \rceil$. So $n > 1/\delta^2$ and $\sqrt{n} > 1/\delta$. Let $x = 1/\sqrt{n} < \delta$. Then

$$\left| \frac{1}{x^2} - L \right| = |n - L| = |L - n| < 1$$

and so $-1 < L - n < 1$. Thus $n - 1 < L < n + 1$. Now, let $y = \frac{1}{\sqrt{n+2}} < x < \delta$. Then

$$\left| \frac{1}{y^2} - L \right| = |L - (n+2)| < 1.$$

Hence $n+1 < L < n+3$. Therefore, $n+1 < L < n+1$, which is a contradiction.

12.27 (a) Proof Let $\epsilon > 0$ be given. Since g is bounded, there exists a positive real number B such that $|g(x)| < B$ for each $x \in \mathbf{R}$. Then $\epsilon/B > 0$. Since $\lim_{x \to a} f(x) = 0$, there exists $\delta > 0$ such that if $0 < |x-a| < \delta$, then $|f(x) - 0| < \frac{\epsilon}{B}$ and so $|f(x)| < \epsilon/B$. Therefore, $|f(x)g(x) - 0| = |f(x)||g(x)| < \frac{\epsilon}{B} \cdot B = \epsilon$.

(b) Since $\lim_{x\to 0} x^2 = 0$ and $|\sin\left(\frac{1}{x^2}\right)| \le 1$ for all $x \in \mathbf{R} - \{0\}$, it follows from (a) that

$$\lim_{x\to 0} x^2 \sin\left(\frac{1}{x^2}\right) = 0.$$

12.29 (a) *Proof* Let $\epsilon > 0$ be given. Since $\lim_{x\to 0} f(x) = L$, there exists $\delta > 0$ such that if $0 < |x - 0| < \delta$, then $|f(x) - L| < \epsilon$. Let $y = x - c$. Because $\lim_{y\to 0} f(y) = L$, given $\epsilon > 0$, there exists $\delta > 0$ such that if $0 < |y - 0| < \delta$, then $|f(y) - L| < \epsilon$. Thus if $0 < |x - c| < \delta$, then $|f(x - c) - L| < \epsilon$. Therefore, $\lim_{x\to c} f(x - c) = L$. $\blacksquare$

(b) *Proof* By assumption, $f(x) = f(x - c + c) = f(x - c) + f(c)$. Since $\lim_{x\to 0} f(x) = L$, it follows by (a) that $\lim_{x-c\to 0} f(x - c) = \lim_{x\to c} f(x - c) = L$. Thus

$$\lim_{x\to c} f(x) = \lim_{x\to c}[f(x - c) + f(c)] = \lim_{x\to c} f(x - c) + \lim_{x\to c} f(c) = L + f(c)$$

and so $\lim_{x\to c} f(x)$ exists for each $x \in \mathbf{R}$. $\blacksquare$

Section 12.4: Fundamental Properties of Limits of Functions

12.31 (a) $\lim_{x\to 1}(x^3 - 2x^2 - 5x + 8) = 1^3 - 2(1)^2 - 5 \cdot 1 + 8 = 2.$
(b) $\lim_{x\to 1}(4x + 7)(3x^2 - 2) = (4 \cdot 1 + 7)(3 \cdot (1)^2 - 2) = 11 \cdot 1 = 11.$
(c) $\lim_{x\to 2} \frac{2x^2-1}{3x^3+1} = \frac{2 \cdot 2^2-1}{3 \cdot 2^3+1} = \frac{7}{25}.$

12.33 *Proof* First, by Theorem 12.28, $\lim_{x\to a} c_0 = c_0$. For $1 \le k \le n$, it follows by Theorems 12.25, 12.28 and 12.30 that $\lim_{x\to a} (c_k x^k) = (\lim_{x\to a} c_k)(\lim_{x\to a} x^k) = c_k a^k$. By Exercise 12.32, $\lim_{x\to a}(c_n x^n + c_{n-1} x^{n-1} + \cdots + c_1 x + c_0) = c_n a^n + c_{n-1} a^{n-1} + \cdots + c_1 a + c_0 = p(a)$. $\blacksquare$

Section 12.5: Continuity

12.35 Observe that f is not defined at $x = 2$ and

$$\lim_{x\to 2} \frac{x^2 - 4}{x^3 - 2x^2} = 1.$$

(Use an argument similar to that in Result 12.15.) Thus if we define $f(2) = 1$, then $\lim_{x\to 2} f(x) = 1 = f(2)$ and so f is continuous at 2.

12.37 *Proof* Let a be a real number that is not an integer. Then $n < a < n + 1$ for some $n \in \mathbf{Z}$ and $f(a) = \lceil a \rceil = n + 1$. We show that $\lim_{x\to a} f(x) = f(a) = n + 1$. Let $\epsilon > 0$ be given and choose

$$\delta = \min(a - n, (n + 1) - a).$$

Let $x \in \mathbf{R}$ such that $0 < |x - a| < \delta$. Thus $n \le a - \delta < x < a + \delta \le n + 1$ and so $f(x) = \lceil x \rceil = n + 1$. Therefore,

$$|f(x) - f(a)| = |(n + 1) - (n + 1)| = 0 < \epsilon,$$

completing the proof. $\blacksquare$

12.39 We show that $\lim_{x\to 10} \sqrt{x - 1} = f(10) = 3$.
Proof Let $\epsilon > 0$ be given and choose $\delta = \min(1, 5\epsilon)$. Let $x \in \mathbf{R}$ such that $0 < |x - 10| < \delta$. Since $|x - 10| < 1$, it follows that $9 < x < 11$ and so $\sqrt{x - 1} + 3 > 5$. Therefore, $1/(\sqrt{x - 1} + 3) < 1/5$. Hence

$$|\sqrt{x - 1} - 3| = \left|\frac{(\sqrt{x - 1} - 3)(\sqrt{x - 1} + 3)}{\sqrt{x - 1} + 3}\right| = \frac{|x - 10|}{\sqrt{x - 1} + 3} < \frac{1}{5}(5\epsilon) = \epsilon,$$

completing the proof. $\blacksquare$

Section 12.6: Differentiability

12.41 $f'(3) = 6.$

Proof   Let $\epsilon > 0$ be given and choose $\delta = \epsilon$. Let $x \in \mathbf{R}$ such that $0 < |x - 3| < \delta = \epsilon$. Then

$$\left| \frac{f(x) - f(3)}{x - 3} - 6 \right| = \left| \frac{x^2 - 9}{x - 3} - 6 \right| = \left| \frac{(x - 3)(x + 3)}{x - 3} - 6 \right|$$

$$= |(x + 3) - 6| = |x - 3| < \epsilon.$$

Thus $f'(3) = 6$.

12.43   Claim: $f'(a) = 3a^2$.

Proof   Let $\epsilon > 0$ be given and choose $\delta = \min\left\{ \frac{\epsilon}{1+3a}, 1 \right\}$. Let $x \in \mathbf{R}$ such that $0 < |x - a| < \delta$. Then

$$|x + 2a| = |(x - a) + 3a| \le |x - a| + 3|a| < 1 + 3a.$$

Observe that

$$\left| \frac{f(x) - f(a)}{x - a} - 3a^2 \right| = \left| \frac{(x - a)(x^2 + ax + a^2)}{x - a} - 3a^2 \right| = |x^2 + ax + a^2 - 3a^2|$$

$$= |x^2 + ax - 2a^2| = |x - a||x + 2a| < \delta(1 + 3a)$$

$$\le \left(\frac{\epsilon}{1 + 3a} \right)(1 + 3a) = \epsilon.$$

Thus $f'(a) = 3a^2$.   ∎

EXERCISES FOR CHAPTER 13

Section 13.1: Binary Operations

13.1   **(a)** $x * (y * z) = x * x = y$ and $(x * y) * z = z * z = y$. So $x * (y * z) = (x * y) * z$.

  (b) $x * (x * x) = x * y = z$ and $(x * x) * x = y * x = y$.

  (c) $y * (y * y) = y * x = y$ and $(y * y) * y = x * y = z$.

  (d) The binary operation $*$ is neither associative nor commutative.

13.3   **(a)** Let $A_1, A_2 \in T$. Then $A_1 = \begin{bmatrix} a_1 & -b_1 \\ b_1 & a_1 \end{bmatrix}$ and $A_2 = \begin{bmatrix} a_2 & -b_2 \\ b_2 & a_2 \end{bmatrix}$ for some $a_1, b_1, a_2, b_2 \in \mathbf{R}$. Then

  $A_1 + A_2 = \begin{bmatrix} a_1 + a_2 & -(b_1 + b_2) \\ b_1 + b_2 & a_1 + a_2 \end{bmatrix}$. Since $A_1 + A_2 \in T$, it follows that T is closed under addition.

  (b) Since $A_1 A_2 = \begin{bmatrix} a_1 & -b_1 \\ b_1 & a_1 \end{bmatrix}\begin{bmatrix} a_2 & -b_2 \\ b_2 & a_2 \end{bmatrix} = \begin{bmatrix} a_1a_2 - b_1b_2 & -(a_1b_2 + b_1a_2) \\ a_1b_2 + b_1a_2 & a_1a_2 - b_1b_2 \end{bmatrix} \in T$, it follows that T is

  closed under matrix multiplication.

13.5   **Proof**   Let $a, b \in T$. Thus $a * a = a$ and $b * b = b$. Hence

$$(a * b) * (a * b) = (a * b) * (b * a) = a * (b * (b * a)) = a * ((b * b) * a)$$

$$= a * (b * a) = a * (a * b) = (a * a) * b = a * b,$$

as desired.   ∎

13.7   All four properties G1–G4 are satisfied. In this case, $[-1] = [n - 1]$ is an identity element and $[-a - 2]$ is an inverse for $[a]$.

Section 13.2: Groups

13.9   See the table.

$*$	a	b	c	d
a	d	c	b	a
b	c	d	a	b
c	b	a	d	c
d	a	b	c	d

13.11 **(a)** First, observe that $[1][b] = [b]$ for each $[b] \in \mathbf{Z}_6^*$ and that $[5][-b] = [b]$ for each $[b] \in \mathbf{Z}_6^*$. There is no $[x] \in \mathbf{Z}_6^*$ such that $[2][x] = [1]$ or that $[3][x] = [1]$ or that $[4][x] = [1]$.

(b) Because of Theorem 13.5, the answer to (a) is not surprising.

13.13 ***Proof***   First, we show that if g is any element of G such that $g * g = g$, then $g = e$. Suppose then that $g * g = g$. By (ii), there exists $g' \in G$ such that $g * g' = e$. Thus

$$e = g * g' = (g * g) * g' = g * (g * g') = g * e = g$$

by (i). We now show that $g' * g = e$. Since

$$(g' * g) * (g' * g) = ((g' * g) * g') * g = (g' * (g * g')) * g$$
$$= (g' * e) * g = g' * g,$$

it follows that $g' * g = e$.

Next, we show that $e * g = g$ for every $g \in G$. Let $g \in G$. Then, as we just showed, there is $g' \in G$ such that $g' * g = e$. Thus $e * g = (g * g') * g = g * (g' * g) = g * e = g$ by (i). ∎

Section 13.3: Permutation Groups

13.15 Let $a, b, c \in A$. Let $\alpha, \beta \in S_A$ such that $\alpha(a) = b$, $\alpha(b) = a$ and $\alpha(x) = x$ for $x \neq a, b$; while $\beta(b) = c$, $\beta(c) = b$ and $\beta(x) = x$ for $x \neq b, c$. Then $(\alpha \circ \beta)(b) = \alpha(\beta(b)) = \alpha(c) = c$; while $(\beta \circ \alpha)(b) = \beta(\alpha(b)) = \beta(a) = a$. Thus $\alpha \circ \beta \neq \beta \circ \alpha$.

13.17 $x^2 = \alpha_1$ for all $x \in \{\alpha_1, \alpha_2, \alpha_3, \alpha_4\}$, $x^3 = \alpha_1$ for all $x \in \{\alpha_1, \alpha_5, \alpha_6\}$.

13.19 Consider the operation table shown below. Thus $\circ$ is a binary operation on G. Since composition of permutations on A is associative, property G1 is satisfied. In addition, β_1 is an identity and the elements $\beta_1, \beta_2, \beta_3, \beta_4, \beta_5, \beta_6$ are inverses of $\beta_1, \beta_3, \beta_2, \beta_4, \beta_6, \beta_5$, respectively. Therefore, properties G2 and G3 are satisfied and so $(G, \circ)$ is a group.

$\circ$	β_1	β_2	β_3	β_4	β_5	β_6
β_1	β_1	β_2	β_3	β_4	β_5	β_6
β_2	β_2	β_3	β_1	β_5	β_6	β_4
β_3	β_3	β_1	β_2	β_6	β_4	β_5
β_4	β_4	β_5	β_6	β_1	β_2	β_3
β_5	β_5	β_6	β_4	β_2	β_3	β_1
β_6	β_6	β_4	β_5	β_3	β_1	β_2

Section 13.4: Fundamental Properties of Groups

13.21 ***Proof***   Assume that $b * a = c * a$. Let s be an inverse for a. Then $(b * a) * s = (c * a) * s$. Thus
$$b = b * e = b * (a * s) = (b * a) * s = (c * a) * s = c * (a * s) = c * e = c$$
and so $b = c$. ∎

13.23 **(a)** $x = a^{-1} * c * b^{-1}$. (If x_1 and x_2 are two solutions, then $a * x_1 * b = a * x_2 * b = c$. An application of the Left and Right Cancellation Laws yields $x_1 = x_2$.)

(b) $x = b^{-1} * a^{-1} * c$. (Verifying the uniqueness is similar to (a).)

13.25 ***Proof***   Assume that G is abelian. Let $a, b \in G$. By Theorem 13.11, $(ab)^{-1} = b^{-1}a^{-1}$. Since G is abelian, $b^{-1}a^{-1} = a^{-1}b^{-1}$. For the converse, assume that G is a group such that $b^{-1}a^{-1} = a^{-1}b^{-1}$ for every pair a, b of elements of G. We show that G is abelian. Let $x, y \in G$. Then $x^{-1}, y^{-1} \in G$. By assumption, $\left(x^{-1}\right)^{-1} \left(y^{-1}\right)^{-1} = \left(y^{-1}\right)^{-1} \left(x^{-1}\right)^{-1}$ and so $xy = yx$. Thus G is abelian. ∎

13.27 Claim: e' is the identity of G.

Proof   Since $e'b = b$ and $eb = b$, it follows that $e'b = eb$. Applying the Right Cancellation Law in Theorem 13.7, we have $e' = e$. ∎

13.29 Since G has even order, $G - \{e\}$ has an odd number of elements. Consider those elements $g \in G$ for which $g \neq g^{-1}$ and let $S_g = \{g, g^{-1}\}$. Hence $S_g = S_{g^{-1}}$. If we take the union of all such sets S_g for which $g \neq g^{-1}$, then $\cup S_g \subset G - \{e\}$. Hence there exists an element $h \in G - \{e\}$ such that $h \notin \cup S_g$ and so $h = h^{-1}$. Thus $h^2 = e$.

Section 13.5: Subgroups

13.31 **Proof** Let $a, b \in n\mathbf{Z}$. Then $a = nk$ and $b = n\ell$ for $k, \ell \in \mathbf{Z}$. Since $a + b = nk + n\ell = n(k + \ell)$ and $k + \ell \in \mathbf{Z}$, it follows that $a + b \in n\mathbf{Z}$. The identity of $n\mathbf{Z}$ is $0 = 0n$. For $a = nk$, the integer $-a = (-k)n$ is the inverse of a since $a + (-a) = kn + (-k)n = 0$. Since $-k \in \mathbf{Z}$, $-a \in n\mathbf{Z}$. By the Subgroup Test, $(n\mathbf{Z}, +)$ is a subgroup of $(\mathbf{Z}, +)$.   ∎

13.33 **(a)** The statement is true.

   Proof Since H and K are subgroups of G, it follows that $e \in H$ and $e \in K$. So $e \in H \cap K$ and $H \cap K \neq \emptyset$. Let $a, b \in H \cap K$. Then $a, b \in H$ and $a, b \in K$. Since H and K are subgroups of G, it follows that $ab \in H$ and $ab \in K$. So $ab \in H \cap K$. Let $a \in H \cap K$. It remains to show that $a^{-1} \in H \cap K$. Since $a \in H$, $a \in K$ and H and K are subgroups of G, it follows that $a^{-1} \in H$ and $a^{-1} \in K$. So $a^{-1} \in H \cap K$. By the Subgroup Test, $H \cap K$ is a subgroup of G.   ∎

   (b) The statement is false. For example, $H = \{[0], [3]\}$ and $K = \{[0], [2], [4]\}$ are subgroups of $(\mathbf{Z}_6, +)$, but $H \cup K$ is not a subgroup of $(\mathbf{Z}_6, +)$.

13.35 **Proof** Since $\sqrt{3} \in H$, it follows that $H \neq \emptyset$. First, we show that H is closed under multiplication. Let $r = a + b\sqrt{3}$ and $s = c + d\sqrt{3}$ be elements of H, where at least one of a and b is nonzero and at least one of c and d is nonzero. Therefore, $r \neq 0$ and $s \neq 0$. Hence $rs \neq 0$ and

$$rs = (ac + 3bd) + (ad + bc)\sqrt{3}.$$

Thus at least one of $ac + 3bd$ and $ad + bc$ is nonzero. Since $ac + 3bd, ad + bc \in \mathbf{Q}$, it follows that $rs \in H$, and so H is closed under multiplication.

   Next, we show that every element of H has an inverse in H. Let $r = a + b\sqrt{3} \in H$, where at least one of a and b is nonzero. Then

$$\frac{1}{r} = \frac{1}{a + b\sqrt{3}} = \frac{1}{a + b\sqrt{3}} \cdot \frac{a - b\sqrt{3}}{a - b\sqrt{3}}$$

$$= -\frac{a}{3b^2 - a^2} + \frac{b}{3b^2 - a^2}\sqrt{3}.$$

Observe that $3b^2 - a^2 \neq 0$; for if $3b^2 - a^2 = 0$, then $a/b = \pm\sqrt{3}$, which is impossible since $a/b \in \mathbf{Q}$ and $\sqrt{3}$ is irrational. Hence $1/r \in H$.

   By the Subgroup Test, H is a subgroup.   ∎

13.37 **Proof** Let $A_1, A_2 \in H$. Then each of $\det(A_1)$ and $\det(A_2)$ is 1 or -1. Since the determinant of $A_1 A_2$ is the product of the determinants of A_1 and A_2, it follows that $\det(A_1 A_2)$ is 1 or -1. Thus $A_1 A_2 \in H$. Next, let $A \in H$. So $\det(A)$ is 1 or -1. Since $\det(A) \neq 0$, it follows that A^{-1} exists. We show that $A^{-1} \in H$. Since $AA^{-1} = I$ (the identity in $M_2^*(\mathbf{R})$) and $\det(I) = 1$, it follows that $\det(A^{-1})\det(A) = 1$. Because $\det(A)$ is 1 or -1, so is $\det(A^{-1})$. Thus $A^{-1} \in H$. By the Subgroup Test, $(H, \cdot)$ is a subgroup of $(M_2^*(\mathbf{R}), \cdot)$.   ∎

13.39 **Proof** For the identity e of G, it follows that $e^2 = e \in H$ and so $H \neq \emptyset$. Let $a, b \in H$. Then $a^2 = b^2 = e$. Then $(ab)^2 = a^2 b^2 = e \cdot e = e$ and so $ab \in H$. Therefore, H is closed under multiplication. Let $a \in H$. Then $a^2 = e$. Thus $(a^2)^{-1} = e$. However, $(a^2)^{-1} = (a^{-1})^2 = e$ and so $a^{-1} \in H$. By the Subgroup Test, H is a subgroup of G.   ∎

13.41 The statement is false. Since $22 \nmid 372$, no group of order 372 contains a subgroup of order 22.

13.43 **(a)** **Proof** Since H is closed under $*$, it suffices to show $g^{-1} \in H$ for each $g \in H$. Let $H = \{g_1, g_2, \ldots, g_k\}$ and let $g \in H$. We claim that $g * g_1, g * g_2, \ldots, g * g_k$ are k distinct elements in H, for suppose this is not the case. Then $g * g_s = g * g_t$ for distinct elements $g_s, g_t \in H$. By the Left Cancellation Law, $g_s = g_t$, which is impossible. Thus, as claimed, $g * g_1, g * g_2, \ldots, g * g_k$ are k distinct elements in H and so

$$H = \{g * g_1, g * g_2, \ldots, g * g_k\}.$$

Since $g \in H$, it follows that $g = g * g_i$ for some integer i with $1 \leq i \leq k$. Hence $g = g * g_i = g * e$ for the identity e of G. By the Left Cancellation Law, $g_i = e$ and so $e \in H$. Therefore, $g * g_j = e$ for some integer j with $1 \leq j \leq k$ and so $g_j = g^{-1}$, implying that $g^{-1} \in H$. By the Subgroup Test, H is a subgroup of G.   ∎

   (b) The set $\mathbf{N}$ is a subset of the infinite group $(\mathbf{Z}, +)$. Note that $\mathbf{N}$ is closed under $+$, but $\mathbf{N}$ is not a subgroup of $(\mathbf{Z}, +)$ by Exercise 13.32(a).

13.45 Since there are six distinct left cosets of H in G, one of which is H and every two left cosets have the same number of elements, it follows that the order of H is $48/6 = 8$.

13.47 The statement is false. Suppose that $g^2 \in gH$. Then $g^2 = gh$ for some $h \in H$. Thus $g = h$, contradicting the fact that $gH \neq H$.

Section 13.6: Isomorphic Groups

13.49 (a) Since 1 is not the image of any integer under ϕ, the function ϕ is not onto and so ϕ is not an isomorphism.

(b) Since $\phi(0) = 1$, the image of the identity 0 in $(\mathbf{Z}, +)$ is not the identity in $(\mathbf{Z}, +)$. By Theorem 13.18(a), ϕ is not an isomorphism.

(c) The function ϕ is an isomorphism.

Proof First, we show that ϕ is one-to-one. Suppose that $\phi(a) = \phi(b)$, where $a, b \in \mathbf{R}$. Then $2^a = 2^b$. Thus $a = \log_2 2^a = \log_2 2^b = b$ and so ϕ is one-to-one. Next, we show that ϕ is onto. Let $r \in \mathbf{R}^+$. Then $\log_2 r \in \mathbf{R}$. Hence $\phi(\log_2 r) = 2^{\log_2 r} = r$ and so ϕ is onto. Finally, we show that ϕ is operation-preserving. For $a, b \in \mathbf{R}$,

$$\phi(a + b) = 2^{a+b} = 2^a \cdot 2^b = \phi(a) \cdot \phi(b).$$

Therefore, ϕ is an isomorphism. ∎

(d) Let $A = \begin{bmatrix} 1 & 0 \\ 0 & 1 \end{bmatrix}$ and $B = \begin{bmatrix} 2 & 0 \\ 0 & \frac{1}{2} \end{bmatrix}$. Then $\phi(A) = \phi(B) = 1$, but $A \neq B$. Thus ϕ is not one-to-one and so ϕ is not an isomorphism.

13.51 ***Proof*** Assume that $\phi : G \to H$ is an isomorphism. Since ϕ is a bijection, ϕ^{-1} is a bijection by Theorem 9.15. It remains to show that ϕ^{-1} is operation-preserving. Let $h_1, h_2 \in H$. Then there exist $g_1, g_2 \in G$ such that $\phi(g_1) = h_1$ and $\phi(g_2) = h_2$. Thus $\phi^{-1}(h_1) = g_1$ and $\phi^{-1}(h_2) = g_2$. Furthermore, $\phi(g_1 * g_2) = \phi(g_1) \circ \phi(g_2) = h_1 \circ h_2$. Hence $\phi^{-1}(h_1 \circ h_2) = g_1 * g_2 = \phi^{-1}(h_1) * \phi^{-1}(h_2)$. Thus ϕ^{-1} is operation-preserving and so ϕ^{-1} is an isomorphism. ∎

13.53 (a) ***Proof*** Let $a, b \in G$. Since $a \circ b = b * a \in G$, it follows that $\circ$ is a binary operation on G. Let $a, b, c \in G$. Then $(a \circ b) \circ c = c * (a \circ b) = c * (b * a) = (c * b) * a = (b \circ c) * a = a \circ (b \circ c)$. Thus $\circ$ is an associative operation. Let e be the identity of $(G, *)$. Then

$$a \circ e = e * a = a = a * e = e \circ a$$

and so e is the identity of $(G, \circ)$. Let $g \in (G, \circ)$ and let g^{-1} be the inverse of g in $(G, *)$. Then

$$g \circ g^{-1} = g^{-1} * g = e = g * g^{-1} = g^{-1} \circ g.$$

Thus g^{-1} is the inverse of g in $(G, \circ)$. Therefore, $(G, \circ)$ is a group. ∎

(b) ***Proof*** Consider the function $\phi : (G, *) \to (G, \circ)$ defined by $\phi(g) = g^{-1}$ for each $g \in G$. We show that ϕ is an isomorphism. First, we show that ϕ is bijective. Let $\phi(g_1) = \phi(g_2)$, where $g_1, g_2 \in (G, *)$. Then $g_1^{-1} = g_2^{-1}$. Since $\left(g_1^{-1}\right)^{-1} = \left(g_2^{-1}\right)^{-1}$ in $(G, \circ)$, it follows that $g_1 = g_2$ in $(G, *)$. Thus ϕ is one-to-one. Let $h \in (G, \circ)$. Then $\phi(h^{-1}) = \left(h^{-1}\right)^{-1} = h$ and so ϕ is onto. It remains to show that ϕ is operation-preserving. Let $g_1, g_2 \in (G, *)$. Then $\phi(g_1 * g_2) = (g_1 * g_2)^{-1} = (g_2 \circ g_1)^{-1} = g_1^{-1} \circ g_2^{-1} = \phi(g_1) \circ \phi(g_2)$ and so ϕ is operation-preserving. Therefore, ϕ is an isomorphism, implying that $(G, *)$ and $(G, \circ)$ are isomorphic. ∎

13.55 ***Proof*** Define $\phi : \mathbf{Z} \to \mathbf{Z}$ by $\phi(n) = n + 1$ for each $n \in \mathbf{Z}$. Then ϕ is bijective. We show that ϕ is an isomorphism from $(\mathbf{Z}, +)$ to $(\mathbf{Z}, *)$. For $m, n \in \mathbf{Z}$, $\phi(m + n) = m + n + 1$. Since $\phi(m) * \phi(n) = (m + 1) * (n + 1) = (m + 1) + (n + 1) - 1 = m + n + 1$, it follows that $\phi(m + n) = \phi(m) * \phi(n)$. Thus ϕ is an isomorphism. ∎

13.57 (a) ***Proof*** First, we show that each f_n, $n \in \mathbf{Z}$, is one-to-one. Suppose that $f_n(a) = f_n(b)$, where $a, b \in A$. Hence $\frac{a}{1+na} = \frac{b}{1+nb}$. Then $a(1 + nb) = b(1 + na)$ and so $a = b$. Thus f_n is one-to-one. Next, we show that each f_n, $n \in \mathbf{Z}$, is onto. Let $c \in A$. Then $\frac{c}{1-nc} \in A$ and $f_n\left(\frac{c}{1-nc}\right) = c$. Hence f_n is onto. Since f_n is one-to-one and onto, it is bijective. ∎

(b) ***Proof***　Let f_n, $f_m \in P$. For $a \in A$, $(f_n \circ f_m)(a) = f_n(f_m(a)) = f_n(\frac{a}{1+ma}) = \frac{a}{1+(m+n)a}$. Thus $f_n \circ f_m \in P$. Let $f_n \in P$. Then f_n^{-1} exists and $f_n^{-1}(x) = \frac{x}{1-nx}$ for $x \in A$. Thus $f_n^{-1} \in P$. By the Subgroup Test, $(P, \circ)$ is a subgroup of $(S_A, \circ)$.　■

(c) ***Proof***　Define $\phi : \mathbf{Z} \to P$ by $\phi(n) = f_n$ for each $n \in \mathbf{Z}$. Thus ϕ is a bijection. For $n, m \in \mathbf{Z}$, $\phi(n + m) = f_{n+m}$. For each $a \in A$, $(f_n \circ f_m)(a) = f_n(f_m(a)) = f_n(\frac{a}{1+ma}) = \frac{a}{1+(m+n)a}$ and so $f_n \circ f_m = f_{n+m}$. Thus $\phi(n + m) = \phi(n) \circ \phi(m)$ and so ϕ is an isomorphism.　■

13.59 (a) ***Proof***　Let $x, y \in A$. Then $x = \frac{m}{n}$ and $y = \frac{p}{q}$, where m, n, p, q are odd integers. Then $xy = \frac{mp}{nq}$, where mp and nq are odd integers. Reducing $\frac{mp}{nq}$ to lowest terms results in an element of A. Next, let $x \in A$. Then $x = \frac{m}{n}$, where m and n are odd integers. Then $x^{-1} = \frac{n}{m} \in A$. By the Subgroup Test, $(A, \cdot)$ is a subgroup of $(\mathbf{Q}^*, \cdot)$.　■

(b) ***Proof***　Let $a \in A$. Then $a = \frac{m}{n}$, where m and n are odd integers. We show that f_a is one-to-one. Assume that $f_a(x) = f_a(y)$, where $x, y \in \mathbf{R}^*$. Then $x^a = y^a$ and so $x^{\frac{m}{n}} = y^{\frac{m}{n}}$. Thus $(x^{\frac{m}{n}})^n = (y^{\frac{m}{n}})^n$ and $x^m = y^m$. Hence $x = (x^m)^{\frac{1}{m}} = (y^m)^{\frac{1}{m}} = y$. Hence f_a is one-to-one. Next, we show that f_a is onto. Let $r \in \mathbf{R}^*$. Then $f(r^{\frac{n}{m}}) = (r^{\frac{n}{m}})^a = (r^{\frac{n}{m}})^{\frac{m}{n}} = r$. Hence f_a is onto. Since f_a is one-to-one and onto, it is a permutation.　■

(c) ***Proof***　Let f_a, $f_b \in F$. For $x \in \mathbf{R}^*$, $(f_b \circ f_a)(x) = f_b(f_a(x)) = f_b(x^a) = (x^a)^b = x^{ab} = f_{ba}(x)$. Since $a = \frac{m}{n}$ and $b = \frac{p}{q}$, where m, n, p, q are odd integers, $ab = \frac{mp}{nq}$, where mp and nq are odd integers, and $ab = \frac{mp}{nq}$ (reduced to lowest terms), it follows that $f_b \circ f_a \in F$. Next, let $f_a \in F$. So $a = \frac{m}{n}$ where m and n are odd integers. Since $f_a^{-1}(x) = x^{\frac{n}{m}}$, it follows that $f_a^{-1} \in F$. By the Subgroup Test, $(F, \circ)$ is a subgroup of $(S_{\mathbf{R}^*}, \circ)$.　■

(d) ***Proof***　Define $\phi : A \to F$ by $\phi(a) = f_a$ for each $a \in A$. Thus ϕ is a bijection. Also, for $a, b \in A$, $\phi(ab) = f_{ab}$ and $\phi(a) \circ \phi(b) = f_a \circ f_b$. Since $(f_a \circ f_b)(x) = f_a(f_b(x)) = f_a(x^b) = (x^b)^a = x^{ab}$, ϕ is an isomorphism.　■

References

1. Carl B. Boyer, *The History of the Calculus and its Conceptual Development*. Dover Publications, New York (1959).
2. Carl B. Boyer, *A History of Mathematics* (Second Edition). John Wiley & Sons, New York (1989).
3. Raymond H. Cox, *A Proof of the Schröder–Bernstein Theorem*. American Mathematical Monthly **75** (1968) 508.
4. Joseph W. Dauben, *Georg Cantor: His Mathematics and Philosophy of the Infinite*. Harvard University Press, Cambridge, MA (1979).
5. John Derbyshire, *Prime Obsession: Bernhard Riemann and the Greatest Unsolved Problems in Mathematics*. Joseph Henry Press, Washington, DC (2003).
6. William Dunham, *Euler: The Master of Us All*. The Dolciani Mathematical Expositions, Vol. 22, Mathematical Association of America, Washington, DC (1999).
7. Ross Honsberger, *More Mathematical Morsels*. The Dolciani Mathematical Expositions, Vol. 10. Mathematical Association of America, Washington, DC (1991).
8. Victor J. Katz, *A History of Mathematics: An Introduction* (Second Edition). Addison Wesley Longman, Reading, MA (1998).
9. Alan Levine, *Discovering Higher Mathematics: Four Habits of Highly Effective Mathematicians*. Academic Press, San Diego (2000).
10. Albert D. Polimeni and H. Joseph Straight, *Foundations of Discrete Mathematics* (Second Edition). Brooks/Cole, Pacific Grove, CA (1989).
11. Laura Toti Rigatelli, *Évariste Galois*, 1811–1832. Springer-Verlag, Berlin (1996).
12. George F. Simmons, *Calculus Gems: Brief Lives and Memorable Mathematics*. McGraw-Hill, New York (1992).
13. Dava Sobel, *Galileo's Daughter: A Historical Memoir of Science, Faith, and Love*. Walker Publishing, New York (1999).
14. Dirk Struik, *A Concise History of Mathematics* (Fourth Edition). Dover Publications, New York (1987).
15. Robert M. Young, *Excursions in Calculus: An Interplay of the Continuous and the Discrete*. The Dolciani Mathematical Expositions, Vol. 13. Mathematical Association of America, Washington, DC (1992).

Index of Symbols

Index